Information Theory
and Network Coding

Information Technology: Transmission, Processing, and Storage

Series Editors:
Robert Gallager
Massachusetts Institute of Technology
Cambridge, Massachusetts

Jack Keil Wolf
University of California at San Diego
La Jolla, California

(continued after index)

Raymond W. Yeung

Information Theory and Network Coding

Raymond W. Yeung
The Chinese University of Hong Kong
Department of Information Engineering
Hong Kong
People's Republic of China
whyeung@ie.cuhk.edu.hk

Series Editors
Robert Gallager
Massachusetts Institute of Technology
Department of Electrical Engineering
  and Computer Science
Cambridge, MA
USA

Jack Keil Wolf
University of California, San Diego
Department of Electrical
  and Computer Engineering
La Jolla, CA
USA

ISBN: 978-1-4419-4630-0   e-ISBN: 978-0-387-79234-7

Printed on acid-free paper

9 8 7 6 5 4 3 2 1

springer.com

information theory are derived and explained. Extra care is taken in handling joint distributions with zero probability masses. There is a section devoted to the discussion of maximum entropy distributions. The chapter ends with a section on the entropy rate of a stationary information source.

Chapter 3 is an introduction to the theory of I-Measure which establishes a one-to-one correspondence between Shannon's information measures and set theory. A number of examples are given to show how the use of information diagrams can simplify the proofs of many results in information theory. Such diagrams are becoming standard tools for solving information theory problems.

Chapter 4 is a discussion of zero-error data compression by uniquely decodable codes, with prefix codes as a special case. A proof of the entropy bound for prefix codes which involves neither the Kraft inequality nor the fundamental inequality is given. This proof facilitates the discussion of the redundancy of prefix codes.

Chapter 5 is a thorough treatment of weak typicality. The weak asymptotic equipartition property and the source coding theorem are discussed. An explanation of the fact that a good data compression scheme produces almost i.i.d. bits is given. There is also an introductory discussion of the Shannon–McMillan–Breiman theorem. The concept of weak typicality will be further developed in Chapter 10 for continuous random variables.

Chapter 6 contains a detailed discussion of strong typicality which applies to random variables with finite alphabets. The results developed in this chapter will be used for proving the channel coding theorem and the rate-distortion theorem in the next two chapters.

The discussion in Chapter 7 of the discrete memoryless channel is an enhancement of the discussion in the previous book. In particular, the new definition of the discrete memoryless channel enables rigorous formulation and analysis of coding schemes for such channels with or without feedback. The proof of the channel coding theorem uses a graphical model approach that helps explain the conditional independence of the random variables.

Chapter 8 is an introduction to rate-distortion theory. The version of the distortion theorem here, proved by using strong typicality, is a stronger n of the original theorem obtained by Shannon.

Chapter 9, the Blahut–Arimoto algorithms for computing the channel y and the rate-distortion function are discussed, and a simplified proof ergence is given. Great care is taken in handling distributions with bability masses.

ers 10 and 11 are devoted to the discussion of information theory for s random variables. Chapter 10 introduces differential entropy and ormation measures, and their basic properties are discussed. The equipartition property for continuous random variables is proved. tion on maximum differential entropy distributions echoes the apter 2 on maximum entropy distributions.

To my parents and my family

Preface

This book is an evolution from my book *A First Course in Information Theory* published in 2002 when network coding was still at its infancy. The last few years have witnessed the rapid development of network coding into a research field of its own in information science. With its root in information theory, network coding has not only brought about a paradigm shift in network communications at large, but also had significant influence on such specific research fields as coding theory, networking, switching, wireless communications, distributed data storage, cryptography, and optimization the[...] While new applications of network coding keep emerging, the fundament[...] sults that lay the foundation of the subject are more or less mature. the main goals of this book therefore is to present these results in a [...] and coherent manner.

While the previous book focused only on information theory [...] random variables, the current book contains two new chapters o[...] theory for continuous random variables, namely the chapter [...] entropy and the chapter on continuous-valued channels. W[...] included, the book becomes more comprehensive and is m[...] used as a textbook for a course in an electrical engineerin[...]

What Is in This book

Out of the 21 chapters in this book, the first 16 c[...] *Components of Information Theory*, and the last 5 c[...] *Fundamentals of Network Coding*. Part I covers the [...] theory and prepares the reader for the discussion[...] of the chapters will give a better idea of what i[...]

Chapter 1 contains a high-level introducti[...] First, there is a discussion on the nature of [...] results in Shannon's original paper in 1948[...] also pointers to Shannon's biographies a[...]

Chapter 2 introduces Shannon's in[...] dom variables and their basic properti[...]

Chapter 11 discusses a variety of continuous-valued channels, with the continuous memoryless channel being the basic building block. In proving the capacity of the memoryless Gaussian channel, a careful justification is given for the existence of the differential entropy of the output random variable. Based on this result, the capacity of a system of parallel/correlated Gaussian channels is obtained. Heuristic arguments leading to the formula for the capacity of the bandlimited white/colored Gaussian channel are given. The chapter ends with a proof of the fact that zero-mean Gaussian noise is the worst additive noise.

Chapter 12 explores the structure of the I-Measure for Markov structures. Set-theoretic characterizations of full conditional independence and Markov random field are discussed. The treatment of Markov random field here maybe too specialized for the average reader, but the structure of the I-Measure and the simplicity of the information diagram for a Markov chain are best explained as a special case of a Markov random field.

Information inequalities are sometimes called the laws of information theory because they govern the impossibilities in information theory. In Chapter 13, the geometrical meaning of information inequalities and the relation between information inequalities and conditional independence are explained in depth. The framework for information inequalities discussed here is the basis of the next two chapters.

Chapter 14 explains how the problem of proving information inequalities can be formulated as a linear programming problem. This leads to a complete characterization of all information inequalities provable by conventional techniques. These inequalities, called Shannon-type inequalities, can be proved by the World Wide Web available software package ITIP. It is also shown how Shannon-type inequalities can be used to tackle the implication problem of conditional independence in probability theory.

Shannon-type inequalities are all the information inequalities known during the first half century of information theory. In the late 1990s, a few new inequalities, called non-Shannon-type inequalities, were discovered. These inequalities imply the existence of laws in information theory beyond those laid down by Shannon. In Chapter 15, we discuss these inequalities and their applications.

Chapter 16 explains an intriguing relation between information theory and group theory. Specifically, for every information inequality satisfied by any joint probability distribution, there is a corresponding group inequality satisfied by any finite group and its subgroups and vice versa. Inequalities of the latter type govern the orders of any finite group and their subgroups. Group-theoretic proofs of Shannon-type information inequalities are given. At the end of the chapter, a group inequality is obtained from a non-Shannon-type inequality discussed in Chapter 15. The meaning and the implication of this inequality are yet to be understood.

Chapter 17 starts Part II of the book with a discussion of the butterfly network, the primary example in network coding. Variations of the butterfly network are analyzed in detail. The advantage of network coding over store-and-forward in wireless and satellite communications is explained through a simple example. We also explain why network coding with multiple information sources is substantially different from network coding with a single information source.

In Chapter 18, the fundamental bound for single-source network coding, called the max-flow bound, is explained in detail. The bound is established for a general class of network codes.

In Chapter 19, we discuss various classes of linear network codes on acyclic networks that achieve the max-flow bound to different extents. Static network codes, a special class of linear network codes that achieves the max-flow bound in the presence of channel failure, are also discussed. Polynomial-time algorithms for constructing these codes are presented.

In Chapter 20, we formulate and analyze convolutional network codes on cyclic networks. The existence of such codes that achieve the max-flow bound is proved.

Network coding theory is further developed in Chapter 21. The scenario when more than one information source are multicast in a point-to-point acyclic network is discussed. An implicit characterization of the achievable information rate region which involves the framework for information inequalities developed in Part I is proved.

How to Use This book

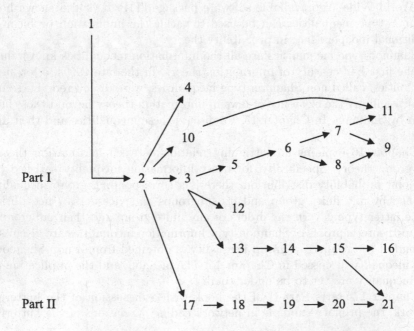

To my parents and my family

Preface

This book is an evolution from my book *A First Course in Information Theory* published in 2002 when network coding was still at its infancy. The last few years have witnessed the rapid development of network coding into a research field of its own in information science. With its root in information theory, network coding has not only brought about a paradigm shift in network communications at large, but also had significant influence on such specific research fields as coding theory, networking, switching, wireless communications, distributed data storage, cryptography, and optimization theory. While new applications of network coding keep emerging, the fundamental results that lay the foundation of the subject are more or less mature. One of the main goals of this book therefore is to present these results in a unifying and coherent manner.

While the previous book focused only on information theory for discrete random variables, the current book contains two new chapters on information theory for continuous random variables, namely the chapter on differential entropy and the chapter on continuous-valued channels. With these topics included, the book becomes more comprehensive and is more suitable to be used as a textbook for a course in an electrical engineering department.

What Is in This book

Out of the 21 chapters in this book, the first 16 chapters belong to Part I, *Components of Information Theory*, and the last 5 chapters belong to Part II, *Fundamentals of Network Coding*. Part I covers the basic topics in information theory and prepares the reader for the discussions in Part II. A brief rundown of the chapters will give a better idea of what is in this book.

Chapter 1 contains a high-level introduction to the contents of this book. First, there is a discussion on the nature of information theory and the main results in Shannon's original paper in 1948 which founded the field. There are also pointers to Shannon's biographies and his works.

Chapter 2 introduces Shannon's information measures for discrete random variables and their basic properties. Useful identities and inequalities in information theory are derived and explained. Extra care is taken in handling joint distributions with zero probability masses. There is a section devoted to the discussion of maximum entropy distributions. The chapter ends with a section on the entropy rate of a stationary information source.

Chapter 3 is an introduction to the theory of I-Measure which establishes a one-to-one correspondence between Shannon's information measures and set theory. A number of examples are given to show how the use of information diagrams can simplify the proofs of many results in information theory. Such diagrams are becoming standard tools for solving information theory problems.

Chapter 4 is a discussion of zero-error data compression by uniquely decodable codes, with prefix codes as a special case. A proof of the entropy bound for prefix codes which involves neither the Kraft inequality nor the fundamental inequality is given. This proof facilitates the discussion of the redundancy of prefix codes.

Chapter 5 is a thorough treatment of weak typicality. The weak asymptotic equipartition property and the source coding theorem are discussed. An explanation of the fact that a good data compression scheme produces almost i.i.d. bits is given. There is also an introductory discussion of the Shannon–McMillan–Breiman theorem. The concept of weak typicality will be further developed in Chapter 10 for continuous random variables.

Chapter 6 contains a detailed discussion of strong typicality which applies to random variables with finite alphabets. The results developed in this chapter will be used for proving the channel coding theorem and the rate-distortion theorem in the next two chapters.

The discussion in Chapter 7 of the discrete memoryless channel is an enhancement of the discussion in the previous book. In particular, the new definition of the discrete memoryless channel enables rigorous formulation and analysis of coding schemes for such channels with or without feedback. The proof of the channel coding theorem uses a graphical model approach that helps explain the conditional independence of the random variables.

Chapter 8 is an introduction to rate-distortion theory. The version of the rate-distortion theorem here, proved by using strong typicality, is a stronger version of the original theorem obtained by Shannon.

In Chapter 9, the Blahut–Arimoto algorithms for computing the channel capacity and the rate-distortion function are discussed, and a simplified proof for convergence is given. Great care is taken in handling distributions with zero probability masses.

Chapters 10 and 11 are devoted to the discussion of information theory for continuous random variables. Chapter 10 introduces differential entropy and related information measures, and their basic properties are discussed. The asymptotic equipartition property for continuous random variables is proved. The last section on maximum differential entropy distributions echoes the section in Chapter 2 on maximum entropy distributions.

Chapter 11 discusses a variety of continuous-valued channels, with the continuous memoryless channel being the basic building block. In proving the capacity of the memoryless Gaussian channel, a careful justification is given for the existence of the differential entropy of the output random variable. Based on this result, the capacity of a system of parallel/correlated Gaussian channels is obtained. Heuristic arguments leading to the formula for the capacity of the bandlimited white/colored Gaussian channel are given. The chapter ends with a proof of the fact that zero-mean Gaussian noise is the worst additive noise.

Chapter 12 explores the structure of the I-Measure for Markov structures. Set-theoretic characterizations of full conditional independence and Markov random field are discussed. The treatment of Markov random field here maybe too specialized for the average reader, but the structure of the I-Measure and the simplicity of the information diagram for a Markov chain are best explained as a special case of a Markov random field.

Information inequalities are sometimes called the laws of information theory because they govern the impossibilities in information theory. In Chapter 13, the geometrical meaning of information inequalities and the relation between information inequalities and conditional independence are explained in depth. The framework for information inequalities discussed here is the basis of the next two chapters.

Chapter 14 explains how the problem of proving information inequalities can be formulated as a linear programming problem. This leads to a complete characterization of all information inequalities provable by conventional techniques. These inequalities, called Shannon-type inequalities, can be proved by the World Wide Web available software package ITIP. It is also shown how Shannon-type inequalities can be used to tackle the implication problem of conditional independence in probability theory.

Shannon-type inequalities are all the information inequalities known during the first half century of information theory. In the late 1990s, a few new inequalities, called non-Shannon-type inequalities, were discovered. These inequalities imply the existence of laws in information theory beyond those laid down by Shannon. In Chapter 15, we discuss these inequalities and their applications.

Chapter 16 explains an intriguing relation between information theory and group theory. Specifically, for every information inequality satisfied by any joint probability distribution, there is a corresponding group inequality satisfied by any finite group and its subgroups and vice versa. Inequalities of the latter type govern the orders of any finite group and their subgroups. Group-theoretic proofs of Shannon-type information inequalities are given. At the end of the chapter, a group inequality is obtained from a non-Shannon-type inequality discussed in Chapter 15. The meaning and the implication of this inequality are yet to be understood.

Chapter 17 starts Part II of the book with a discussion of the butterfly network, the primary example in network coding. Variations of the butterfly network are analyzed in detail. The advantage of network coding over store-and-forward in wireless and satellite communications is explained through a simple example. We also explain why network coding with multiple information sources is substantially different from network coding with a single information source.

In Chapter 18, the fundamental bound for single-source network coding, called the max-flow bound, is explained in detail. The bound is established for a general class of network codes.

In Chapter 19, we discuss various classes of linear network codes on acyclic networks that achieve the max-flow bound to different extents. Static network codes, a special class of linear network codes that achieves the max-flow bound in the presence of channel failure, are also discussed. Polynomial-time algorithms for constructing these codes are presented.

In Chapter 20, we formulate and analyze convolutional network codes on cyclic networks. The existence of such codes that achieve the max-flow bound is proved.

Network coding theory is further developed in Chapter 21. The scenario when more than one information source are multicast in a point-to-point acyclic network is discussed. An implicit characterization of the achievable information rate region which involves the framework for information inequalities developed in Part I is proved.

How to Use This book

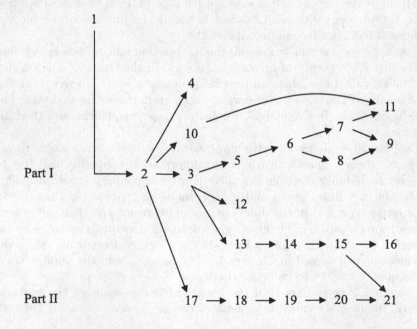

Part I of this book by itself may be regarded as a comprehensive textbook in information theory. The main reason why the book is in the present form is because in my opinion, the discussion of network coding in Part II is incomplete without Part I. Nevertheless, except for Chapter 21 on multi-source network coding, Part II by itself may be used satisfactorily as a textbook on single-source network coding.

An elementary course on probability theory and an elementary course on linear algebra are prerequisites to Part I and Part II, respectively. For Chapter 11, some background knowledge on digital communication systems would be helpful, and for Chapter 20, some prior exposure to discrete-time linear systems is necessary. The reader is recommended to read the chapters according to the above chart. However, one will not have too much difficulty jumping around in the book because there should be sufficient references to the previous relevant sections.

This book inherits the writing style from the previous book, namely that all the derivations are from the first principle. The book contains a large number of examples, where important points are very often made. To facilitate the use of the book, there is a summary at the end of each chapter.

This book can be used as a textbook or a reference book. As a textbook, it is ideal for a two-semester course, with the first and second semesters covering selected topics from Part I and Part II, respectively. A comprehensive instructor's manual is available upon request. Please contact the author at whyeung@ie.cuhk.edu.hk for information and access.

Just like any other lengthy document, this book for sure contains errors and omissions. To alleviate the problem, an errata will be maintained at the book homepage http://www.ie.cuhk.edu.hk/IT_book2/.

Hong Kong, China   *Raymond W. Yeung*
December, 2007

Acknowledgments

The current book, an expansion of my previous book *A First Course in Information Theory*, was written within the year 2007. Thanks to the generous support of the Friedrich Wilhelm Bessel Research Award from the Alexander von Humboldt Foundation of Germany, I had the luxury of working on the project full-time from January to April when I visited Munich University of Technology. I would like to thank Joachim Hagenauer and Ralf Koetter for nominating me for the award and for hosting my visit. I would also like to thank the Department of Information Engineering, The Chinese University of Hong Kong, for making this arrangement possible.

There are many individuals who have directly or indirectly contributed to this book. First, I am indebted to Toby Berger who taught me information theory and writing. I am most thankful to Zhen Zhang, Ning Cai, and Bob Li for their friendship and inspiration. Without the results obtained through our collaboration, the book cannot possibly be in its current form. I would also like to thank Venkat Anantharam, Vijay Bhargava, Dick Blahut, Agnes and Vincent Chan, Tom Cover, Imre Csiszár, Tony Ephremides, Bob Gallager, Bruce Hajek, Te Sun Han, Jim Massey, Prakash Narayan, Alon Orlitsky, Shlomo Shamai, Sergio Verdú, Victor Wei, Frans Willems, and Jack Wolf for their support and encouragement throughout the years. I would also like to thank all the collaborators of my work for their contribution and all the anonymous reviewers for their useful comments.

I would like to thank a number of individuals who helped in the project. I benefited tremendously from the discussions with David Tse who gave a lot of suggestions for writing the chapters on differential entropy and continuous-valued channels. Terence Chan, Ka Wo Cheung, Bruce Hajek, Siu-Wai Ho, Siu Ting Ho, Tat Ming Lok, Prakash Narayan, Will Ng, Sagar Shenvi, Xiang-Gen Xia, Shaohua Yang, Ken Zeger, and Zhixue Zhang gave many valuable comments at different stages of the writing. My graduate students Silas Fong, Min Tan, and Shenghao Yang proofread the chapters on network coding in great detail. Silas Fong also helped compose the figures throughout the book.

On the domestic side, I am most grateful to my wife Rebecca for her love. During our stay in Munich, she took good care of the whole family so that I was able to concentrate on my writing. We are most thankful to our family friend Ms. Pui Yee Wong for taking care of Rebecca when she was ill during the final stage of the project and to my sister Georgiana for her moral support. In this regard, we are indebted to Dr. Yu Lap Yip for his timely diagnosis. I would also like to thank my sister-in-law Ophelia Tsang who comes over during the weekends to help taking care of our daughter Shannon, who continues to be the sweetheart of the family and was most supportive during the time her mom was ill.

Contents

Part II Fundamentals of Network Coding

1

The Science of Information

In a communication system, we try to convey information from one point to another, very often in a noisy environment. Consider the following scenario. A secretary needs to send facsimiles regularly and she wants to convey as much information as possible on each page. She has a choice of the font size, which means that more characters can be squeezed onto a page if a smaller font size is used. In principle, she can squeeze as many characters as desired on a page by using a small enough font size. However, there are two factors in the system which may cause errors. First, the fax machine has a finite resolution. Second, the characters transmitted may be received incorrectly due to noise in the telephone line. Therefore, if the font size is too small, the characters may not be recognizable on the facsimile. On the other hand, although some characters on the facsimile may not be recognizable, the recipient can still figure out the words from the context provided that the number of such characters is not excessive. In other words, it is not necessary to choose a font size such that all the characters on the facsimile are recognizable almost surely. Then we are motivated to ask: What is the maximum amount of meaningful information which can be conveyed on one page of facsimile?

This question may not have a definite answer because it is not very well posed. In particular, we do not have a precise measure of meaningful information. Nevertheless, this question is an illustration of the kind of fundamental questions we can ask about a communication system.

Information, which is not a physical entity but an abstract concept, is hard to quantify in general. This is especially the case if human factors are involved when the information is utilized. For example, when we play Beethoven's violin concerto from an audio compact disc, we receive the musical information from the loudspeakers. We enjoy this information because it arouses certain kinds of emotion within ourselves. While we receive the same information every time we play the same piece of music, the kinds of emotion aroused may be different from time to time because they depend on our mood at that particular moment. In other words, we can derive utility from the same information every time in a different way. For this reason, it is extremely difficult to devise a measure which can quantify the amount of information contained in a piece of music.

In 1948, Bell Telephone Laboratories scientist Claude E. Shannon (1916–2001) published a paper entitled "The Mathematical Theory of Communication" [322] which laid the foundation of an important field now known as information theory. In his paper, the model of a point-to-point communication system depicted in Figure 1.1 is considered. In this model, a message is generated by the information source. The message is converted by the transmitter into a signal which is suitable for transmission. In the course of transmission, the signal may be contaminated by a noise source, so that the received signal may be different from the transmitted signal. Based on the received signal, the receiver then makes an estimate on the message and delivers it to the destination.

In this abstract model of a point-to-point communication system, one is only concerned about whether the message generated by the source can be delivered correctly to the receiver without worrying about how the message is actually used by the receiver. In a way, Shannon's model does not cover all possible aspects of a communication system. However, in order to develop a precise and useful theory of information, the scope of the theory has to be restricted.

In [322], Shannon introduced two fundamental concepts about "information" from the communication point of view. First, information is *uncertainty*. More specifically, if a piece of information we are interested in is deterministic, then it has no value at all because it is already known with no uncertainty. From this point of view, for example, the continuous transmission of a still picture on a television broadcast channel is superfluous. Consequently, an information source is naturally modeled as a random variable or a random process, and probability is employed to develop the theory of information. Second, information to be transmitted is *digital*. This means that the information source should first be converted into a stream of 0s and 1s called *bits*, and the remaining task is to deliver these bits to the receiver correctly with no reference to their actual meaning. This is the foundation of all modern digital

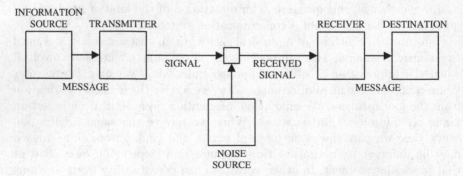

Fig. 1.1. Schematic diagram for a general point-to-point communication system.

communication systems. In fact, this work of Shannon appears to contain the first published use of the term "bit," which stands for *binary* digit.

In the same work, Shannon also proved two important theorems. The first theorem, called the *source coding theorem*, introduces *entropy* as the fundamental measure of information which characterizes the minimum rate of a source code representing an information source essentially free of error. The source coding theorem is the theoretical basis for *lossless* data compression.[1] The second theorem, called the *channel coding theorem*, concerns communication through a noisy channel. It was shown that associated with every noisy channel is a parameter, called the *capacity*, which is strictly positive except for very special channels, such that information can be communicated reliably through the channel as long as the information rate is less than the capacity. These two theorems, which give fundamental limits in point-to-point communication, are the two most important results in information theory.

In science, we study the laws of Nature which must be obeyed by any physical system. These laws are used by engineers to design systems to achieve specific goals. Therefore, science is the foundation of engineering. Without science, engineering can only be done by trial and error.

In information theory, we study the fundamental limits in communication regardless of the technologies involved in the actual implementation of the communication systems. These fundamental limits not only are used as guidelines by communication engineers, but also give insights into what optimal coding schemes are like. Information theory is therefore the science of information.

Since Shannon published his original paper in 1948, information theory has been developed into a major research field in both communication theory and applied probability.

For a non-technical introduction to information theory, we refer the reader to *Encyclopedia Britannica* [103]. In fact, we strongly recommend the reader to first read this excellent introduction before starting this book. For biographies of Claude Shannon, a legend of the twentieth century who had made fundamental contribution to the Information Age, we refer the readers to [55] and [340]. The latter is also a complete collection of Shannon's papers.

Unlike most branches of applied mathematics in which physical systems are studied, abstract systems of communication are studied in information theory. In reading this book, it is not unusual for a beginner to be able to understand all the steps in a proof but has no idea what the proof is leading to. The best way to learn information theory is to study the materials first and come back at a later time. Many results in information theory are rather subtle, to the extent that an expert in the subject may from time to time realize that his/her understanding of certain basic results has been inadequate or even incorrect. While a novice should expect to raise his/her level of understanding of the

---

[1] A data compression scheme is lossless if the data can be recovered with an arbitrarily small probability of error.

subject by reading this book, he/she should not be discouraged to find after finishing the book that there are actually more things yet to be understood. In fact, this is exactly the challenge and the beauty of information theory.

Part I

Components of Information Theory

2

Information Measures

Shannon's information measures refer to entropy, conditional entropy, mutual information, and conditional mutual information. They are the most important measures of information in information theory. In this chapter, we introduce these measures and establish some basic properties they possess. The physical meanings of these measures will be discussed in depth in subsequent chapters. We then introduce the informational divergence which measures the "distance" between two probability distributions and prove some useful inequalities in information theory. The chapter ends with a section on the entropy rate of a stationary information source.

2.1 Independence and Markov Chains

We begin our discussion in this chapter by reviewing two basic concepts in probability: independence of random variables and Markov chain. All the random variables in this book except for Chapters 10 and 11 are assumed to be discrete unless otherwise specified.

Let X be a random variable taking values in an alphabet $\mathcal{X}$. The probability distribution for X is denoted as $\{p_X(x), x \in \mathcal{X}\}$, with $p_X(x) = \Pr\{X = x\}$. When there is no ambiguity, $p_X(x)$ will be abbreviated as $p(x)$, and $\{p(x)\}$ will be abbreviated as $p(x)$. The *support* of X, denoted by $\mathcal{S}_X$, is the set of all $x \in \mathcal{X}$ such that $p(x) > 0$. If $\mathcal{S}_X = \mathcal{X}$, we say that p is *strictly positive*. Otherwise, we say that p is not strictly positive, or p contains zero probability masses. All the above notations naturally extend to two or more random variables. As we will see, probability distributions with zero probability masses are very delicate, and they need to be handled with great care.

Definition 2.1. *Two random variables X and Y are independent, denoted by $X \perp Y$, if*

$$p(x, y) = p(x)p(y) \tag{2.1}$$

for all x and y (i.e., for all $(x, y) \in \mathcal{X} \times \mathcal{Y}$).

For more than two random variables, we distinguish between two types of independence.

Definition 2.2 (Mutual Independence). *For $n \geq 3$, random variables $X_1, X_2, \cdots, X_n$ are mutually independent if*

$$p(x_1, x_2, \cdots, x_n) = p(x_1)p(x_2) \cdots p(x_n) \tag{2.2}$$

for all $x_1, x_2, \cdots, x_n$.

Definition 2.3 (Pairwise Independence). *For $n \geq 3$, random variables $X_1, X_2, \cdots, X_n$ are pairwise independent if X_i and X_j are independent for all $1 \leq i < j \leq n$.*

Note that mutual independence implies pairwise independence. We leave it as an exercise for the reader to show that the converse is not true.

Definition 2.4 (Conditional Independence). *For random variables X, Y, and Z, X is independent of Z conditioning on Y, denoted by $X \perp Z|Y$, if*

$$p(x, y, z)p(y) = p(x, y)p(y, z) \tag{2.3}$$

for all x, y, and z, or equivalently,

$$p(x, y, z) = \begin{cases} \frac{p(x,y)p(y,z)}{p(y)} = p(x,y)p(z|y) & \text{if } p(y) > 0 \\ 0 & \text{otherwise} \end{cases} \tag{2.4}$$

The first definition of conditional independence above is sometimes more convenient to use because it is not necessary to distinguish between the cases $p(y) > 0$ and $p(y) = 0$. However, the physical meaning of conditional independence is more explicit in the second definition.

Proposition 2.5. *For random variables X, Y, and Z, $X \perp Z|Y$ if and only if*

$$p(x, y, z) = a(x, y)b(y, z) \tag{2.5}$$

for all x, y, and z such that $p(y) > 0$.

Proof. The 'only if' part follows immediately from the definition of conditional independence in (2.4), so we will only prove the 'if' part. Assume

$$p(x, y, z) = a(x, y)b(y, z) \tag{2.6}$$

for all x, y, and z such that $p(y) > 0$. Then for such x, y, and z, we have

$$p(x, y) = \sum_z p(x, y, z) = \sum_z a(x, y)b(y, z) = a(x, y)\sum_z b(y, z) \tag{2.7}$$

and

$$p(y,z) = \sum_x p(x,y,z) = \sum_x a(x,y)b(y,z) = b(y,z)\sum_x a(x,y). \qquad (2.8)$$

Furthermore,

$$p(y) = \sum_z p(y,z) = \left(\sum_x a(x,y)\right)\left(\sum_z b(y,z)\right) > 0. \qquad (2.9)$$

Therefore,

$$\frac{p(x,y)p(y,z)}{p(y)} = \frac{\left(a(x,y)\sum_z b(y,z)\right)\left(b(y,z)\sum_x a(x,y)\right)}{\left(\sum_x a(x,y)\right)\left(\sum_z b(y,z)\right)} \qquad (2.10)$$

$$= a(x,y)b(y,z) \qquad (2.11)$$

$$= p(x,y,z). \qquad (2.12)$$

For x, y, and z such that $p(y) = 0$, since

$$0 \le p(x,y,z) \le p(y) = 0, \qquad (2.13)$$

we have

$$p(x,y,z) = 0. \qquad (2.14)$$

Hence, $X \perp Z|Y$ according to (2.4). The proof is accomplished. $\square$

Definition 2.6 (Markov Chain). *For random variables* $X_1, X_2, \cdots, X_n$, *where* $n \ge 3$, $X_1 \to X_2 \to \cdots \to X_n$ *forms a Markov chain if*

$$p(x_1, x_2, \cdots, x_n)p(x_2)p(x_3)\cdots p(x_{n-1})$$
$$= p(x_1, x_2)p(x_2, x_3)\cdots p(x_{n-1}, x_n) \qquad (2.15)$$

for all $x_1, x_2, \cdots, x_n$, *or equivalently,*

$$p(x_1, x_2, \cdots, x_n) =$$

$$\begin{cases} p(x_1, x_2)p(x_3|x_2)\cdots p(x_n|x_{n-1}) & \text{if } p(x_2), p(x_3), \cdots, p(x_{n-1}) > 0 \\ 0 & \text{otherwise} \end{cases}. \qquad (2.16)$$

We note that $X \perp Z|Y$ is equivalent to the Markov chain $X \to Y \to Z$.

Proposition 2.7. $X_1 \to X_2 \to \cdots \to X_n$ *forms a Markov chain if and only if* $X_n \to X_{n-1} \to \cdots \to X_1$ *forms a Markov chain.*

Proof. This follows directly from the symmetry in the definition of a Markov chain in (2.15). □

In the following, we state two basic properties of a Markov chain. The proofs are left as an exercise.

Proposition 2.8. $X_1 \to X_2 \to \cdots \to X_n$ *forms a Markov chain if and only if*

$$
\begin{aligned}
&X_1 \to X_2 \to X_3 \\
&(X_1, X_2) \to X_3 \to X_4 \\
&\quad\vdots \\
&(X_1, X_2, \cdots, X_{n-2}) \to X_{n-1} \to X_n
\end{aligned}
\tag{2.17}
$$

form Markov chains.

Proposition 2.9. $X_1 \to X_2 \to \cdots \to X_n$ *forms a Markov chain if and only if*

$$
p(x_1, x_2, \cdots, x_n) = f_1(x_1, x_2) f_2(x_2, x_3) \cdots f_{n-1}(x_{n-1}, x_n)
\tag{2.18}
$$

for all $x_1, x_2, \cdots, x_n$ *such that* $p(x_2), p(x_3), \cdots, p(x_{n-1}) > 0$.

Note that Proposition 2.9 is a generalization of Proposition 2.5. From Proposition 2.9, one can prove the following important property of a Markov chain. Again, the details are left as an exercise.

Proposition 2.10 (Markov Subchains). *Let* $\mathcal{N}_n = \{1, 2, \cdots, n\}$ *and let* $X_1 \to X_2 \to \cdots \to X_n$ *form a Markov chain. For any subset* α *of* $\mathcal{N}_n$, *denote* $(X_i, i \in \alpha)$ *by* X_α. *Then for any disjoint subsets* $\alpha_1, \alpha_2, \cdots, \alpha_m$ *of* $\mathcal{N}_n$ *such that*

$$
k_1 < k_2 < \cdots < k_m
\tag{2.19}
$$

for all $k_j \in \alpha_j$, $j = 1, 2, \cdots, m$,

$$
X_{\alpha_1} \to X_{\alpha_2} \to \cdots \to X_{\alpha_m}
\tag{2.20}
$$

forms a Markov chain. That is, a subchain of $X_1 \to X_2 \to \cdots \to X_n$ *is also a Markov chain.*

Example 2.11. Let $X_1 \to X_2 \to \cdots \to X_{10}$ form a Markov chain and $\alpha_1 = \{1, 2\}$, $\alpha_2 = \{4\}$, $\alpha_3 = \{6, 8\}$, and $\alpha_4 = \{10\}$ be subsets of $\mathcal{N}_{10}$. Then Proposition 2.10 says that

$$
(X_1, X_2) \to X_4 \to (X_6, X_8) \to X_{10}
\tag{2.21}
$$

also forms a Markov chain.

We have been very careful in handling probability distributions with zero probability masses. In the rest of the section, we show that such distributions are very delicate in general. We first prove the following property of a strictly positive probability distribution involving four random variables.[1]

Proposition 2.12. *Let* $X_1, X_2, X_3,$ *and* X_4 *be random variables such that* $p(x_1, x_2, x_3, x_4)$ *is strictly positive. Then*

$$\left.\begin{array}{c} X_1 \perp X_4 | (X_2, X_3) \\ X_1 \perp X_3 | (X_2, X_4) \end{array}\right\} \Rightarrow X_1 \perp (X_3, X_4) | X_2. \tag{2.22}$$

Proof. If $X_1 \perp X_4 | (X_2, X_3)$, then

$$p(x_1, x_2, x_3, x_4) = \frac{p(x_1, x_2, x_3) p(x_2, x_3, x_4)}{p(x_2, x_3)}. \tag{2.23}$$

On the other hand, if $X_1 \perp X_3 | (X_2, X_4)$, then

$$p(x_1, x_2, x_3, x_4) = \frac{p(x_1, x_2, x_4) p(x_2, x_3, x_4)}{p(x_2, x_4)}. \tag{2.24}$$

Equating (2.23) and (2.24), we have

$$p(x_1, x_2, x_3) = \frac{p(x_2, x_3) p(x_1, x_2, x_4)}{p(x_2, x_4)}. \tag{2.25}$$

Therefore,

$$p(x_1, x_2) = \sum_{x_3} p(x_1, x_2, x_3) \tag{2.26}$$

$$= \sum_{x_3} \frac{p(x_2, x_3) p(x_1, x_2, x_4)}{p(x_2, x_4)} \tag{2.27}$$

$$= \frac{p(x_2) p(x_1, x_2, x_4)}{p(x_2, x_4)} \tag{2.28}$$

or

$$\frac{p(x_1, x_2, x_4)}{p(x_2, x_4)} = \frac{p(x_1, x_2)}{p(x_2)}. \tag{2.29}$$

Hence from (2.24),

$$p(x_1, x_2, x_3, x_4) = \frac{p(x_1, x_2, x_4) p(x_2, x_3, x_4)}{p(x_2, x_4)} = \frac{p(x_1, x_2) p(x_2, x_3, x_4)}{p(x_2)} \tag{2.30}$$

for all $x_1, x_2, x_3,$ and x_4, i.e., $X_1 \perp (X_3, X_4) | X_2$. □

---

[1] Proposition 2.12 is called the *intersection* axiom in Bayesian networks. See [287].

If $p(x_1, x_2, x_3, x_4) = 0$ for some x_1, x_2, x_3, and x_4, i.e., p is not strictly positive, the arguments in the above proof are not valid. In fact, the proposition may not hold in this case. For instance, let $X_1 = Y$, $X_2 = Z$, and $X_3 = X_4 = (Y, Z)$, where Y and Z are independent random variables. Then $X_1 \perp X_4|(X_2, X_3)$, $X_1 \perp X_3|(X_2, X_4)$, but $X_1 \not\perp (X_3, X_4)|X_2$. Note that for this construction, p is not strictly positive because $p(x_1, x_2, x_3, x_4) = 0$ if $x_3 \neq (x_1, x_2)$ or $x_4 \neq (x_1, x_2)$.

The above example is somewhat counter-intuitive because it appears that Proposition 2.12 should hold for all probability distributions via a continuity argument[2] which would go like this. For any distribution p, let $\{p_k\}$ be a sequence of strictly positive distributions such that $p_k \to p$ and p_k satisfies (2.23) and (2.24) for all k, i.e.,

$$p_k(x_1, x_2, x_3, x_4)p_k(x_2, x_3) = p_k(x_1, x_2, x_3)p_k(x_2, x_3, x_4) \qquad (2.31)$$

and

$$p_k(x_1, x_2, x_3, x_4)p_k(x_2, x_4) = p_k(x_1, x_2, x_4)p_k(x_2, x_3, x_4). \qquad (2.32)$$

Then by the proposition, p_k also satisfies (2.30), i.e.,

$$p_k(x_1, x_2, x_3, x_4)p_k(x_2) = p_k(x_1, x_2)p_k(x_2, x_3, x_4). \qquad (2.33)$$

Letting $k \to \infty$, we have

$$p(x_1, x_2, x_3, x_4)p(x_2) = p(x_1, x_2)p(x_2, x_3, x_4) \qquad (2.34)$$

for all x_1, x_2, x_3, and x_4, i.e., $X_1 \perp (X_3, X_4)|X_2$. Such an argument would be valid if there always exists a sequence $\{p_k\}$ as prescribed. However, the existence of the distribution $p(x_1, x_2, x_3, x_4)$ constructed immediately after Proposition 2.12 simply says that it is not always possible to find such a sequence $\{p_k\}$.

Therefore, probability distributions which are not strictly positive can be very delicate. For strictly positive distributions, we see from Proposition 2.5 that their conditional independence structures are closely related to the factorization problem of such distributions, which has been investigated by Chan [59].

2.2 Shannon's Information Measures

We begin this section by introducing the *entropy* of a random variable. As we will see shortly, all Shannon's information measures can be expressed as linear combinations of entropies.

---

[2] See Section 2.3 for a more detailed discussion on continuous functionals.

Definition 2.13. *The entropy $H(X)$ of a random variable X is defined as*

$$H(X) = -\sum_x p(x) \log p(x). \tag{2.35}$$

In the definitions of all information measures, we adopt the convention that summation is taken over the corresponding support. Such a convention is necessary because $p(x) \log p(x)$ in (2.13) is undefined if $p(x) = 0$.

The base of the logarithm in (2.13) can be chosen to be any convenient real number greater than 1. We write $H(X)$ as $H_\alpha(X)$ when the base of the logarithm is α. When the base of the logarithm is 2, the unit for entropy is the *bit*. When the base of the logarithm is e, the unit for entropy is the *nat*. When the base of the logarithm is an integer $D \geq 2$, the unit for entropy is the *D-it* (*D*-ary digit). In the context of source coding, the base is usually taken to be the size of the code alphabet. This will be discussed in Chapter 4.

In computer science, a bit means an entity which can take the value 0 or 1. In information theory, the entropy of a random variable is measured in bits. The reader should distinguish these two meanings of a bit from each other carefully.

Let $g(X)$ be any function of a random variable X. We will denote the expectation of $g(X)$ by $Eg(X)$, i.e.,

$$Eg(X) = \sum_x p(x)g(x), \tag{2.36}$$

where the summation is over S_X. Then the definition of the entropy of a random variable X can be written as

$$H(X) = -E \log p(X). \tag{2.37}$$

Expressions of Shannon's information measures in terms of expectations will be useful in subsequent discussions.

The entropy $H(X)$ of a random variable X is a functional of the probability distribution $p(x)$ which measures the average amount of information contained in X or, equivalently, the average amount of *uncertainty* removed upon revealing the outcome of X. Note that $H(X)$ depends only on $p(x)$, not on the actual values in $\mathcal{X}$. Occasionally, we also denote $H(X)$ by $H(p)$.

For $0 \leq \gamma \leq 1$, define

$$h_b(\gamma) = -\gamma \log \gamma - (1 - \gamma) \log(1 - \gamma) \tag{2.38}$$

with the convention $0 \log 0 = 0$, so that $h_b(0) = h_b(1) = 0$. With this convention, $h_b(\gamma)$ is continuous at $\gamma = 0$ and $\gamma = 1$. h_b is called the *binary entropy function*. For a binary random variable X with distribution $\{\gamma, 1 - \gamma\}$,

$$H(X) = h_b(\gamma). \tag{2.39}$$

Figure 2.1 is the plot of $h_b(\gamma)$ versus γ in the base 2. Note that $h_b(\gamma)$ achieves the maximum value 1 when $\gamma = \frac{1}{2}$.

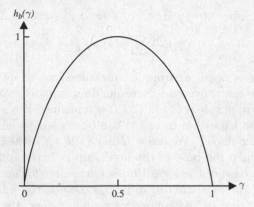

Fig. 2.1. $h_b(\gamma)$ versus γ in the base 2.

The definition of the *joint entropy* of two random variables is similar to the definition of the entropy of a single random variable. Extension of this definition to more than two random variables is straightforward.

Definition 2.14. *The joint entropy $H(X,Y)$ of a pair of random variables X and Y is defined as*

$$H(X,Y) = -\sum_{x,y} p(x,y) \log p(x,y) = -E \log p(X,Y). \qquad (2.40)$$

For two random variables, we define in the following the *conditional entropy* of one random variable when the other random variable is given.

Definition 2.15. *For random variables X and Y, the conditional entropy of Y given X is defined as*

$$H(Y|X) = -\sum_{x,y} p(x,y) \log p(y|x) = -E \log p(Y|X). \qquad (2.41)$$

From (2.41), we can write

$$H(Y|X) = \sum_{x} p(x) \left[-\sum_{y} p(y|x) \log p(y|x) \right]. \qquad (2.42)$$

The inner sum is the entropy of Y conditioning on a fixed $x \in \mathcal{S}_X$. Thus we are motivated to express $H(Y|X)$ as

$$H(Y|X) = \sum_{x} p(x) H(Y|X = x), \qquad (2.43)$$

where

$$H(Y|X = x) = -\sum_{y} p(y|x) \log p(y|x). \tag{2.44}$$

Observe that the right-hand sides of (2.13) and (2.44) have exactly the same form. Similarly, for $H(Y|X, Z)$, we write

$$H(Y|X, Z) = \sum_{z} p(z)H(Y|X, Z = z), \tag{2.45}$$

where

$$H(Y|X, Z = z) = -\sum_{x,y} p(x, y|z) \log p(y|x, z). \tag{2.46}$$

Proposition 2.16.

$$H(X, Y) = H(X) + H(Y|X) \tag{2.47}$$

and

$$H(X, Y) = H(Y) + H(X|Y). \tag{2.48}$$

Proof. Consider

$$H(X, Y) = -E \log p(X, Y) \tag{2.49}$$
$$= -E \log[p(X)p(Y|X)] \tag{2.50}$$
$$= -E \log p(X) - E \log p(Y|X) \tag{2.51}$$
$$= H(X) + H(Y|X). \tag{2.52}$$

Note that (2.50) is justified because the summation of the expectation is over $\mathcal{S}_{XY}$, and we have used the linearity of expectation[3] to obtain (2.51). This proves (2.47), and (2.48) follows by symmetry. □

This proposition has the following interpretation. Consider revealing the outcome of a pair of random variables X and Y in two steps: first the outcome of X and then the outcome of Y. Then the proposition says that the total amount of uncertainty removed upon revealing both X and Y is equal to the sum of the uncertainty removed upon revealing X (uncertainty removed in the first step) and the uncertainty removed upon revealing Y once X is known (uncertainty removed in the second step).

Definition 2.17. *For random variables X and Y, the mutual information between X and Y is defined as*

$$I(X; Y) = \sum_{x,y} p(x, y) \log \frac{p(x, y)}{p(x)p(y)} = E \log \frac{p(X, Y)}{p(X)p(Y)}. \tag{2.53}$$

---

[3] See Problem 5 at the end of the chapter.

Remark $I(X;Y)$ is symmetrical in X and Y.

Proposition 2.18. *The mutual information between a random variable X and itself is equal to the entropy of X, i.e., $I(X;X) = H(X)$.*

Proof. This can be seen by considering

$$I(X;X) = E \log \frac{p(X)}{p(X)^2} \tag{2.54}$$

$$= -E \log p(X) \tag{2.55}$$

$$= H(X). \tag{2.56}$$

The proposition is proved. □

Remark The entropy of X is sometimes called the self-information of X.

Proposition 2.19.

$$I(X;Y) = H(X) - H(X|Y), \tag{2.57}$$

$$I(X;Y) = H(Y) - H(Y|X), \tag{2.58}$$

and

$$I(X;Y) = H(X) + H(Y) - H(X,Y), \tag{2.59}$$

provided that all the entropies and conditional entropies are finite (see Example 2.46 in Section 2.8).

The proof of this proposition is left as an exercise.

From (2.57), we can interpret $I(X;Y)$ as the reduction in uncertainty about X when Y is given or, equivalently, the amount of information about X provided by Y. Since $I(X;Y)$ is symmetrical in X and Y, from (2.58), we can as well interpret $I(X;Y)$ as the amount of information about Y provided by X.

The relations between the (joint) entropies, conditional entropies, and mutual information for two random variables X and Y are given in Propositions 2.16 and 2.19. These relations can be summarized by the diagram in Figure 2.2 which is a variation of the Venn diagram.[4] One can check that all the relations between Shannon's information measures for X and Y which are shown in Figure 2.2 are consistent with the relations given in Propositions 2.16 and 2.19. This one-to-one correspondence between Shannon's information measures and set theory is not just a coincidence for two random variables. We will discuss this in depth when we introduce the I-Measure in Chapter 3.

Analogous to entropy, there is a conditional version of mutual information called conditional mutual information.

---

[4] The rectangle representing the universal set in a usual Venn diagram is missing in Figure 2.2.

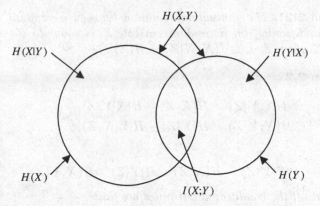

Fig. 2.2. Relationship between entropies and mutual information for two random variables.

Definition 2.20. *For random variables X, Y, and Z, the mutual information between X and Y conditioning on Z is defined as*

$$I(X;Y|Z) = \sum_{x,y,z} p(x,y,z) \log \frac{p(x,y|z)}{p(x|z)p(y|z)} = E \log \frac{p(X,Y|Z)}{p(X|Z)p(Y|Z)}. \quad (2.60)$$

Remark $I(X;Y|Z)$ is symmetrical in X and Y.

Analogous to conditional entropy, we write

$$I(X;Y|Z) = \sum_z p(z)I(X;Y|Z=z), \quad (2.61)$$

where

$$I(X;Y|Z=z) = \sum_{x,y} p(x,y|z) \log \frac{p(x,y|z)}{p(x|z)p(y|z)}. \quad (2.62)$$

Similarly, when conditioning on two random variables, we write

$$I(X;Y|Z,T) = \sum_t p(t)I(X;Y|Z,T=t), \quad (2.63)$$

where

$$I(X;Y|Z,T=t) = \sum_{x,y,z} p(x,y,z|t) \log \frac{p(x,y|z,t)}{p(x|z,t)p(y|z,t)}. \quad (2.64)$$

Conditional mutual information satisfies the same set of relations given in Propositions 2.18 and 2.19 for mutual information except that all the terms are now conditioned on a random variable Z. We state these relations in the next two propositions. The proofs are omitted.

Proposition 2.21. *The mutual information between a random variable X and itself conditioning on a random variable Z is equal to the conditional entropy of X given Z, i.e., $I(X;X|Z) = H(X|Z)$.*

Proposition 2.22.

$$I(X;Y|Z) = H(X|Z) - H(X|Y,Z), \tag{2.65}$$
$$I(X;Y|Z) = H(Y|Z) - H(Y|X,Z), \tag{2.66}$$

and

$$I(X;Y|Z) = H(X|Z) + H(Y|Z) - H(X,Y|Z), \tag{2.67}$$

provided that all the conditional entropies are finite.

Remark All Shannon's information measures are finite if the random variables involved have finite alphabets. Therefore, Propositions 2.19 and 2.22 apply provided that all the random variables therein have finite alphabets.

To conclude this section, we show that all Shannon's information measures are special cases of conditional mutual information. Let Φ be a degenerate random variable, i.e., Φ takes a constant value with probability 1. Consider the mutual information $I(X;Y|Z)$. When $X = Y$ and $Z = \Phi$, $I(X;Y|Z)$ becomes the entropy $H(X)$. When $X = Y$, $I(X;Y|Z)$ becomes the conditional entropy $H(X|Z)$. When $Z = \Phi$, $I(X;Y|Z)$ becomes the mutual information $I(X;Y)$. Thus all Shannon's information measures are special cases of conditional mutual information.

2.3 Continuity of Shannon's Information Measures for Fixed Finite Alphabets

In this section, we prove that for fixed finite alphabets, all Shannon's information measures are continuous functionals of the joint distribution of the random variables involved. To formulate the notion of continuity, we first introduce the *variational distance*[5] as a distance measure between two probability distributions on a common alphabet.

Definition 2.23. *Let p and q be two probability distributions on a common alphabet $\mathcal{X}$. The variational distance between p and q is defined as*

$$V(p,q) = \sum_{x \in \mathcal{X}} |p(x) - q(x)|. \tag{2.68}$$

---

[5] The variational distance is also referred to as the $\mathcal{L}^1$ distance in mathematics.

For a fixed finite alphabet $\mathcal{X}$, let $\mathcal{P}_{\mathcal{X}}$ be the set of all distributions on $\mathcal{X}$. Then according to (2.13), the entropy of a distribution p on an alphabet $\mathcal{X}$ is defined as

$$H(p) = -\sum_{x \in \mathcal{S}_p} p(x) \log p(x), \tag{2.69}$$

where $\mathcal{S}_p$ denotes the support of p and $\mathcal{S}_p \subset \mathcal{X}$. In order for $H(p)$ to be continuous with respect to convergence in variational distance[6] at a particular distribution $p \in \mathcal{P}_{\mathcal{X}}$, for any $\epsilon > 0$, there exists $\delta > 0$ such that

$$|H(p) - H(q)| < \epsilon \tag{2.70}$$

for all $q \in \mathcal{P}_{\mathcal{X}}$ satisfying

$$V(p, q) < \delta, \tag{2.71}$$

or equivalently,

$$\lim_{p' \to p} H(p') = H\left(\lim_{p' \to p} p'\right) = H(p), \tag{2.72}$$

where the convergence $p' \to p$ is in variational distance.

Since $a \log a \to 0$ as $a \to 0$, we define a function $l : [0, \infty) \to \Re$ by

$$l(a) = \begin{cases} a \log a & \text{if } a > 0 \\ 0 & \text{if } a = 0 \end{cases}, \tag{2.73}$$

i.e., $l(a)$ is a continuous extension of $a \log a$. Then (2.69) can be rewritten as

$$H(p) = -\sum_{x \in \mathcal{X}} l(p(x)), \tag{2.74}$$

where the summation above is over all x in $\mathcal{X}$ instead of $\mathcal{S}_p$. Upon defining a function $l_x : \mathcal{P}_{\mathcal{X}} \to \Re$ for all $x \in \mathcal{X}$ by

$$l_x(p) = l(p(x)), \tag{2.75}$$

(2.74) becomes

$$H(p) = -\sum_{x \in \mathcal{X}} l_x(p). \tag{2.76}$$

Evidently, $l_x(p)$ is continuous in p (with respect to convergence in variational distance). Since the summation in (2.76) involves a finite number of terms, we conclude that $H(p)$ is a continuous functional of p.

We now proceed to prove the continuity of conditional mutual information which covers all cases of Shannon's information measures. Consider $I(X; Y|Z)$ and let p_{XYZ} be the joint distribution of X, Y, and Z, where the alphabets $\mathcal{X}$, $\mathcal{Y}$, and $\mathcal{Z}$ are assumed to be finite. From (2.47) and (2.67), we obtain

$$I(X; Y|Z) = H(X, Z) + H(Y, Z) - H(X, Y, Z) - H(Z). \tag{2.77}$$

---

[6] Convergence in variational distance is the same as $\mathcal{L}^1$-convergence.

Note that each term on the right-hand side above is the unconditional entropy of the corresponding marginal distribution. Then (2.77) can be rewritten as

$$I_{X;Y|Z}(p_{XYZ}) = H(p_{XZ}) + H(p_{YZ}) - H(p_{XYZ}) - H(p_Z), \qquad (2.78)$$

where we have used $I_{X;Y|Z}(p_{XYZ})$ to denote $I(X;Y|Z)$. It follows that

$$\lim_{p'_{XYZ} \to p_{XYZ}} I_{X;Y|Z}(p'_{XYZ})$$

$$= \lim_{p'_{XYZ} \to p_{XYZ}} [H(p'_{XZ}) + H(p'_{YZ}) - H(p'_{XYZ}) - H(p'_Z)] \qquad (2.79)$$

$$= \lim_{p'_{XYZ} \to p_{XYZ}} H(p'_{XZ}) + \lim_{p'_{XYZ} \to p_{XYZ}} H(p'_{YZ})$$

$$- \lim_{p'_{XYZ} \to p_{XYZ}} H(p'_{XYZ}) - \lim_{p'_{XYZ} \to p_{XYZ}} H(p'_Z). \qquad (2.80)$$

It can readily be proved, for example, that

$$\lim_{p'_{XYZ} \to p_{XYZ}} p'_{XZ} = p_{XZ}, \qquad (2.81)$$

so that

$$\lim_{p'_{XYZ} \to p_{XYZ}} H(p'_{XZ}) = H\left(\lim_{p'_{XYZ} \to p_{XYZ}} p'_{XZ} \right) = H(p_{XZ}) \qquad (2.82)$$

by the continuity of $H(\cdot)$ when the alphabets involved are fixed and finite. The details are left as an exercise. Hence, we conclude that

$$\lim_{p'_{XYZ} \to p_{XYZ}} I_{X;Y|Z}(p'_{XYZ})$$

$$= H(p_{XZ}) + H(p_{YZ}) - H(p_{XYZ}) - H(p_Z) \qquad (2.83)$$

$$= I_{X;Y|Z}(p_{XYZ}), \qquad (2.84)$$

i.e., $I_{X;Y|Z}(p_{XYZ})$ is a continuous functional of p_{XYZ}.

Since conditional mutual information covers all cases of Shannon's information measures, we have proved that all Shannon's information measures are continuous with respect to convergence in variational distance under the assumption that the alphabets are fixed and finite. It is not difficult to show that under this assumption, convergence in variational distance is equivalent to $\mathcal{L}^2$-convergence, i.e., convergence in Euclidean distance (see Problem 8). It follows that Shannon's information measures are also continuous with respect to $\mathcal{L}^2$-convergence. The variational distance, however, is more often used as a distance measure between two probability distributions because it can be directly related with the informational divergence to be discussed in Section 2.5.

The continuity of Shannon's information measures we have proved in this section is rather restrictive and needs to be applied with caution. In fact, if the alphabets are not fixed, Shannon's information measures are everywhere discontinuous with respect to convergence in a number of commonly used distance measures. We refer the readers to Problems 28–31 for a discussion of these issues.

2.4 Chain Rules

In this section, we present a collection of information identities known as the chain rules which are often used in information theory.

Proposition 2.24 (Chain Rule for Entropy).

$$H(X_1, X_2, \cdots, X_n) = \sum_{i=1}^{n} H(X_i | X_1, \cdots, X_{i-1}). \tag{2.85}$$

Proof. The chain rule for $n = 2$ has been proved in Proposition 2.16. We prove the chain rule by induction on n. Assume (2.85) is true for $n = m$, where $m \geq 2$. Then

$$H(X_1, \cdots, X_m, X_{m+1})$$
$$= H(X_1, \cdots, X_m) + H(X_{m+1} | X_1, \cdots, X_m) \tag{2.86}$$

$$= \sum_{i=1}^{m} H(X_i | X_1, \cdots, X_{i-1}) + H(X_{m+1} | X_1, \cdots, X_m) \tag{2.87}$$

$$= \sum_{i=1}^{m+1} H(X_i | X_1, \cdots, X_{i-1}), \tag{2.88}$$

where in (2.86) we have used (2.47) by letting $X = (X_1, \cdots, X_m)$ and $Y = X_{m+1}$, and in (2.87) we have used (2.85) for $n = m$. This proves the chain rule for entropy. $\square$

The chain rule for entropy has the following conditional version.

Proposition 2.25 (Chain Rule for Conditional Entropy).

$$H(X_1, X_2, \cdots, X_n | Y) = \sum_{i=1}^{n} H(X_i | X_1, \cdots, X_{i-1}, Y). \tag{2.89}$$

Proof. The proposition can be proved by considering

$$H(X_1, X_2, \cdots, X_n | Y)$$
$$= H(X_1, X_2, \cdots, X_n, Y) - H(Y) \tag{2.90}$$
$$= H((X_1, Y), X_2, \cdots, X_n) - H(Y) \tag{2.91}$$

$$= H(X_1, Y) + \sum_{i=2}^{n} H(X_i | X_1, \cdots, X_{i-1}, Y) - H(Y) \tag{2.92}$$

$$= H(X_1 | Y) + \sum_{i=2}^{n} H(X_i | X_1, \cdots, X_{i-1}, Y) \tag{2.93}$$

$$= \sum_{i=1}^{n} H(X_i | X_1, \cdots, X_{i-1}, Y), \tag{2.94}$$

where (2.90) and (2.93) follow from Proposition 2.16, while (2.92) follows from Proposition 2.24.

Alternatively, the proposition can be proved by considering

$$H(X_1, X_2, \cdots, X_n | Y)$$

$$= \sum_y p(y) H(X_1, X_2, \cdots, X_n | Y = y) \qquad (2.95)$$

$$= \sum_y p(y) \sum_{i=1}^{n} H(X_i | X_1, \cdots, X_{i-1}, Y = y) \qquad (2.96)$$

$$= \sum_{i=1}^{n} \sum_y p(y) H(X_i | X_1, \cdots, X_{i-1}, Y = y) \qquad (2.97)$$

$$= \sum_{i=1}^{n} H(X_i | X_1, \cdots, X_{i-1}, Y), \qquad (2.98)$$

where (2.95) and (2.98) follow from (2.43) and (2.45), respectively, and (2.96) follows from an application of Proposition 2.24 to the joint distribution of $X_1, X_2, \cdots, X_n$ conditioning on $\{Y = y\}$. This proof offers an explanation to the observation that (2.89) can be obtained directly from (2.85) by conditioning on Y in every term. $\square$

Proposition 2.26 (Chain Rule for Mutual Information).

$$I(X_1, X_2, \cdots, X_n; Y) = \sum_{i=1}^{n} I(X_i; Y | X_1, \cdots, X_{i-1}). \qquad (2.99)$$

Proof. Consider

$$I(X_1, X_2, \cdots, X_n; Y)$$

$$= H(X_1, X_2, \cdots, X_n) - H(X_1, X_2, \cdots, X_n | Y) \qquad (2.100)$$

$$= \sum_{i=1}^{n} [H(X_i | X_1, \cdots, X_{i-1}) - H(X_i | X_1, \cdots, X_{i-1}, Y)] \qquad (2.101)$$

$$= \sum_{i=1}^{n} I(X_i; Y | X_1, \cdots, X_{i-1}), \qquad (2.102)$$

where in (2.101), we have invoked both Propositions 2.24 and 2.25. The chain rule for mutual information is proved. $\square$

Proposition 2.27 (Chain Rule for Conditional Mutual Information).
For random variables $X_1, X_2, \cdots, X_n$, Y, and Z,

$$I(X_1, X_2, \cdots, X_n; Y | Z) = \sum_{i=1}^{n} I(X_i; Y | X_1, \cdots, X_{i-1}, Z). \qquad (2.103)$$

Proof. This is the conditional version of the chain rule for mutual information. The proof is similar to that for Proposition 2.25. The details are omitted.   □

2.5 Informational Divergence

Let p and q be two probability distributions on a common alphabet $\mathcal{X}$. We very often want to measure how much p is different from q and vice versa. In order to be useful, this measure must satisfy the requirements that it is always nonnegative and it takes the zero value if and only if $p = q$. We denote the support of p and q by $\mathcal{S}_p$ and $\mathcal{S}_q$, respectively. The *informational divergence* defined below serves this purpose.

Definition 2.28. *The informational divergence between two probability distributions p and q on a common alphabet $\mathcal{X}$ is defined as*

$$D(p\|q) = \sum_x p(x) \log \frac{p(x)}{q(x)} = E_p \log \frac{p(X)}{q(X)}, \tag{2.104}$$

where E_p denotes expectation with respect to p.

In the above definition, in addition to the convention that the summation is taken over $\mathcal{S}_p$, we further adopt the convention $c \log \frac{c}{0} = \infty$ for $c > 0$. With this convention, if $D(p\|q) < \infty$, then $p(x) = 0$ whenever $q(x) = 0$, i.e., $\mathcal{S}_p \subset \mathcal{S}_q$.

In the literature, the informational divergence is also referred to as *relative entropy* or the *Kullback-Leibler distance*. We note that $D(p\|q)$ is not symmetrical in p and q, so it is not a true *metric* or "distance." Moreover, $D(\cdot\|\cdot)$ does not satisfy the triangular inequality (see Problem 14).

In the rest of the book, the informational divergence will be referred to as *divergence* for brevity. Before we prove that divergence is always nonnegative, we first establish the following simple but important inequality called the *fundamental inequality* in information theory.

Lemma 2.29 (Fundamental Inequality). *For any $a > 0$,*

$$\ln a \leq a - 1 \tag{2.105}$$

with equality if and only if $a = 1$.

Proof. Let $f(a) = \ln a - a + 1$. Then $f'(a) = 1/a - 1$ and $f''(a) = -1/a^2$. Since $f(1) = 0$, $f'(1) = 0$, and $f''(1) = -1 < 0$, we see that $f(a)$ attains its maximum value 0 when $a = 1$. This proves (2.105). It is also clear that equality holds in (2.105) if and only if $a = 1$. Figure 2.3 is an illustration of the fundamental inequality.   □

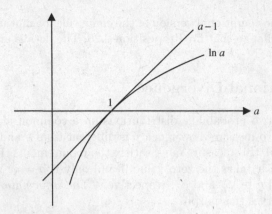

Fig. 2.3. The fundamental inequality $\ln a \leq a - 1$.

Corollary 2.30. *For any $a > 0$,*

$$\ln a \geq 1 - \frac{1}{a} \tag{2.106}$$

with equality if and only if $a = 1$.

Proof. This can be proved by replacing a by $1/a$ in (2.105). $\square$

We can see from Figure 2.3 that the fundamental inequality results from the convexity of the logarithmic function. In fact, many important results in information theory are also direct or indirect consequences of the convexity of the logarithmic function!

Theorem 2.31 (Divergence Inequality). *For any two probability distributions p and q on a common alphabet $\mathcal{X}$,*

$$D(p\|q) \geq 0 \tag{2.107}$$

with equality if and only if $p = q$.

Proof. If $q(x) = 0$ for some $x \in \mathcal{S}_p$, then $D(p\|q) = \infty$ and the theorem is trivially true. Therefore, we assume that $q(x) > 0$ for all $x \in \mathcal{S}_p$. Consider

$$D(p\|q) = (\log e) \sum_{x \in \mathcal{S}_p} p(x) \ln \frac{p(x)}{q(x)} \tag{2.108}$$

$$\geq (\log e) \sum_{x \in \mathcal{S}_p} p(x) \left(1 - \frac{q(x)}{p(x)} \right) \tag{2.109}$$

$$= (\log e) \left[\sum_{x \in \mathcal{S}_p} p(x) - \sum_{x \in \mathcal{S}_p} q(x) \right] \tag{2.110}$$

$$\geq 0, \tag{2.111}$$

where (2.109) results from an application of (2.106), and (2.111) follows from

$$\sum_{x \in \mathcal{S}_p} q(x) \leq 1 = \sum_{x \in \mathcal{S}_p} p(x). \tag{2.112}$$

This proves (2.107).

For equality to hold in (2.107), equality must hold in (2.109) for all $x \in \mathcal{S}_p$ and also in (2.111). For the former, we see from Lemma 2.29 that this is the case if and only if

$$p(x) = q(x) \quad \text{for all } x \in \mathcal{S}_p, \tag{2.113}$$

which implies

$$\sum_{x \in \mathcal{S}_p} q(x) = \sum_{x \in \mathcal{S}_p} p(x) = 1, \tag{2.114}$$

i.e., (2.111) holds with equality. Thus (2.113) is a necessary and sufficient condition for equality to hold in (2.107).

It is immediate that $p = q$ implies (2.113), so it remains to prove the converse. Since $\sum_x q(x) = 1$ and $q(x) \geq 0$ for all x, $p(x) = q(x)$ for all $x \in \mathcal{S}_p$ implies $q(x) = 0$ for all $x \notin \mathcal{S}_p$, and therefore $p = q$. The theorem is proved. $\square$

We now prove a very useful consequence of the divergence inequality called the *log-sum inequality*.

Theorem 2.32 (Log-Sum Inequality). *For positive numbers $a_1, a_2, \cdots$ and nonnegative numbers $b_1, b_2, \cdots$ such that $\sum_i a_i < \infty$ and $0 < \sum_i b_i < \infty$,*

$$\sum_i a_i \log \frac{a_i}{b_i} \geq \left(\sum_i a_i \right) \log \frac{\sum_i a_i}{\sum_i b_i} \tag{2.115}$$

with the convention that $\log \frac{a_i}{0} = \infty$. Moreover, equality holds if and only if $a_i/b_i = \text{constant}$ for all i.

The log-sum inequality can easily be understood by writing it out for the case when there are two terms in each of the summations:

$$a_1 \log \frac{a_1}{b_1} + a_2 \log \frac{a_2}{b_2} \geq (a_1 + a_2) \log \frac{a_1 + a_2}{b_1 + b_2}. \tag{2.116}$$

Proof. Let $a_i' = a_i / \sum_j a_j$ and $b_i' = b_i / \sum_j b_j$. Then $\{a_i'\}$ and $\{b_i'\}$ are probability distributions. Using the divergence inequality, we have

$$0 \leq \sum_i a_i' \log \frac{a_i'}{b_i'} \tag{2.117}$$

$$= \sum_i \frac{a_i}{\sum_j a_j} \log \frac{a_i / \sum_j a_j}{b_i / \sum_j b_j} \tag{2.118}$$

$$= \frac{1}{\sum_j a_j} \left[\sum_i a_i \log \frac{a_i}{b_i} - \left(\sum_i a_i \right) \log \frac{\sum_j a_j}{\sum_j b_j} \right], \tag{2.119}$$

which implies (2.115). Equality holds if and only if $a_i' = b_i'$ for all i or $a_i/b_i = constant$ for all i. The theorem is proved.   □

One can also prove the divergence inequality by using the log-sum inequality (see Problem 20), so the two inequalities are in fact equivalent. The log-sum inequality also finds application in proving the next theorem which gives a lower bound on the divergence between two probability distributions on a common alphabet in terms of the variational distance between them. We will see further applications of the log-sum inequality when we discuss the convergence of some iterative algorithms in Chapter 9.

Theorem 2.33 (Pinsker's Inequality).

$$D(p\|q) \geq \frac{1}{2\ln 2}V^2(p,q). \tag{2.120}$$

Both divergence and the variational distance can be used as measures of the difference between two probability distributions defined on the same alphabet. Pinsker's inequality has the important implication that for two probability distributions p and q defined on the same alphabet, if $D(p\|q)$ or $D(q\|p)$ is small, then so is $V(p,q)$. Furthermore, for a sequence of probability distributions q_k, as $k \to \infty$, if $D(p\|q_k) \to 0$ or $D(q_k\|p) \to 0$, then $V(p,q_k) \to 0$. In other words, convergence in divergence is a stronger notion of convergence than convergence in variational distance.

The proof of Pinsker's inequality as well as its consequence discussed above is left as an exercise (see Problems 23 and 24).

2.6 The Basic Inequalities

In this section, we prove that all Shannon's information measures, namely entropy, conditional entropy, mutual information, and conditional mutual information, are always nonnegative. By this, we mean that these quantities are nonnegative for all joint distributions for the random variables involved.

Theorem 2.34. *For random variables X, Y, and Z,*

$$I(X;Y|Z) \geq 0, \tag{2.121}$$

with equality if and only if X and Y are independent when conditioning on Z.

Proof. Observe that

$$I(X;Y|Z)$$

$$= \sum_{x,y,z} p(x,y,z) \log \frac{p(x,y|z)}{p(x|z)p(y|z)} \tag{2.122}$$

$$= \sum_{z} p(z) \sum_{x,y} p(x,y|z) \log \frac{p(x,y|z)}{p(x|z)p(y|z)} \tag{2.123}$$

$$= \sum_{z} p(z) D(p_{XY|z} \| p_{X|z} p_{Y|z}), \tag{2.124}$$

where we have used $p_{XY|z}$ to denote $\{p(x,y|z), (x,y) \in \mathcal{X} \times \mathcal{Y}\}$, etc. Since for a fixed z, both $p_{XY|z}$ and $p_{X|z}p_{Y|z}$ are joint probability distributions on $\mathcal{X} \times \mathcal{Y}$, we have

$$D(p_{XY|z} \| p_{X|z} p_{Y|z}) \geq 0. \tag{2.125}$$

Therefore, we conclude that $I(X;Y|Z) \geq 0$. Finally, we see from Theorem 2.31 that $I(X;Y|Z) = 0$ if and only if for all $z \in \mathcal{S}_z$,

$$p(x,y|z) = p(x|z)p(y|z) \tag{2.126}$$

or

$$p(x,y,z) = p(x,z)p(y|z) \tag{2.127}$$

for all x and y. Therefore, X and Y are independent conditioning on Z. The proof is accomplished. $\square$

As we have seen in Section 2.2 that all Shannon's information measures are special cases of conditional mutual information, we already have proved that all Shannon's information measures are always nonnegative. The nonnegativity of all Shannon's information measures is called the *basic inequalities*.

For entropy and conditional entropy, we offer the following more direct proof for their nonnegativity. Consider the entropy $H(X)$ of a random variable X. For all $x \in \mathcal{S}_X$, since $0 < p(x) \leq 1$, $\log p(x) \leq 0$. It then follows from the definition in (2.13) that $H(X) \geq 0$. For the conditional entropy $H(Y|X)$ of random variable Y given random variable X, since $H(Y|X = x) \geq 0$ for each $x \in \mathcal{S}_X$, we see from (2.43) that $H(Y|X) \geq 0$.

Proposition 2.35. $H(X) = 0$ *if and only if X is deterministic.*

Proof. If X is deterministic, i.e., there exists $x^* \in \mathcal{X}$ such that $p(x^*) = 1$ and $p(x) = 0$ for all $x \neq x^*$, then $H(X) = -p(x^*) \log p(x^*) = 0$. On the other hand, if X is not deterministic, i.e., there exists $x^* \in \mathcal{X}$ such that $0 < p(x^*) < 1$, then $H(X) \geq -p(x^*) \log p(x^*) > 0$. Therefore, we conclude that $H(X) = 0$ if and only if X is deterministic. $\square$

Proposition 2.36. $H(Y|X) = 0$ *if and only if* Y *is a function of* X.

Proof. From (2.43), we see that $H(Y|X) = 0$ if and only if $H(Y|X = x) = 0$ for each $x \in \mathcal{S}_X$. Then from the last proposition, this happens if and only if Y is deterministic for each given x. In other words, Y is a function of X. $\square$

Proposition 2.37. $I(X;Y) = 0$ *if and only if* X *and* Y *are independent.*

Proof. This is a special case of Theorem 2.34 with Z being a degenerate random variable. $\square$

One can regard (conditional) mutual information as a measure of (conditional) dependency between two random variables. When the (conditional) mutual information is exactly equal to 0, the two random variables are (conditionally) independent.

We refer to inequalities involving Shannon's information measures only (possibly with constant terms) as *information inequalities*. The basic inequalities are important examples of information inequalities. Likewise, we refer to identities involving Shannon's information measures only as *information identities*. From the information identities (2.47), (2.57), and (2.65), we see that all Shannon's information measures can be expressed as linear combinations of entropies provided that the latter are all finite. Specifically,

$$H(Y|X) = H(X,Y) - H(X), \tag{2.128}$$
$$I(X;Y) = H(X) + H(Y) - H(X,Y), \tag{2.129}$$

and

$$I(X;Y|Z) = H(X,Z) + H(Y,Z) - H(X,Y,Z) - H(Z). \tag{2.130}$$

Therefore, an information inequality is an inequality which involves only entropies.

As we will see later in the book, information inequalities form the most important set of tools for proving converse coding theorems in information theory. Except for a number of so-called non-Shannon-type inequalities, all known information inequalities are implied by the basic inequalities. Information inequalities will be studied systematically in Chapters 13, 14, and 15. In the next section, we will prove some consequences of the basic inequalities which are often used in information theory.

2.7 Some Useful Information Inequalities

In this section, we prove some useful consequences of the basic inequalities introduced in the last section. Note that the conditional versions of these inequalities can be proved by techniques similar to those used in the proof of Proposition 2.25.

Theorem 2.38 (Conditioning Does Not Increase Entropy).

$$H(Y|X) \leq H(Y) \tag{2.131}$$

with equality if and only if X and Y are independent.

Proof. This can be proved by considering

$$H(Y|X) = H(Y) - I(X;Y) \leq H(Y), \tag{2.132}$$

where the inequality follows because $I(X;Y)$ is always nonnegative. The inequality is tight if and only if $I(X;Y) = 0$, which is equivalent by Proposition 2.37 to X and Y being independent. □

Similarly, it can be shown that

$$H(Y|X,Z) \leq H(Y|Z), \tag{2.133}$$

which is the conditional version of the above proposition. These results have the following interpretation. Suppose Y is a random variable we are interested in, and X and Z are side-information about Y. Then our uncertainty about Y cannot be increased on the average upon receiving side-information X. Once we know X, our uncertainty about Y again cannot be increased on the average upon further receiving side-information Z.

Remark Unlike entropy, the mutual information between two random variables can be increased by conditioning on a third random variable. We refer the reader to Section 3.4 for a discussion.

Theorem 2.39 (Independence Bound for Entropy).

$$H(X_1, X_2, \cdots, X_n) \leq \sum_{i=1}^{n} H(X_i) \tag{2.134}$$

with equality if and only if X_i, $i = 1, 2, \cdots, n$ are mutually independent.

Proof. By the chain rule for entropy,

$$H(X_1, X_2, \cdots, X_n) = \sum_{i=1}^{n} H(X_i|X_1, \cdots, X_{i-1}) \tag{2.135}$$

$$\leq \sum_{i=1}^{n} H(X_i), \tag{2.136}$$

where the inequality follows because we have proved in the last theorem that conditioning does not increase entropy. The inequality is tight if and only if it is tight for each i, i.e.,

$$H(X_i|X_1, \cdots, X_{i-1}) = H(X_i) \tag{2.137}$$

for $1 \leq i \leq n$. From the last theorem, this is equivalent to X_i being independent of $X_1, X_2, \cdots, X_{i-1}$ for each i. Then

$$
\begin{aligned}
p(x_1, x_2, \cdots, x_n) & \\
&= p(x_1, x_2, \cdots, x_{n-1})p(x_n) & (2.138) \\
&= p(p(x_1, x_2, \cdots, x_{n-2})p(x_{n-1})p(x_n) & (2.139)
\end{aligned}
$$

$$\vdots$$

$$= p(x_1)p(x_2) \cdots p(x_n) \tag{2.140}$$

for all $x_1, x_2, \cdots, x_n$, i.e., $X_1, X_2, \cdots, X_n$ are mutually independent.

Alternatively, we can prove the theorem by considering

$$\sum_{i=1}^{n} H(X_i) - H(X_1, X_2, \cdots, X_n)$$

$$= -\sum_{i=1}^{n} E \log p(X_i) + E \log p(X_1, X_2, \cdots, X_n) \tag{2.141}$$

$$= -E \log[p(X_1)p(X_2) \cdots p(X_n)] + E \log p(X_1, X_2, \cdots, X_n) \tag{2.142}$$

$$= E \log \frac{p(X_1, X_2, \cdots, X_n)}{p(X_1)p(X_2) \cdots p(X_n)} \tag{2.143}$$

$$= D(p_{X_1 X_2 \cdots X_n} \| p_{X_1} p_{X_2} \cdots p_{X_n}) \tag{2.144}$$

$$\geq 0, \tag{2.145}$$

where equality holds if and only if

$$p(x_1, x_2, \cdots, x_n) = p(x_1)p(x_2) \cdots p(x_n) \tag{2.146}$$

for all $x_1, x_2, \cdots, x_n$, i.e., $X_1, X_2, \cdots, X_n$ are mutually independent. $\square$

Theorem 2.40.
$$I(X; Y, Z) \geq I(X; Y), \tag{2.147}$$
with equality if and only if $X \to Y \to Z$ forms a Markov chain.

Proof. By the chain rule for mutual information, we have

$$I(X; Y, Z) = I(X; Y) + I(X; Z|Y) \geq I(X; Y). \tag{2.148}$$

The above inequality is tight if and only if $I(X; Z|Y) = 0$, or $X \to Y \to Z$ forms a Markov chain. The theorem is proved. $\square$

Lemma 2.41. *If $X \to Y \to Z$ forms a Markov chain, then*

$$I(X; Z) \leq I(X; Y) \tag{2.149}$$

and

$$I(X; Z) \leq I(Y; Z). \tag{2.150}$$

Before proving this inequality, we first discuss its meaning. Suppose X is a random variable we are interested in, and Y is an observation of X. If we infer X via Y, our uncertainty about X on the average is $H(X|Y)$. Now suppose we process Y (either deterministically or probabilistically) to obtain a random variable Z. If we infer X via Z, our uncertainty about X on the average is $H(X|Z)$. Since $X \to Y \to Z$ forms a Markov chain, from (2.149), we have

$$H(X|Z) = H(X) - I(X;Z) \tag{2.151}$$
$$\geq H(X) - I(X;Y) \tag{2.152}$$
$$= H(X|Y), \tag{2.153}$$

i.e., further processing of Y can only increase our uncertainty about X on the average.

Proof. Assume $X \to Y \to Z$, i.e., $X \perp Z|Y$. By Theorem 2.34, we have

$$I(X;Z|Y) = 0. \tag{2.154}$$

Then

$$I(X;Z) = I(X;Y,Z) - I(X;Y|Z) \tag{2.155}$$
$$\leq I(X;Y,Z) \tag{2.156}$$
$$= I(X;Y) + I(X;Z|Y) \tag{2.157}$$
$$= I(X;Y). \tag{2.158}$$

In (2.155) and (2.157), we have used the chain rule for mutual information. The inequality in (2.156) follows because $I(X;Y|Z)$ is always nonnegative, and (2.158) follows from (2.154). This proves (2.149).

Since $X \to Y \to Z$ is equivalent to $Z \to Y \to X$, we also have proved (2.150). This completes the proof of the lemma. □

From Lemma 2.41, we can prove the more general data processing theorem.

Theorem 2.42 (Data Processing Theorem). *If $U \to X \to Y \to V$ forms a Markov chain, then*

$$I(U;V) \leq I(X;Y). \tag{2.159}$$

Proof. Assume $U \to X \to Y \to V$. Then by Proposition 2.10, we have $U \to X \to Y$ and $U \to Y \to V$. From the first Markov chain and Lemma 2.41, we have

$$I(U;Y) \leq I(X;Y). \tag{2.160}$$

From the second Markov chain and Lemma 2.41, we have

$$I(U;V) \leq I(U;Y). \tag{2.161}$$

Combining (2.160) and (2.161), we obtain (2.159), proving the theorem. □

2.8 Fano's Inequality

In the last section, we have proved a few information inequalities involving only Shannon's information measures. In this section, we first prove an upper bound on the entropy of a random variable in terms of the size of the alphabet. This inequality is then used in the proof of Fano's inequality, which is extremely useful in proving converse coding theorems in information theory.

Theorem 2.43. *For any random variable X,*

$$H(X) \leq \log |\mathcal{X}|, \tag{2.162}$$

where $|\mathcal{X}|$ denotes the size of the alphabet $\mathcal{X}$. This upper bound is tight if and only if X is distributed uniformly on $\mathcal{X}$.

Proof. Let u be the uniform distribution on $\mathcal{X}$, i.e., $u(x) = |\mathcal{X}|^{-1}$ for all $x \in \mathcal{X}$. Then

$$\log |\mathcal{X}| - H(X)$$

$$= -\sum_{x \in \mathcal{S}_X} p(x) \log |\mathcal{X}|^{-1} + \sum_{x \in \mathcal{S}_X} p(x) \log p(x) \tag{2.163}$$

$$= -\sum_{x \in \mathcal{S}_X} p(x) \log u(x) + \sum_{x \in \mathcal{S}_X} p(x) \log p(x) \tag{2.164}$$

$$= \sum_{x \in \mathcal{S}_X} p(x) \log \frac{p(x)}{u(x)} \tag{2.165}$$

$$= D(p\|u) \tag{2.166}$$

$$\geq 0, \tag{2.167}$$

proving (2.162). This upper bound is tight if and if only $D(p\|u) = 0$, which from Theorem 2.31 is equivalent to $p(x) = u(x)$ for all $x \in \mathcal{X}$, completing the proof. □

Corollary 2.44. *The entropy of a random variable may take any nonnegative real value.*

Proof. Consider a random variable X defined on a fixed finite alphabet $\mathcal{X}$. We see from the last theorem that $H(X) = \log |\mathcal{X}|$ is achieved when X is distributed uniformly on $\mathcal{X}$. On the other hand, $H(X) = 0$ is achieved when X is deterministic. For $0 \leq a \leq |\mathcal{X}|^{-1}$, let

$$g(a) = H\left(\{1 - (|\mathcal{X}| - 1)a, a, \cdots, a\}\right) \tag{2.168}$$

$$= -l(1 - (|\mathcal{X}| - 1)a) - (|\mathcal{X}| - 1)l(a), \tag{2.169}$$

where $l(\cdot)$ is defined in (2.73). Note that $g(a)$ is continuous in a, with $g(0) = 0$ and $g(|\mathcal{X}|^{-1}) = \log |\mathcal{X}|$. For any value $0 < b < \log |\mathcal{X}|$, by the intermediate value theorem of continuous functions, there exists a distribution for X such that $H(X) = b$. Then we see that $H(X)$ can take any positive value by letting $|\mathcal{X}|$ be sufficiently large. This accomplishes the proof. □

Remark Let $|\mathcal{X}| = D$, or the random variable X is a D-ary symbol. When the base of the logarithm is D, (2.162) becomes

$$H_D(X) \le 1. \tag{2.170}$$

Recall that the unit of entropy is the D-it when the logarithm is in the base D. This inequality says that a D-ary symbol can carry at most 1 D-it of information. This maximum is achieved when X has a uniform distribution. We already have seen the binary case when we discuss the binary entropy function $h_b(p)$ in Section 2.2.

We see from Theorem 2.43 that the entropy of a random variable is finite as long as it has a finite alphabet. However, if a random variable has a countable alphabet,[7] its entropy may or may not be finite. This will be shown in the next two examples.

Example 2.45. Let X be a random variable such that

$$\Pr\{X = i\} = 2^{-i}, \tag{2.171}$$

$i = 1, 2, \cdots$. Then

$$H_2(X) = \sum_{i=1}^{\infty} i2^{-i} = 2, \tag{2.172}$$

which is finite.

For a random variable X with a countable alphabet and finite entropy, we show in Appendix 2.A that the entropy of X can be approximated by the entropy of a truncation of the distribution of X.

Example 2.46. Let Y be a random variable which takes values in the subset of pairs of integers

$$\left\{ (i,j) : 1 \le i < \infty \text{ and } 1 \le j \le \frac{2^{2^i}}{2^i} \right\} \tag{2.173}$$

such that

$$\Pr\{Y = (i,j)\} = 2^{-2^i} \tag{2.174}$$

for all i and j. First, we check that

$$\sum_{i=1}^{\infty} \sum_{j=1}^{2^{2^i}/2^i} \Pr\{Y = (i,j)\} = \sum_{i=1}^{\infty} 2^{-2^i} \left(\frac{2^{2^i}}{2^i} \right) = 1. \tag{2.175}$$

---

[7] An alphabet is countable means that it is either finite or countably infinite.

Then

$$H_2(Y) = -\sum_{i=1}^{\infty} \sum_{j=1}^{2^{2^i}/2^i} 2^{-2^i} \log_2 2^{-2^i} = \sum_{i=1}^{\infty} 1, \qquad (2.176)$$

which does not converge.

Let X be a random variable and $\hat{X}$ be an estimate on X which takes value in the same alphabet $\mathcal{X}$. Let the probability of error P_e be

$$P_e = \Pr\{X \neq \hat{X}\}. \qquad (2.177)$$

If $P_e = 0$, i.e., $X = \hat{X}$ with probability 1, then $H(X|\hat{X}) = 0$ by Proposition 2.36. Intuitively, if P_e is small, i.e., $X = \hat{X}$ with probability close to 1, then $H(X|\hat{X})$ should be close to 0. Fano's inequality makes this intuition precise.

Theorem 2.47 (Fano's Inequality). *Let X and $\hat{X}$ be random variables taking values in the same alphabet $\mathcal{X}$. Then*

$$H(X|\hat{X}) \leq h_b(P_e) + P_e \log(|\mathcal{X}| - 1), \qquad (2.178)$$

where h_b is the binary entropy function.

Proof. Define a random variable

$$Y = \begin{cases} 0 \text{ if } X = \hat{X} \\ 1 \text{ if } X \neq \hat{X} \end{cases}. \qquad (2.179)$$

The random variable Y is an indicator of the error event $\{X \neq \hat{X}\}$, with $\Pr\{Y = 1\} = P_e$ and $H(Y) = h_b(P_e)$. Since Y is a function X and $\hat{X}$,

$$H(Y|X, \hat{X}) = 0. \qquad (2.180)$$

Then

$$H(X|\hat{X})$$
$$= H(X|\hat{X}) + H(Y|X, \hat{X}) \qquad (2.181)$$
$$= H(X, Y|\hat{X}) \qquad (2.182)$$
$$= H(Y|\hat{X}) + H(X|\hat{X}, Y) \qquad (2.183)$$
$$\leq H(Y) + H(X|\hat{X}, Y) \qquad (2.184)$$
$$= H(Y) + \sum_{\hat{x} \in \mathcal{X}} \left[\Pr\{\hat{X} = \hat{x}, Y = 0\} H(X|\hat{X} = \hat{x}, Y = 0) \right.$$
$$\left. + \Pr\{\hat{X} = \hat{x}, Y = 1\} H(X|\hat{X} = \hat{x}, Y = 1) \right]. \qquad (2.185)$$

In the above, (2.181) follows from (2.180), (2.184) follows because conditioning does not increase entropy, and (2.185) follows from an application of

(2.43). Now X must take the value $\hat{x}$ if $\hat{X} = \hat{x}$ and $Y = 0$. In other words, X is conditionally deterministic given $\hat{X} = \hat{x}$ and $Y = 0$. Therefore, by Proposition 2.35,

$$H(X|\hat{X} = \hat{x}, Y = 0) = 0. \tag{2.186}$$

If $\hat{X} = \hat{x}$ and $Y = 1$, then X must take a value in the set $\{x \in \mathcal{X} : x \neq \hat{x}\}$ which contains $|\mathcal{X}| - 1$ elements. By Theorem 2.43, we have

$$H(X|\hat{X} = \hat{x}, Y = 1) \leq \log(|\mathcal{X}| - 1), \tag{2.187}$$

where this upper bound does not depend on $\hat{x}$. Hence,

$$H(X|\hat{X})$$

$$\leq h_b(P_e) + \left(\sum_{\hat{x} \in \mathcal{X}} \Pr\{\hat{X} = \hat{x}, Y = 1\} \right) \log(|\mathcal{X}| - 1) \tag{2.188}$$

$$= h_b(P_e) + \Pr\{Y = 1\} \log(|\mathcal{X}| - 1) \tag{2.189}$$

$$= h_b(P_e) + P_e \log(|\mathcal{X}| - 1), \tag{2.190}$$

which completes the proof. $\square$

Very often, we only need the following simplified version when we apply Fano's inequality. The proof is omitted.

Corollary 2.48. $H(X|\hat{X}) < 1 + P_e \log |\mathcal{X}|.$

Fano's inequality has the following implication. If the alphabet $\mathcal{X}$ is finite, as $P_e \to 0$, the upper bound in (2.178) tends to 0, which implies $H(X|\hat{X})$ also tends to 0. However, this is not necessarily the case if $\mathcal{X}$ is countable, which is shown in the next example.

Example 2.49. Let $\hat{X}$ take the value 0 with probability 1. Let Z be an independent binary random variable taking values in $\{0, 1\}$. Define the random variable X by

$$X = \begin{cases} 0 & \text{if } Z = 0 \\ Y & \text{if } Z = 1 \end{cases}, \tag{2.191}$$

where Y is the random variable in Example 2.46 whose entropy is infinity. Let

$$P_e = \Pr\{X \neq \hat{X}\} = \Pr\{Z = 1\}. \tag{2.192}$$

Then

$$H(X|\hat{X}) \tag{2.193}$$

$$= H(X) \tag{2.194}$$

$$\geq H(X|Z) \tag{2.195}$$

$$= \Pr\{Z = 0\} H(X|Z = 0) + \Pr\{Z = 1\} H(X|Z = 1) \tag{2.196}$$

$$= (1 - P_e) \cdot 0 + P_e \cdot H(Y) \tag{2.197}$$

$$= \infty \tag{2.198}$$

for any $P_e > 0$. Therefore, $H(X|\hat{X})$ does not tend to 0 as $P_e \to 0$.

2.9 Maximum Entropy Distributions

In Theorem 2.43, we have proved that for any random variable X,

$$H(X) \leq \log |\mathcal{X}|, \tag{2.199}$$

with equality when X is distributed uniformly over $\mathcal{X}$. In this section, we revisit this result in the context that X is a real random variable.

To simplify our discussion, all the logarithms are in the base e. Consider the following problem:

Maximize $H(p)$ over all probability distributions p defined on a countable subset $\mathcal{S}$ of the set of real numbers, subject to

$$\sum_{x \in \mathcal{S}_p} p(x) r_i(x) = a_i \quad \text{for } 1 \leq i \leq m, \tag{2.200}$$

where $\mathcal{S}_p \subset \mathcal{S}$ and $r_i(x)$ is defined for all $x \in \mathcal{S}$.

The following theorem renders a solution to this problem.

Theorem 2.50. *Let*

$$p^*(x) = e^{-\lambda_0 - \sum_{i=1}^m \lambda_i r_i(x)} \tag{2.201}$$

for all $x \in \mathcal{S}$, where $\lambda_0, \lambda_1, \cdots, \lambda_m$ are chosen such that the constraints in (2.200) are satisfied. Then p^* maximizes $H(p)$ over all probability distribution p on $\mathcal{S}$, subject to the constraints in (2.200).

Proof. For any p satisfying the constraints in (2.200), consider

$$H(p^*) - H(p)$$

$$= -\sum_{x \in \mathcal{S}} p^*(x) \ln p^*(x) + \sum_{x \in \mathcal{S}_p} p(x) \ln p(x) \tag{2.202}$$

$$= -\sum_{x \in \mathcal{S}} p^*(x) \left(-\lambda_0 - \sum_i \lambda_i r_i(x) \right) + \sum_{x \in \mathcal{S}_p} p(x) \ln p(x) \tag{2.203}$$

$$= \lambda_0 \left(\sum_{x \in \mathcal{S}} p^*(x) \right) + \sum_i \lambda_i \left(\sum_{x \in \mathcal{S}} p^*(x) r_i(x) \right) + \sum_{x \in \mathcal{S}_p} p(x) \ln p(x) \tag{2.204}$$

$$= \lambda_0 \cdot 1 + \sum_i \lambda_i a_i + \sum_{x \in \mathcal{S}_p} p(x) \ln p(x) \tag{2.205}$$

$$= \lambda_0 \left(\sum_{x \in \mathcal{S}_p} p(x) \right) + \sum_i \lambda_i \left(\sum_{x \in \mathcal{S}_p} p(x) r_i(x) \right) + \sum_{x \in \mathcal{S}_p} p(x) \ln p(x) \tag{2.206}$$

$$= -\sum_{x \in \mathcal{S}_p} p(x) \left(-\lambda_0 - \sum_i \lambda_i r_i(x) \right) + \sum_{x \in \mathcal{S}_p} p(x) \ln p(x) \tag{2.207}$$

$$= - \sum_{x \in \mathcal{S}_p} p(x) \ln p^*(x) + \sum_{x \in \mathcal{S}_p} p(x) \ln p(x) \tag{2.208}$$

$$= \sum_{x \in \mathcal{S}_p} p(x) \ln \frac{p(x)}{p^*(x)} \tag{2.209}$$

$$= D(p \| p^*) \tag{2.210}$$

$$\geq 0. \tag{2.211}$$

In the above, (2.207) is obtained from (2.203) by replacing $p^*(x)$ by $p(x)$ and $x \in \mathcal{S}$ by $x \in \mathcal{S}_p$ in the first summation, while the intermediate steps (2.204)–(2.206) are justified by noting that both p^* and p satisfy the constraints in (2.200). The last step is an application of the divergence inequality (Theorem 2.31). The proof is accomplished. □

Remark For all $x \in \mathcal{S}$, $p^*(x) > 0$, so that $\mathcal{S}_{p^*} = \mathcal{S}$.

The following corollary of Theorem 2.50 is rather subtle.

Corollary 2.51. *Let p^* be a probability distribution defined on $\mathcal{S}$ with*

$$p^*(x) = e^{-\lambda_0 - \sum_{i=1}^{m} \lambda_i r_i(x)} \tag{2.212}$$

for all $x \in \mathcal{S}$. Then p^* maximizes $H(p)$ over all probability distribution p defined on $\mathcal{S}$, subject to the constraints

$$\sum_{x \in \mathcal{S}_p} p(x) r_i(x) = \sum_{x \in \mathcal{S}} p^*(x) r_i(x) \quad \text{for } 1 \leq i \leq m. \tag{2.213}$$

Example 2.52. Let $\mathcal{S}$ be finite and let the set of constraints in (2.200) be empty. Then

$$p^*(x) = e^{-\lambda_0}, \tag{2.214}$$

a constant that does not depend on x. Therefore, p^* is simply the uniform distribution over $\mathcal{S}$, i.e., $p^*(x) = |\mathcal{S}|^{-1}$ for all $x \in \mathcal{S}$. This is consistent with Theorem 2.43.

Example 2.53. Let $\mathcal{S} = \{0, 1, 2, \cdots\}$, and let the set of constraints in (2.200) be

$$\sum_x p(x) x = a, \tag{2.215}$$

where $a \geq 0$, i.e., the mean of the distribution p is fixed at some nonnegative value a. We now determine p^* using the prescription in Theorem 2.50. Let

$$q_i = e^{-\lambda_i} \tag{2.216}$$

for $i = 0, 1$. Then by (2.201),

$$p^*(x) = q_0 q_1^x. \tag{2.217}$$

Evidently, p^* is a geometric distribution, so that

$$q_0 = 1 - q_1. \tag{2.218}$$

Finally, we invoke the constraint (2.200) on p to obtain $q_1 = (a+1)^{-1}$. The details are omitted.

2.10 Entropy Rate of a Stationary Source

In the previous sections, we have discussed various properties of the entropy of a finite collection of random variables. In this section, we discuss the *entropy rate* of a discrete-time information source.

A discrete-time information source $\{X_k, k \geq 1\}$ is an infinite collection of random variables indexed by the set of positive integers. Since the index set is ordered, it is natural to regard the indices as time indices. We will refer to the random variables X_k as *letters*.

We assume that $H(X_k) < \infty$ for all k. Then for any finite subset A of the index set $\{k : k \geq 1\}$, we have

$$H(X_k, k \in A) \leq \sum_{k \in A} H(X_k) < \infty. \tag{2.219}$$

However, it is not meaningful to discuss $H(X_k, k \geq 1)$ because the joint entropy of an infinite collection of letters is infinite except for very special cases. On the other hand, since the indices are ordered, we can naturally define the *entropy rate* of an information source, which gives the average entropy per letter of the source.

Definition 2.54. *The entropy rate of an information source $\{X_k\}$ is defined as*

$$H_X = \lim_{n \to \infty} \frac{1}{n} H(X_1, X_2, \cdots, X_n) \tag{2.220}$$

when the limit exists.

We show in the next two examples that the entropy rate of a source may or may not exist.

Example 2.55. Let $\{X_k\}$ be an i.i.d. source with generic random variable X. Then

$$\lim_{n \to \infty} \frac{1}{n} H(X_1, X_2, \cdots, X_n) = \lim_{n \to \infty} \frac{nH(X)}{n} \tag{2.221}$$

$$= \lim_{n \to \infty} H(X) \tag{2.222}$$

$$= H(X), \tag{2.223}$$

i.e., the entropy rate of an i.i.d. source is the entropy of any of its single letters.

Example 2.56. Let $\{X_k\}$ be a source such that X_k are mutually independent and $H(X_k) = k$ for $k \geq 1$. Then

$$\frac{1}{n}H(X_1, X_2, \cdots, X_n) = \frac{1}{n}\sum_{k=1}^{n} k \qquad (2.224)$$

$$= \frac{1}{n}\frac{n(n+1)}{2} \qquad (2.225)$$

$$= \frac{1}{2}(n+1), \qquad (2.226)$$

which does not converge as $n \to \infty$ although $\check{H}(X_k) < \infty$ for all k. Therefore, the entropy rate of $\{X_k\}$ does not exist.

Toward characterizing the asymptotic behavior of $\{X_k\}$, it is natural to consider the limit

$$H'_X = \lim_{n\to\infty} H(X_n|X_1, X_2, \cdots, X_{n-1}) \qquad (2.227)$$

if it exists. The quantity $H(X_n|X_1, X_2, \cdots, X_{n-1})$ is interpreted as the conditional entropy of the next letter given that we know all the past history of the source, and H'_X is the limit of this quantity after the source has been run for an indefinite amount of time.

Definition 2.57. *An information source $\{X_k\}$ is stationary if*

$$X_1, X_2, \cdots, X_m \qquad (2.228)$$

and

$$X_{1+l}, X_{2+l}, \cdots, X_{m+l} \qquad (2.229)$$

have the same joint distribution for any $m, l \geq 1$.

In the rest of the section, we will show that stationarity is a sufficient condition for the existence of the entropy rate of an information source.

Lemma 2.58. *Let $\{X_k\}$ be a stationary source. Then H'_X exists.*

Proof. Since $H(X_n|X_1, X_2, \cdots, X_{n-1})$ is lower bounded by zero for all n, it suffices to prove that $H(X_n|X_1, X_2, \cdots, X_{n-1})$ is non-increasing in n to conclude that the limit H'_X exists. Toward this end, for $n \geq 2$, consider

$$H(X_n|X_1, X_2, \cdots, X_{n-1})$$
$$\leq H(X_n|X_2, X_3, \cdots, X_{n-1}) \qquad (2.230)$$
$$= H(X_{n-1}|X_1, X_2, \cdots, X_{n-2}), \qquad (2.231)$$

where the last step is justified by the stationarity of $\{X_k\}$. The lemma is proved. $\square$.

Lemma 2.59 (Cesáro Mean). *Let a_k and b_k be real numbers. If $a_n \to a$ as $n \to \infty$ and $b_n = \frac{1}{n}\sum_{k=1}^n a_k$, then $b_n \to a$ as $n \to \infty$.*

Proof. The idea of the lemma is the following. If $a_n \to a$ as $n \to \infty$, then the average of the first n terms in $\{a_k\}$, namely b_n, also tends to a as $n \to \infty$.

The lemma is formally proved as follows. Since $a_n \to a$ as $n \to \infty$, for every $\epsilon > 0$, there exists $N(\epsilon)$ such that $|a_n - a| < \epsilon$ for all $n > N(\epsilon)$. For $n > N(\epsilon)$, consider

$$|b_n - a| = \left| \frac{1}{n} \sum_{i=1}^n a_i - a \right| \tag{2.232}$$

$$= \left| \frac{1}{n} \sum_{i=1}^n (a_i - a) \right| \tag{2.233}$$

$$\leq \frac{1}{n} \sum_{i=1}^n |a_i - a| \tag{2.234}$$

$$= \frac{1}{n} \left(\sum_{i=1}^{N(\epsilon)} |a_i - a| + \sum_{i=N(\epsilon)+1}^n |a_i - a| \right) \tag{2.235}$$

$$< \frac{1}{n} \sum_{i=1}^{N(\epsilon)} |a_i - a| + \frac{(n - N(\epsilon))\epsilon}{n} \tag{2.236}$$

$$< \frac{1}{n} \sum_{i=1}^{N(\epsilon)} |a_i - a| + \epsilon. \tag{2.237}$$

The first term tends to 0 as $n \to \infty$. Therefore, for any $\epsilon > 0$, by taking n to be sufficiently large, we can make $|b_n - a| < 2\epsilon$. Hence $b_n \to a$ as $n \to \infty$, proving the lemma. $\square$

We now prove that H'_X is an alternative definition/interpretation of the entropy rate of $\{X_k\}$ when $\{X_k\}$ is stationary.

Theorem 2.60. *The entropy rate H_X of a stationary source $\{X_k\}$ exists and is equal to H'_X.*

Proof. Since we have proved in Lemma 2.58 that H'_X always exists for a stationary source $\{X_k\}$, in order to prove the theorem, we only have to prove that $H_X = H'_X$. By the chain rule for entropy,

$$\frac{1}{n} H(X_1, X_2, \cdots, X_n) = \frac{1}{n} \sum_{k=1}^n H(X_k | X_1, X_2, \cdots, X_{k-1}). \tag{2.238}$$

Since

$$\lim_{k \to \infty} H(X_k | X_1, X_2, \cdots, X_{k-1}) = H'_X \tag{2.239}$$

from (2.227), it follows from Lemma 2.59 that

$$H_X = \lim_{n \to \infty} \frac{1}{n} H(X_1, X_2, \cdots, X_n) = H'_X. \qquad (2.240)$$

The theorem is proved.   □

In this theorem, we have proved that the entropy rate of a random source $\{X_k\}$ exists under the fairly general assumption that $\{X_k\}$ is stationary. However, the entropy rate of a stationary source $\{X_k\}$ may not carry any physical meaning unless $\{X_k\}$ is also ergodic. This will be explained when we discuss the Shannon–McMillan–Breiman Theorem in Section 5.4.

Appendix 2.A: Approximation of Random Variables with Countably Infinite Alphabets by Truncation

Let X be a random variable with a countable alphabet $\mathcal{X}$ such that $H(X) < \infty$. Without loss of generality, $\mathcal{X}$ is taken to be the set of positive integers. Define a random variable $X(m)$ which takes values in

$$\mathcal{N}_m = \{1, 2, \cdots, m\} \qquad (2.241)$$

such that

$$\Pr\{X(m) = k\} = \frac{\Pr\{X = k\}}{\Pr\{X \in \mathcal{N}_m\}} \qquad (2.242)$$

for all $k \in \mathcal{N}_m$, i.e., the distribution of $X(m)$ is the truncation of the distribution of X up to m.

It is intuitively correct that $H(X(m)) \to H(X)$ as $m \to \infty$, which we formally prove in this appendix. For every $m \geq 1$, define the binary random variable

$$B(m) = \begin{cases} 1 \text{ if } X \leq m \\ 0 \text{ if } X > m \end{cases}. \qquad (2.243)$$

Consider

$$H(X) = -\sum_{k=1}^{m} \Pr\{X = k\} \log \Pr\{X = k\}$$

$$- \sum_{k=m+1}^{\infty} \Pr\{X = k\} \log \Pr\{X = k\}. \qquad (2.244)$$

As $m \to \infty$,

$$-\sum_{k=1}^{m} \Pr\{X = k\} \log \Pr\{X = k\} \to H(X). \qquad (2.245)$$

Since $H(X) < \infty$,

$$- \sum_{k=m+1}^{\infty} \Pr\{X = k\} \log \Pr\{X = k\} \to 0 \qquad (2.246)$$

as $k \to \infty$. Now consider

$$
\begin{aligned}
H(X) & \\
&= H(X|B(m)) + I(X; B(m)) & (2.247) \\
&= H(X|B(m) = 1)\Pr\{B(m) = 1\} + H(X|B(m) = 0) \\
&\quad \times \Pr\{B(m) = 0\} + I(X; B(m)) & (2.248) \\
&= H(X(m))\Pr\{B(m) = 1\} + H(X|B(m) = 0) \\
&\quad \times \Pr\{B(m) = 0\} + I(X; B(m)). & (2.249)
\end{aligned}
$$

As $m \to \infty$, $H(B(m)) \to 0$ since $\Pr\{B(m) = 1\} \to 1$. This implies $I(X; B(m)) \to 0$ because

$$I(X; B(m)) \leq H(B(m)). \qquad (2.250)$$

In (2.249), we further consider

$$
\begin{aligned}
&H(X|B(m) = 0)\Pr\{B(m) = 0\} \\
&= - \sum_{k=m+1}^{\infty} \Pr\{X = k\} \log \frac{\Pr\{X = k\}}{\Pr\{B(m) = 0\}} & (2.251) \\
&= - \sum_{k=m+1}^{\infty} \Pr\{X = k\}(\log \Pr\{X = k\} \\
&\quad - \log \Pr\{B(m) = 0\}) & (2.252) \\
&= - \sum_{k=m+1}^{\infty} (\Pr\{X = k\} \log \Pr\{X = k\}) \\
&\quad + \left(\sum_{k=m+1}^{\infty} \Pr\{X = k\} \right) \log \Pr\{B(m) = 0\} & (2.253) \\
&= - \sum_{k=m+1}^{\infty} \Pr\{X = k\} \log \Pr\{X = k\} \\
&\quad + \Pr\{B(m) = 0\} \log \Pr\{B(m) = 0\}. & (2.254)
\end{aligned}
$$

As $m \to \infty$, the summation above tends to 0 by (2.246). Since $\Pr\{B(m) = 0\} \to 0$, $\Pr\{B(m) = 0\} \log \Pr\{B(m) = 0\} \to 0$. Therefore,

$$H(X|B(m) = 0)\Pr\{B(m) = 0\} \to 0, \qquad (2.255)$$

and we see from (2.249) that $H(X(m)) \to H(X)$ as $m \to \infty$.

Chapter Summary

Markov Chain: $X \to Y \to Z$ forms a Markov chain if and only if

$$p(x, y, z) = a(x, y)b(y, z)$$

for all x, y, and z such that $p(y) > 0$.

Shannon's Information Measures:

$$H(X) = -\sum_x p(x) \log p(x) = -E \log p(X),$$

$$I(X; Y) = \sum_{x,y} p(x, y) \log \frac{p(x, y)}{p(x)p(y)} = E \log \frac{p(X, Y)}{p(X)p(Y)},$$

$$H(Y|X) = -\sum_{x,y} p(x, y) \log p(y|x) = -E \log p(Y|X),$$

$$I(X; Y|Z) = \sum_{x,y,z} p(x, y, z) \log \frac{p(x, y|z)}{p(x|z)p(y|z)} = E \log \frac{p(X, Y|Z)}{p(X|Z)p(Y|Z)}.$$

Some Useful Identities:

$$H(X) = I(X; X),$$
$$H(Y|X) = H(X, Y) - H(X),$$
$$I(X; Y) = H(X) - H(X|Y),$$
$$I(X; Y|Z) = H(X|Z) - H(X|Y, Z).$$

Chain Rule for Entropy:

$$H(X_1, X_2, \cdots, X_n) = \sum_{i=1}^{n} H(X_i|X_1, \cdots, X_{i-1}).$$

Chain Rule for Mutual Information:

$$I(X_1, X_2, \cdots, X_n; Y) = \sum_{i=1}^{n} I(X_i; Y|X_1, \cdots, X_{i-1}).$$

Informational Divergence: For two probability distributions p and q on a common alphabet $\mathcal{X}$,

$$D(p\|q) = \sum_x p(x) \log \frac{p(x)}{q(x)} = E_p \log \frac{p(X)}{q(X)}.$$

Fundamental Inequality: For any $a > 0$, $\ln a \le a - 1$, with equality if and only if $a = 1$.

Divergence Inequality: $D(p\|q) \geq 0$, with equality if and only if $p = q$.

Log-Sum Inequality: For positive numbers $a_1, a_2, \cdots$ and nonnegative numbers $b_1, b_2, \cdots$ such that $\sum_i a_i < \infty$ and $0 < \sum_i b_i < \infty$,

$$\sum_i a_i \log \frac{a_i}{b_i} \geq \left(\sum_i a_i \right) \log \frac{\sum_i a_i}{\sum_i b_i}.$$

Equality holds if and only if $\frac{a_i}{b_i} = constant$ for all i.

The Basic Inequalities: All Shannon's information measures are nonnegative.

Some Useful Properties of Shannon's Information Measures:

1. $H(X) \leq \log |\mathcal{X}|$ with equality if and only if X is uniform.
2. $H(X) = 0$ if and only if X is deterministic.
3. $H(Y|X) = 0$ if and only if Y is a function of X.
4. $I(X;Y) = 0$ if and only X and Y are independent.

Fano's Inequality: Let X and $\hat{X}$ be random variables taking values in the same alphabet $\mathcal{X}$. Then

$$H(X|\hat{X}) \leq h_b(P_e) + P_e \log(|\mathcal{X}| - 1).$$

Conditioning Does Not Increase Entropy: $H(Y|X) \leq H(Y)$, with equality if and only if X and Y are independent.

Independence Bound for Entropy:

$$H(X_1, X_2, \cdots, X_n) \leq \sum_{i=1}^{n} H(X_i)$$

with equality if and only if X_i, $i = 1, 2, \cdots, n$ are mutually independent.

Data Processing Theorem: If $U \rightarrow X \rightarrow Y \rightarrow V$ forms a Markov chain, then $I(U;V) \leq I(X;Y)$.

Maximum Entropy Distributions: Let

$$p^*(x) = e^{-\lambda_0 - \sum_{i=1}^{m} \lambda_i r_i(x)}$$

for all $x \in \mathcal{S}$, where $\lambda_0, \lambda_1, \cdots, \lambda_m$ are chosen such that the constraints

$$\sum_{x \in \mathcal{S}_p} p(x) r_i(x) = a_i \quad \text{for } 1 \leq i \leq m$$

are satisfied. Then p^* maximizes $H(p)$ over all probability distributions p on $\mathcal{S}$ subject to the above constraints.

Entropy Rate of a Stationary Source:

1. The entropy rate of an information source $\{X_k\}$ is defined as

$$H_X = \lim_{n \to \infty} \frac{1}{n} H(X_1, X_2, \cdots, X_n)$$

when the limit exists.
2. The entropy rate H_X of a stationary source $\{X_k\}$ exists and is equal to

$$H'_X = \lim_{n \to \infty} H(X_n | X_1, X_2, \cdots, X_{n-1}).$$

Problems

1. Let X and Y be random variables with alphabets $\mathcal{X} = \mathcal{Y} = \{1, 2, 3, 4, 5\}$ and joint distribution $p(x, y)$ given by

$$\frac{1}{25}\begin{bmatrix} 1 & 1 & 1 & 1 & 1 \\ 2 & 1 & 2 & 0 & 0 \\ 2 & 0 & 1 & 1 & 1 \\ 0 & 3 & 0 & 2 & 0 \\ 0 & 0 & 1 & 1 & 3 \end{bmatrix}.$$

Calculate $H(X), H(Y), H(X|Y), H(Y|X)$, and $I(X;Y)$.
2. Prove Propositions 2.8, 2.9, 2.10, 2.19, 2.21, and 2.22.
3. Give an example which shows that pairwise independence does not imply mutual independence.
4. Verify that $p(x, y, z)$ as defined in Definition 2.4 is a probability distribution. You should exclude all the zero probability masses from the summation carefully.
5. *Linearity of expectation* It is well known that expectation is linear, i.e., $E[f(X) + g(Y)] = Ef(X) + Eg(Y)$, where the summation in an expectation is taken over the corresponding alphabet. However, we adopt in information theory the convention that the summation in an expectation is taken over the corresponding support. Justify carefully the linearity of expectation under this convention.
6. The identity $I(X;Y) = H(X) - H(X|Y)$ is invalid if $H(X|Y)$ (and hence $H(X)$) is equal to infinity. Give an example such that $I(X;Y)$ has a finite value but both $H(X)$ and $H(Y|X)$ are equal to infinity.
7. Let p'_{XY} and p_{XY} be probability distributions defined on $\mathcal{X} \times \mathcal{Y}$, where $\mathcal{X}$ and $\mathcal{Y}$ are fixed finite alphabets. Prove that

$$\lim_{p'_{XY} \to p_{XY}} p'_x = p_X,$$

where the limit is taken with respect to the variational distance.

8. Let p_k and p be probability distributions defined on a common finite alphabet. Show that as $k \to \infty$, if $p_k \to p$ in variational distance, then $p_k \to p$ in $\mathcal{L}^2$ and vice versa.

9. Consider any probability distribution $p(x,y,z)$ and let

$$q(x,y,z) = \begin{cases} p(x)p(y)p(z|x,y) & \text{if } p(x,y) > 0 \\ 0 & \text{otherwise} \end{cases}.$$

a) Show that $q(x,y,z)$ is in general not a probability distribution.

b) By ignoring the fact that $q(x,y,z)$ may not be a probability distribution, application of the divergence inequality $D(p\|q) \geq 0$ would yield the inequality

$$H(X) + H(Y) + H(Z|X,Y) \geq H(X,Y,Z),$$

which indeed holds for all jointly distributed random variables X, Y, and Z. Explain.

10. Let $C_\alpha = \sum_{n=2}^{\infty} \frac{1}{n(\log n)^\alpha}$.

a) Prove that

$$C_\alpha \begin{cases} < \infty \text{ if } \alpha > 1 \\ = \infty \text{ if } 0 \leq \alpha \leq 1 \end{cases}.$$

Then

$$p_\alpha(n) = [C_\alpha n(\log n)^\alpha]^{-1}, \quad n = 2, 3, \cdots$$

is a probability distribution for $\alpha > 1$.

b) Prove that

$$H(p_\alpha) \begin{cases} < \infty \text{ if } \alpha > 2 \\ = \infty \text{ if } 1 < \alpha \leq 2 \end{cases}.$$

11. Prove that $H(p)$ is concave in p, i.e., for $0 \leq \lambda \leq 1$ and $\bar{\lambda} = 1 - \lambda$,

$$\lambda H(p_1) + \bar{\lambda} H(p_2) \leq H(\lambda p_1 + \bar{\lambda} p_2).$$

12. Let $(X, Y) \sim p(x, y) = p(x)p(y|x)$.

a) Prove that for fixed $p(x)$, $I(X;Y)$ is a convex functional of $p(y|x)$.

b) Prove that for fixed $p(y|x)$, $I(X;Y)$ is a concave functional of $p(x)$.

13. Do $I(X;Y) = 0$ and $I(X;Y|Z) = 0$ imply each other? If so, give a proof. If not, give a counterexample.

14. Give an example for which $D(\cdot\|\cdot)$ does not satisfy the triangular inequality.

15. Let X be a function of Y. Prove that $H(X) \leq H(Y)$. Interpret this result.

16. Prove that for any $n \geq 2$,

$$H(X_1, X_2, \cdots, X_n) \geq \sum_{i=1}^{n} H(X_i|X_j, j \neq i).$$

17. Prove that

$$H(X_1, X_2) + H(X_2, X_3) + H(X_1, X_3) \geq 2H(X_1, X_2, X_3).$$

Hint: Sum the identities

$$H(X_1, X_2, X_3) = H(X_j, j \neq i) + H(X_i | X_j, j \neq i)$$

for $i = 1, 2, 3$ and apply the result in Problem 16.

18. For a subset α of $\mathcal{N}_n = \{1, 2, \cdots, n\}$, denote $(X_i, i \in \alpha)$ by X_α. For $1 \leq k \leq n$, let

$$H_k = \frac{1}{\binom{n}{k}} \sum_{\alpha : |\alpha| = k} \frac{H(X_\alpha)}{k}.$$

Here H_k is interpreted as the average entropy per random variable when k random variables are taken from $X_1, X_2, \cdots, X_n$ at a time. Prove that

$$H_1 \geq H_2 \geq \cdots \geq H_n.$$

This sequence of inequalities, due to Han [147], is a generalization of the independence bound for entropy (Theorem 2.39). See Problem 6 in Chapter 21 for an application of these inequalities.

19. For a subset α of $\mathcal{N}_n = \{1, 2, \cdots, n\}$, let $\overline{\alpha} = \mathcal{N}_n \backslash \alpha$ and denote $(X_i, i \in \alpha)$ by X_α. For $1 \leq k \leq n$, let

$$H_k' = \frac{1}{\binom{n}{k}} \sum_{\alpha : |\alpha| = k} \frac{H(X_\alpha | X_{\overline{\alpha}})}{k}.$$

Prove that

$$H_1' \leq H_2' \leq \cdots \leq H_n'.$$

Note that H_n' is equal to H_n in the last problem. This sequence of inequalities is again due to Han [147]. See Yeung and Cai [406] for an application of these inequalities.

20. Prove the divergence inequality by using the log-sum inequality.

21. Prove that $D(p\|q)$ is convex in the pair (p, q), i.e., if (p_1, q_1) and (p_2, q_2) are two pairs of probability distributions on a common alphabet, then

$$D(\lambda p_1 + \overline{\lambda} p_2 \| \lambda q_1 + \overline{\lambda} q_2) \leq \lambda D(p_1 \| q_1) + \overline{\lambda} D(p_2 \| q_2)$$

for all $0 \leq \lambda \leq 1$, where $\overline{\lambda} = 1 - \lambda$.

22. Let p_{XY} and q_{XY} be two probability distributions on $\mathcal{X} \times \mathcal{Y}$. Prove that $D(p_{XY} \| q_{XY}) \geq D(p_X \| q_X)$.

23. *Pinsker's inequality.* Let $V(p, q)$ denote the variational distance between two probability distributions p and q on a common alphabet $\mathcal{X}$. We will determine the largest c which satisfies

$$D(p\|q) \geq c d^2(p, q).$$

a) Let $A = \{x : p(x) \geq q(x)\}$, $\hat{p} = \{p(A), 1 - p(A)\}$, and $\hat{q} = \{q(A), 1 - q(A)\}$. Show that $D(p\|q) \geq D(\hat{p}\|\hat{q})$ and $V(p, q) = V(\hat{p}, \hat{q})$.

b) Show that toward determining the largest value of c, we only have to consider the case when $\mathcal{X}$ is binary.

c) By virtue of (b), it suffices to determine the largest c such that

$$p \log \frac{p}{q} + (1 - p) \log \frac{1 - p}{1 - q} - 4c(p - q)^2 \geq 0$$

for all $0 \leq p, q \leq 1$, with the convention that $0 \log \frac{0}{b} = 0$ for $b \geq 0$ and $a \log \frac{a}{0} = \infty$ for $a > 0$. By observing that equality in the above holds if $p = q$ and considering the derivative of the left-hand side with respect to q, show that the largest value of c is equal to $(2 \ln 2)^{-1}$.

24. Let p and $q_k, k \geq 1$ be probability distributions on a common alphabet. Show that if q_k converges to p in divergence, then it also converges to p in variational distance.

25. Find a necessary and sufficient condition for Fano's inequality to be tight.

26. Determine the probability distribution defined on $\{0, 1, \cdots, n\}$ that maximizes the entropy subject to the constraint that the mean is equal to m, where $0 \leq m \leq n$.

27. Show that for a stationary source $\{X_k\}$, $\frac{1}{n} H(X_1, X_2, \cdots, X_n)$ is non-increasing in n.

28. For real numbers $\alpha > 1$ and $\beta > 0$ and an integer $n \geq \alpha$, define the probability distribution

$$\mathcal{D}_n^{(\alpha,\beta)} = \left\{ 1 - \left(\frac{\log \alpha}{\log n} \right)^\beta, \underbrace{\frac{1}{n} \left(\frac{\log \alpha}{\log n} \right)^\beta, \cdots, \frac{1}{n} \left(\frac{\log \alpha}{\log n} \right)^\beta}_{n}, 0, 0, \cdots \right\}.$$

Let $\nu = \{1, 0, 0, \ldots\}$ be the deterministic distribution.

a) Show that $\lim_{n \to \infty} D\left(\nu \| \mathcal{D}_n^{(\alpha,\beta)} \right) = 0$.

b) Determine $\lim_{n \to \infty} H\left(\mathcal{D}_n^{(\alpha,\beta)} \right)$.

29. *Discontinuity of entropy with respect to convergence in divergence.* Let $\mathcal{P}$ be the set of all probability distributions on a countable alphabet. A function $f : \mathcal{P} \to \Re$ is continuous with respect to convergence in divergence at $P \in \mathcal{P}$ if for any $\epsilon > 0$, there exists $\delta > 0$ such that $|f(P) - f(Q)| < \epsilon$ for all $Q \in \mathcal{P}$ satisfying $D(P\|Q) < \delta$; otherwise, f is discontinuous at P.

a) Let $H : \mathcal{P} \to \Re$ be the entropy function. Show that H is discontinuous at the deterministic distribution $\nu = \{1, 0, 0, \cdots, \}$. Hint: Use the results in Problem 28.

b) Show that H is discontinuous at $P = \{p_0, p_1, p_2, \cdots\}$ for all P such that $H(P) < \infty$. Hint: Consider the probability distribution

$$Q_n = \left\{ p_0 - \frac{p_0}{\sqrt{\log n}}, p_1 + \frac{p_0}{n\sqrt{\log n}}, p_2 + \frac{p_0}{n\sqrt{\log n}}, \cdots, \right.$$

$$\left. p_n + \frac{p_0}{n\sqrt{\log n}}, p_{n+1}, p_{n+2}, \cdots \right\}$$

for large n.

30. *Discontinuity of entropy with respect to convergence in variational distance.* Refer to Problem 29. The continuity of a function $f : \mathcal{P} \to \Re$ with respect to convergence in variational distance can be defined similarly.

   a) Show that if a function f is continuous with respect to convergence in variational distance, then it is also continuous with respect to convergence in divergence. Hint: Use Pinsker's inequality.

   b) Repeat (b) in Problem 29 with continuity defined with respect to convergence in variational distance.

31. *Continuity of the entropy function for a fixed finite alphabet.* Refer to Problems 29 and 30. Suppose the domain of H is confined to $\mathcal{P}'$, the set of all probability distributions on a fixed finite alphabet. Show that H is continuous with respect to convergence in divergence.

32. Let $\mathbf{p} = \{p_1, p_2, \cdots, p_n\}$ and $\mathbf{q} = \{q_1, q_2, \cdots, q_n\}$ be two sets of real numbers such that $p_i \geq p_{i'}$ and $q_i \geq q_{i'}$ for all $i < i'$. We say that $\mathbf{p}$ is *majorized* by $\mathbf{q}$ if $\sum_{i=1}^{m} p_i \leq \sum_{j=1}^{m} q_j$ for all $m = 1, 2, \ldots, n$, where equality holds when $m = n$. A function $f : \Re^n \to \Re$ is *Schur-concave* if $f(\mathbf{p}) \geq f(\mathbf{q})$ whenever $\mathbf{p}$ is majorized by $\mathbf{q}$. Now let $\mathbf{p}$ and $\mathbf{q}$ be probability distributions. We will show in the following steps that $H(\cdot)$ is Schur-concave.

   a) Show that for $\mathbf{p} \neq \mathbf{q}$, there exist $1 \leq j < k \leq n$ which satisfy the following:
      i) j is the largest index i such that $p_i < q_i$.
      ii) k is the smallest index i such that $i > j$ and $p_i > q_i$.
      iii) $p_i = q_i$ for all $j < i < k$.

   b) Consider the distribution $\mathbf{q}^* = \{q_1^*, q_2^*, \cdots, q_n^*\}$ defined by $q_i^* = q_i$ for $i \neq j, k$ and

   $$(q_j^*, q_k^*) = \begin{cases} (p_j, q_k + (q_j - p_j)) & \text{if } p_k - q_k \geq q_j - p_j \\ (q_j - (p_k - q_k), p_k) & \text{if } p_k - q_k < q_j - p_j \end{cases}.$$

   Note that either $q_j^* = p_j$ or $q_k^* = p_k$. Show that
      i) $q_i^* \geq q_{i'}^*$ for all $i \leq i'$.
      ii) $\sum_{i=1}^{m} p_i \leq \sum_{i=1}^{m} q_i^*$ for all $m = 1, 2, \cdots, n$.
      iii) $H(\mathbf{q}^*) \geq H(\mathbf{q})$.

   c) Prove that $H(\mathbf{p}) \geq H(\mathbf{q})$ by induction on the Hamming distance between $\mathbf{p}$ and $\mathbf{q}$, i.e., the number of places where $\mathbf{p}$ and $\mathbf{q}$ differ.

   In general, if a concave function f is symmetric, i.e., $f(\mathbf{p}) = f(\mathbf{p}')$ where $\mathbf{p}'$ is a permutation of $\mathbf{p}$, then f is Schur-concave. We refer the reader to [246] for the theory of majorization. (Hardy, Littlewood, and Pólya [154].)

Historical Notes

The concept of entropy has its root in thermodynamics. Shannon [322] was the first to use entropy as a measure of information. Informational divergence was introduced by Kullback and Leibler [214], and it has been studied extensively by Csiszár [80] and Amari [14].

Most of the materials in this chapter can be found in standard textbooks in information theory. The main concepts and results are due to Shannon [322]. Pinsker's inequality is due to Pinsker [292]. Fano's inequality has its origin in the converse proof of the channel coding theorem (to be discussed in Chapter 7) by Fano [107]. Generalizations of Fano's inequality which apply to random variables with countable alphabets have been obtained by Han and Verdú [153] and by Ho [165] (see also [168]). Maximum entropy, a concept in statistical mechanics, was expounded in Jaynes [186].

3

The I-Measure

In Chapter 2, we have illustrated the relationship between Shannon's information measures for two random variables by the diagram in Figure 2.2. For convenience, Figure 2.2 is reproduced in Figure 3.1 with the random variables X and Y replaced by X_1 and X_2, respectively. This diagram suggests that Shannon's information measures for any $n \geq 2$ random variables may have a set-theoretic structure.

In this chapter, we develop a theory which establishes a one-to-one correspondence between Shannon's information measures and set theory in full generality. With this correspondence, manipulations of Shannon's information measures can be viewed as set operations, thus allowing the rich suite of tools in set theory to be used in information theory. Moreover, the structure of Shannon's information measures can easily be visualized by means of an *information diagram* if four or fewer random variables are involved. The use of information diagrams simplifies many difficult proofs in information theory

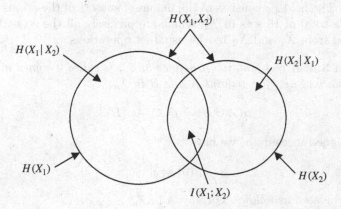

Fig. 3.1. Relationship between entropies and mutual information for two random variables.

problems. More importantly, these results, which may be difficult to discover in the first place, can easily be obtained by inspection of an information diagram.

The main concepts to be used in this chapter are from measure theory. However, it is not necessary for the reader to know measure theory to read this chapter.

3.1 Preliminaries

In this section, we introduce a few basic concepts in measure theory which will be used subsequently. These concepts will be illustrated by simple examples.

Definition 3.1. *The field $\mathcal{F}_n$ generated by sets $\tilde{X}_1, \tilde{X}_2, \cdots, \tilde{X}_n$ is the collection of sets which can be obtained by any sequence of usual set operations (union, intersection, complement, and difference) on $\tilde{X}_1, \tilde{X}_2, \cdots, \tilde{X}_n$.*

Definition 3.2. *The atoms of $\mathcal{F}_n$ are sets of the form $\cap_{i=1}^n Y_i$, where Y_i is either $\tilde{X}_i$ or $\tilde{X}_i^c$, the complement of $\tilde{X}_i$.*

There are 2^n atoms and 2^{2^n} sets in $\mathcal{F}_n$. Evidently, all the atoms in $\mathcal{F}_n$ are disjoint, and each set in $\mathcal{F}_n$ can be expressed uniquely as the union of a subset of the atoms of $\mathcal{F}_n$.[1] We assume that the sets $\tilde{X}_1, \tilde{X}_2, \cdots, \tilde{X}_n$ intersect with each other generically, i.e., all the atoms of $\mathcal{F}_n$ are nonempty unless otherwise specified.

Example 3.3. The sets $\tilde{X}_1$ and $\tilde{X}_2$ generate the field $\mathcal{F}_2$. The atoms of $\mathcal{F}_2$ are

$$\tilde{X}_1 \cap \tilde{X}_2, \tilde{X}_1^c \cap \tilde{X}_2, \tilde{X}_1 \cap \tilde{X}_2^c, \tilde{X}_1^c \cap \tilde{X}_2^c, \tag{3.1}$$

which are represented by the four distinct regions in the Venn diagram in Figure 3.2. The field $\mathcal{F}_2$ consists of the unions of subsets of the atoms in (3.1). There are a total of 16 sets in $\mathcal{F}_2$, which are precisely all the sets which can be obtained from $\tilde{X}_1$ and $\tilde{X}_2$ by the usual set operations.

Definition 3.4. *A real function μ defined on $\mathcal{F}_n$ is called a signed measure if it is set-additive, i.e., for disjoint A and B in $\mathcal{F}_n$,*

$$\mu(A \cup B) = \mu(A) + \mu(B). \tag{3.2}$$

For a signed measure μ, we have

$$\mu(\emptyset) = 0, \tag{3.3}$$

which can be seen as follows. For any A in $\mathcal{F}_n$,

---

[1] We adopt the convention that the union of the empty subset of the atoms of $\mathcal{F}_n$ is the empty set.

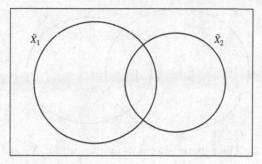

Fig. 3.2. The Venn diagram for $\tilde{X}_1$ and $\tilde{X}_2$.

$$\mu(A) = \mu(A \cup \emptyset) = \mu(A) + \mu(\emptyset) \tag{3.4}$$

by set-additivity because A and $\emptyset$ are disjoint, which implies (3.3).

A signed measure μ on $\mathcal{F}_n$ is completely specified by its values on the atoms of $\mathcal{F}_n$. The values of μ on the other sets in $\mathcal{F}_n$ can be obtained via set-additivity.

Example 3.5. A signed measure μ on $\mathcal{F}_2$ is completely specified by the values

$$\mu(\tilde{X}_1 \cap \tilde{X}_2), \mu(\tilde{X}_1^c \cap \tilde{X}_2), \mu(\tilde{X}_1 \cap \tilde{X}_2^c), \mu(\tilde{X}_1^c \cap \tilde{X}_2^c). \tag{3.5}$$

The value of μ on $\tilde{X}_1$, for example, can be obtained as

$$\mu(\tilde{X}_1) = \mu((\tilde{X}_1 \cap \tilde{X}_2) \cup (\tilde{X}_1 \cap \tilde{X}_2^c)) \tag{3.6}$$
$$= \mu(\tilde{X}_1 \cap \tilde{X}_2) + \mu(\tilde{X}_1 \cap \tilde{X}_2^c). \tag{3.7}$$

3.2 The *I*-Measure for Two Random Variables

To fix ideas, we first formulate in this section the one-to-one correspondence between Shannon's information measures and set theory for two random variables. For random variables X_1 and X_2, let $\tilde{X}_1$ and $\tilde{X}_2$ be sets corresponding to X_1 and X_2, respectively. The sets $\tilde{X}_1$ and $\tilde{X}_2$ generate the field $\mathcal{F}_2$ whose atoms are listed in (3.1). In our formulation, we set the universal set Ω to $\tilde{X}_1 \cup \tilde{X}_2$ for reasons which will become clear later. With this choice of Ω, the Venn diagram for $\tilde{X}_1$ and $\tilde{X}_2$ is represented by the diagram in Figure 3.3. For simplicity, the sets $\tilde{X}_1$ and $\tilde{X}_2$ are, respectively, labeled by X_1 and X_2 in the diagram. We call this the *information diagram* for the random variables X_1 and X_2. In this diagram, the universal set, which is the union of $\tilde{X}_1$ and $\tilde{X}_2$, is not shown explicitly just as in a usual Venn diagram. Note that with our choice of the universal set, the atom $\tilde{X}_1^c \cap \tilde{X}_2^c$ degenerates to the empty set, because

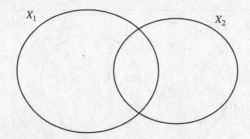

Fig. 3.3. The generic information diagram for X_1 and X_2.

$$\tilde{X}_1^c \cap \tilde{X}_2^c = (\tilde{X}_1 \cup \tilde{X}_2)^c = \Omega^c = \emptyset. \tag{3.8}$$

Thus this atom is not shown in the information diagram in Figure 3.3.

For random variables X_1 and X_2, the Shannon's information measures are

$$H(X_1), H(X_2), H(X_1|X_2), H(X_2|X_1), H(X_1, X_2), I(X_1; X_2). \tag{3.9}$$

Writing $A \cap B^c$ as $A - B$, we now define a signed measure[2] μ^* by

$$\mu^*(\tilde{X}_1 - \tilde{X}_2) = H(X_1|X_2), \tag{3.10}$$
$$\mu^*(\tilde{X}_2 - \tilde{X}_1) = H(X_2|X_1), \tag{3.11}$$

and

$$\mu^*(\tilde{X}_1 \cap \tilde{X}_2) = I(X_1; X_2). \tag{3.12}$$

These are the values of μ^* on the nonempty atoms of $\mathcal{F}_2$ (i.e., atoms of $\mathcal{F}_2$ other than $\tilde{X}_1^c \cap \tilde{X}_2^c$). The values of μ^* on the other sets in $\mathcal{F}_2$ can be obtained via set-additivity. In particular, the relations

$$\mu^*(\tilde{X}_1 \cup \tilde{X}_2) = H(X_1, X_2), \tag{3.13}$$
$$\mu^*(\tilde{X}_1) = H(X_1), \tag{3.14}$$

and

$$\mu^*(\tilde{X}_2) = H(X_2) \tag{3.15}$$

can readily be verified. For example, (3.13) is seen to be true by considering

$$\mu^*(\tilde{X}_1 \cup \tilde{X}_2)$$
$$= \mu^*(\tilde{X}_1 - \tilde{X}_2) + \mu^*(\tilde{X}_2 - \tilde{X}_1) + \mu^*(\tilde{X}_1 \cap \tilde{X}_2) \tag{3.16}$$
$$= H(X_1|X_2) + H(X_2|X_1) + I(X_1; X_2) \tag{3.17}$$
$$= H(X_1, X_2). \tag{3.18}$$

The right-hand sides of (3.10)–(3.15) are the six Shannon's information measures for X_1 and X_2 in (3.9). Now observe that (3.10)–(3.15) are consistent with how the Shannon's information measures on the right-hand side

---

[2] It happens that μ^* defined here for $n = 2$ assumes only nonnegative values, but we will see in Section 3.4 that μ^* can assume negative values for $n \geq 3$.

are identified in Figure 3.1, with the left circle and the right circle representing the sets $\tilde{X}_1$ and $\tilde{X}_2$, respectively. Specifically, in each of these equations, the left-hand side and the right-hand side correspond to each other via the following substitution of symbols:

$$H/I \;\leftrightarrow\; \mu^*,$$
$$, \;\leftrightarrow\; \cup,$$
$$; \;\leftrightarrow\; \cap, \qquad\qquad (3.19)$$
$$| \;\leftrightarrow\; -.$$

Note that we make no distinction between the symbols H and I in this substitution. Thus, for two random variables X_1 and X_2, Shannon's information measures can be regarded formally as a signed measure on $\mathcal{F}_2$. We will refer to μ^* as the *I-Measure* for the random variables X_1 and X_2.[3]

Upon realizing that Shannon's information measures can be viewed as a signed measure, we can apply the rich family of operations in set theory to information theory. This explains why Figure 3.1 or Figure 3.3 represents the relationships among all Shannon's information measures for two random variables correctly. As an example, consider the following set identity which is readily identified in Figure 3.3:

$$\mu^*(\tilde{X}_1 \cup \tilde{X}_2) = \mu^*(\tilde{X}_1) + \mu^*(\tilde{X}_2) - \mu^*(\tilde{X}_1 \cap \tilde{X}_2). \qquad (3.20)$$

This identity is a special case of the inclusion-exclusion formula in set theory. By means of the substitution of symbols in (3.19), we immediately obtain the information identity

$$H(X_1, X_2) = H(X_1) + H(X_2) - I(X_1; X_2). \qquad (3.21)$$

We end this section with a remark. The value of μ^* on the atom $\tilde{X}_1^c \cap \tilde{X}_2^c$ has no apparent information-theoretic meaning. In our formulation, we set the universal set Ω to $\tilde{X}_1 \cup \tilde{X}_2$ so that the atom $\tilde{X}_1^c \cap \tilde{X}_2^c$ degenerates to the empty set. Then $\mu^*(\tilde{X}_1^c \cap \tilde{X}_2^c)$ naturally vanishes because μ^* is a measure, so that μ^* is completely specified by all Shannon's information measures involving the random variables X_1 and X_2.

3.3 Construction of the I-Measure μ^*

We have constructed the I-Measure for two random variables in the last section. We now construct the I-Measure for any $n \geq 2$ random variables.

Consider n random variables $X_1, X_2, \cdots, X_n$. For any random variable X, let $\tilde{X}$ be a set corresponding to X. Let

---

[3] The reader should not confuse μ^* with the probability measure defining the random variables X_1 and X_2. The former, however, is determined by the latter.

$$\mathcal{N}_n = \{1, 2, \cdots, n\}. \tag{3.22}$$

Define the universal set Ω to be the union of the sets $\tilde{X}_1, \tilde{X}_2, \cdots, \tilde{X}_n$, i.e.,

$$\Omega = \bigcup_{i \in \mathcal{N}_n} \tilde{X}_i. \tag{3.23}$$

We use $\mathcal{F}_n$ to denote the field generated by $\tilde{X}_1, \tilde{X}_2, \cdots, \tilde{X}_n$. The set

$$A_0 = \bigcap_{i \in \mathcal{N}_n} \tilde{X}_i^c \tag{3.24}$$

is called the *empty atom* of $\mathcal{F}_n$ because

$$\bigcap_{i \in \mathcal{N}_n} \tilde{X}_i^c = \left(\bigcup_{i \in \mathcal{N}_n} \tilde{X}_i \right)^c = \Omega^c = \emptyset. \tag{3.25}$$

All the atoms of $\mathcal{F}_n$ other than A_0 are called *nonempty atoms*.

Let $\mathcal{A}$ be the set of all nonempty atoms of $\mathcal{F}_n$. Then $|\mathcal{A}|$, the cardinality of $\mathcal{A}$, is equal to $2^n - 1$. A signed measure μ on $\mathcal{F}_n$ is completely specified by the values of μ on the nonempty atoms of $\mathcal{F}_n$.

To simplify notation, we will use X_G to denote $(X_i, i \in G)$ and $\tilde{X}_G$ to denote $\cup_{i \in G} \tilde{X}_i$ for any nonempty subset G of $\mathcal{N}_n$.

Theorem 3.6. *Let*

$$\mathcal{B} = \left\{ \tilde{X}_G : G \text{ is a nonempty subset of } \mathcal{N}_n \right\}. \tag{3.26}$$

Then a signed measure μ on $\mathcal{F}_n$ is completely specified by $\{\mu(B), B \in \mathcal{B}\}$, which can be any set of real numbers.

Proof. The number of elements in $\mathcal{B}$ is equal to the number of nonempty subsets of $\mathcal{N}_n$, which is $2^n - 1$. Thus $|\mathcal{A}| = |\mathcal{B}| = 2^n - 1$. Let $k = 2^n - 1$. Let $\mathbf{u}$ be a column k-vector of $\mu(A), A \in \mathcal{A}$, and $\mathbf{h}$ be a column k-vector of $\mu(B), B \in \mathcal{B}$. Since all the sets in $\mathcal{B}$ can expressed uniquely as the union of some nonempty atoms in $\mathcal{A}$, by the set-additivity of μ, for each $B \in \mathcal{B}$, $\mu(B)$ can be expressed uniquely as the sum of some components of $\mathbf{u}$. Thus

$$\mathbf{h} = C_n \mathbf{u}, \tag{3.27}$$

where C_n is a *unique* $k \times k$ matrix. On the other hand, it can be shown (see Appendix 3.A) that for each $A \in \mathcal{A}$, $\mu(A)$ can be expressed as a linear combination of $\mu(B), B \in \mathcal{B}$ by applications, if necessary, of the following two identities:

$$\mu(A \cap B - C) = \mu(A - C) + \mu(B - C) - \mu(A \cup B - C), \tag{3.28}$$

$$\mu(A - B) = \mu(A \cup B) - \mu(B). \tag{3.29}$$

However, the existence of the said expression does not imply its uniqueness. Nevertheless, we can write

$$\mathbf{u} = D_n \mathbf{h} \qquad (3.30)$$

for some $k \times k$ matrix D_n. Upon substituting (3.27) into (3.30), we obtain

$$\mathbf{u} = (D_n C_n)\mathbf{u}, \qquad (3.31)$$

which implies that D_n is the inverse of C_n as (3.31) holds regardless of the choice of μ. Since C_n is unique, so is D_n. Therefore, $\mu(A), A \in \mathcal{A}$ are uniquely determined once $\mu(B), B \in \mathcal{B}$ are specified. Hence, a signed measure μ on $\mathcal{F}_n$ is completely specified by $\{\mu(B), B \in \mathcal{B}\}$, which can be any set of real numbers. The theorem is proved.   □

We now prove the following two lemmas which are related by the substitution of symbols in (3.19).

Lemma 3.7.

$$\mu(A \cap B - C) = \mu(A \cup C) + \mu(B \cup C) - \mu(A \cup B \cup C) - \mu(C). \qquad (3.32)$$

Proof. From (3.28) and (3.29), we have

$$
\begin{aligned}
&\mu(A \cap B - C) \\
&= \mu(A - C) + \mu(B - C) - \mu(A \cup B - C) & (3.33) \\
&= (\mu(A \cup C) - \mu(C)) + (\mu(B \cup C) - \mu(C)) \\
&\quad -(\mu(A \cup B \cup C) - \mu(C)) & (3.34) \\
&= \mu(A \cup C) + \mu(B \cup C) - \mu(A \cup B \cup C) - \mu(C). & (3.35)
\end{aligned}
$$

The lemma is proved.   □

Lemma 3.8.

$$I(X;Y|Z) = H(X,Z) + H(Y,Z) - H(X,Y,Z) - H(Z). \qquad (3.36)$$

Proof. Consider

$$
\begin{aligned}
&I(X;Y|Z) \\
&= H(X|Z) - H(X|Y,Z) & (3.37) \\
&= H(X,Z) - H(Z) - (H(X,Y,Z) - H(Y,Z)) & (3.38) \\
&= H(X,Z) + H(Y,Z) - H(X,Y,Z) - H(Z). & (3.39)
\end{aligned}
$$

The lemma is proved.   □

We now construct the I-Measure μ^* on $\mathcal{F}_n$ using Theorem 3.6 by defining

$$\mu^*(\tilde{X}_G) = H(X_G) \tag{3.40}$$

for all nonempty subsets G of $\mathcal{N}_n$. In order for μ^* to be meaningful, it has to be consistent with all Shannon's information measures (via the substitution of symbols in (3.19)). In that case, the following must hold for all (not necessarily disjoint) subsets G, G', G'' of $\mathcal{N}_n$ where G and G' are nonempty:

$$\mu^*(\tilde{X}_G \cap \tilde{X}_{G'} - \tilde{X}_{G''}) = I(X_G; X_{G'}|X_{G''}). \tag{3.41}$$

When $G'' = \emptyset$, (3.41) becomes

$$\mu^*(\tilde{X}_G \cap \tilde{X}_{G'}) = I(X_G; X_{G'}). \tag{3.42}$$

When $G = G'$, (3.41) becomes

$$\mu^*(\tilde{X}_G - \tilde{X}_{G''}) = H(X_G|X_{G''}). \tag{3.43}$$

When $G = G'$ and $G'' = \emptyset$, (3.41) becomes

$$\mu^*(\tilde{X}_G) = H(X_G). \tag{3.44}$$

Thus (3.41) covers all the four cases of Shannon's information measures, and it is the necessary and sufficient condition for μ^* to be consistent with all Shannon's information measures.

Theorem 3.9. μ^* *is the unique signed measure on* $\mathcal{F}_n$ *which is consistent with all Shannon's information measures.*

Proof. Consider

$$
\begin{aligned}
\mu^*(\tilde{X}_G \cap \tilde{X}_{G'} - \tilde{X}_{G''}) & \\
= \mu^*(\tilde{X}_{G\cup G''}) + \mu^*(\tilde{X}_{G'\cup G''}) - \mu^*(\tilde{X}_{G\cup G'\cup G''}) - \mu^*(\tilde{X}_{G''}) & \tag{3.45} \\
= H(X_{G\cup G''}) + H(X_{G'\cup G''}) - H(X_{G\cup G'\cup G''}) - H(X_{G''}) & \tag{3.46} \\
= I(X_G; X_{G'}|X_{G''}), & \tag{3.47}
\end{aligned}
$$

where (3.45) and (3.47) follow from Lemmas 3.7 and 3.8, respectively, and (3.46) follows from (3.40), the definition of μ^*. Thus we have proved (3.41), i.e., μ^* is consistent with all Shannon's information measures.

In order that μ^* is consistent with all Shannon's information measures, for all nonempty subsets G of $\mathcal{N}_n$, μ^* has to satisfy (3.44), which in fact is the definition of μ^* in (3.40). Therefore, μ^* is the unique signed measure on $\mathcal{F}_n$ which is consistent with all Shannon's information measures. $\square$

3.4 μ^* Can Be Negative

In the previous sections, we have been cautious in referring to the I-Measure μ^* as a signed measure instead of a measure.[4] In this section, we show that μ^* in fact can take negative values for $n \geq 3$.

For $n = 2$, the three nonempty atoms of $\mathcal{F}_2$ are

$$\tilde{X}_1 \cap \tilde{X}_2, \tilde{X}_1 - \tilde{X}_2, \tilde{X}_2 - \tilde{X}_1. \tag{3.48}$$

The values of μ^* on these atoms are, respectively,

$$I(X_1; X_2), H(X_1|X_2), H(X_2|X_1). \tag{3.49}$$

These quantities are Shannon's information measures and hence nonnegative by the basic inequalities. Therefore, μ^* is always nonnegative for $n = 2$.

For $n = 3$, the seven nonempty atoms of $\mathcal{F}_3$ are

$$\tilde{X}_i - \tilde{X}_{\{j,k\}}, \tilde{X}_i \cap \tilde{X}_j - \tilde{X}_k, \tilde{X}_1 \cap \tilde{X}_2 \cap \tilde{X}_3, \tag{3.50}$$

where $1 \leq i < j < k \leq 3$. The values of μ^* on the first two types of atoms are

$$\mu^*(\tilde{X}_i - \tilde{X}_{\{j,k\}}) = H(X_i|X_j, X_k) \tag{3.51}$$

and

$$\mu^*(\tilde{X}_i \cap \tilde{X}_j - \tilde{X}_k) = I(X_i; X_j|X_k), \tag{3.52}$$

respectively, which are Shannon's information measures and therefore nonnegative. However, $\mu^*(\tilde{X}_1 \cap \tilde{X}_2 \cap \tilde{X}_3)$ does not correspond to a Shannon's information measure. In the next example, we show that $\mu^*(\tilde{X}_1 \cap \tilde{X}_2 \cap \tilde{X}_3)$ can actually be negative.

Example 3.10. In this example, all entropies are in the base 2. Let X_1 and X_2 be independent binary random variables with

$$\Pr\{X_i = 0\} = \Pr\{X_i = 1\} = 0.5, \tag{3.53}$$

$i = 1, 2$. Let

$$X_3 = (X_1 + X_2) \bmod 2. \tag{3.54}$$

It is easy to check that X_3 has the same marginal distribution as X_1 and X_2. Thus,

$$H(X_i) = 1 \tag{3.55}$$

for $i = 1, 2, 3$. Moreover, X_1, X_2, and X_3 are pairwise independent. Therefore,

$$H(X_i, X_j) = 2 \tag{3.56}$$

---

[4] A measure can assume only nonnegative values.

and

$$I(X_i; X_j) = 0 \tag{3.57}$$

for $1 \leq i < j \leq 3$. We further see from (3.54) that each random variable is a function of the other two random variables. Then by the chain rule for entropy, we have

$$H(X_1, X_2, X_3) = H(X_1, X_2) + H(X_3|X_1, X_2) \tag{3.58}$$
$$= 2 + 0 \tag{3.59}$$
$$= 2. \tag{3.60}$$

Now for $1 \leq i < j < k \leq 3$,

$$I(X_i; X_j|X_k)$$
$$= H(X_i, X_k) + H(X_j, X_k) - H(X_1, X_2, X_3) - H(X_k) \tag{3.61}$$
$$= 2 + 2 - 2 - 1 \tag{3.62}$$
$$= 1, \tag{3.63}$$

where we have invoked Lemma 3.8. It then follows that

$$\mu^*(\tilde{X}_1 \cap \tilde{X}_2 \cap \tilde{X}_3) = \mu^*(\tilde{X}_1 \cap \tilde{X}_2) - \mu^*(\tilde{X}_1 \cap \tilde{X}_2 - \tilde{X}_3) \tag{3.64}$$
$$= I(X_1; X_2) - I(X_1; X_2|X_3) \tag{3.65}$$
$$= 0 - 1 \tag{3.66}$$
$$= -1. \tag{3.67}$$

Thus μ^* takes a negative value on the atom $\tilde{X}_1 \cap \tilde{X}_2 \cap \tilde{X}_3$.

Motivated by the substitution of symbols in (3.19) for Shannon's information measures, we will write $\mu^*(\tilde{X}_1 \cap \tilde{X}_2 \cap \tilde{X}_3)$ as $I(X_1; X_2; X_3)$. In general, we will write

$$\mu^*(\tilde{X}_{G_1} \cap \tilde{X}_{G_2} \cap \cdots \cap \tilde{X}_{G_m} - \tilde{X}_F) \tag{3.68}$$

as

$$I(X_{G_1}; X_{G_2}; \cdots; X_{G_m}|X_F) \tag{3.69}$$

and refer to it as the *mutual information* between $X_{G_1}, X_{G_2}, \cdots, X_{G_m}$ conditioning on X_F. Then (3.64) in the above example can be written as

$$I(X_1; X_2; X_3) = I(X_1; X_2) - I(X_1; X_2|X_3). \tag{3.70}$$

For this example, $I(X_1; X_2; X_3) < 0$, which implies

$$I(X_1; X_2|X_3) > I(X_1; X_2). \tag{3.71}$$

Therefore, unlike entropy, the mutual information between two random variables can be increased by conditioning on a third random variable. Also, we note in (3.70) that although the expression on the right-hand side is not symbolically symmetrical in X_1, X_2, and X_3, we see from the left-hand side that it is in fact symmetrical in X_1, X_2, and X_3.

3.5 Information Diagrams

We have established in Section 3.3 a one-to-one correspondence between Shannon's information measures and set theory. Therefore, it is valid to use an *information diagram*, which is a variation of a Venn diagram, to represent the relationship between Shannon's information measures.

For simplicity, a set $\tilde{X}_i$ will be labeled by X_i in an information diagram. We have seen the generic information diagram for $n = 2$ in Figure 3.3. A generic information diagram for $n = 3$ is shown in Figure 3.4. The information-theoretic labeling of the values of μ^* on some of the sets in $\mathcal{F}_3$ is shown in the diagram. As an example, the information diagram for the I-Measure for random variables X_1, X_2, and X_3 discussed in Example 3.10 is shown in Figure 3.5.

For $n \geq 4$, it is not possible to display an information diagram perfectly in two dimensions. In general, an information diagram for n random variables needs $n - 1$ dimensions to be displayed perfectly. Nevertheless, for $n = 4$, an information diagram can be displayed in two dimensions almost perfectly as

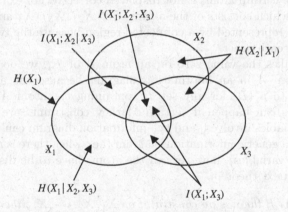

Fig. 3.4. The generic information diagram for X_1, X_2, and X_3.

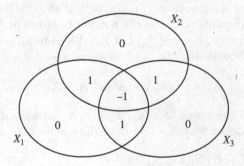

Fig. 3.5. The information diagram for X_1, X_2, and X_3 in Example 3.10.

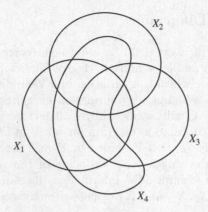

Fig. 3.6. The generic information diagram for X_1, X_2, X_3, and X_4.

shown in Figure 3.6. This information diagram is correct in that the region representing the set $\tilde{X}_4$ splits each atom in Figure 3.4 into two atoms. However, the adjacency of certain atoms is not displayed correctly. For example, the set $\tilde{X}_1 \cap \tilde{X}_2 \cap \tilde{X}_4^c$, which consists of the atoms $\tilde{X}_1 \cap \tilde{X}_2 \cap \tilde{X}_3 \cap \tilde{X}_4^c$ and $\tilde{X}_1 \cap \tilde{X}_2 \cap \tilde{X}_3^c \cap \tilde{X}_4^c$, is not represented by a connected region because the two atoms are not adjacent to each other.

When μ^* takes the value zero on an atom A of $\mathcal{F}_n$, we do not need to display the atom A in an information diagram because the atom A does not contribute to $\mu^*(B)$ for any set B containing the atom A. As we will see shortly, this can happen if certain Markov constraints are imposed on the random variables involved, and the information diagram can be simplified accordingly. In a generic information diagram (i.e., when there is no constraint on the random variables), however, all the atoms have to be displayed, as is implied by the next theorem.

Theorem 3.11. *If there is no constraint on $X_1, X_2, \cdots, X_n$, then μ^* can take any set of nonnegative values on the nonempty atoms of $\mathcal{F}_n$.*

Proof. We will prove the theorem by constructing an I-Measure μ^* which can take any set of nonnegative values on the nonempty atoms of $\mathcal{F}_n$. Recall that $\mathcal{A}$ is the set of all nonempty atoms of $\mathcal{F}_n$. Let $Y_A, A \in \mathcal{A}$ be mutually independent random variables. Now define the random variables $X_i, i = 1, 2, \cdots, n$ by

$$X_i = (Y_A : A \in \mathcal{A} \text{ and } A \subset \tilde{X}_i). \tag{3.72}$$

We determine the I-Measure μ^* for $X_1, X_2, \cdots, X_n$ so defined as follows. Since Y_A are mutually independent, for any nonempty subsets G of $\mathcal{N}_n$, we have

$$H(X_G) = H(X_i, i \in G) \tag{3.73}$$
$$= H((Y_A : A \in \mathcal{A} \text{ and } A \subset \tilde{X}_i), i \in G) \tag{3.74}$$

$$= H(Y_A : A \in \mathcal{A} \text{ and } A \subset \tilde{X}_G) \tag{3.75}$$

$$= \sum_{A \in \mathcal{A}: A \subset \tilde{X}_G} H(Y_A). \tag{3.76}$$

On the other hand,

$$H(X_G) = \mu^*(\tilde{X}_G) = \sum_{A \in \mathcal{A}: A \subset \tilde{X}_G} \mu^*(A). \tag{3.77}$$

Equating the right-hand sides of (3.76) and (3.77), we have

$$\sum_{A \in \mathcal{A}: A \subset \tilde{X}_G} H(Y_A) = \sum_{A \in \mathcal{A}: A \subset \tilde{X}_G} \mu^*(A). \tag{3.78}$$

Evidently, we can make the above equality hold for all nonempty subsets G of $\mathcal{N}_n$ by taking

$$\mu^*(A) = H(Y_A) \tag{3.79}$$

for all $A \in \mathcal{A}$. By the uniqueness of μ^*, this is also the only possibility for μ^*. Since $H(Y_A)$ can take any nonnegative value by Corollary 2.44, μ^* can take any set of nonnegative values on the nonempty atoms of $\mathcal{F}_n$. The theorem is proved. $\square$

In the rest of this section, we explore the structure of Shannon's information measures when $X_1 \to X_2 \to \cdots \to X_n$ forms a Markov chain. To start with, we consider $n = 3$, i.e., $X_1 \to X_2 \to X_3$ forms a Markov chain. Since

$$\mu^*(\tilde{X}_1 \cap \tilde{X}_2^c \cap \tilde{X}_3) = I(X_1; X_3 | X_2) = 0, \tag{3.80}$$

the atom $\tilde{X}_1 \cap \tilde{X}_2^c \cap \tilde{X}_3$ does not have to be displayed in an information diagram. Therefore, in constructing the information diagram, the regions representing the random variables X_1, X_2, and X_3 should overlap with each other such that the region corresponding to the atom $\tilde{X}_1 \cap \tilde{X}_2^c \cap \tilde{X}_3$ is empty, while the regions corresponding to all other nonempty atoms are nonempty. Figure 3.7 shows such a construction, in which each random variable is represented by a "mountain." From Figure 3.7, we see that $\tilde{X}_1 \cap \tilde{X}_2 \cap \tilde{X}_3$, as the only atom on which μ^* may take a negative value, now becomes identical to the atom $\tilde{X}_1 \cap \tilde{X}_3$. Therefore, we have

$$I(X_1; X_2; X_3) = \mu^*(\tilde{X}_1 \cap \tilde{X}_2 \cap \tilde{X}_3) \tag{3.81}$$

$$= \mu^*(\tilde{X}_1 \cap \tilde{X}_3) \tag{3.82}$$

$$= I(X_1; X_3) \tag{3.83}$$

$$\geq 0. \tag{3.84}$$

Hence, we conclude that when $X_1 \to X_2 \to X_3$ forms a Markov chain, μ^* is always nonnegative.

Next, we consider $n = 4$, i.e., $X_1 \to X_2 \to X_3 \to X_4$ forms a Markov chain. With reference to Figure 3.6, we first show that under this Markov constraint, μ^* always vanishes on certain nonempty atoms:

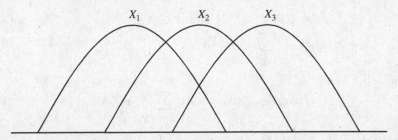

Fig. 3.7. The information diagram for the Markov chain $X_1 \to X_2 \to X_3$.

1. The Markov chain $X_1 \to X_2 \to X_3$ implies

$$I(X_1; X_3; X_4|X_2) + I(X_1; X_3|X_2, X_4) = I(X_1; X_3|X_2) = 0. \qquad (3.85)$$

2. The Markov chain $X_1 \to X_2 \to X_4$ implies

$$I(X_1; X_3; X_4|X_2) + I(X_1; X_4|X_2, X_3) = I(X_1; X_4|X_2) = 0. \qquad (3.86)$$

3. The Markov chain $X_1 \to X_3 \to X_4$ implies

$$I(X_1; X_2; X_4|X_3) + I(X_1; X_4|X_2, X_3) = I(X_1; X_4|X_3) = 0. \qquad (3.87)$$

4. The Markov chain $X_2 \to X_3 \to X_4$ implies

$$I(X_1; X_2; X_4|X_3) + I(X_2; X_4|X_1, X_3) = I(X_2; X_4|X_3) = 0. \qquad (3.88)$$

5. The Markov chain $(X_1, X_2) \to X_3 \to X_4$ implies

$$I(X_1; X_2; X_4|X_3) + I(X_1; X_4|X_2, X_3) + I(X_2; X_4|X_1, X_3)$$
$$= I(X_1, X_2; X_4|X_3) \qquad (3.89)$$
$$= 0. \qquad (3.90)$$

Now (3.85) and (3.86) imply

$$I(X_1; X_4|X_2, X_3) = I(X_1; X_3|X_2, X_4), \qquad (3.91)$$

(3.87) and (3.91) imply

$$I(X_1; X_2; X_4|X_3) = -I(X_1; X_3|X_2, X_4), \qquad (3.92)$$

and (3.88) and (3.92) imply

$$I(X_2; X_4|X_1, X_3) = I(X_1; X_3|X_2, X_4). \qquad (3.93)$$

The terms on the left-hand sides of (3.91), (3.92), and (3.93) are the three terms on the left-hand side of (3.90). Then we substitute (3.91), (3.92), and (3.93) in (3.90) to obtain

$$\mu^*(\tilde{X}_1 \cap \tilde{X}_2^c \cap \tilde{X}_3 \cap \tilde{X}_4^c) = I(X_1; X_3 | X_2, X_4) = 0. \tag{3.94}$$

From (3.85), (3.91), (3.92), and (3.93), (3.94) implies

$$\mu^*(\tilde{X}_1 \cap \tilde{X}_2^c \cap \tilde{X}_3 \cap \tilde{X}_4) = I(X_1; X_3; X_4 | X_2) = 0, \tag{3.95}$$

$$\mu^*(\tilde{X}_1 \cap \tilde{X}_2^c \cap \tilde{X}_3^c \cap \tilde{X}_4) = I(X_1; X_4 | X_2, X_3) = 0, \tag{3.96}$$

$$\mu^*(\tilde{X}_1 \cap \tilde{X}_2 \cap \tilde{X}_3^c \cap \tilde{X}_4) = I(X_1; X_2; X_4 | X_3) = 0, \tag{3.97}$$

$$\mu^*(\tilde{X}_1^c \cap \tilde{X}_2 \cap \tilde{X}_3^c \cap \tilde{X}_4) = I(X_2; X_4 | X_1, X_3) = 0. \tag{3.98}$$

From (3.94) to (3.98), we see that μ^* always vanishes on the atoms

$$\begin{aligned}
&\tilde{X}_1 \cap \tilde{X}_2^c \cap \tilde{X}_3 \cap \tilde{X}_4^c, \\
&\tilde{X}_1 \cap \tilde{X}_2^c \cap \tilde{X}_3 \cap \tilde{X}_4, \\
&\tilde{X}_1 \cap \tilde{X}_2^c \cap \tilde{X}_3^c \cap \tilde{X}_4, \\
&\tilde{X}_1 \cap \tilde{X}_2 \cap \tilde{X}_3^c \cap \tilde{X}_4, \\
&\tilde{X}_1^c \cap \tilde{X}_2 \cap \tilde{X}_3^c \cap \tilde{X}_4
\end{aligned} \tag{3.99}$$

of $\mathcal{F}_4$, which we mark by an asterisk in the information diagram in Figure 3.8.

In fact, the reader can gain a lot of insight by letting $I(X_1; X_3 | X_2, X_4) = a \geq 0$ in (3.85) and tracing the subsequent steps leading to the above conclusion in the information diagram in Figure 3.6.

It is not necessary to display the five atoms in (3.99) in an information diagram because μ^* always vanishes on these atoms. Therefore, in constructing the information diagram, the regions representing the random variables should overlap with each other such that the regions corresponding to these five nonempty atoms are empty, while the regions corresponding to the other ten nonempty atoms, namely

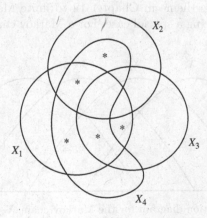

Fig. 3.8. The atoms of $\mathcal{F}_4$ on which μ^* vanishes when $X_1 \to X_2 \to X_3 \to X_4$ forms a Markov chain.

$$\tilde{X}_1 \cap \tilde{X}_2^c \cap \tilde{X}_3^c \cap \tilde{X}_4^c, \; \tilde{X}_1 \cap \tilde{X}_2 \cap \tilde{X}_3^c \cap \tilde{X}_4^c,$$
$$\tilde{X}_1 \cap \tilde{X}_2 \cap \tilde{X}_3 \cap \tilde{X}_4^c, \; \tilde{X}_1 \cap \tilde{X}_2 \cap \tilde{X}_3 \cap \tilde{X}_4,$$
$$\tilde{X}_1^c \cap \tilde{X}_2 \cap \tilde{X}_3^c \cap \tilde{X}_4^c, \; \tilde{X}_1^c \cap \tilde{X}_2 \cap \tilde{X}_3 \cap \tilde{X}_4^c, \tag{3.100}$$
$$\tilde{X}_1^c \cap \tilde{X}_2 \cap \tilde{X}_3 \cap \tilde{X}_4, \; \tilde{X}_1^c \cap \tilde{X}_2^c \cap \tilde{X}_3 \cap \tilde{X}_4^c,$$
$$\tilde{X}_1^c \cap \tilde{X}_2^c \cap \tilde{X}_3 \cap \tilde{X}_4, \; \tilde{X}_1^c \cap \tilde{X}_2^c \cap \tilde{X}_3^c \cap \tilde{X}_4,$$

are nonempty. Figure 3.9 shows such a construction. The reader should compare the information diagrams in Figures 3.7 and 3.9 and observe that the latter is an extension of the former.

From Figure 3.9, we see that the values of μ^* on the ten nonempty atoms in (3.100) are equivalent to

$$H(X_1|X_2, X_3, X_4), \; I(X_1; X_2|X_3, X_4),$$
$$I(X_1; X_3|X_4), \; I(X_1; X_4),$$
$$H(X_2|X_1, X_3, X_4), \; I(X_2; X_3|X_1; X_4), \tag{3.101}$$
$$I(X_2; X_4|X_1), \; H(X_3|X_1, X_2, X_4),$$
$$I(X_3; X_4|X_1, X_2), \; H(X_4|X_1, X_2, X_3),$$

respectively.[5] Since these are all Shannon's information measures and thus nonnegative, we conclude that μ^* is always nonnegative.

When $X_1 \to X_2 \to \cdots \to X_n$ forms a Markov chain, for $n = 3$, there is only one nonempty atom, namely $\tilde{X}_1 \cap \tilde{X}_2^c \cap \tilde{X}_3$, on which μ^* always vanishes. This atom can be determined directly from the Markov constraint $I(X_1; X_3|X_2) = 0$. For $n = 4$, the five nonempty atoms on which μ^* always vanishes are listed in (3.99). The determination of these atoms, as we have seen, is not straightforward. We have also shown that for $n = 3$ and $n = 4$, μ^* is always nonnegative.

We will extend this theme in Chapter 12 to finite Markov random fields with Markov chains being a special case. For a Markov chain, the information

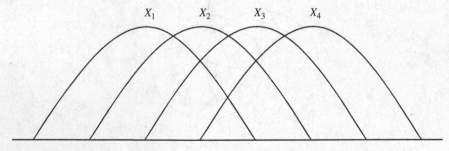

Fig. 3.9. The information diagram for the Markov chain $X_1 \to X_2 \to X_3 \to X_4$.

---

[5] A formal proof will be given in Theorem 12.30.

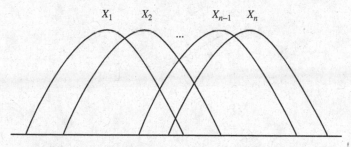

Fig. 3.10. The information diagram for the Markov chain $X_1 \to X_2 \to \cdots \to X_n$.

diagram can always be displayed in two dimensions as in Figure 3.10, and μ^* is always nonnegative. These will be explained in Chapter 12.

3.6 Examples of Applications

In this section, we give a few examples of applications of information diagrams. These examples show how information diagrams can help solve information theory problems.

The use of an information diagram is highly intuitive. To obtain an information identity from an information diagram is WYSIWYG.[6] However, how to obtain an information inequality from an information diagram needs some explanation.

Very often, we use a Venn diagram to represent a measure μ which takes nonnegative values. If we see in the Venn diagram two sets A and B such that A is a subset of B, then we can immediately conclude that $\mu(A) \leq \mu(B)$ because

$$\mu(B) - \mu(A) = \mu(B - A) \geq 0. \tag{3.102}$$

However, an *I*-Measure μ^* can take negative values. Therefore, when we see in an information diagram that A is a subset of B, we cannot conclude from this fact alone that $\mu^*(A) \leq \mu^*(B)$ unless we know from the setup of the problem that μ^* is nonnegative. (For example, μ^* is nonnegative if the random variables involved form a Markov chain.) Instead, information inequalities can be obtained from an information diagram in conjunction with the basic inequalities. The following examples illustrate how it works.

Example 3.12 (Concavity of Entropy). Let $X_1 \sim p_1(x)$ and $X_2 \sim p_2(x)$. Let

$$X \sim p(x) = \lambda p_1(x) + \bar\lambda p_2(x), \tag{3.103}$$

where $0 \leq \lambda \leq 1$ and $\bar\lambda = 1 - \lambda$. We will show that

$$H(X) \geq \lambda H(X_1) + \bar\lambda H(X_2). \tag{3.104}$$

---

[6] What you see is what you get.

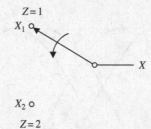

$Z = 1$

X_1

X

X_2

$Z = 2$

Fig. 3.11. The schematic diagram for Example 3.12.

Consider the system in Fig. 3.11 in which the position of the switch is determined by a random variable Z with

$$\Pr\{Z = 1\} = \lambda \quad \text{and} \quad \Pr\{Z = 2\} = \bar{\lambda}, \tag{3.105}$$

where Z is independent of X_1 and X_2. The switch takes position i if $Z = i$, $i = 1, 2$. The random variable Z is called a *mixing random variable* for the probability distributions $p_1(x)$ and $p_2(x)$. Figure 3.12 shows the information diagram for X and Z. From the diagram, we see that $\tilde{X} - \tilde{Z}$ is a subset of $\tilde{X}$. Since μ^* is nonnegative for two random variables, we can conclude that

$$\mu^*(\tilde{X}) \geq \mu^*(\tilde{X} - \tilde{Z}), \tag{3.106}$$

which is equivalent to

$$H(X) \geq H(X|Z). \tag{3.107}$$

Then

$$H(X) \geq H(X|Z) \tag{3.108}$$

$$= \Pr\{Z = 1\}H(X|Z = 1) + \Pr\{Z = 2\}H(X|Z = 2) \tag{3.109}$$

$$= \lambda H(X_1) + \bar{\lambda}H(X_2), \tag{3.110}$$

proving (3.104). This shows that $H(X)$ is a concave functional of $p(x)$.

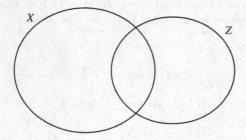

X

Z

Fig. 3.12. The information diagram for Example 3.12.

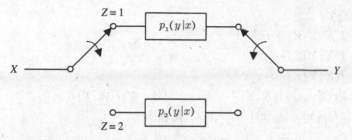

Fig. 3.13. The schematic diagram for Example 3.13.

Example 3.13 (Convexity of Mutual Information). Let

$$(X, Y) \sim p(x, y) = p(x)p(y|x). \tag{3.111}$$

We will show that for fixed $p(x)$, $I(X; Y)$ is a convex functional of $p(y|x)$.

Let $p_1(y|x)$ and $p_2(y|x)$ be two transition matrices. Consider the system in Figure 3.13 in which the position of the switch is determined by a random variable Z as in the last example, where Z is independent of X, i.e.,

$$I(X; Z) = 0. \tag{3.112}$$

In the information diagram for X, Y, and Z in Figure 3.14, let

$$I(X; Z|Y) = a \geq 0. \tag{3.113}$$

Since $I(X; Z) = 0$, we see that

$$I(X; Y; Z) = -a, \tag{3.114}$$

because

$$I(X; Z) = I(X; Z|Y) + I(X; Y; Z). \tag{3.115}$$

Then

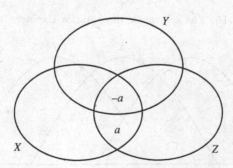

Fig. 3.14. The information diagram for Example 3.13.

$$I(X;Y)$$
$$= I(X;Y|Z) + I(X;Y;Z) \tag{3.116}$$
$$= I(X;Y|Z) - a \tag{3.117}$$
$$\leq I(X;Y|Z) \tag{3.118}$$
$$= \Pr\{Z=1\}I(X;Y|Z=1) + \Pr\{Z=2\}I(X;Y|Z=2) \tag{3.119}$$
$$= \lambda I(p(x), p_1(y|x)) + \bar{\lambda}I(p(x), p_2(y|x)), \tag{3.120}$$

where $I(p(x), p_i(y|x))$ denotes the mutual information between the input and output of a channel with input distribution $p(x)$ and transition matrix $p_i(y|x)$. This shows that for fixed $p(x)$, $I(X;Y)$ is a convex functional of $p(y|x)$.

Example 3.14 (Concavity of Mutual Information). Let

$$(X,Y) \sim p(x,y) = p(x)p(y|x). \tag{3.121}$$

We will show that for fixed $p(y|x)$, $I(X;Y)$ is a concave functional of $p(x)$.

Consider the system in Figure 3.15, where the position of the switch is determined by a random variable Z as in the last example. In this system, when X is given, Y is independent of Z, or $Z \to X \to Y$ forms a Markov chain. Then μ^* is nonnegative, and the information diagram for X, Y, and Z is shown in Figure 3.16.

From Figure 3.16, since $\tilde{X} \cap \tilde{Y} - \tilde{Z}$ is a subset of $\tilde{X} \cap \tilde{Y}$ and μ^* is nonnegative, we immediately see that

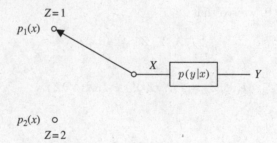

Fig. 3.15. The schematic diagram for Example 3.14.

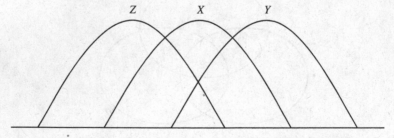

Fig. 3.16. The information diagram for Example 3.14.

$$I(X;Y)$$
$$\geq I(X;Y|Z) \tag{3.122}$$
$$= \Pr\{Z = 1\}I(X;Y|Z = 1) + \Pr\{Z = 2\}I(X;Y|Z = 2) \tag{3.123}$$
$$= \lambda I(p_1(x), p(y|x)) + \bar{\lambda}I(p_2(x), p(y|x)). \tag{3.124}$$

This shows that for fixed $p(y|x)$, $I(X;Y)$ is a concave functional of $p(x)$.

Example 3.15 (Imperfect Secrecy Theorem). Let X be the plain text, Y be the cipher text, and Z be the key in a secret key cryptosystem. Since X can be recovered from Y and Z, we have

$$H(X|Y,Z) = 0. \tag{3.125}$$

We will show that this constraint implies

$$I(X;Y) \geq H(X) - H(Z). \tag{3.126}$$

The quantity $I(X;Y)$ is a measure of the security level of the cryptosystem. In general, we want to make $I(X;Y)$ small so that the eavesdropper cannot obtain too much information about the plain text X by observing the cipher text Y. The inequality in (3.126) says that the system can attain a certain level of security only if $H(Z)$ (often called the key length) is sufficiently large. In particular, if perfect secrecy is required, i.e., $I(X;Y) = 0$, then $H(Z)$ must be at least equal to $H(X)$. This special case is known as Shannon's perfect secrecy theorem [323].[7]
We now prove (3.126). Let

$$I(X;Y|Z) = a \geq 0, \tag{3.127}$$
$$I(Y;Z|X) = b \geq 0, \tag{3.128}$$
$$H(Z|X,Y) = c \geq 0, \tag{3.129}$$

and

$$I(X;Y;Z) = d. \tag{3.130}$$

(See Figure 3.17.) Since $I(Y;Z) \geq 0$,

$$b + d \geq 0. \tag{3.131}$$

In comparing $H(X)$ with $H(Z)$, we do not have to consider $I(X;Z|Y)$ and $I(X;Y;Z)$ since they belong to both $H(X)$ and $H(Z)$. Then we see from Figure 3.17 that

$$H(X) - H(Z) = a - b - c. \tag{3.132}$$

---

[7] Shannon used a combinatorial argument to prove this theorem. An information-theoretic proof can be found in Massey [251].

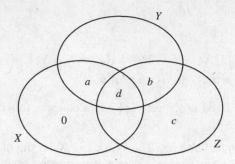

Fig. 3.17. The information diagram for Example 3.15.

Therefore,

$$I(X;Y) = a + d \qquad (3.133)$$
$$\geq a - b \qquad (3.134)$$
$$\geq a - b - c \qquad (3.135)$$
$$= H(X) - H(Z), \qquad (3.136)$$

where (3.134) and (3.135) follow from (3.131) and (3.129), respectively, proving (3.126).

Note that in deriving our result, the assumptions that $H(Y|X,Z) = 0$, i.e., the cipher text is a function of the plain text and the key, and $I(X;Z) = 0$, i.e., the plain text and the key are independent, are not necessary.

Example 3.16. Figure 3.18 shows the information diagram for the Markov chain $X \rightarrow Y \rightarrow Z$. From this diagram, we can identify the following two information identities:

$$I(X;Y) = I(X;Y,Z), \qquad (3.137)$$
$$H(X|Y) = H(X|Y,Z). \qquad (3.138)$$

Since μ^* is nonnegative and $\tilde{X} \cap \tilde{Z}$ is a subset of $\tilde{X} \cap \tilde{Y}$, we have

$$I(X;Z) \leq I(X;Y), \qquad (3.139)$$

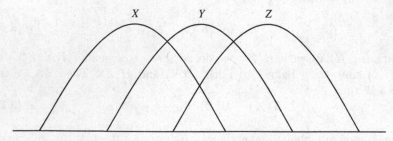

Fig. 3.18. The information diagram for the Markov chain $X \rightarrow Y \rightarrow Z$.

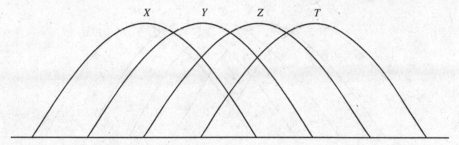

Fig. 3.19. The information diagram for the Markov chain $X \rightarrow Y \rightarrow Z \rightarrow T$.

which has already been obtained in Lemma 2.41. Similarly, we can also obtain

$$H(X|Y) \leq H(X|Z). \tag{3.140}$$

Example 3.17 (Data Processing Theorem). Figure 3.19 shows the information diagram for the Markov chain $X \rightarrow Y \rightarrow Z \rightarrow T$. Since μ^* is nonnegative and $\tilde{X} \cap \tilde{T}$ is a subset of $\tilde{Y} \cap \tilde{Z}$, we have

$$I(X;T) \leq I(Y;Z), \tag{3.141}$$

which is the data processing theorem (Theorem 2.42).

We end this chapter by giving an application of the information diagram for a Markov chain with five random variables.

Example 3.18. In this example, we prove with the help of an information diagram that for five random variables $X, Y, Z, T,$ and U such that $X \rightarrow Y \rightarrow Z \rightarrow T \rightarrow U$ forms a Markov chain,

$$H(Y) + H(T) =$$
$$I(Z; X, Y, T, U) + I(X, Y; T, U) + H(Y|Z) + H(T|Z). \tag{3.142}$$

In the information diagram for $X, Y, Z, T,$ and U in Figure 3.20, we first identify the atoms of $H(Y)$ and then the atoms of $H(T)$ by marking each of

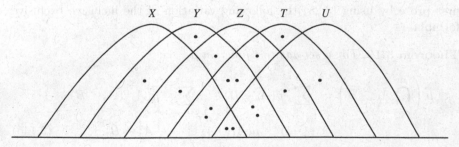

Fig. 3.20. The atoms of $H(Y) + H(T)$.

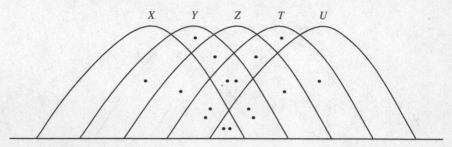

Fig. 3.21. The atoms of $I(Z; X, Y, T, U) + I(X, Y; T, U) + H(Y|Z) + H(T|Z)$.

them by a dot. If an atom belongs to both $H(Y)$ and $H(T)$, it receives two dots. The resulting diagram represents

$$H(Y) + H(T).   \tag{3.143}$$

By repeating the same procedure for

$$I(Z; X, Y, T, U) + I(X, Y; T, U) + H(Y|Z) + H(T|Z),   \tag{3.144}$$

we obtain the information diagram in Figure 3.21. Comparing these two information diagrams, we find that they are identical. Hence, the information identity in (3.142) always holds conditioning on the Markov chain $X \to Y \to Z \to T \to U$. This identity is critical in proving an outer bound on the achievable coding rate region of the multiple descriptions problem in Fu et al. [125]. It is virtually impossible to discover this identity without the help of an information diagram!

Appendix 3.A: A Variation of the Inclusion–Exclusion Formula

In this appendix, we show that for each $A \in \mathcal{A}$, $\mu(A)$ can be expressed as a linear combination of $\mu(B), B \in \mathcal{B}$ via applications of (3.28) and (3.29). We first prove by using (3.28) the following variation of the inclusive–exclusive formula.

Theorem 3.19. *For a set-additive function* μ,

$$\mu\left(\bigcap_{k=1}^{n} A_k - B\right) = \sum_{1 \le i \le n} \mu(A_i - B) - \sum_{1 \le i < j \le n} \mu(A_i \cup A_j - B) + \cdots$$
$$+ (-1)^{n+1} \mu(A_1 \cup A_2 \cup \cdots \cup A_n - B).   \tag{3.145}$$

Proof. The theorem will be proved by induction on n. First, (3.145) is obviously true for $n = 1$. Assume (3.145) is true for some $n \geq 1$. Now consider

$$\mu\left(\bigcap_{k=1}^{n+1} A_k - B\right)$$

$$= \mu\left(\left(\bigcap_{k=1}^{n} A_k\right) \cap A_{n+1} - B\right) \tag{3.146}$$

$$= \mu\left(\bigcap_{k=1}^{n} A_k - B\right) + \mu(A_{n+1} - B) - \mu\left(\left(\bigcap_{k=1}^{n} A_k\right) \cup A_{n+1} - B\right) \tag{3.147}$$

$$= \left\{ \sum_{1 \leq i \leq n} \mu(A_i - B) - \sum_{1 \leq i < j \leq n} \mu(A_i \cup A_j - B) + \cdots \right.$$

$$\left. + (-1)^{n+1}\mu(A_1 \cup A_2 \cup \cdots \cup A_n - B) \right\} + \mu(A_{n+1} - B)$$

$$- \mu\left(\bigcap_{k=1}^{n}(A_k \cup A_{n+1}) - B\right) \tag{3.148}$$

$$= \left\{ \sum_{1 \leq i \leq n} \mu(A_i - B) - \sum_{1 \leq i < j \leq n} \mu(A_i \cup A_j - B) + \cdots \right.$$

$$\left. + (-1)^{n+1}\mu(A_1 \cup A_2 \cup \cdots \cup A_n - B) \right\} + \mu(A_{n+1} - B)$$

$$- \left\{ \sum_{1 \leq i \leq n} \mu(A_i \cup A_{n+1} - B) - \sum_{1 \leq i < j \leq n} \mu(A_i \cup A_j \cup A_{n+1} - B) \right.$$

$$\left. + \cdots + (-1)^{n+1}\mu(A_1 \cup A_2 \cup \cdots \cup A_n \cup A_{n+1} - B) \right\} \tag{3.149}$$

$$= \sum_{1 \leq i \leq n+1} \mu(A_i - B) - \sum_{1 \leq i < j \leq n+1} \mu(A_i \cup A_j - B) + \cdots$$

$$+ (-1)^{n+2}\mu(A_1 \cup A_2 \cup \cdots \cup A_{n+1} - B). \tag{3.150}$$

In the above, (3.28) was used in obtaining (3.147), and the induction hypothesis was used in obtaining (3.148) and (3.149). The theorem is proved.
□

Now a nonempty atom of $\mathcal{F}_n$ has the form

$$\bigcap_{i=1}^{n} Y_i, \tag{3.151}$$

where Y_i is either $\tilde{X}_i$ or $\tilde{X}_i^c$, and there exists at least one i such that $Y_i = \tilde{X}_i$. Then we can write the atom in (3.151) as

$$\bigcap_{i:Y_i=\tilde{X}_i} \tilde{X}_i - \left(\bigcup_{j:Y_j=\tilde{X}_j^c} \tilde{X}_j \right). \tag{3.152}$$

Note that the intersection above is always nonempty. Then using (3.145) and (3.29), we see that for each $A \in \mathcal{A}$, $\mu(A)$ can be expressed as a linear combination of $\mu(B), B \in \mathcal{B}$.

Chapter Summary

Definition: The field $\mathcal{F}_n$ generated by sets $\tilde{X}_1, \tilde{X}_2, \cdots, \tilde{X}_n$ is the collection of sets which can be obtained by any sequence of usual set operations (union, intersection, complement, and difference) on $\tilde{X}_1, \tilde{X}_2, \cdots, \tilde{X}_n$.

Definition: A real function μ defined on $\mathcal{F}_n$ is called a signed measure if it is set-additive, i.e., for disjoint A and B in $\mathcal{F}_n$,

$$\mu(A \cup B) = \mu(A) + \mu(B).$$

I-Measure μ^*: There exists a unique signed measure μ^* on $\mathcal{F}_n$ which is consistent with all Shannon's information measures.

μ^* Can Be Negative: Let X_1 and X_2 be i.i.d. uniform on $\{0,1\}$, and let $X_3 = X_1 + X_2$ mod 2. Then

$$\mu^*(\tilde{X}_1 \cap \tilde{X}_2 \cap \tilde{X}_3) = I(X_1; X_2; X_3) = -1.$$

Information Diagrams for Two, Three, and Four Random Variables:

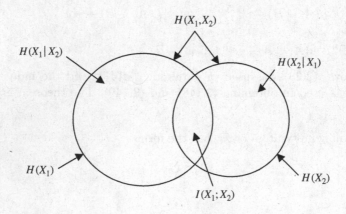

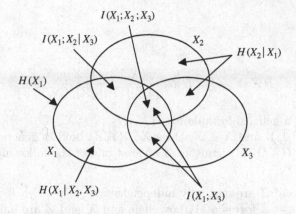

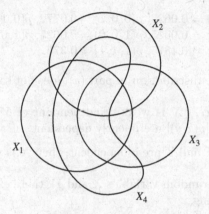

Information Diagram for Markov Chain $X_1 \to X_2 \to \cdots \to X_n$:

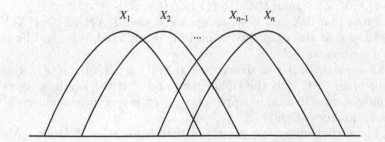

Problems

1. Show that

$$I(X;Y;Z) = E\log\frac{p(X,Y)p(Y,Z)p(X,Z)}{p(X)p(Y)p(Z)p(X,Y,Z)}$$

and obtain a general formula for $I(X_1;X_2,;\cdots;X_n)$.

2. Suppose $X \perp Y$ and $X \perp Z$. Does $X \perp (Y,Z)$ hold in general?

3. Show that $I(X;Y;Z)$ vanishes if at least one of the following conditions hold:

   a) X, Y, and Z are mutually independent;
   b) $X \to Y \to Z$ forms a Markov chain and X and Z are independent.

4. a) Verify that $I(X;Y;Z)$ vanishes for the distribution $p(x,y,z)$ given by

$$p(0,0,0) = 0.0625, \; p(0,0,1) = 0.0772, \; p(0,1,0) = 0.0625,$$
$$p(0,1,1) = 0.0625, \; p(1,0,0) = 0.0625, \; p(1,0,1) = 0.1103,$$
$$p(1,1,0) = 0.1875, \; p(1,1,1) = 0.375.$$

   b) Verify that the distribution in part (a) does not satisfy the conditions in Problem 3.

5. *Weak independence.* X is weakly independent of Y if the rows of the transition matrix $[p(x|y)]$ are linearly dependent.

   a) Show that if X and Y are independent, then X is weakly independent of Y.
   b) Show that for random variables X and Y, there exists a random variable Z satisfying

   i) $X \to Y \to Z$
   ii) X and Z are independent
   iii) Y and Z are not independent
   if and only if X is weakly independent of Y.
   (Berger and Yeung [29].)

6. Prove that
   a) $I(X;Y;Z) \geq -\min\{I(X;Y|Z), I(Y;Z|X), I(X,Z|Y)\}$.
   b) $I(X;Y;Z) \leq \min\{I(X;Y), I(Y;Z), I(X;Z)\}$.

7. a) Prove that if X and Y are independent, then $I(X,Y;Z) \geq I(X;Y|Z)$.
   b) Show that the inequality in part (a) is not valid in general by giving a counterexample.

8. In Example 3.15, it was shown that $I(X;Y) \geq H(X) - H(Z)$, where X is the plain text, Y is the cipher text, and Z is the key in a secret key cryptosystem. Give an example of a secret key cryptosystem such that this inequality is tight.

9. a) Prove that under the constraint that $X \to Y \to Z$ forms a Markov chain, $X \perp Y|Z$ and $X \perp Z$ imply $X \perp Y$.

b) Prove that the implication in (a) continues to be valid without the Markov chain constraint.

10. a) Show that $Y \perp Z|T$ does not imply $Y \perp Z|(X,T)$ by giving a counterexample.

   b) Prove that $Y \perp Z|T$ implies $Y \perp Z|(X,T)$ conditioning on $X \to Y \to Z \to T$.

11. a) Let $X \to Y \to (Z,T)$ form a Markov chain. Prove that $I(X;Z) + I(X;T) \leq I(X;Y) + I(Z;T)$.

   b) Let $X \to Y \to Z \to T$ form a Markov chain. Determine which of the following inequalities always hold:

   i) $I(X;T) + I(Y;Z) \geq I(X;Z) + I(Y;T)$.

   ii) $I(X;T) + I(Y;Z) \geq I(X;Y) + I(Z;T)$.

   iii) $I(X;Y) + I(Z;T) \geq I(X;Z) + I(Y;T)$.

12. *Secret sharing.* For a given finite set $\mathcal{P}$ and a collection $\mathcal{A}$ of subsets of $\mathcal{P}$, a secret sharing scheme is a random variable S and a family of random variables $\{X_p : p \in \mathcal{P}\}$ such that for all $A \in \mathcal{A}$,

$$H(S|X_A) = 0,$$

and for all $B \notin \mathcal{A}$,

$$H(S|X_B) = H(S).$$

Here, S is the *secret* and $\mathcal{P}$ is the set of *participants* of the scheme. A participant p of the scheme possesses a *share* X_p of the secret. The set $\mathcal{A}$ specifies the *access structure* of the scheme: For a subset A of $\mathcal{P}$, by pooling their shares, if $A \in \mathcal{A}$, the participants in A can reconstruct S, otherwise they can know nothing about S.

   a) i) Prove that for $A, B \subset \mathcal{P}$, if $B \notin \mathcal{A}$ and $A \cup B \in \mathcal{A}$, then

$$H(X_A|X_B) = H(S) + H(X_A|X_B, S).$$

   ii) Prove that if $B \in \mathcal{A}$, then

$$H(X_A|X_B) = H(X_A|X_B, S).$$

   (Capocelli et al. [56].)

   b) Prove that for $A, B, C \subset \mathcal{P}$ such that $A \cup C \in \mathcal{A}$, $B \cup C \in \mathcal{A}$, and $C \notin \mathcal{A}$, then

$$I(X_A; X_B|X_C) \geq H(S).$$

   (van Dijk [363].)

13. Consider four random variables X, Y, Z, and T which satisfy the following constraints: $H(T|X) = H(T)$, $H(T|X,Y) = 0$, $H(T|Y) = H(T)$, $H(Y|Z) = 0$, and $H(T|Z) = 0$. Prove that

   a) $H(T|X,Y,Z) = I(Z;T|X,Y) = 0$.

   b) $I(X;T|Y,Z) = I(X;Y;T|Z) = I(Y;T|X,Z) = 0$.

   c) $I(X;Z;T) = I(Y;Z;T) = 0$.

d) $H(Y|X, Z, T) = I(X; Y|Z, T) = 0.$
e) $I(X; Y; Z) \geq 0.$
f) $I(X; Z) \geq H(T).$

The inequality in (f) finds application in a secret sharing problem studied by Blundo et al. [43].

14. Prove that for random variables X, Y, Z, and T,

$$\left. \begin{array}{l} X \perp Z|Y \\ (X,Y) \perp T|Z \\ Y \perp Z|T \\ Y \perp Z|X \\ X \perp T \end{array} \right\} \Rightarrow Y \perp Z.$$

Hint: Observe that $X \perp Z|Y$ and $(X,Y) \perp T|Z$ are equivalent to $X \to Y \to Z \to T$ and use an information diagram.

15. Prove that

$$\left. \begin{array}{l} X \perp Y \\ X \perp Y|(Z,T) \\ Z \perp T|X \\ Z \perp T|Y \end{array} \right\} \Leftrightarrow \left\{ \begin{array}{l} Z \perp T \\ Z \perp T|(X,Y) \\ X \perp Y|Z \\ X \perp Y|T. \end{array} \right.$$

(Studený [346].)

Historical Notes

The original work on the set-theoretic structure of Shannon's information measures is due to Hu [173]. It was established in this paper that every information identity implies a set identity via a substitution of symbols. This allows the tools for proving information identities to be used in proving set identities. Since the paper was published in Russian, it was largely unknown to the West until it was described in Csiszár and Körner [83]. Throughout the years, the use of Venn diagrams to represent the structure of Shannon's information measures for two or three random variables has been suggested by various authors, for example, Reza [301], Abramson [2], and Papoulis [286], but no formal justification was given until Yeung [398] introduced the I-Measure.

McGill [265] proposed a multiple mutual information for any number of random variables which is equivalent to the mutual information between two or more random variables discussed here. Properties of this quantity have been investigated by Kawabata [196] and Yeung [398].

Along a related direction, Han [146] viewed the linear combination of entropies as a vector space and developed a lattice-theoretic description of Shannon's information measures.

4

Zero-Error Data Compression

In a random experiment, a coin is tossed n times. Let X_i be the outcome of the ith toss, with

$$\Pr\{X_i = \text{HEAD}\} = p \ \text{ and } \ \Pr\{X_i = \text{TAIL}\} = 1 - p, \qquad (4.1)$$

where $0 \le p \le 1$. It is assumed that X_i are i.i.d., and the value of p is known. We are asked to describe the outcome of the random experiment without error (with zero error) by using binary symbols. One way to do this is to encode a HEAD by a "0" and a TAIL by a "1." Then the outcome of the random experiment is encoded into a binary codeword of length n. When the coin is fair, i.e., $p = 0.5$, this is the best we can do because the probability of every outcome of the experiment is equal to 2^{-n}. In other words, all the outcomes are equally likely.

However, if the coin is biased, i.e., $p \ne 0.5$, the probability of an outcome of the experiment depends on the number of HEADs and the number of TAILs in the outcome. In other words, the probabilities of the outcomes are no longer uniform. It turns out that we can take advantage of this by encoding more likely outcomes into shorter codewords and less likely outcomes into longer codewords. By doing so, it is possible to use fewer than n bits *on the average* to describe the outcome of the random experiment. In particular, in the extreme case when $p = 0$ or 1, we actually do not need to describe the outcome of the experiment because it is deterministic.

At the beginning of Chapter 2, we mentioned that the entropy $H(X)$ measures the amount of information contained in a random variable X. In this chapter, we substantiate this claim by exploring the role of entropy in the context of zero-error data compression.

4.1 The Entropy Bound

In this section, we establish that $H(X)$ is a fundamental lower bound on the expected length of the number of symbols needed to describe the outcome of a random variable X with zero error. This is called the *entropy bound.*

Definition 4.1. *A D-ary source code $\mathcal{C}$ for a source random variable X is a mapping from $\mathcal{X}$ to $\mathcal{D}^*$, the set of all finite length sequences of symbols taken from a D-ary code alphabet.*

Consider an information source $\{X_k, k \geq 1\}$, where X_k are discrete random variables which take values in the same alphabet. We apply a source code $\mathcal{C}$ to each X_k and concatenate the codewords. Once the codewords are concatenated, the boundaries of the codewords are no longer explicit. In other words, when the code $\mathcal{C}$ is applied to a source sequence, a sequence of code symbols are produced, and the codewords may no longer be distinguishable. We are particularly interested in *uniquely decodable codes* which are defined as follows.

Definition 4.2. *A code $\mathcal{C}$ is uniquely decodable if for any finite source sequence, the sequence of code symbols corresponding to this source sequence is different from the sequence of code symbols corresponding to any other (finite) source sequence.*

Suppose we use a code $\mathcal{C}$ to encode a source file into a coded file. If $\mathcal{C}$ is uniquely decodable, then we can always recover the source file from the coded file. An important class of uniquely decodable codes, called *prefix codes*, are discussed in the next section. But we first look at an example of a code which is not uniquely decodable.

Example 4.3. Let $\mathcal{X} = \{A, B, C, D\}$. Consider the code $\mathcal{C}$ defined by

x	$\mathcal{C}(x)$
A	0
B	1
C	01
D	10.

Then all the three source sequences AAD, ACA, and AABA produce the code sequence 0010. Thus from the code sequence 0010, we cannot tell which of the three source sequences it comes from. Therefore, $\mathcal{C}$ is not uniquely decodable.

In the next theorem, we prove that for any uniquely decodable code, the lengths of the codewords have to satisfy an inequality called the *Kraft inequality.*

Theorem 4.4 (Kraft Inequality). *Let $\mathcal{C}$ be a D-ary source code, and let $l_1, l_2, \cdots, l_m$ be the lengths of the codewords. If $\mathcal{C}$ is uniquely decodable, then*

$$\sum_{k=1}^{m} D^{-l_k} \leq 1. \tag{4.2}$$

Proof. Let N be an arbitrary positive integer, and consider

$$\left(\sum_{k=1}^{m} D^{-l_k}\right)^N = \sum_{k_1=1}^{m} \sum_{k_2=1}^{m} \cdots \sum_{k_N=1}^{m} D^{-(l_{k_1}+l_{k_2}+\cdots+l_{k_N})}. \qquad (4.3)$$

By collecting terms on the right-hand side, we write

$$\left(\sum_{k=1}^{m} D^{-l_k}\right)^N = \sum_{i=1}^{Nl_{max}} A_i D^{-i}, \qquad (4.4)$$

where

$$l_{max} = \max_{1\leq k\leq m} l_k \qquad (4.5)$$

and A_i is the coefficient of D^{-i} in $\left(\sum_{k=1}^{m} D^{-l_k}\right)^N$. Now observe that A_i gives the total number of sequences of N codewords with a total length of i code symbols (see Example 4.5 below). Since the code is uniquely decodable, these code sequences must be distinct, and therefore

$$A_i \leq D^i \qquad (4.6)$$

because there are D^i distinct sequences of i code symbols. Substituting this inequality into (4.4), we have

$$\left(\sum_{k=1}^{m} D^{-l_k}\right)^N \leq \sum_{i=1}^{Nl_{max}} 1 = Nl_{max}, \qquad (4.7)$$

or

$$\sum_{k=1}^{m} D^{-l_k} \leq (Nl_{max})^{1/N}. \qquad (4.8)$$

Since this inequality holds for any N, upon letting $N \to \infty$, we obtain (4.2), completing the proof. $\square$

Example 4.5. In this example, we illustrate the quantity A_i in the proof of Theorem 4.4 for the code $\mathcal{C}$ in Example 4.3. Let $l_1 = l_2 = 1$ and $l_3 = l_4 = 2$. Let $N = 2$ and consider

$$\left(\sum_{k=1}^{4} 2^{-l_k}\right)^2 = (2\cdot 2^{-1} + 2\cdot 2^{-2})^2 \qquad (4.9)$$

$$= 4\cdot 2^{-2} + 8\cdot 2^{-3} + 4\cdot 2^{-4}. \qquad (4.10)$$

Then $A_2 = 4$, $A_3 = 8$, and $A_4 = 8$, i.e., the total number of sequences of 2 codewords with a total length of 2, 3, and 4 code symbols are 4, 8, and 4, respectively. For a total length of 3, for instance, the 8 code sequences are $001(AC)$, $010(AD)$, $101(BC)$, $110(BD)$, $010(CA)$, $011(CB)$, $100(DA)$, and $101(DB)$.

Let X be a source random variable with probability distribution

$$\{p_1, p_2, \cdots, p_m\}, \tag{4.11}$$

where $m \geq 2$. When we use a uniquely decodable code $\mathcal{C}$ to encode the outcome of X, we are naturally interested in the expected length of a codeword, which is given by

$$L = \sum_i p_i l_i. \tag{4.12}$$

We will also refer to L as the expected length of the code $\mathcal{C}$. The quantity L gives the average number of symbols we need to describe the outcome of X when the code $\mathcal{C}$ is used, and it is a measure of the efficiency of the code $\mathcal{C}$. Specifically, the smaller the expected length L is, the better the code $\mathcal{C}$ is.

In the next theorem, we will prove a fundamental lower bound on the expected length of any uniquely decodable D-ary code. We first explain why this is the lower bound we should expect. In a uniquely decodable code, we use L D-ary symbols on the average to describe the outcome of X. Recall from the remark following Theorem 2.43 that a D-ary symbol can carry at most one D-it of information. Then the maximum amount of information which can be carried by the codeword on the average is $L \cdot 1 = L$ D-its. Since the code is uniquely decodable, the amount of entropy carried by the codeword on the average is $H_D(X)$. Therefore, we have

$$H_D(X) \leq L. \tag{4.13}$$

In other words, the expected length of a uniquely decodable code is at least the entropy of the source. This argument is rigorized in the proof of the next theorem.

Theorem 4.6 (Entropy Bound). *Let $\mathcal{C}$ be a D-ary uniquely decodable code for a source random variable X with entropy $H_D(X)$. Then the expected length of $\mathcal{C}$ is lower bounded by $H_D(X)$, i.e.,*

$$L \geq H_D(X). \tag{4.14}$$

This lower bound is tight if and only if $l_i = -\log_D p_i$ for all i.

Proof. Since $\mathcal{C}$ is uniquely decodable, the lengths of its codewords satisfy the Kraft inequality. Write

$$L = \sum_i p_i \log_D D^{l_i} \tag{4.15}$$

and recall from Definition 2.13 that

$$H_D(X) = -\sum_i p_i \log_D p_i. \tag{4.16}$$

Then

$$L - H_D(X) = \sum_i p_i \log_D(p_i D^{l_i}) \tag{4.17}$$

$$= (\ln D)^{-1} \sum_i p_i \ln(p_i D^{l_i}) \tag{4.18}$$

$$\geq (\ln D)^{-1} \sum_i p_i \left(1 - \frac{1}{p_i D^{l_i}}\right) \tag{4.19}$$

$$= (\ln D)^{-1} \left[\sum_i p_i - \sum_i D^{-l_i}\right] \tag{4.20}$$

$$\geq (\ln D)^{-1}(1 - 1) \tag{4.21}$$

$$= 0, \tag{4.22}$$

where we have invoked the fundamental inequality in (4.19) and the Kraft inequality in (4.21). This proves (4.14). In order for this lower bound to be tight, both (4.19) and (4.21) have to be tight simultaneously. Now (4.19) is tight if and only if $p_i D^{l_i} = 1$ or $l_i = -\log_D p_i$ for all i. If this holds, we have

$$\sum_i D^{-l_i} = \sum_i p_i = 1, \tag{4.23}$$

i.e., (4.21) is also tight. This completes the proof of the theorem.   □

The entropy bound can be regarded as a generalization of Theorem 2.43, as is seen from the following corollary.

Corollary 4.7. $H(X) \leq \log |\mathcal{X}|$.

Proof. Consider encoding each outcome of a random variable X by a distinct symbol in $\{1, 2, \cdots, |\mathcal{X}|\}$. This is obviously a $|\mathcal{X}|$-ary uniquely decodable code with expected length 1. Then by the entropy bound, we have

$$H_{|\mathcal{X}|}(X) \leq 1, \tag{4.24}$$

which becomes

$$H(X) \leq \log |\mathcal{X}| \tag{4.25}$$

when the base of the logarithm is not specified.   □

Motivated by the entropy bound, we now introduce the *redundancy* of a uniquely decodable code.

Definition 4.8. *The redundancy R of a D-ary uniquely decodable code is the difference between the expected length of the code and the entropy of the source.*

We see from the entropy bound that the redundancy of a uniquely decodable code is always nonnegative.

4.2 Prefix Codes

4.2.1 Definition and Existence

Definition 4.9. *A code is called a prefix-free code if no codeword is a prefix of any other codeword. For brevity, a prefix-free code will be referred to as a prefix code.*

Example 4.10. The code $\mathcal{C}$ in Example 4.3 is not a prefix code because the codeword 0 is a prefix of the codeword 01, and the codeword 1 is a prefix of the codeword 10. It can easily be checked that the following code $\mathcal{C}'$ is a prefix code.

x	$\mathcal{C}'(x)$
A	0
B	10
C	110
D	1111

A D-ary tree is a graphical representation of a collection of finite sequences of D-ary symbols. In a D-ary tree, each node has at most D children. If a node has at least one child, it is called an *internal node*, otherwise it is called a *leaf*. The children of an internal node are labeled by the D symbols in the code alphabet.

A D-ary prefix code can be represented by a D-ary tree with the leaves of the tree being the codewords. Such a tree is called the *code tree* for the prefix code. Figure 4.1 shows the code tree for the prefix code $\mathcal{C}'$ in Example 4.10.

As we have mentioned in Section 4.1, once a sequence of codewords are concatenated, the boundaries of the codewords are no longer explicit. Prefix codes have the desirable property that the end of a codeword can be recognized instantaneously so that it is not necessary to make reference to the future codewords during the decoding process. For example, for the source sequence $BCDAC \cdots$, the code $\mathcal{C}'$ in Example 4.10 produces the code sequence

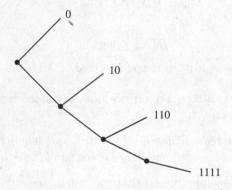

Fig. 4.1. The code tree for the code $\mathcal{C}'$.

1011011110110 $\cdots$. Based on this binary sequence, the decoder can reconstruct the source sequence as follows. The first bit 1 cannot form the first codeword because 1 is not a valid codeword. The first two bits 10 must form the first codeword because it is a valid codeword and it is not the prefix of any other codeword. The same procedure is repeated to locate the end of the next codeword, and the code sequence is parsed as $10, 110, 1111, 0, 110, \cdots$. Then the source sequence $BCDAC \cdots$ can be reconstructed correctly.

Since a prefix code can always be decoded correctly, it is a uniquely decodable code. Therefore, by Theorem 4.4, the codeword lengths of a prefix code also satisfy the Kraft inequality. In the next theorem, we show that the Kraft inequality fully characterizes the existence of a prefix code.

Theorem 4.11. *There exists a D-ary prefix code with codeword lengths l_1, $l_2, \cdots, l_m$ if and only if the Kraft inequality*

$$\sum_{k=1}^{m} D^{-l_k} \leq 1 \qquad (4.26)$$

is satisfied.

Proof. We only need to prove the existence of a D-ary prefix code with codeword lengths $l_1, l_2, \cdots, l_m$ if these lengths satisfy the Kraft inequality. Without loss of generality, assume that $l_1 \leq l_2 \leq \cdots \leq l_m$.

Consider all the D-ary sequences of lengths less than or equal to l_m and regard them as the nodes of the full D-ary tree of depth l_m. We will refer to a sequence of length l as a node of *order l*. Our strategy is to choose nodes as codewords in non-decreasing order of the codeword lengths. Specifically, we choose a node of order l_1 as the first codeword, then a node of order l_2 as the second codeword, so on and so forth, such that each newly chosen codeword is not prefixed by any of the previously chosen codewords. If we can successfully choose all the m codewords, then the resultant set of codewords forms a prefix code with the desired set of lengths.

There are $D^{l_1} > 1$ (since $l_1 \geq 1$) nodes of order l_1 which can be chosen as the first codeword. Thus choosing the first codeword is always possible. Assume that the first i codewords have been chosen successfully, where $1 \leq i \leq m-1$, and we want to choose a node of order l_{i+1} as the $(i+1)$st codeword such that it is not prefixed by any of the previously chosen codewords. In other words, the $(i+1)$st node to be chosen cannot be a descendant of any of the previously chosen codewords. Observe that for $1 \leq j \leq i$, the codeword with length l_j has $D^{l_{i+1}-l_j}$ descendents of order l_{i+1}. Since all the previously chosen codewords are not prefeces of each other, their descendents of order l_{i+1} do not overlap. Therefore, upon noting that the total number of nodes of order l_{i+1} is $D^{l_{i+1}}$, the number of nodes which can be chosen as the $(i+1)$st codeword is

$$D^{l_{i+1}} - D^{l_{i+1}-l_1} - \cdots - D^{l_{i+1}-l_i}. \qquad (4.27)$$

If $l_1, l_2, \cdots, l_m$ satisfy the Kraft inequality, we have

$$D^{-l_1} + \cdots + D^{-l_i} + D^{-l_{i+1}} \leq 1. \tag{4.28}$$

Multiplying by $D^{l_{i+1}}$ and rearranging the terms, we have

$$D^{l_{i+1}} - D^{l_{i+1}-l_1} - \cdots - D^{l_{i+1}-l_i} \geq 1. \tag{4.29}$$

The left-hand side is the number of nodes which can be chosen as the $(i+1)$st codeword as given in (4.27). Therefore, it is possible to choose the $(i+1)$st codeword. Thus we have shown the existence of a prefix code with codeword lengths $l_1, l_2, \cdots, l_m$, completing the proof.   □

A probability distribution $\{p_i\}$ such that for all i, $p_i = D^{-t_i}$, where t_i is a positive integer, is called a *D-adic* distribution. When $D = 2$, $\{p_i\}$ is called a *dyadic* distribution. From Theorem 4.6 and the above theorem, we can obtain the following result as a corollary.

Corollary 4.12. *There exists a D-ary prefix code which achieves the entropy bound for a distribution $\{p_i\}$ if and only if $\{p_i\}$ is D-adic.*

Proof. Consider a D-ary prefix code which achieves the entropy bound for a distribution $\{p_i\}$. Let l_i be the length of the codeword assigned to the probability p_i. By Theorem 4.6, for all i, $l_i = -\log_D p_i$ or $p_i = D^{-l_i}$. Thus $\{p_i\}$ is D-adic.

Conversely, suppose $\{p_i\}$ is D-adic, and let $p_i = D^{-t_i}$ for all i. Let $l_i = t_i$ for all i. Then by the Kraft inequality, there exists a prefix code with codeword lengths $\{l_i\}$, because

$$\sum_i D^{-l_i} = \sum_i D^{-t_i} = \sum_i p_i = 1. \tag{4.30}$$

Assigning the codeword with length l_i to the probability p_i for all i, we see from Theorem 4.6 that this code achieves the entropy bound.   □

4.2.2 Huffman Codes

As we have mentioned, the efficiency of a uniquely decodable code is measured by its expected length. Thus for a given source X, we are naturally interested in prefix codes which have the minimum expected length. Such codes, called optimal codes, can be constructed by the *Huffman procedure*, and these codes are referred to as *Huffman codes*. In general, there exists more than one optimal code for a source, and some optimal codes cannot be constructed by the Huffman procedure.

For simplicity, we first discuss binary Huffman codes. A binary prefix code for a source X with distribution $\{p_i\}$ is represented by a binary code tree, with each leaf in the code tree corresponding to a codeword. The Huffman procedure is to form a code tree such that the expected length is minimum. The procedure is described by a very simple rule:

Keep merging the two smallest probability masses until one probability mass (i.e., 1) is left.

The merging of two probability masses corresponds to the formation of an internal node of the code tree. We now illustrate the Huffman procedure by the following example.

Example 4.13. Let X be the source with $\mathcal{X} = \{A, B, C, D, E\}$, and the probabilities are 0.35, 0.1, 0.15, 0.2, 0.2, respectively. The Huffman procedure is shown in Figure 4.2. In the first step, we merge probability masses 0.1 and 0.15 into a probability mass 0.25. In the second step, we merge probability masses 0.2 and 0.2 into a probability mass 0.4. In the third step, we merge probability masses 0.35 and 0.25 into a probability mass 0.6. Finally, we merge probability masses 0.6 and 0.4 into a probability mass 1. A code tree is then formed. Upon assigning 0 and 1 (in any convenient way) to each pair of branches at an internal node, we obtain the codeword assigned to each source symbol.

In the Huffman procedure, sometimes there is more than one choice of merging the two smallest probability masses. We can take any one of these choices without affecting the optimality of the code eventually obtained.

For an alphabet of size m, it takes $m - 1$ steps to complete the Huffman procedure for constructing a binary code, because we merge two probability masses in each step. In the resulting code tree, there are m leaves and $m - 1$ internal nodes.

In the Huffman procedure for constructing a D-ary code, the smallest D probability masses are merged in each step. If the resulting code tree is formed in $k + 1$ steps, where $k \geq 0$, then there will be $k + 1$ internal nodes and $D + k(D - 1)$ leaves, where each leaf corresponds to a source symbol in the alphabet. If the alphabet size m has the form $D + k(D - 1)$, then we can apply the Huffman procedure directly. Otherwise, we need to add a few dummy symbols with probability 0 to the alphabet in order to make the total number of symbols have the form $D + k(D - 1)$.

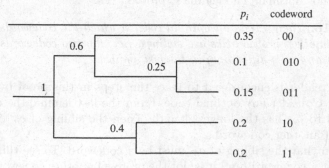

Fig. 4.2. The Huffman procedure.

Example 4.14. If we want to construct a quaternary Huffman code ($D = 4$) for the source in the last example, we need to add 2 dummy symbols so that the total number of symbols becomes $7 = 4 + (1)3$, where $k = 1$. In general, we need to add at most $D - 2$ dummy symbols.

In Section 4.1, we have proved the entropy bound for a uniquely decodable code. This bound also applies to a prefix code since a prefix code is uniquely decodable. In particular, it applies to a Huffman code, which is a prefix code by construction. Thus the expected length of a Huffman code is at least the entropy of the source. In Example 4.13, the entropy $H(X)$ is 2.202 bits, while the expected length of the Huffman code is

$$0.35(2) + 0.1(3) + 0.15(3) + 0.2(2) + 0.2(2) = 2.25. \qquad (4.31)$$

We now turn to proving the optimality of a Huffman code. For simplicity, we will only prove the optimality of a binary Huffman code. Extension of the proof to the general case is straightforward.

Without loss of generality, assume that

$$p_1 \geq p_2 \geq \cdots \geq p_m. \qquad (4.32)$$

Denote the codeword assigned to p_i by c_i, and denote its length by l_i. To prove that a Huffman code is actually optimal, we make the following observations.

Lemma 4.15. *In an optimal code, shorter codewords are assigned to larger probabilities.*

Proof. Consider $1 \leq i < j \leq m$ such that $p_i > p_j$. Assume that in a code, the codewords c_i and c_j are such that $l_i > l_j$, i.e., a shorter codeword is assigned to a smaller probability. Then by exchanging c_i and c_j, the expected length of the code is changed by

$$(p_i l_j + p_j l_i) - (p_i l_i + p_j l_j) = (p_i - p_j)(l_j - l_i) < 0 \qquad (4.33)$$

since $p_i > p_j$ and $l_i > l_j$. In other words, the code can be improved and therefore is not optimal. The lemma is proved.   □

Lemma 4.16. *There exists an optimal code in which the codewords assigned to the two smallest probabilities are siblings, i.e., the two codewords have the same length and they differ only in the last symbol.*

Proof. The reader is encouraged to trace the steps in this proof by drawing a code tree. Consider any optimal code. From the last lemma, the codeword c_m assigned to p_m has the longest length. Then the sibling of c_m cannot be the prefix of another codeword.

We claim that the sibling of c_m must be a codeword. To see this, assume that it is not a codeword (and it is not the prefix of another codeword). Then we can replace c_m by its parent to improve the code because the length of the codeword assigned to p_m is reduced by 1, while all the other codewords remain unchanged. This is a contradiction to the assumption that the code is optimal. Therefore, the sibling of c_m must be a codeword.

If the sibling of c_m is assigned to p_{m-1}, then the code already has the desired property, i.e., the codewords assigned to the two smallest probabilities are siblings. If not, assume that the sibling of c_m is assigned to p_i, where $i < m - 1$. Since $p_i \geq p_{m-1}$, $l_{m-1} \geq l_i = l_m$. On the other hand, by Lemma 4.15, l_{m-1} is always less than or equal to l_m, which implies that $l_{m-1} = l_m = l_i$. Then we can exchange the codewords for p_i and p_{m-1} without changing the expected length of the code (i.e., the code remains optimal) to obtain the desired code. The lemma is proved. $\square$

Suppose c_i and c_j are siblings in a code tree. Then $l_i = l_j$. If we replace c_i and c_j by a common codeword at their parent, call it c_{ij}, then we obtain a reduced code tree, and the probability of c_{ij} is $p_i + p_j$. Accordingly, the probability set becomes a reduced probability set with p_i and p_j replaced by a probability $p_i + p_j$. Let L and L' be the expected lengths of the original code and the reduced code, respectively. Then

$$L - L' = (p_i l_i + p_j l_j) - (p_i + p_j)(l_i - 1) \tag{4.34}$$

$$= (p_i l_i + p_j l_i) - (p_i + p_j)(l_i - 1) \tag{4.35}$$

$$= p_i + p_j, \tag{4.36}$$

which implies

$$L = L' + (p_i + p_j). \tag{4.37}$$

This relation says that the difference between the expected length of the original code and the expected length of the reduced code depends only on the values of the two probabilities merged but not on the structure of the reduced code tree.

Theorem 4.17. *The Huffman procedure produces an optimal prefix code.*

Proof. Consider an optimal code in which c_m and c_{m-1} are siblings. Such an optimal code exists by Lemma 4.16. Let $\{p_i'\}$ be the reduced probability set obtained from $\{p_i\}$ by merging p_m and p_{m-1}. From (4.37), we see that L' is the expected length of an optimal code for $\{p_i'\}$ if and only if L is the expected length of an optimal code for $\{p_i\}$. Therefore, if we can find an optimal code for $\{p_i'\}$, we can use it to construct an optimal code for $\{p_i\}$. Note that by merging p_m and p_{m-1}, the size of the problem, namely the total number of probability masses, is reduced by one. To find an optimal code for $\{p_i'\}$, we again merge the two smallest probabilities in $\{p_i'\}$. This is repeated until the size of the problem is eventually reduced to 2, as we know that an optimal code has two codewords of length 1. In the last step of the Huffman procedure, two probability masses are merged, which corresponds to the formation of a code with two codewords of length 1. Thus the Huffman procedure indeed produces an optimal code. $\square$

We have seen that the expected length of a Huffman code is lower bounded by the entropy of the source. On the other hand, it would be desirable to obtain an upper bound in terms of the entropy of the source. This is given in the next theorem.

Theorem 4.18. *The expected length of a Huffman code, denoted by L_{Huff}, satisfies*

$$L_{\text{Huff}} < H_D(X) + 1. \tag{4.38}$$

This bound is the tightest among all the upper bounds on L_{Huff} which depends only on the source entropy.

Proof. We will construct a prefix code with expected length less than $H(X) + 1$. Then, because a Huffman code is an optimal prefix code, its expected length L_{Huff} is upper bounded by $H(X) + 1$.

Consider constructing a prefix code with codeword lengths $\{l_i\}$, where

$$l_i = \lceil -\log_D p_i \rceil. \tag{4.39}$$

Then

$$-\log_D p_i \le l_i < -\log_D p_i + 1 \tag{4.40}$$

or

$$p_i \ge D^{-l_i} > D^{-1} p_i. \tag{4.41}$$

Thus

$$\sum_i D^{-l_i} \le \sum_i p_i = 1, \tag{4.42}$$

i.e., $\{l_i\}$ satisfies the Kraft inequality, which implies that it is possible to construct a prefix code with codeword lengths $\{l_i\}$.

It remains to show that L, the expected length of this code, is less than $H(X) + 1$. Toward this end, consider

$$L = \sum_i p_i l_i \tag{4.43}$$

$$< \sum_i p_i (-\log_D p_i + 1) \tag{4.44}$$

$$= -\sum_i p_i \log_D p_i + \sum_i p_i \tag{4.45}$$

$$= H(X) + 1, \tag{4.46}$$

where (4.44) follows from the upper bound in (4.40). Thus we conclude that

$$L_{\text{Huff}} \le L < H(X) + 1. \tag{4.47}$$

To see that this upper bound is the tightest possible, we have to show that there exists a sequence of distributions P_k such that L_{Huff} approaches $H(X)+1$ as $k \to \infty$. This can be done by considering the sequence of D-ary distributions

$$P_k = \left\{ 1 - \frac{D-1}{k}, \frac{1}{k}, \cdots, \frac{1}{k} \right\}, \tag{4.48}$$

where $k \geq D$. The Huffman code for each P_k consists of D codewords of length 1. Thus L_{Huff} is equal to 1 for all k. As $k \to \infty$, $H(X) \to 0$, and hence L_{Huff} approaches $H(X) + 1$. The theorem is proved. □

The code constructed in the above proof is known as the *Shannon code*. The idea is that in order for the code to be near-optimal, we should choose l_i close to $-\log p_i$ for all i. When $\{p_i\}$ is D-adic, l_i can be chosen to be exactly $-\log p_i$ because the latter are integers. In this case, the entropy bound is tight.

From the entropy bound and the above theorem, we have

$$H(X) \leq L_{\text{Huff}} < H(X) + 1. \tag{4.49}$$

Now suppose we use a Huffman code to encode $X_1, X_2, \cdots, X_n$ which are n i.i.d. copies of X. Let us denote the length of this Huffman code by L_{Huff}^n. Then (4.49) becomes

$$nH(X) \leq L_{\text{Huff}}^n < nH(X) + 1. \tag{4.50}$$

Dividing by n, we obtain

$$H(X) \leq \frac{1}{n} L_{\text{Huff}}^n < H(X) + \frac{1}{n}. \tag{4.51}$$

As $n \to \infty$, the upper bound approaches the lower bound. Therefore, $n^{-1} L_{\text{Huff}}^n$, the coding rate of the code, namely the average number of code symbols needed to encode a source symbol, approaches $H(X)$ as $n \to \infty$. But of course, as n becomes large, constructing a Huffman code becomes very complicated. Nevertheless, this result indicates that entropy is a fundamental measure of information.

4.3 Redundancy of Prefix Codes

The entropy bound for a uniquely decodable code has been proved in Section 4.1. In this section, we present an alternative proof specifically for prefix codes which offers much insight into the redundancy of such codes.

Let X be a source random variable with probability distribution

$$\{p_1, p_2, \cdots, p_m\}, \tag{4.52}$$

where $m \geq 2$. A D-ary prefix code for X can be represented by a D-ary code tree with m leaves, where each leaf corresponds to a codeword. We denote the leaf corresponding to p_i by c_i and the order of c_i by l_i, and assume that the alphabet is

$$\{0, 1, \cdots, D-1\}. \tag{4.53}$$

Let $\mathcal{I}$ be the index set of all the internal nodes (including the root) in the code tree.

Instead of matching codewords by brute force, we can use the code tree of a prefix code for more efficient decoding. To decode a codeword, we trace the path specified by the codeword from the root of the code tree until it terminates at the leaf corresponding to that codeword. Let q_k be the probability of reaching an internal node $k \in \mathcal{I}$ during the decoding process. The probability q_k is called the *reaching probability* of internal node k. Evidently, q_k is equal to the sum of the probabilities of all the leaves descending from node k.

Let $\tilde{p}_{k,j}$ be the probability that the jth branch of node k is taken during the decoding process. The probabilities $\tilde{p}_{k,j}, 0 \le j \le D - 1$, are called the *branching probabilities* of node k, and

$$q_k = \sum_j \tilde{p}_{k,j}. \tag{4.54}$$

Once node k is reached, the *conditional branching distribution* is

$$\left\{ \frac{\tilde{p}_{k,0}}{q_k}, \frac{\tilde{p}_{k,1}}{q_k}, \cdots, \frac{\tilde{p}_{k,D-1}}{q_k} \right\}. \tag{4.55}$$

Then define the *conditional entropy* of node k by

$$h_k = H_D \left(\left\{ \frac{\tilde{p}_{k,0}}{q_k}, \frac{\tilde{p}_{k,1}}{q_k}, \cdots, \frac{\tilde{p}_{k,D-1}}{q_k} \right\} \right), \tag{4.56}$$

where with a slight abuse of notation, we have used $H_D(\cdot)$ to denote the entropy in the base D of the conditional branching distribution in the parenthesis. By Theorem 2.43, $h_k \le 1$. The following lemma relates the entropy of X with the structure of the code tree.

Lemma 4.19. $H_D(X) = \sum_{k \in \mathcal{I}} q_k h_k.$

Proof. We prove the lemma by induction on the number of internal nodes of the code tree. If there is only one internal node, it must be the root of the tree. Then the lemma is trivially true upon observing that the reaching probability of the root is equal to 1.

Assume the lemma is true for all code trees with n internal nodes. Now consider a code tree with $n + 1$ internal nodes. Let k be an internal node such that k is the parent of a leaf c with maximum order. Each sibling of c may or may not be a leaf. If it is not a leaf, then it cannot be the ascendant of another leaf because we assume that c is a leaf with maximum order. Now consider revealing the outcome of X in two steps. In the first step, if the outcome of X is not a leaf descending from node k, we identify the outcome exactly, otherwise we identify the outcome to be a child of node k. We call this random variable V. If we do not identify the outcome exactly in the first step, which happens with probability q_k, we further identify in the second step which of the children (child) of node k the outcome is (there is only one child of node k which can be the outcome if all the siblings of c are not leaves). We call this random variable W. If the second step is not necessary, we assume that W takes a constant value with probability 1. Then $X = (V, W)$.

The outcome of V can be represented by a code tree with n internal nodes which is obtained by pruning the original code tree at node k. Then by the induction hypothesis,

$$H(V) = \sum_{k' \in \mathcal{I} \setminus \{k\}} q_{k'} h_{k'}. \tag{4.57}$$

By the chain rule for entropy, we have

$$H(X) = H(V) + H(W|V) \tag{4.58}$$

$$= \sum_{k' \in \mathcal{I} \setminus \{k\}} q_{k'} h_{k'} + (1 - q_k) \cdot 0 + q_k h_k \tag{4.59}$$

$$= \sum_{k' \in \mathcal{I}} q_{k'} h_{k'}. \tag{4.60}$$

The lemma is proved. $\square$

The next lemma expresses the expected length L of a prefix code in terms of the reaching probabilities of the internal nodes of the code tree.

Lemma 4.20. $L = \sum_{k \in \mathcal{I}} q_k.$

Proof. Define

$$a_{ki} = \begin{cases} 1 \text{ if leaf } c_i \text{ is a descendent of internal node } k \\ 0 \text{ otherwise.} \end{cases} \tag{4.61}$$

Then

$$l_i = \sum_{k \in \mathcal{I}} a_{ki}, \tag{4.62}$$

because there are exactly l_i internal nodes of which c_i is a descendent if the order of c_i is l_i. On the other hand,

$$q_k = \sum_i a_{ki} p_i. \tag{4.63}$$

Then

$$L = \sum_i p_i l_i \tag{4.64}$$

$$= \sum_i p_i \sum_{k \in \mathcal{I}} a_{ki} \tag{4.65}$$

$$= \sum_{k \in \mathcal{I}} \sum_i p_i a_{ki} \tag{4.66}$$

$$= \sum_{k \in \mathcal{I}} q_k, \tag{4.67}$$

proving the lemma.   □

Define the *local redundancy* of an internal node k by

$$r_k = q_k(1 - h_k). \tag{4.68}$$

This quantity is local to node k in the sense that it depends only on the branching probabilities of node k, and it vanishes if and only if $\tilde{p}_{k,j} = q_k/D$ for all j, i.e., if and only if the node is *balanced*. Note that $r_k \geq 0$ because $h_k \leq 1$.

The next theorem says that the redundancy R of a prefix code is equal to the sum of the local redundancies of all the internal nodes of the code tree.

Theorem 4.21 (Local Redundancy Theorem). *Let L be the expected length of a D-ary prefix code for a source random variable X, and R be the redundancy of the code. Then*

$$R = \sum_{k \in \mathcal{I}} r_k. \tag{4.69}$$

Proof. By Lemmas 4.19 and 4.20, we have

$$R = L - H_D(X) \tag{4.70}$$

$$= \sum_{k \in \mathcal{I}} q_k - \sum_k q_k h_k \tag{4.71}$$

$$= \sum_{k \in \mathcal{I}} q_k(1 - h_k) \tag{4.72}$$

$$= \sum_{k \in \mathcal{I}} r_k. \tag{4.73}$$

The theorem is proved.   □

We now present an slightly different version of the entropy bound.

Corollary 4.22 (Entropy Bound). *Let R be the redundancy of a prefix code. Then $R \geq 0$ with equality if and only if all the internal nodes in the code tree are balanced.*

Proof. Since $r_k \geq 0$ for all k, it is evident from the local redundancy theorem that $R \geq 0$. Moreover $R = 0$ if and only if $r_k = 0$ for all k, which means that all the internal nodes in the code tree are balanced.   □

Remark Before the entropy bound was stated in Theorem 4.6, we gave the intuitive explanation that the entropy bound results from the fact that a D-ary symbol can carry at most one D-it of information. Therefore, when the entropy bound is tight, each code symbol has to carry exactly one D-it of information. Now consider revealing a random codeword one symbol after another. The above corollary states that in order for the entropy bound to be tight, all the internal nodes in the code tree must be balanced. That is, as long as the codeword is not completed, the next code symbol to be revealed always carries one D-it of information because it is distributed uniformly on the alphabet. This is consistent with the intuitive explanation we gave for the entropy bound.

Example 4.23. The local redundancy theorem allows us to lower bound the redundancy of a prefix code based on partial knowledge on the structure of the code tree. More specifically,

$$R \geq \sum_{k \in \mathcal{I}'} r_k \qquad (4.74)$$

for any subset $\mathcal{I}'$ of $\mathcal{I}$.

Let p_{m-1}, p_m be the two smallest probabilities in the source distribution. In constructing a binary Huffman code, p_{m-1} and p_m are merged. Then the redundancy of a Huffman code is lower bounded by

$$(p_{m-1} + p_m) \left[1 - H_2 \left(\left\{ \frac{p_{m-1}}{p_{m-1} + p_m}, \frac{p_m}{p_{m-1} + p_m} \right\} \right) \right], \qquad (4.75)$$

the local redundancy of the parent of the two leaves corresponding to p_{m-1} and p_m. See Yeung [399] for progressive lower and upper bounds on the redundancy of a Huffman code.

Chapter Summary

Kraft Inequality: For a D-ary uniquely decodable source code,

$$\sum_{k=1}^{m} D^{-l_k} \leq 1.$$

Entropy Bound:
$$L = \sum_k p_k l_k \geq H_D(X),$$

with equality if and only if the distribution of X is D-adic.

Definition: A code is called a prefix code if no codeword is a prefix of any other codeword.

Existence of Prefix Code: A D-ary prefix code with codeword lengths $l_1, l_2, \cdots, l_m$ exists if and only if the Kraft inequality is satisfied.

Huffman Code:

1. A Huffman code is a prefix code with the shortest expected length for a given source.
2. $H_D(X) \leq L_{\text{Huff}} < H_D(X) + 1$.

Huffman Procedure: Keep merging the D smallest probability masses.

Redundancy of Prefix Code:

$$L - H_D(X) = R = \sum_{k \in \mathcal{I}} r_k,$$

where $r_k = q_k(1 - h_k)$ is the local redundancy of an internal node k.

Problems

1. Construct a binary Huffman code for the distribution $\{0.25, 0.05, 0.1, 0.13, 0.2, 0.12, 0.08, 0.07\}$.
2. Construct a ternary Huffman code for the source distribution in Problem 1.
3. Show that a Huffman code is an optimal uniquely decodable code for a given source distribution.
4. Construct an optimal binary prefix code for the source distribution in Problem 1 such that all the codewords have even lengths.
5. Prove directly that the codeword lengths of a prefix code satisfy the Kraft inequality without using Theorem 4.4.
6. Prove that if $p_1 > 0.4$, then the shortest codeword of a binary Huffman code has length equal to 1. Then prove that the redundancy of such a Huffman code is lower bounded by $1 - h_b(p_1)$. (Johnsen [192].)
7. *Suffix codes.* A code is a suffix code if no codeword is a suffix of any other codeword. Show that a suffix code is uniquely decodable.
8. *Fix-free codes.* A code is a fix-free code if it is both a prefix code and a suffix code. Let $l_1, l_2, \cdots, l_m$ be m positive integers. Prove that if

$$\sum_{k=1}^{m} 2^{-l_k} \leq \frac{1}{2},$$

then there exists a binary fix-free code with codeword lengths $l_1, l_2, \cdots, l_m$. (Ahlswede et al. [5].)

9. *Random coding for prefix codes.* Construct a binary prefix code with code-word lengths $l_1 \leq l_2 \leq \cdots \leq l_m$ as follows. For each $1 \leq k \leq m$, the code-word with length l_k is chosen independently from the set of all 2^{l_k} possible binary strings with length l_k according to the uniform distribution. Let $P_m(good)$ be the probability that the code so constructed is a prefix code.

a) Prove that $P_2(good) = (1 - 2^{-l_1})^+$, where

$$(x)^+ = \begin{cases} x \text{ if } x \geq 0 \\ 0 \text{ if } x < 0 \end{cases}.$$

b) Prove by induction on m that

$$P_m(good) = \prod_{k=1}^{m} \left(1 - \sum_{j=1}^{k-1} s^{-l_j}\right)^+.$$

c) Observe that there exists a prefix code with codeword lengths $l_1, l_2, \cdots,$ l_m if and only if $P_m(good) > 0$. Show that $P_m(good) > 0$ is equivalent to the Kraft inequality.

By using this random coding method, one can derive the Kraft inequality without knowing the inequality ahead of time. (Ye and Yeung [395].)

10. Let X be a source random variable. Suppose a certain probability mass p_k in the distribution of X is given. Let

$$l_j = \begin{cases} \lceil -\log p_j \rceil & \text{if } j = k \\ \lceil -\log(p_j + x_j) \rceil & \text{if } j \neq k \end{cases},$$

where

$$x_j = p_j \left(\frac{p_k - 2^{-\lceil -\log p_k \rceil}}{1 - p_k}\right)$$

for all $j \neq k$.

a) Show that $1 \leq l_j \leq \lceil -\log p_j \rceil$ for all j.

b) Show that $\{l_j\}$ satisfies the Kraft inequality.

c) Obtain an upper bound on L_{Huff} in terms of $H(X)$ and p_k which is tighter than $H(X) + 1$. This shows that when partial knowledge about the source distribution in addition to the source entropy is available, tighter upper bounds on L_{Huff} can be obtained.

(Ye and Yeung [396].)

Historical Notes

The foundation for the material in this chapter can be found in Shannon's original paper [322]. The Kraft inequality for uniquely decodable codes was first proved by McMillan [267]. The proof given here is due to Karush [195]. The Huffman coding procedure was devised and proved to be optimal by

Huffman [175]. The same procedure was devised independently by Zimmerman [418]. Linder et al. [236] have proved the existence of an optimal prefix code for an infinite source alphabet which can be constructed from Huffman codes for truncations of the source distribution. The local redundancy theorem is due to Yeung [399]. A comprehensive survey of code trees for lossless data compression can be found in Abrahams [1].

5

Weak Typicality

In the last chapter, we have discussed the significance of entropy in the context of zero-error data compression. In this chapter and the next, we explore entropy in terms of the asymptotic behavior of i.i.d. sequences. Specifically, two versions of the *asymptotic equipartition property* (AEP), namely the weak AEP and the strong AEP, are discussed. The role of these AEPs in information theory is analogous to the role of the weak law of large numbers in probability theory. In this chapter, the weak AEP and its relation with the source coding theorem are discussed. All the logarithms are in the base 2 unless otherwise specified.

5.1 The Weak AEP

We consider an information source $\{X_k, k \geq 1\}$ where X_k are i.i.d. with distribution $p(x)$. We use X to denote the generic random variable and $H(X)$ to denote the common entropy for all X_k, where $H(X) < \infty$. Let $\mathbf{X} = (X_1, X_2, \cdots, X_n)$. Since X_k are i.i.d.,

$$p(\mathbf{X}) = p(X_1)p(X_2)\cdots p(X_n). \tag{5.1}$$

Note that $p(\mathbf{X})$ is a random variable because it is a function of the random variables $X_1, X_2, \cdots, X_n$. We now prove an asymptotic property of $p(\mathbf{X})$ called the *weak asymptotic equipartition property* (weak AEP).

Theorem 5.1 (Weak AEP I).

$$-\frac{1}{n}\log p(\mathbf{X}) \to H(X) \tag{5.2}$$

in probability as $n \to \infty$, i.e., for any $\epsilon > 0$, for n sufficiently large,

$$\Pr\left\{\left|-\frac{1}{n}\log p(\mathbf{X}) - H(X)\right| \leq \epsilon\right\} > 1 - \epsilon. \tag{5.3}$$

Proof. Since $X_1, X_2, \cdots, X_n$ are i.i.d., by (5.1),

$$-\frac{1}{n}\log p(\mathbf{X}) = -\frac{1}{n}\sum_{k=1}^{n}\log p(X_k). \tag{5.4}$$

The random variables $\log p(X_k)$ are also i.i.d. Then by the weak law of large numbers, the right-hand side of (5.4) tends to

$$-E\log p(X) = H(X), \tag{5.5}$$

in probability, proving the theorem.   □

The weak AEP is nothing more than a straightforward application of the weak law of large numbers. However, as we will see shortly, this property has significant implications.

Definition 5.2. *The weakly typical set* $W_{[X]\epsilon}^n$ *with respect to* $p(x)$ *is the set of sequences* $\mathbf{x} = (x_1, x_2, \cdots, x_n) \in \mathcal{X}^n$ *such that*

$$\left|-\frac{1}{n}\log p(\mathbf{x}) - H(X)\right| \leq \epsilon, \tag{5.6}$$

or equivalently,

$$H(X) - \epsilon \leq -\frac{1}{n}\log p(\mathbf{x}) \leq H(X) + \epsilon, \tag{5.7}$$

where ϵ *is an arbitrarily small positive real number. The sequences in* $W_{[X]\epsilon}^n$ *are called weakly* ϵ-*typical sequences.*

The quantity

$$-\frac{1}{n}\log p(\mathbf{x}) = -\frac{1}{n}\sum_{k=1}^{n}\log p(x_k) \tag{5.8}$$

is called the *empirical entropy* of the sequence $\mathbf{x}$. The empirical entropy of a weakly typical sequence is close to the true entropy $H(X)$. The important properties of the set $W_{[X]\epsilon}^n$ are summarized in the next theorem which we will see is equivalent to the weak AEP.

Theorem 5.3 (Weak AEP II). *The following hold for any* $\epsilon > 0$:

1) If $\mathbf{x} \in W_{[X]\epsilon}^n$, *then*

$$2^{-n(H(X)+\epsilon)} \leq p(\mathbf{x}) \leq 2^{-n(H(X)-\epsilon)}. \tag{5.9}$$

2) For n *sufficiently large,*

$$\Pr\{\mathbf{X} \in W_{[X]\epsilon}^n\} > 1 - \epsilon. \tag{5.10}$$

3) For n sufficiently large,

$$(1 - \epsilon)2^{n(H(X)-\epsilon)} \leq |W^n_{[X]\epsilon}| \leq 2^{n(H(X)+\epsilon)}. \tag{5.11}$$

Proof. Property 1 follows immediately from the definition of $W^n_{[X]\epsilon}$ in (5.7). Property 2 is equivalent to Theorem 5.1. To prove Property 3, we use the lower bound in (5.9) and consider

$$|W^n_{[X]\epsilon}|2^{-n(H(X)+\epsilon)} \leq \Pr\{W^n_{[X]\epsilon}\} \leq 1, \tag{5.12}$$

which implies

$$|W^n_{[X]\epsilon}| \leq 2^{n(H(X)+\epsilon)}. \tag{5.13}$$

Note that this upper bound holds for any $n \geq 1$. On the other hand, using the upper bound in (5.9) and Theorem 5.1, for n sufficiently large, we have

$$1 - \epsilon \leq \Pr\{W^n_{[X]\epsilon}\} \leq |W^n_{[X]\epsilon}|2^{-n(H(X)-\epsilon)}. \tag{5.14}$$

Then

$$|W^n_{[X]\epsilon}| \geq (1 - \epsilon)2^{n(H(X)-\epsilon)}. \tag{5.15}$$

Combining (5.13) and (5.15) gives Property 3. The theorem is proved. $\square$

Remark Theorem 5.3 is a consequence of Theorem 5.1. However, Property 2 in Theorem 5.3 is equivalent to Theorem 5.1. Therefore, Theorem 5.1 and Theorem 5.3 are equivalent, and they will both be referred to as the weak AEP.

The weak AEP has the following interpretation. Suppose $\mathbf{X} = (X_1, X_2, \cdots, X_n)$ is drawn i.i.d. according to $p(x)$, where n is large. After the sequence is drawn, we ask what the probability of occurrence of the sequence is. The weak AEP says that the probability of occurrence of the sequence drawn is close to $2^{-nH(X)}$ with very high probability. Such a sequence is called a weakly typical sequence. Moreover, the total number of weakly typical sequences is approximately equal to $2^{nH(X)}$. The weak AEP, however, does not say that most of the sequences in $\mathcal{X}^n$ are weakly typical. In fact, the number of weakly typical sequences is in general insignificant compared with the total number of sequences, because

$$\frac{|W^n_{[X]\delta}|}{|\mathcal{X}|^n} \approx \frac{2^{nH(X)}}{2^{n \log|\mathcal{X}|}} = 2^{-n(\log|\mathcal{X}|-H(X))} \to 0 \tag{5.16}$$

as $n \to \infty$ as long as $H(X)$ is strictly less than $\log|\mathcal{X}|$. The idea is that although the size of the weakly typical set may be insignificant compared with the size of the set of all sequences, the former has almost all the probability.

When n is large, one can almost think of the sequence $\mathbf{X}$ as being obtained by choosing a sequence from the weakly typical set according to the uniform distribution. Very often, we concentrate on the properties of typical sequences because any property which is proved to be true for typical sequences will then be true with high probability. This in turn determines the average behavior of a large sample.

Remark The most likely sequence is in general not weakly typical although the probability of the weakly typical set is close to 1 when n is large. For example, for X_k i.i.d. with $p(0) = 0.1$ and $p(1) = 0.9$, $(1, 1, \cdots, 1)$ is the most likely sequence, but it is not weakly typical because its empirical entropy is not close to the true entropy. The idea is that as $n \to \infty$, the probability of every sequence, including that of the most likely sequence, tends to 0. Therefore, it is not necessary for a weakly typical set to include the most likely sequence in order to possess a probability close to 1.

5.2 The Source Coding Theorem

To encode a random sequence $\mathbf{X} = (X_1, X_2, \cdots, X_n)$ drawn i.i.d. according to $p(x)$ by a *block code*, we construct a one-to-one mapping from a subset $\mathcal{A}$ of $\mathcal{X}^n$ to an index set

$$\mathcal{I} = \{1, 2, \cdots, M\}, \tag{5.17}$$

where $|\mathcal{A}| = M \leq |\mathcal{X}|^n$. We do not have to assume that $|\mathcal{X}|$ is finite. The indices in $\mathcal{I}$ are called *codewords*, and the integer n is called the *block length* of the code. If a sequence $\mathbf{x} \in \mathcal{A}$ occurs, the encoder outputs the corresponding codeword which is specified by approximately $\log M$ bits. If a sequence $\mathbf{x} \notin \mathcal{A}$ occurs, the encoder outputs the constant codeword 1. In either case, the codeword output by the encoder is decoded to the sequence in $\mathcal{A}$ corresponding to that codeword by the decoder. If a sequence $\mathbf{x} \in \mathcal{A}$ occurs, then $\mathbf{x}$ is decoded correctly by the decoder. If a sequence $\mathbf{x} \notin \mathcal{A}$ occurs, then $\mathbf{x}$ is not decoded correctly by the decoder. For such a code, its performance is measured by the coding rate defined as $n^{-1} \log M$ (in bits per source symbol), and the probability of error is given by

$$P_e = \Pr\{\mathbf{X} \notin \mathcal{A}\}. \tag{5.18}$$

If the code is not allowed to make any error, i.e., $P_e = 0$, it is clear that M must be taken to be $|\mathcal{X}|^n$, or $\mathcal{A} = \mathcal{X}^n$. In that case, the coding rate is equal to $\log |\mathcal{X}|$. However, if we allow P_e to be any small quantity, Shannon [322] showed that there exists a block code whose coding rate is arbitrarily close to $H(X)$ when n is sufficiently large. This is the direct part of Shannon's *source coding theorem*, and in this sense the source sequence $\mathbf{X}$ is said to be reconstructed *almost perfectly*.

We now prove the direct part of the source coding theorem by constructing a desired code. First, we fix $\epsilon > 0$ and take

$$\mathcal{A} = W_{[X]\epsilon}^n \tag{5.19}$$

and

$$M = |\mathcal{A}|. \tag{5.20}$$

For sufficiently large n, by the weak AEP,

$$(1 - \epsilon)2^{n(H(X)-\epsilon)} \leq M = |\mathcal{A}| = |W_{[X]\epsilon}^n| \leq 2^{n(H(X)+\epsilon)}. \tag{5.21}$$

Therefore, the coding rate $n^{-1} \log M$ satisfies

$$\frac{1}{n} \log(1 - \epsilon) + H(X) - \epsilon \leq \frac{1}{n} \log M \leq H(X) + \epsilon. \tag{5.22}$$

Also by the weak AEP,

$$P_e = \Pr\{\mathbf{X} \notin \mathcal{A}\} = \Pr\{\mathbf{X} \notin W_{[X]\epsilon}^n\} < \epsilon. \tag{5.23}$$

Letting $\epsilon \to 0$, the coding rate tends to $H(X)$, while P_e tends to 0. This proves the direct part of the source coding theorem.

The converse part of the source coding theorem says that if we use a block code with block length n and coding rate less than $H(X) - \zeta$, where $\zeta > 0$ does not change with n, then $P_e \to 1$ as $n \to \infty$. To prove this, consider any code with block length n and coding rate less than $H(X) - \zeta$, so that M, the total number of codewords, is at most $2^{n(H(X)-\zeta)}$. We can use some of these codewords for the typical sequences $\mathbf{x} \in W_{[X]\epsilon}^n$, and some for the non-typical sequences $\mathbf{x} \notin W_{[X]\epsilon}^n$. The total probability of the typical sequences covered by the code, by the weak AEP, is upper bounded by

$$2^{n(H(X)-\zeta)}2^{-n(H(X)-\epsilon)} = 2^{-n(\zeta-\epsilon)}. \tag{5.24}$$

Therefore, the total probability covered by the code is upper bounded by

$$2^{-n(\zeta-\epsilon)} + \Pr\{\mathbf{X} \notin W_{[X]\epsilon}^n\} < 2^{-n(\zeta-\epsilon)} + \epsilon \tag{5.25}$$

for n sufficiently large, again by the weak AEP. This probability is equal to $1 - P_e$ because P_e is the probability that the source sequence $\mathbf{X}$ is not covered by the code. Thus

$$1 - P_e < 2^{-n(\zeta-\epsilon)} + \epsilon \tag{5.26}$$

or

$$P_e > 1 - (2^{-n(\zeta-\epsilon)} + \epsilon). \tag{5.27}$$

This inequality holds when n is sufficiently large for any $\epsilon > 0$, in particular for $\epsilon < \zeta$. Then for any $\epsilon < \zeta$, $P_e > 1 - 2\epsilon$ when n is sufficiently large. Hence, $P_e \to 1$ as $n \to \infty$ and then $\epsilon \to 0$. This proves the converse part of the source coding theorem.

5.3 Efficient Source Coding

Theorem 5.4. *Let* $\mathbf{Y} = (Y_1, Y_2, \cdots, Y_m)$ *be a random binary sequence of length* m. *Then* $H(\mathbf{Y}) \leq m$ *with equality if and only if* Y_i *are drawn i.i.d. according to the uniform distribution on* $\{0, 1\}$.

Proof. By the independence bound for entropy,

$$H(\mathbf{Y}) \leq \sum_{i=1}^{m} H(Y_i) \tag{5.28}$$

with equality if and only if Y_i are mutually independent. By Theorem 2.43,

$$H(Y_i) \leq \log 2 = 1 \tag{5.29}$$

with equality if and only if Y_i is distributed uniformly on $\{0, 1\}$. Combining (5.28) and (5.29), we have

$$H(\mathbf{Y}) \leq \sum_{i=1}^{m} H(Y_i) \leq m, \tag{5.30}$$

where this upper bound is tight if and only if Y_i are mutually independent and each of them is distributed uniformly on $\{0, 1\}$. The theorem is proved. $\square$

Let $\mathbf{Y} = (Y_1, Y_2, \cdots, Y_n)$ be a sequence of length n such that Y_i are drawn i.i.d. according to the uniform distribution on $\{0, 1\}$, and let Y denote the generic random variable. Then $H(Y) = 1$. According to the source coding theorem, for almost perfect reconstruction of $\mathbf{Y}$, the coding rate of the source code must be at least 1. It turns out that in this case it is possible to use a source code with coding rate exactly equal to 1 while the source sequence $\mathbf{Y}$ can be reconstructed with zero error. This can be done by simply encoding all the 2^n possible binary sequences of length n, i.e., by taking $M = 2^n$. Then the coding rate is given by

$$n^{-1} \log M = n^{-1} \log 2^n = 1. \tag{5.31}$$

Since each symbol in $\mathbf{Y}$ is a bit and the rate of the best possible code describing $\mathbf{Y}$ is 1 bit per symbol, $Y_1, Y_2, \cdots, Y_n$ are called *fair bits*, with the connotation that they are incompressible.

It turns out that the whole idea of efficient source coding by a block code is to describe the information source by a binary sequence consisting of "almost fair" bits. Consider a sequence of block codes which encodes $\mathbf{X} = (X_1, X_2, \cdots, X_n)$ into $\mathbf{Y} = (Y_1, Y_2, \cdots, Y_m)$, where X_k are i.i.d. with generic random variable X, $\mathbf{Y}$ is a binary sequence with length

$$m \approx nH(X), \tag{5.32}$$

and $n \to \infty$. For simplicity, we assume that the common alphabet $\mathcal{X}$ is finite. Let $\hat{\mathbf{X}} \in \mathcal{X}^n$ be the reconstruction of $\mathbf{X}$ by the decoder and P_e be the probability of error, i.e.,

$$P_e = \Pr\{\mathbf{X} \neq \hat{\mathbf{X}}\}. \tag{5.33}$$

Further assume $P_e \to 0$ as $n \to \infty$. We will show that $\mathbf{Y}$ consists of almost fair bits.

By Fano's inequality,

$$H(\mathbf{X}|\hat{\mathbf{X}}) \leq 1 + P_e \log |\mathcal{X}|^n = 1 + nP_e \log |\mathcal{X}|. \tag{5.34}$$

Since $\hat{\mathbf{X}}$ is a function of $\mathbf{Y}$,

$$H(\mathbf{Y}) = H(\mathbf{Y}, \hat{\mathbf{X}}) \geq H(\hat{\mathbf{X}}). \tag{5.35}$$

It follows that

$$H(\mathbf{Y}) \geq H(\hat{\mathbf{X}}) \tag{5.36}$$
$$\geq I(\mathbf{X}; \hat{\mathbf{X}}) \tag{5.37}$$
$$= H(\mathbf{X}) - H(\mathbf{X}|\hat{\mathbf{X}}) \tag{5.38}$$
$$\geq nH(X) - (1 + nP_e \log |\mathcal{X}|) \tag{5.39}$$
$$= n(H(X) - P_e \log |\mathcal{X}|) - 1. \tag{5.40}$$

On the other hand, by Theorem 5.4,

$$H(\mathbf{Y}) \leq m. \tag{5.41}$$

Combining (5.40) and (5.41), we have

$$n(H(X) - P_e \log |\mathcal{X}|) - 1 \leq H(\mathbf{Y}) \leq m. \tag{5.42}$$

Since $P_e \to 0$ as $n \to \infty$, the above lower bound on $H(\mathbf{Y})$ is approximately equal to

$$nH(X) \approx m \tag{5.43}$$

when n is large (cf. (5.32)). Therefore,

$$H(\mathbf{Y}) \approx m. \tag{5.44}$$

In light of Theorem 5.4, $\mathbf{Y}$ almost attains the maximum possible entropy. In this sense, we say that $\mathbf{Y}$ consists of almost fair bits.

5.4 The Shannon–McMillan–Breiman Theorem

For an i.i.d. information source $\{X_k\}$ with generic random variable X and generic distribution $p(x)$, the weak AEP states that

$$-\frac{1}{n} \log p(\mathbf{X}) \to H(X) \tag{5.45}$$

in probability as $n \to \infty$, where $\mathbf{X} = (X_1, X_2, \cdots, X_n)$. Here $H(X)$ is the entropy of the generic random variables X as well as the entropy rate of the source $\{X_k\}$.

In Section 2.10, we showed that the entropy rate H of a source $\{X_k\}$ exists if the source is stationary. The *Shannon–McMillan–Breiman theorem* states that if $\{X_k\}$ is also *ergodic*, then

$$\Pr\left\{-\lim_{n\to\infty}\frac{1}{n}\log\Pr\{\mathbf{X}\} = H\right\} = 1. \tag{5.46}$$

This means that if $\{X_k\}$ is stationary and ergodic, then $-\frac{1}{n}\log\Pr\{\mathbf{X}\}$ not only almost always converges, but also almost always converges to H. For this reason, the Shannon–McMillan–Breiman theorem is also referred to as the weak AEP for ergodic stationary sources.

The formal definition of an ergodic source and the statement of the Shannon–McMillan–Breiman theorem require the use of measure theory which is beyond the scope of this book. We point out that the event in (5.46) involves an infinite collection of random variables which cannot be described by a joint distribution except in very special cases. Without measure theory, the probability of this event in general cannot be properly defined. However, this does not prevent us from developing some appreciation of the Shannon–McMillan–Breiman theorem.

Let $\mathcal{X}$ be the common alphabet for a stationary source $\{X_k\}$. Roughly speaking, a stationary source $\{X_k\}$ is ergodic if the time average exhibited by a single realization of the source is equal to the ensemble average with probability 1. More specifically, for any $k_1, k_2, \cdots, k_m$,

$$\Pr\left\{\lim_{n\to\infty}\frac{1}{n}\sum_{l=0}^{n-1}f(X_{k_1+l}, X_{k_2+l}, \cdots, X_{k_m+l})\right.$$

$$\left. = Ef(X_{k_1}, X_{k_2}, \cdots, X_{k_m})\right\} = 1, \tag{5.47}$$

where f is a function defined on $\mathcal{X}^m$ which satisfies suitable conditions. For the special case that $\{X_k\}$ satisfies

$$\Pr\left\{\lim_{n\to\infty}\frac{1}{n}\sum_{l=1}^{n}X_l = EX_k\right\} = 1, \tag{5.48}$$

we say that $\{X_k\}$ is *mean ergodic*.

Example 5.5. The i.i.d. source $\{X_k\}$ is mean ergodic under suitable conditions because the strong law of the large numbers states that (5.48) is satisfied.

Example 5.6. Consider the source $\{X_k\}$ defined as follows. Let Z be a binary random variable uniformly distributed on $\{0, 1\}$. For all k, let $X_k = Z$. Then

$$\Pr\left\{\lim_{n\to\infty}\frac{1}{n}\sum_{l=1}^{n}X_l=0\right\}=\frac{1}{2}\tag{5.49}$$

and

$$\Pr\left\{\lim_{n\to\infty}\frac{1}{n}\sum_{l=1}^{n}X_l=1\right\}=\frac{1}{2}.\tag{5.50}$$

Since $EX_k=\frac{1}{2}$,

$$\Pr\left\{\lim_{n\to\infty}\frac{1}{n}\sum_{l=1}^{n}X_l=EX_k\right\}=0.\tag{5.51}$$

Therefore, $\{X_k\}$ is not mean ergodic and hence not ergodic.

If an information source $\{X_k\}$ is stationary and ergodic, by the Shannon–McMillan–Breiman theorem,

$$-\frac{1}{n}\log\Pr\{\mathbf{X}\}\approx H\tag{5.52}$$

when n is large. That is, with probability close to 1, the probability of the sequence $\mathbf{X}$ which occurs is approximately equal to 2^{-nH}. Then by means of arguments similar to the proof of Theorem 5.3, we see that there exist approximately 2^{nH} sequences in $\mathcal{X}^n$ whose probabilities are approximately equal to 2^{-nH}, and the total probability of these sequences is almost 1. Therefore, by encoding these sequences with approximately nH bits, the source sequence $\mathbf{X}$ can be recovered with an arbitrarily small probability of error when the block length n is sufficiently large. This is a generalization of the direct part of the source coding theorem which gives a physical meaning to the entropy rate of ergodic stationary sources. We remark that if a source is stationary but not ergodic, although the entropy rate always exists, it may not carry any physical meaning.

As an example, by regarding printed English as a stationary ergodic process, Shannon [325] estimated by a guessing game that its entropy rate is about 1.3 bits per letter. Cover and King [77] described a gambling estimate of the entropy rate of printed English which gives 1.34 bits per letter. These results show that it is not possible to describe printed English accurately by using less than about 1.3 bits per letter.

Chapter Summary

Weak AEP I:
$$-\frac{1}{n}\log p(\mathbf{X})\to H(X)\text{ in probability.}$$

Weakly Typical Set:
$$W_{[X]\epsilon}^n=\left\{\mathbf{x}\in\mathcal{X}^n:\left|-n^{-1}\log p(\mathbf{x})-H(X)\right|\le\epsilon\right\}.$$

Weak AEP II:

1. $2^{-n(H(X)+\epsilon)} \leq p(\mathbf{x}) \leq 2^{-n(H(X)-\epsilon)}$ for $\mathbf{x} \in W^n_{[X]\epsilon}$.
2. $\Pr\{\mathbf{X} \in W^n_{[X]\epsilon}\} > 1 - \epsilon$ for n sufficiently large.
3. $(1 - \epsilon)2^{n(H(X)-\epsilon)} \leq |W^n_{[X]\epsilon}| \leq 2^{n(H(X)+\epsilon)}$ for n sufficiently large.

Source Coding Theorem: An i.i.d. random sequence $X_1, X_2, \cdots, X_n$ with generic random variable X can be compressed at rate $H(X) + \epsilon$ while $P_e \to 0$ as $n \to \infty$. If a rate less than $H(X)$ is used, then $P_e \to 1$ as $n \to \infty$.

Shannon–McMillan–Breiman Theorem: For a stationary source $\{X_k\}$ with entropy rate H,

$$\Pr\left\{-\lim_{n\to\infty}\frac{1}{n}\log\Pr\{\mathbf{X}\} = H\right\} = 1.$$

Problems

1. Show that for any $\epsilon > 0$, $W^n_{[X]\epsilon}$ is nonempty for sufficiently large n.
2. *The source coding theorem with a general block code.* In proving the converse of the source coding theorem, we assume that each codeword in $\mathcal{I}$ corresponds to a unique sequence in $\mathcal{X}^n$. More generally, a block code with block length n is defined by an encoding function $f : \mathcal{X}^n \to \mathcal{I}$ and a decoding function $g : \mathcal{I} \to \mathcal{X}^n$. Prove that $P_e \to 1$ as $n \to \infty$ even if we are allowed to use a general block code.
3. Following Problem 2, we further assume that we can use a block code with probabilistic encoding and decoding. For such a code, encoding is defined by a transition matrix F from $\mathcal{X}^n$ to $\mathcal{I}$ and decoding is defined by a transition matrix G from $\mathcal{I}$ to $\mathcal{X}^n$. Prove that $P_e \to 1$ as $n \to \infty$ even if we are allowed to use such a code.
4. In the discussion in Section 5.3, we made the assumption that the common alphabet $\mathcal{X}$ is finite. Can you draw the same conclusion when $\mathcal{X}$ is countable but $H(X)$ is finite? Hint: use Problem 2.
5. *Alternative definition of weak typicality.* Let $\mathbf{X} = (X_1, X_2, \cdots, X_n)$ be an i.i.d. sequence whose generic random variable X is distributed with $p(x)$. Let $q_{\mathbf{x}}$ be the empirical distribution of the sequence $\mathbf{x}$, i.e., $q_{\mathbf{x}}(x) = n^{-1}N(x; \mathbf{x})$ for all $x \in \mathcal{X}$, where $N(x; \mathbf{x})$ is the number of occurrence of x in $\mathbf{x}$.
    a) Show that for any $\mathbf{x} \in \mathcal{X}^n$,

    $$-\frac{1}{n}\log p(\mathbf{x}) = D(q_{\mathbf{x}}\|p) + H(q_{\mathbf{x}}).$$

    b) Show that for any $\epsilon > 0$, the weakly typical set $W^n_{[X]\epsilon}$ with respect to $p(x)$ is the set of sequences $\mathbf{x} \in \mathcal{X}^n$ such that

    $$|D(q_{\mathbf{x}}\|p) + H(q_{\mathbf{x}}) - H(p)| \leq \epsilon.$$

c) Show that for sufficiently large n,

$$\Pr\{|D(q_{\mathbf{x}}\|p) + H(q_{\mathbf{x}}) - H(p)| \le \epsilon\} > 1 - \epsilon.$$

(Ho and Yeung [167].)

6. Verify that the empirical entropy of a sequence is different from the entropy of the empirical distribution of the sequence (see Problem 5 for definition).

7. Let p and q be two probability distributions on the same alphabet $\mathcal{X}$ such that $H(p) \ne H(q)$. Show that there exists an $\epsilon > 0$ such that

$$p^n\left(\left\{\mathbf{x} \in \mathcal{X}^n : \left|-\frac{1}{n}\log p^n(\mathbf{x}) - H(q)\right| < \epsilon\right\}\right) \to 0$$

as $n \to \infty$. Give an example that $p \ne q$ but the above convergence does not hold.

8. Let p and q be two probability distributions on the same alphabet $\mathcal{X}$ with the same support.
   a) Prove that for any $\delta > 0$,

$$p^n\left(\left\{\mathbf{x} \in \mathcal{X}^n : \left|-\frac{1}{n}\log q^n(\mathbf{x}) - (H(p) + D(p\|q))\right| < \delta\right\}\right) \to 1$$

as $n \to \infty$.
   b) Prove that for any $\delta > 0$,

$$\left|\left\{\mathbf{x} \in \mathcal{X}^n : \left|-\frac{1}{n}\log q^n(\mathbf{x}) - (H(p) + D(p\|q))\right| < \delta\right\}\right| \le 2^{n(H(p)+D(p\|q)+\delta)}.$$

9. *Universal source coding.* Let $\mathcal{F} = \{\{X_k^{(s)}, k \ge 1\} : s \in \mathcal{S}\}$ be a family of i.i.d. information sources indexed by a finite set $\mathcal{S}$ with a common alphabet $\mathcal{X}$. Define

$$\bar{H} = \max_{s \in \mathcal{S}} H(X^{(s)}),$$

where $X^{(s)}$ is the generic random variable for $\{X_k^{(s)}, k \ge 1\}$, and

$$A_\epsilon^n(\mathcal{S}) = \bigcup_{s \in \mathcal{S}} W_{[X^{(s)}]\epsilon}^n,$$

where $\epsilon > 0$.
   a) Prove that for all $s \in \mathcal{S}$,

$$\Pr\{\mathbf{X}^{(s)} \in A_\epsilon^n(\mathcal{S})\} \to 1$$

as $n \to \infty$, where $\mathbf{X}^{(s)} = (X_1^{(s)}, X_2^{(s)}, \cdots, X_n^{(s)})$.
   b) Prove that for any $\epsilon' > \epsilon$,

$$|A_\epsilon^n(\mathcal{S})| \le 2^{n(\bar{H}+\epsilon')}$$

for sufficiently large n.

c) Suppose we know that an information source is in the family $\mathcal{F}$ but we do not know which one it is. Devise a compression scheme for the information source such that it is asymptotically optimal for every possible source in $\mathcal{F}$.

10. Let $\{X_k, k \geq 1\}$ be an i.i.d. information source with generic random variable X and alphabet $\mathcal{X}$. Assume

$$\sum_x p(x)[\log p(x)]^2 < \infty$$

and define

$$Z_n = -\frac{\log p(\mathbf{X})}{\sqrt{n}} - \sqrt{n}H(X)$$

for $n = 1, 2, \cdots$. Prove that $Z_n \to Z$ in distribution, where Z is a Gaussian random variable with mean 0 and variance $\sum_x p(x)[\log p(x)]^2 - H(X)^2$.

Historical Notes

The weak asymptotic equipartition property (AEP), which is instrumental in proving the source coding theorem, was first proved by Shannon in his original paper [322]. In this paper, he also stated that this property can be extended to an ergodic stationary source. Subsequently, McMillan [266] and Breiman [48] proved this property for an ergodic stationary source with a finite alphabet. Chung [70] extended the theme to a countable alphabet.

6

Strong Typicality

Weak typicality requires that the empirical entropy of a sequence is close to the true entropy. In this chapter, we introduce a stronger notion of typicality which requires that the relative frequency of each possible outcome is close to the corresponding probability. As we will see later, strong typicality is more powerful and flexible than weak typicality as a tool for theorem proving for memoryless problems. However, strong typicality can be used only for random variables with finite alphabets. Throughout this chapter, typicality refers to strong typicality and all the logarithms are in the base 2 unless otherwise specified.

6.1 Strong AEP

We consider an information source $\{X_k, k \geq 1\}$ where X_k are i.i.d. with distribution $p(x)$. We use X to denote the generic random variable and $H(X)$ to denote the common entropy for all X_k, where $H(X) < \infty$. Let $\mathbf{X} = (X_1, X_2, \cdots, X_n)$.

Definition 6.1. *The strongly typical set $T^n_{[X]\delta}$ with respect to $p(x)$ is the set of sequences $\mathbf{x} = (x_1, x_2, \cdots, x_n) \in \mathcal{X}^n$ such that $N(x; \mathbf{x}) = 0$ for $x \notin \mathcal{S}_X$, and*

$$\sum_x \left| \frac{1}{n} N(x; \mathbf{x}) - p(x) \right| \leq \delta, \qquad (6.1)$$

where $N(x; \mathbf{x})$ is the number of occurrences of x in the sequence $\mathbf{x}$ and δ is an arbitrarily small positive real number. The sequences in $T^n_{[X]\delta}$ are called strongly δ-typical sequences.

Throughout this chapter, we adopt the convention that all the summations, products, unions, etc., are taken over the corresponding supports unless otherwise specified. The strongly typical set $T^n_{[X]\delta}$ shares similar properties with its weakly typical counterpart, which is summarized as the *strong asymptotic equipartition property* (strong AEP) below. The interpretation of the strong AEP is similar to that of the weak AEP.

Theorem 6.2 (Strong AEP). *There exists $\eta > 0$ such that $\eta \to 0$ as $\delta \to 0$, and the following hold:*

1) If $\mathbf{x} \in T_{[X]\delta}^n$, then

$$2^{-n(H(X)+\eta)} \leq p(\mathbf{x}) \leq 2^{-n(H(X)-\eta)}. \tag{6.2}$$

2) For n sufficiently large,

$$\Pr\{\mathbf{X} \in T_{[X]\delta}^n\} > 1 - \delta. \tag{6.3}$$

3) For n sufficiently large,

$$(1-\delta)2^{n(H(X)-\eta)} \leq |T_{[X]\delta}^n| \leq 2^{n(H(X)+\eta)}. \tag{6.4}$$

Proof. To prove Property 1, for $\mathbf{x} \in T_{[X]\delta}^n$, we write

$$p(\mathbf{x}) = \prod_x p(x)^{N(x;\mathbf{x})}. \tag{6.5}$$

Then

$$\log p(\mathbf{x})$$
$$= \sum_x N(x;\mathbf{x}) \log p(x) \tag{6.6}$$
$$= \sum_x (N(x;\mathbf{x}) - np(x) + np(x)) \log p(x) \tag{6.7}$$
$$= n \sum_x p(x) \log p(x) - n \sum_x \left(\frac{1}{n}N(x;\mathbf{x}) - p(x)\right)(-\log p(x)) \tag{6.8}$$
$$= -n \left[H(X) + \sum_x \left(\frac{1}{n}N(x;\mathbf{x}) - p(x)\right)(-\log p(x)) \right]. \tag{6.9}$$

Since $\mathbf{x} \in T_{[X]\delta}^n$,

$$\sum_x \left| \frac{1}{n}N(x;\mathbf{x}) - p(x) \right| \leq \delta, \tag{6.10}$$

which implies

$$\left| \sum_x \left(\frac{1}{n}N(x;\mathbf{x}) - p(x)\right)(-\log p(x)) \right|$$

$$\leq \sum_x \left| \frac{1}{n} N(x; \mathbf{x}) - p(x) \right| (-\log p(x)) \tag{6.11}$$

$$\leq -\log \left(\min_x p(x) \right) \sum_x \left| \frac{1}{n} N(x; \mathbf{x}) - p(x) \right| \tag{6.12}$$

$$\leq -\delta \log \left(\min_x p(x) \right) \tag{6.13}$$

$$= \eta, \tag{6.14}$$

where

$$\eta = -\delta \log \left(\min_x p(x) \right) > 0. \tag{6.15}$$

Therefore,

$$-\eta \leq \sum_x \left(\frac{1}{n} N(x; \mathbf{x}) - p(x) \right) (-\log p(x)) \leq \eta. \tag{6.16}$$

It then follows from (6.9) that

$$-n(H(X) + \eta) \leq \log p(\mathbf{x}) \leq -n(H(X) - \eta) \tag{6.17}$$

or

$$2^{-n(H(X)+\eta)} \leq p(\mathbf{x}) \leq 2^{-n(H(X)-\eta)}, \tag{6.18}$$

where $\eta \to 0$ as $\delta \to 0$, proving Property 1.

To prove Property 2, we write

$$N(x; \mathbf{X}) = \sum_{k=1}^{n} B_k(x), \tag{6.19}$$

where

$$B_k(x) = \begin{cases} 1 \text{ if } X_k = x \\ 0 \text{ if } X_k \neq x \end{cases}. \tag{6.20}$$

Then $B_k(x), k = 1, 2, \cdots, n$ are i.i.d. random variables with

$$\Pr\{B_k(x) = 1\} = p(x) \tag{6.21}$$

and

$$\Pr\{B_k(x) = 0\} = 1 - p(x). \tag{6.22}$$

Note that

$$EB_k(x) = (1 - p(x)) \cdot 0 + p(x) \cdot 1 = p(x). \tag{6.23}$$

By the weak law of large numbers, for any $\delta > 0$ and for any $x \in \mathcal{X}$,

$$\Pr \left\{ \left| \frac{1}{n} \sum_{k=1}^{n} B_k(x) - p(x) \right| > \frac{\delta}{|\mathcal{X}|} \right\} < \frac{\delta}{|\mathcal{X}|} \tag{6.24}$$

for n sufficiently large. Then

$$\Pr\left\{\left|\frac{1}{n}N(x;\mathbf{X}) - p(x)\right| > \frac{\delta}{|\mathcal{X}|} \text{ for some } x\right\}$$

$$= \Pr\left\{\left|\frac{1}{n}\sum_{k=1}^{n}B_k(x) - p(x)\right| > \frac{\delta}{|\mathcal{X}|} \text{ for some } x\right\} \qquad (6.25)$$

$$= \Pr\left\{\bigcup_x\left\{\left|\frac{1}{n}\sum_{k=1}^{n}B_k(x) - p(x)\right| > \frac{\delta}{|\mathcal{X}|}\right\}\right\} \qquad (6.26)$$

$$\leq \sum_x \Pr\left\{\left|\frac{1}{n}\sum_{k=1}^{n}B_k(x) - p(x)\right| > \frac{\delta}{|\mathcal{X}|}\right\} \qquad (6.27)$$

$$< \sum_x \frac{\delta}{|\mathcal{X}|} \qquad (6.28)$$

$$= \delta, \qquad (6.29)$$

where we have used the union bound[1] to obtain (6.27). Since

$$\sum_x \left|\frac{1}{n}N(x;\mathbf{x}) - p(x)\right| > \delta \qquad (6.30)$$

implies

$$\left|\frac{1}{n}N(x;\mathbf{x}) - p(x)\right| > \frac{\delta}{|\mathcal{X}|} \quad \text{for some } x \in \mathcal{X}, \qquad (6.31)$$

we have

$$\Pr\left\{\mathbf{X} \in T_{[X]\delta}^n\right\}$$

$$= \Pr\left\{\sum_x \left|\frac{1}{n}N(x;\mathbf{X}) - p(x)\right| \leq \delta\right\} \qquad (6.32)$$

$$= 1 - \Pr\left\{\sum_x \left|\frac{1}{n}N(x;\mathbf{X}) - p(x)\right| > \delta\right\} \qquad (6.33)$$

$$\geq 1 - \Pr\left\{\left|\frac{1}{n}N(x;\mathbf{X}) - p(x)\right| > \frac{\delta}{|\mathcal{X}|} \text{ for some } x \in \mathcal{X}\right\} \qquad (6.34)$$

$$> 1 - \delta, \qquad (6.35)$$

proving Property 2.

Finally, Property 3 follows from Property 1 and Property 2 in exactly the same way as in Theorem 5.3, so the proof is omitted. □

Remark Analogous to weak typicality, we note that the upper bound on $|T_{[X]\delta}^n|$ in Property 3 holds for all $n \geq 1$, and for any $\delta > 0$, there exists at least one strongly typical sequence when n is sufficiently large. See Problem 1 in Chapter 5.

---

[1] The union bound refers to $\Pr\{A \cup B\} \leq \Pr\{A\} + \Pr\{B\}$.

In the rest of the section, we prove an enhancement of Property 2 of the strong AEP which gives an exponential bound on the probability of obtaining a non-typical vector.[2] The reader may skip this part at the first reading.

Theorem 6.3. *For sufficiently large n, there exists $\varphi(\delta) > 0$ such that*

$$\Pr\{\mathbf{X} \notin T^n_{[X]\delta}\} < 2^{-n\varphi(\delta)}. \tag{6.36}$$

The proof of this theorem is based on the Chernoff bound [66] which we prove in the next lemma.

Lemma 6.4 (Chernoff Bound). *Let Y be a real random variable and s be any nonnegative real number. Then for any real number a,*

$$\log \Pr\{Y \geq a\} \leq -sa + \log E\left[2^{sY}\right] \tag{6.37}$$

and

$$\log \Pr\{Y \leq a\} \leq sa + \log E\left[2^{-sY}\right]. \tag{6.38}$$

Proof. Let

$$u(y) = \begin{cases} 1 \text{ if } y \geq 0 \\ 0 \text{ if } y < 0 \end{cases}. \tag{6.39}$$

Then for any $s \geq 0$,

$$u(y - a) \leq 2^{s(y-a)}. \tag{6.40}$$

This is illustrated in Figure 6.1. Then

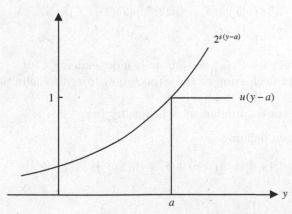

Fig. 6.1. An illustration of $u(y - a) \leq 2^{s(y-a)}$.

---

[2] This result is due to Ning Cai and Raymond W. Yeung. An alternative proof based on Pinsker's inequality (Theorem 2.33) and the method of types has been given by Prakash Narayan (private communication). See also Proposition 1 in Weissman et al. [375].

$$E[u(Y - a)] \leq E\left[2^{s(Y-a)}\right] = 2^{-sa} E\left[2^{sY}\right]. \tag{6.41}$$

Since

$$E[u(Y - a)] = \Pr\{Y \geq a\} \cdot 1 + \Pr\{Y < a\} \cdot 0 = \Pr\{Y \geq a\}, \tag{6.42}$$

we see that

$$\Pr\{Y \geq a\} \leq 2^{-sa} E\left[2^{sY}\right] = 2^{-sa + \log E[2^{sY}]}. \tag{6.43}$$

Then (6.37) is obtained by taking logarithm in the base 2. Upon replacing Y by $-Y$ and a by $-a$ in (6.37), (6.38) is obtained. The lemma is proved. $\square$

Proof of Theorem 6.3. We will follow the notation in the proof of Theorem 6.2. Consider $x \in X$ such that $p(x) > 0$. Applying (6.37), we have

$$\log \Pr\left\{\sum_{k=1}^{n} B_k(x) \geq n\,(p(x) + \delta)\right\}$$

$$\leq -sn\,(p(x) + \delta) + \log E\left[2^{s\sum_{k=1}^{n} B_k(x)}\right] \tag{6.44}$$

$$\overset{a)}{=} -sn\,(p(x) + \delta) + \log\left(\prod_{k=1}^{n} E\left[2^{sB_k(x)}\right]\right) \tag{6.45}$$

$$\overset{b)}{=} -sn\,(p(x) + \delta) + n\log(1 - p(x) + p(x)2^s) \tag{6.46}$$

$$\overset{c)}{\leq} -sn\,(p(x) + \delta) + n(\ln 2)^{-1}(-p(x) + p(x)2^s) \tag{6.47}$$

$$= -n\left[s\,(p(x) + \delta) + (\ln 2)^{-1}p(x)(1 - 2^s)\right], \tag{6.48}$$

where

(a) follows because $B_k(x)$ are mutually independent;
(b) is a direct evaluation of the expectation from the definition of $B_k(x)$ in (6.20);
(c) follows from the fundamental inequality $\ln a \leq a - 1$.

In (6.48), upon defining

$$\beta_x(s, \delta) = s\,(p(x) + \delta) + (\ln 2)^{-1}p(x)(1 - 2^s), \tag{6.49}$$

we have

$$\log \Pr\left\{\sum_{k=1}^{n} B_k(x) \geq n\,(p(x) + \delta)\right\} \leq -n\beta_x(s, \delta) \tag{6.50}$$

or

$$\Pr\left\{\sum_{k=1}^{n} B_k(x) \geq n\,(p(x) + \delta)\right\} \leq 2^{-n\beta_x(s, \delta)}. \tag{6.51}$$

It is readily seen that

$$\beta_x(0, \delta) = 0. \tag{6.52}$$

Regarding δ as fixed and differentiating with respect to s, we have

$$\beta'_x(s, \delta) = p(x)(1 - 2^s) + \delta. \tag{6.53}$$

Then

$$\beta'_x(0, \delta) = \delta > 0 \tag{6.54}$$

and it is readily verified that

$$\beta'_x(s, \delta) \geq 0 \tag{6.55}$$

for

$$0 \leq s \leq \log\left(1 + \frac{\delta}{p(x)}\right). \tag{6.56}$$

Therefore, we conclude that $\beta_x(s, \delta)$ is strictly positive for

$$0 < s \leq \log\left(1 + \frac{\delta}{p(x)}\right). \tag{6.57}$$

On the other hand, by applying (6.38), we can obtain in the same fashion the bound

$$\log \Pr\left\{\sum_{k=1}^n B_k(x) \leq n\left(p(x) - \delta\right)\right\} \leq -n\sigma_x(s, \delta) \tag{6.58}$$

or

$$\Pr\left\{\sum_{k=1}^n B_k(x) \leq n\left(p(x) - \delta\right)\right\} \leq 2^{-n\sigma_x(s,\delta)}, \tag{6.59}$$

where

$$\sigma_x(s, \delta) = -s\left(p(x) - \delta\right) + (\ln 2)^{-1}p(x)(1 - 2^{-s}). \tag{6.60}$$

Then

$$\sigma_x(0, \delta) = 0 \tag{6.61}$$

and

$$\sigma'_x(s, \delta) = p(x)(2^{-s} - 1) + \delta, \tag{6.62}$$

which is nonnegative for

$$0 \leq s \leq -\log\left(1 - \frac{\delta}{p(x)}\right). \tag{6.63}$$

In particular,

$$\sigma'_x(0, \delta) = \delta > 0. \tag{6.64}$$

Therefore, we conclude that $\sigma_x(s, \delta)$ is strictly positive for

$$0 < s \le -\log\left(1 - \frac{\delta}{p(x)}\right). \tag{6.65}$$

By choosing s satisfying

$$0 < s \le \min\left[\log\left(1 + \frac{\delta}{p(x)}\right), -\log\left(1 - \frac{\delta}{p(x)}\right)\right], \tag{6.66}$$

both $\beta_x(s,\delta)$ and $\sigma_x(s,\delta)$ are strictly positive. From (6.51) and (6.59), we have

$$\Pr\left\{\left|\frac{1}{n}\sum_{k=1}^{n}B_k(x) - p(x)\right| \ge \delta\right\}$$

$$= \Pr\left\{\left|\sum_{k=1}^{n}B_k(x) - np(x)\right| \ge n\delta\right\} \tag{6.67}$$

$$\le \Pr\left\{\sum_{k=1}^{n}B_k(x) \ge n\left(p(x) + \delta\right)\right\}$$

$$+ \Pr\left\{\sum_{k=1}^{n}B_k(x) \le n\left(p(x) - \delta\right)\right\} \tag{6.68}$$

$$\le 2^{-n\beta_x(s,\delta)} + 2^{-n\sigma_x(s,\delta)} \tag{6.69}$$

$$\le 2 \cdot 2^{-n\min(\beta_x(s,\delta),\sigma_x(s,\delta))} \tag{6.70}$$

$$= 2^{-n\left[\min(\beta_x(s,\delta),\sigma_x(s,\delta)) - \frac{1}{n}\right]} \tag{6.71}$$

$$= 2^{-n\varphi_x(\delta)}, \tag{6.72}$$

where

$$\varphi_x(\delta) = \min(\beta_x(s,\delta), \sigma_x(s,\delta)) - \frac{1}{n}. \tag{6.73}$$

Then $\varphi_x(\delta)$ is strictly positive for sufficiently large n because both $\beta_x(s,\delta)$ and $\sigma_x(s,\delta)$ are strictly positive.

Finally, consider

$$\Pr\{\mathbf{X} \in T^n_{[X]\delta}\}$$

$$= \Pr\left\{\sum_x\left|\frac{1}{n}N(x;\mathbf{X}) - p(x)\right| \le \delta\right\} \tag{6.74}$$

$$\ge \Pr\left\{\left|\frac{1}{n}N(x;\mathbf{X}) - p(x)\right| \le \frac{\delta}{|\mathcal{X}|} \text{ for all } x \in \mathcal{X}\right\} \tag{6.75}$$

$$= 1 - \Pr\left\{\left|\frac{1}{n}N(x;\mathbf{X}) - p(x)\right| > \frac{\delta}{|\mathcal{X}|} \text{ for some } x \in \mathcal{X}\right\} \tag{6.76}$$

$$\ge 1 - \sum_x \Pr\left\{\left|\frac{1}{n}N(x;\mathbf{X}) - p(x)\right| > \frac{\delta}{|\mathcal{X}|}\right\} \tag{6.77}$$

$$= 1 - \sum_{x} \Pr\left\{ \left| \frac{1}{n} \sum_{k=1}^{n} B_k(x) - p(x) \right| > \frac{\delta}{|\mathcal{X}|} \right\} \tag{6.78}$$

$$= 1 - \sum_{x:p(x)>0} \Pr\left\{ \left| \frac{1}{n} \sum_{k=1}^{n} B_k(x) - p(x) \right| > \frac{\delta}{|\mathcal{X}|} \right\} \tag{6.79}$$

$$\geq 1 - \sum_{x:p(x)>0} 2^{-n\varphi_x\left(\frac{\delta}{|\mathcal{X}|}\right)}, \tag{6.80}$$

where the last step follows from (6.72). Define

$$\varphi(\delta) = \frac{1}{2}\left[\min_{x:p(x)>0} \varphi_x\left(\frac{\delta}{|\mathcal{X}|}\right) \right]. \tag{6.81}$$

Then for sufficiently large n,

$$\Pr\{\mathbf{X} \in T_{[X]\delta}^{n}\} > 1 - 2^{-n\varphi(\delta)} \tag{6.82}$$

or

$$\Pr\{\mathbf{X} \notin T_{[X]\delta}^{n}\} < 2^{-n\varphi(\delta)}, \tag{6.83}$$

where $\varphi(\delta)$ is strictly positive. The theorem is proved. $\square$

6.2 Strong Typicality Versus Weak Typicality

As we have mentioned at the beginning of the chapter, strong typicality is more powerful and flexible than weak typicality as a tool for theorem proving for memoryless problems, but it can be used only for random variables with finite alphabets. We will prove in the next proposition that strong typicality is stronger than weak typicality in the sense that the former implies the latter.

Proposition 6.5. *For any* $\mathbf{x} \in \mathcal{X}^n$, *if* $\mathbf{x} \in T_{[X]\delta}^{n}$, *then* $\mathbf{x} \in W_{[X]\eta}^{n}$, *where* $\eta \to 0$ *as* $\delta \to 0$.

Proof. By Property 1 of strong AEP (Theorem 6.2), if $\mathbf{x} \in T_{[X]\delta}^{n}$, then

$$2^{-n(H(X)+\eta)} \leq p(\mathbf{x}) \leq 2^{-n(H(X)-\eta)} \tag{6.84}$$

or

$$H(X) - \eta \leq -\frac{1}{n}\log p(\mathbf{x}) \leq H(X) + \eta, \tag{6.85}$$

where $\eta \to 0$ as $\delta \to 0$. Then $\mathbf{x} \in W_{[X]\eta}^{n}$ by Definition 5.2. The proposition is proved. $\square$

We have proved in this proposition that strong typicality implies weak typicality, but the converse is not true. This idea can be explained without any detailed analysis. Let X be distributed with p such that $p(0) = 0.5$, $p(1) = 0.25$, and $p(2) = 0.25$. Consider a sequence $\mathbf{x}$ of length n and let $q(i)$ be the relative frequency of occurrence of symbol i in $\mathbf{x}$, i.e., $\frac{1}{n}N(i; \mathbf{x})$, where $i = 0, 1, 2$. In order for the sequence $\mathbf{x}$ to be weakly typical, we need

$$-\frac{1}{n}\log p(\mathbf{x})$$

$$= -q(0)\log 0.5 - q(1)\log 0.25 - q(2)\log 0.25 \tag{6.86}$$

$$\approx H(X) \tag{6.87}$$

$$= -(0.5)\log 0.5 - (0.25)\log 0.25 - (0.25)\log 0.25. \tag{6.88}$$

Obviously, this can be satisfied by choosing $q(i) = p(i)$ for all i. But alternatively, we can choose $q(0) = 0.5$, $q(1) = 0.5$, and $q(2) = 0$. With such a choice of $\{q(i)\}$, the sequence $\mathbf{x}$ is weakly typical with respect to p but obviously not strongly typical with respect to p, because the relative frequency of occurrence of each symbol i is $q(i)$, which is not close to $p(i)$ for $i = 1, 2$.

Therefore, we conclude that strong typicality is indeed stronger than weak typicality. However, as we have pointed out at the beginning of the chapter, strong typicality can only be used for random variables with finite alphabets.

6.3 Joint Typicality

In this section, we discuss strong joint typicality with respect to a bivariate distribution. Generalization to a multivariate distribution is straightforward.

Consider a bivariate information source $\{(X_k, Y_k), k \geq 1\}$ where (X_k, Y_k) are i.i.d. with distribution $p(x, y)$. We use (X, Y) to denote the pair of generic random variables.

Definition 6.6. *The strongly jointly typical set $T^n_{[XY]\delta}$ with respect to $p(x, y)$ is the set of $(\mathbf{x}, \mathbf{y}) \in \mathcal{X}^n \times \mathcal{Y}^n$ such that $N(x, y; \mathbf{x}, \mathbf{y}) = 0$ for $(x, y) \notin \mathcal{S}_{XY}$, and*

$$\sum_x \sum_y \left| \frac{1}{n}N(x, y; \mathbf{x}, \mathbf{y}) - p(x, y) \right| \leq \delta, \tag{6.89}$$

where $N(x, y; \mathbf{x}, \mathbf{y})$ is the number of occurrences of (x, y) in the pair of sequences $(\mathbf{x}, \mathbf{y})$ and δ is an arbitrarily small positive real number. A pair of sequences $(\mathbf{x}, \mathbf{y})$ is called strongly jointly δ-typical if it is in $T^n_{[XY]\delta}$.

Strong typicality satisfies the following *consistency* property.

Theorem 6.7 (Consistency). *If $(\mathbf{x}, \mathbf{y}) \in T^n_{[XY]\delta}$, then $\mathbf{x} \in T^n_{[X]\delta}$ and $\mathbf{y} \in T^n_{[Y]\delta}$.*

Proof. If $(\mathbf{x}, \mathbf{y}) \in T^n_{[XY]\delta}$, then

$$\sum_x \sum_y \left| \frac{1}{n} N(x,y;\mathbf{x},\mathbf{y}) - p(x,y) \right| \leq \delta. \tag{6.90}$$

Upon observing that

$$N(x;\mathbf{x}) = \sum_y N(x,y;\mathbf{x},\mathbf{y}), \tag{6.91}$$

we have

$$\sum_x \left| \frac{1}{n} N(x;\mathbf{x}) - p(x) \right|$$

$$= \sum_x \left| \frac{1}{n} \sum_y N(x,y;\mathbf{x},\mathbf{y}) - \sum_y p(x,y) \right| \tag{6.92}$$

$$= \sum_x \left| \sum_y \left(\frac{1}{n} N(x,y;\mathbf{x},\mathbf{y}) - p(x,y) \right) \right| \tag{6.93}$$

$$\leq \sum_x \sum_y \left| \frac{1}{n} N(x,y;\mathbf{x},\mathbf{y}) - p(x,y) \right| \tag{6.94}$$

$$< \delta. \tag{6.95}$$

Therefore, $\mathbf{x} \in T^n_{[X]\delta}$. Similarly, $\mathbf{y} \in T^n_{[Y]\delta}$. The theorem is proved. $\square$

The following theorem asserts that strong typicality is preserved when a function is applied to a vector componentwise.

Theorem 6.8 (Preservation). *Let* $Y = f(X)$. *If*

$$\mathbf{x} = (x_1, x_2, \cdots, x_n) \in T^n_{[X]\delta}, \tag{6.96}$$

then

$$f(\mathbf{x}) = (y_1, y_2, \cdots, y_n) \in T^n_{[Y]\delta}, \tag{6.97}$$

where $y_i = f(x_i)$ *for* $1 \leq i \leq n$.

Proof. Consider $\mathbf{x} \in T^n_{[X]\delta}$, i.e.,

$$\sum_x \left| \frac{1}{n} N(x;\mathbf{x}) - p(x) \right| < \delta. \tag{6.98}$$

Since $Y = f(X)$,

$$p(y) = \sum_{x \in f^{-1}(y)} p(x) \tag{6.99}$$

for all $y \in \mathcal{Y}$. On the other hand,

$$N(y; f(\mathbf{x})) = \sum_{x \in f^{-1}(y)} N(x; \mathbf{x}) \tag{6.100}$$

for all $y \in \mathcal{Y}$. Then

$$\sum_y \left| \frac{1}{n} N(y; f(\mathbf{x})) - p(y) \right|$$

$$= \sum_y \left| \sum_{x \in f^{-1}(y)} \left(\frac{1}{n} N(x; \mathbf{x}) - p(x) \right) \right| \tag{6.101}$$

$$\leq \sum_y \sum_{x \in f^{-1}(y)} \left| \frac{1}{n} N(x; \mathbf{x}) - p(x) \right| \tag{6.102}$$

$$= \sum_x \left| \frac{1}{n} N(x; \mathbf{x}) - p(x) \right| \tag{6.103}$$

$$< \delta. \tag{6.104}$$

Therefore, $f(\mathbf{x}) \in T_{[Y]\delta}^n$, proving the lemma. $\square$

For a bivariate i.i.d. source $\{(X_k, Y_k)\}$, we have the *strong joint asymptotic equipartition property* (strong JAEP), which can readily be obtained by applying the strong AEP to the source $\{(X_k, Y_k)\}$.

Theorem 6.9 (Strong JAEP). *Let*

$$(\mathbf{X}, \mathbf{Y}) = ((X_1, Y_1), (X_2, Y_2), \cdots, (X_n, Y_n)), \tag{6.105}$$

where (X_i, Y_i) are i.i.d. with generic pair of random variables (X, Y). Then there exists $\lambda > 0$ such that $\lambda \to 0$ as $\delta \to 0$, and the following hold:

1) If $(\mathbf{x}, \mathbf{y}) \in T_{[XY]\delta}^n$, then

$$2^{-n(H(X,Y)+\lambda)} \leq p(\mathbf{x}, \mathbf{y}) \leq 2^{-n(H(X,Y)-\lambda)}. \tag{6.106}$$

2) For n sufficiently large,

$$\Pr\{(\mathbf{X}, \mathbf{Y}) \in T_{[XY]\delta}^n\} > 1 - \delta. \tag{6.107}$$

3) For n sufficiently large,

$$(1 - \delta) 2^{n(H(X,Y)-\lambda)} \leq |T_{[XY]\delta}^n| \leq 2^{n(H(X,Y)+\lambda)}. \tag{6.108}$$

From the strong JAEP, we can see the following. Since there are approximately $2^{nH(X,Y)}$ typical $(\mathbf{x}, \mathbf{y})$ pairs and approximately $2^{nH(X)}$ typical $\mathbf{x}$, for a typical $\mathbf{x}$, the number of $\mathbf{y}$ such that $(\mathbf{x}, \mathbf{y})$ is jointly typical is approximately

$$\frac{2^{nH(X,Y)}}{2^{nH(X)}} = 2^{nH(Y|X)} \tag{6.109}$$

on the average. The next theorem reveals that this is not only true on the average, but it is in fact true for every typical $\mathbf{x}$ as long as there exists at least one $\mathbf{y}$ such that $(\mathbf{x}, \mathbf{y})$ is jointly typical.

Theorem 6.10 (Conditional Strong AEP). *For any $\mathbf{x} \in T_{[X]\delta}^n$, define*

$$T_{[Y|X]\delta}^n(\mathbf{x}) = \{\mathbf{y} \in T_{[Y]\delta}^n : (\mathbf{x}, \mathbf{y}) \in T_{[XY]\delta}^n\}. \tag{6.110}$$

If $|T_{[Y|X]\delta}^n(\mathbf{x})| \geq 1$, then

$$2^{n(H(Y|X)-\nu)} \leq |T_{[Y|X]\delta}^n(\mathbf{x})| \leq 2^{n(H(Y|X)+\nu)}, \tag{6.111}$$

where $\nu \to 0$ as $n \to \infty$ and $\delta \to 0$.

We first prove the following lemma which is along the line of Stirling's approximation [113].

Lemma 6.11. *For any $n > 0$,*

$$n \ln n - n < \ln n! < (n + 1) \ln(n + 1) - n. \tag{6.112}$$

Proof. First, we write

$$\ln n! = \ln 1 + \ln 2 + \cdots + \ln n. \tag{6.113}$$

Since $\ln x$ is a monotonically increasing function of x, we have

$$\int_{k-1}^{k} \ln x \, dx < \ln k < \int_{k}^{k+1} \ln x \, dx. \tag{6.114}$$

Summing over $1 \leq k \leq n$, we have

$$\int_{0}^{n} \ln x \, dx < \ln n! < \int_{1}^{n+1} \ln x \, dx \tag{6.115}$$

or

$$n \ln n - n < \ln n! < (n + 1) \ln(n + 1) - n. \tag{6.116}$$

The lemma is proved. $\square$

Proof of Theorem 6.10. Let δ be a small positive real number and n be a large positive integer to be specified later. Fix an $\mathbf{x} \in T_{[X]\delta}^n$, so that

$$\sum_{x} \left| \frac{1}{n} N(x; \mathbf{x}) - p(x) \right| \leq \delta. \tag{6.117}$$

This implies that for all $x \in \mathcal{X}$,

$$\left| \frac{1}{n} N(x; \mathbf{x}) - p(x) \right| \leq \delta \qquad (6.118)$$

or

$$p(x) - \delta \leq \frac{1}{n} N(x; \mathbf{x}) \leq p(x) + \delta. \qquad (6.119)$$

We first prove the upper bound on $|T^n_{[Y|X]\delta}(\mathbf{x})|$. For any $\nu > 0$, consider

$$2^{-n(H(X)-\nu/2)} \overset{a)}{\geq} p(\mathbf{x}) \qquad (6.120)$$

$$= \sum_{\mathbf{y} \in \mathcal{Y}^n} p(\mathbf{x}, \mathbf{y}) \qquad (6.121)$$

$$\geq \sum_{\mathbf{y} \in T^n_{[Y|X]\delta}(\mathbf{x})} p(\mathbf{x}, \mathbf{y}) \qquad (6.122)$$

$$\overset{b)}{\geq} \sum_{\mathbf{y} \in T^n_{[Y|X]\delta}(\mathbf{x})} 2^{-n(H(X,Y)+\nu/2)} \qquad (6.123)$$

$$= |T^n_{[Y|X]\delta}(\mathbf{x})| 2^{-n(H(X,Y)+\nu/2)}, \qquad (6.124)$$

where (a) and (b) follow from the strong AEP (Theorem 6.2) and the strong joint AEP (Theorem 6.9), respectively. Then we obtain

$$|T^n_{[Y|X]\delta}(\mathbf{x})| \leq 2^{n(H(Y|X)+\nu)}, \qquad (6.125)$$

which is the upper bound to be proved.

Assume that $|T^n_{[Y|X]\delta}(\mathbf{x})| \geq 1$. We now prove the lower bound on $|T^n_{[Y|X]\delta}(\mathbf{x})|$. Let

$$\{K(x,y), (x,y) \in \mathcal{X} \times \mathcal{Y}\} \qquad (6.126)$$

be any set of nonnegative integers such that

1.

$$\sum_y K(x,y) = N(x; \mathbf{x}) \qquad (6.127)$$

for all $x \in \mathcal{X}$, and
2. for any $\mathbf{y} \in \mathcal{Y}^n$, if

$$N(x,y; \mathbf{x}, \mathbf{y}) = K(x,y) \qquad (6.128)$$

for all $(x,y) \in \mathcal{X} \times \mathcal{Y}$, then $(\mathbf{x}, \mathbf{y}) \in T^n_{[XY]\delta}$.

Then by Definition 6.6, $\{K(x,y)\}$ satisfies

$$\sum_x \sum_y \left| \frac{1}{n} K(x,y) - p(x,y) \right| \leq \delta, \qquad (6.129)$$

which implies that for all $(x, y) \in \mathcal{X} \times \mathcal{Y}$,

$$\left| \frac{1}{n} K(x, y) - p(x, y) \right| \le \delta \tag{6.130}$$

or

$$p(x, y) - \delta \le \frac{1}{n} K(x, y) \le p(x, y) + \delta. \tag{6.131}$$

Such a set $\{K(x, y)\}$ exists because $T^n_{[Y|X]\delta}(\mathbf{x})$ is assumed to be nonempty. Straightforward combinatorics reveals that the number of $\mathbf{y}$ which satisfy the constraints in (6.128) is equal to

$$M(K) = \prod_x \frac{N(x; \mathbf{x})!}{\prod_y K(x, y)!}, \tag{6.132}$$

and it is readily seen that

$$|T^n_{[Y|X]\delta}(\mathbf{x})| \ge M(K). \tag{6.133}$$

Using Lemma 6.11, we can lower bound $\ln M(K)$ as follows:

$$\begin{aligned}
\ln M(K) \\
\ge \sum_x \Big\{ & N(x; \mathbf{x}) \ln N(x; \mathbf{x}) - N(x; \mathbf{x}) \\
& - \sum_y [(K(x, y) + 1) \ln(K(x, y) + 1) - K(x, y)] \Big\}
\end{aligned} \tag{6.134}$$

$$\stackrel{a)}{=} \sum_x \Big[N(x; \mathbf{x}) \ln N(x; \mathbf{x}) \\
- \sum_y (K(x, y) + 1) \ln(K(x, y) + 1) \Big] \tag{6.135}$$

$$\stackrel{b)}{\ge} \sum_x \Big\{ N(x; \mathbf{x}) \ln(n \, (p(x) - \delta)) \\
- \sum_y (K(x, y) + 1) \ln \left[n \left(p(x, y) + \delta + \frac{1}{n} \right) \right] \Big\}. \tag{6.136}$$

In the above, (a) follows from (6.127) and (b) is obtained by applying the lower bound on $n^{-1} N(x; \mathbf{x})$ in (6.119) and the upper bound on $n^{-1} K(x, y)$ in (6.131). Also from (6.127), the coefficient of $\ln n$ in (6.136) is given by

$$\sum_x \Big[N(x; \mathbf{x}) - \sum_y (K(x, y) + 1) \Big] = -|\mathcal{X}||\mathcal{Y}|. \tag{6.137}$$

Let δ be sufficiently small and n be sufficiently large so that

$$0 < p(x) - \delta < 1 \tag{6.138}$$

and

$$p(x, y) + \delta + \frac{1}{n} < 1 \tag{6.139}$$

for all x and y. Then in (6.136), both the logarithms

$$\ln(p(x) - \delta) \tag{6.140}$$

and

$$\ln\left(p(x, y) + \delta + \frac{1}{n}\right) \tag{6.141}$$

are negative. Note that the logarithm in (6.140) is well defined by virtue of (6.138). Rearranging the terms in (6.136), applying the upper bound in (6.119) and the lower bound[3] in (6.131), and dividing by n, we have

$$n^{-1} \ln M(K)$$

$$\geq \sum_x (p(x) + \delta) \ln(p(x) - \delta) - \sum_x \sum_y \left(p(x, y) - \delta + \frac{1}{n}\right)$$

$$\times \ln\left(p(x, y) + \delta + \frac{1}{n}\right) - \frac{|\mathcal{X}||\mathcal{Y}| \ln n}{n} \tag{6.142}$$

$$= -H_e(X) + H_e(X, Y) + L_l(n, \delta) \tag{6.143}$$

$$= H_e(Y|X) + L_l(n, \delta), \tag{6.144}$$

where $L_l(n, \delta)$ denotes a function of n and δ which tends to 0 as $n \to \infty$ and $\delta \to 0$. Changing the base of the logarithm to 2, we have

$$n^{-1} \log M(K) \geq H(Y|X) + L_l(n, \delta). \tag{6.145}$$

Then it follows from (6.133) that

$$n^{-1} \log |T^n_{[Y|X]\delta}(\mathbf{x})| \geq H(Y|X) + L_l(n, \delta). \tag{6.146}$$

Upon replacing $L_l(n, \delta)$ by ν, we obtain

$$|T^n_{[Y|X]\delta}(\mathbf{x})| \geq 2^{n(H(Y|X) - \nu)}, \tag{6.147}$$

where $\nu \to 0$ as $n \to \infty$ and $\delta \to 0$ as required. The theorem is proved. □

The above theorem says that for any typical $\mathbf{x}$, as long as there is one typical $\mathbf{y}$ such that $(\mathbf{x}, \mathbf{y})$ is jointly typical, there are approximately $2^{nH(Y|X)}$ $\mathbf{y}$ such that $(\mathbf{x}, \mathbf{y})$ is jointly typical. This theorem has the following corollary that the number of such typical $\mathbf{x}$ grows with n at almost the same rate as the total number of typical $\mathbf{x}$.

---

[3] For the degenerate case when $p(x, y) = 1$ for some x and y, $p(x, y) + \delta + \frac{1}{n} > 1$, and the logarithm in (6.141) is in fact positive. Then the upper bound instead of the lower bound should be applied. The details are omitted.

Corollary 6.12. *For a joint distribution $p(x,y)$ on $\mathcal{X} \times \mathcal{Y}$, let $S_{[X]\delta}^n$ be the set of all sequences $\mathbf{x} \in T_{[X]\delta}^n$ such that $T_{[Y|X]\delta}^n(\mathbf{x})$ is nonempty. Then*

$$|S_{[X]\delta}^n| \geq (1-\delta)2^{n(H(X)-\psi)}, \tag{6.148}$$

where $\psi \to 0$ as $n \to \infty$ and $\delta \to 0$.

Proof. By the consistency of strong typicality (Theorem 6.7), if $(\mathbf{x}, \mathbf{y}) \in T_{[XY]\delta}^n$, then $\mathbf{x} \in T_{[X]\delta}^n$. In particular, $\mathbf{x} \in S_{[X]\delta}^n$. Then

$$T_{[XY]\delta}^n = \bigcup_{\mathbf{x} \in S_{[X]\delta}^n} \{(\mathbf{x}, \mathbf{y}) : \mathbf{y} \in T_{[Y|X]\delta}^n(\mathbf{x})\}. \tag{6.149}$$

Using the lower bound on $|T_{[XY]\delta}^n|$ in Theorem 6.9 and the upper bound on $|T_{[Y|X]\delta}^n(\mathbf{x})|$ in the last theorem, we have

$$(1-\delta)2^{n(H(X,Y)-\lambda)} \leq |T_{[XY]\delta}^n| \leq |S_{[X]\delta}^n| 2^{n(H(Y|X)+\nu)} \tag{6.150}$$

which implies

$$|S_{[X]\delta}^n| \geq (1-\delta)2^{n(H(X)-(\lambda+\nu))}. \tag{6.151}$$

The theorem is proved upon letting $\psi = \lambda + \nu$. $\square$

We have established a rich set of structural properties for strong typicality with respect to a bivariate distribution $p(x,y)$, which is summarized in the two-dimensional *strong joint typicality array* in Figure 6.2. In this array, the rows and the columns are the typical sequences $\mathbf{x} \in S_{[X]\delta}^n$ and $\mathbf{y} \in S_{[Y]\delta}^n$, respectively. The total number of rows and columns are approximately equal to $2^{nH(X)}$ and $2^{nH(Y)}$, respectively. An entry indexed by $(\mathbf{x}, \mathbf{y})$ receives a dot if $(\mathbf{x}, \mathbf{y})$ is strongly jointly typical. The total number of dots is approximately equal to $2^{nH(X,Y)}$. The number of dots in each row is approximately equal to $2^{nH(Y|X)}$, while the number of dots in each column is approximately equal to $2^{nH(X|Y)}$.

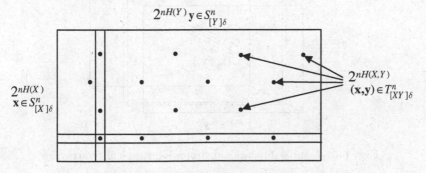

Fig. 6.2. A two-dimensional strong joint typicality array.

For reasons which will become clear in Chapter 16, the strong joint typicality array in Figure 6.2 is said to exhibit an *asymptotic quasi-uniform* structure. By a two-dimensional asymptotic quasi-uniform structure, we mean that in the array all the columns have approximately the same number of dots and all the rows have approximately the same number of dots. The strong joint typicality array for a multivariate distribution continues to exhibit an asymptotic quasi-uniform structure. The three-dimensional strong joint typicality array with respect to a distribution $p(x, y, z)$ is illustrated in Figure 6.3. As before, an entry $(\mathbf{x}, \mathbf{y}, \mathbf{z})$ receives a dot if $(\mathbf{x}, \mathbf{y}, \mathbf{z})$ is strongly jointly typical. This is not shown in the figure otherwise it will be very confusing. The total number of dots in the whole array is approximately equal to $2^{nH(X,Y,Z)}$. These dots are distributed in the array such that all the planes parallel to each other have approximately the same number of dots, and all the cylinders parallel to each other have approximately the same number of dots. More specifically, the total number of dots on the plane for any fixed $\mathbf{z}_0 \in S^n_{[Z]\delta}$ (as shown) is approximately equal to $2^{nH(X,Y|Z)}$, and the total number of dots in the cylinder for any fixed $(\mathbf{x}_0, \mathbf{y}_0)$ pair in $S^n_{[XY]\delta}$ (as shown) is approximately equal to $2^{nH(Z|X,Y)}$, so on and so forth.

We see from the strong AEP and Corollary 6.12 that $S^n_{[X]\delta}$ and $T^n_{[X]\delta}$ grow with n at approximately the same rate. We end this section by stating in the next proposition that $S^n_{[X]\delta}$ indeed contains almost all the probability when n is large. The proof is left as an exercise (see Problem 4).

Proposition 6.13. *With respect to a joint distribution $p(x, y)$ on $\mathcal{X} \times \mathcal{Y}$, for any $\delta > 0$,*

$$\Pr\{\mathbf{X} \in S^n_{[X]\delta}\} > 1 - \delta \tag{6.152}$$

for n sufficiently large.

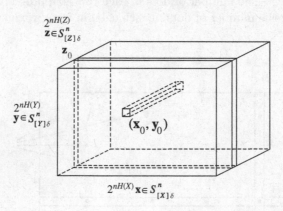

Fig. 6.3. A three-dimensional strong joint typicality array.

6.4 An Interpretation of the Basic Inequalities

The asymptotic quasi-uniform structure exhibited in a strong joint typicality array discussed in the last section is extremely important in information theory. Later in the book, we will see how this structure is involved in proving results such as the channel coding theorem and the rate-distortion theorem. In this section, we show how the basic inequalities can be revealed by examining this structure. It has further been shown by Chan [58] that all unconstrained information inequalities can be obtained from this structure, thus giving a physical meaning to these inequalities.

Consider random variables X, Y, and Z and a fixed $\mathbf{z} \in S^n_{[Z]\delta}$, so that $T^n_{[XY|Z]\delta}(\mathbf{z})$ is nonempty. By the consistency of strong typicality, if $(\mathbf{x}, \mathbf{y}, \mathbf{z}) \in T^n_{[XYZ]\delta}$, then $(\mathbf{x}, \mathbf{z}) \in T^n_{[XZ]\delta}$ and $(\mathbf{y}, \mathbf{z}) \in T^n_{[YZ]\delta}$, or $\mathbf{x} \in T^n_{[X|Z]\delta}(\mathbf{z})$ and $\mathbf{y} \in T^n_{[Y|Z]\delta}(\mathbf{z})$, respectively. Thus

$$T^n_{[XY|Z]\delta}(\mathbf{z}) \subset T^n_{[X|Z]\delta}(\mathbf{z}) \times T^n_{[Y|Z]\delta}(\mathbf{z}), \tag{6.153}$$

which implies

$$|T^n_{[XY|Z]\delta}(\mathbf{z})| \le |T^n_{[X|Z]\delta}(\mathbf{z})||T^n_{[Y|Z]\delta}(\mathbf{z})|. \tag{6.154}$$

Applying the lower bound in Theorem 6.10 to $T^n_{[XY|Z]\delta}(\mathbf{z})$ and the upper bound to $T^n_{[X|Z]\delta}(\mathbf{z})$ and $T^n_{[Y|Z]\delta}(\mathbf{z})$, we have

$$2^{n(H(X,Y|Z)-\zeta)} \le 2^{n(H(X|Z)+\gamma)} 2^{n(H(Y|Z)+\phi)}, \tag{6.155}$$

where $\zeta, \gamma, \phi \to 0$ as $n \to \infty$ and $\delta \to 0$. Taking logarithm to the base 2 and dividing by n, we obtain

$$H(X,Y|Z) \le H(X|Z) + H(Y|Z) \tag{6.156}$$

upon letting $n \to \infty$ and $\delta \to 0$. This inequality is equivalent to

$$I(X;Y|Z) \ge 0. \tag{6.157}$$

Thus we have proved the nonnegativity of conditional mutual information. Since all Shannon's information measures are special cases of conditional mutual information, we have proved the nonnegativity of all Shannon's information measures, namely the basic inequalities.

Chapter Summary

Strong typicality implies weak typicality but can be used only for random variables with finite alphabets.

Strongly Typical Set:

$$T_{[X]\delta}^n = \left\{ \mathbf{x} \in \mathcal{X}^n : \sum_x \left| n^{-1} N(x; \mathbf{x}) - p(x) \right| \leq \delta \right\}.$$

Strong AEP:

1. $2^{-n(H(X)+\eta)} \leq p(\mathbf{x}) \leq 2^{-n(H(X)-\eta)}$ for $\mathbf{x} \in T_{[X]\delta}^n$:

2. $\Pr\{\mathbf{X} \in T_{[X]\delta}^n\} > 1 - \delta$ for n sufficiently large.

3. $(1-\delta)2^{n(H(X)-\eta)} \leq |T_{[X]\delta}^n| \leq 2^{n(H(X)+\eta)}$ for n sufficiently large.

Theorem: For sufficiently large n,

$$\Pr\{\mathbf{X} \notin T_{[X]\delta}^n\} < 2^{-n\varphi(\delta)}.$$

Consistency: If $(\mathbf{x}, \mathbf{y}) \in T_{[XY]\delta}^n$, then $\mathbf{x} \in T_{[X]\delta}^n$ and $\mathbf{y} \in T_{[Y]\delta}^n$.

Preservation: If $\mathbf{x} \in T_{[X]\delta}^n$, then $f(\mathbf{x}) \in T_{[f(X)]\delta}^n$.

Conditional Strong AEP: For $\mathbf{x} \in T_{[X]\delta}^n$, let

$$T_{[Y|X]\delta}^n(\mathbf{x}) = \{\mathbf{y} \in T_{[Y]\delta}^n : (\mathbf{x}, \mathbf{y}) \in T_{[XY]\delta}^n\}.$$

If $|T_{[Y|X]\delta}^n(\mathbf{x})| \geq 1$, then

$$2^{n(H(Y|X)-\nu)} \leq |T_{[Y|X]\delta}^n(\mathbf{x})| \leq 2^{n(H(Y|X)+\nu)}.$$

Problems

1. Show that $(\mathbf{x}, \mathbf{y}) \in T_{[X,Y]\delta}^n$ and $(\mathbf{y}, \mathbf{z}) \in T_{[Y,Z]\delta}^n$ do not imply $(\mathbf{x}, \mathbf{z}) \in T_{[X,Z]\delta}^n$.

2. Let $\mathbf{X} = (X_1, X_2, \cdots, X_n)$, where X_k are i.i.d. with generic random variable X. Prove that

$$\Pr\{\mathbf{X} \in T_{[X]\delta}^n\} \geq 1 - \frac{|\mathcal{X}|^3}{n\delta^2}$$

for any n and $\delta > 0$. This shows that $\Pr\{\mathbf{X} \in T_{[X]\delta}^n\} \to 1$ as $\delta \to 0$ and $n \to \infty$ if $\sqrt{n}\delta \to \infty$.

3. Prove that for a random variable X with a countable alphabet, Property 2 of the strong AEP holds, while Properties 1 and 3 do not hold.

4. Prove Proposition 6.13. Hint: Use the fact that if $(\mathbf{X}, \mathbf{Y}) \in T_{[XY]\delta}^n$, then $\mathbf{X} \in S_{[X]\delta}^n$.

5. Let $\mathcal{P}(\mathcal{X})$ be the set of all probability distributions over a finite alphabet $\mathcal{X}$. Find a polynomial $Q(n)$ such that for any integer n, there exists a subset $\mathcal{P}_n(\mathcal{X})$ of $\mathcal{P}(\mathcal{X})$ such that
   a) $|\mathcal{P}_n(\mathcal{X})| \leq Q(n)$;
   b) for all $P \in \mathcal{P}(\mathcal{X})$, there exists $P_n \in \mathcal{P}_n(\mathcal{X})$ such that

   $$|P_n(x) - P(x)| < \frac{1}{n}$$

   for all $x \in \mathcal{X}$.
   Hint: Let $\mathcal{P}_n(\mathcal{X})$ be the set of all probability distributions over $\mathcal{X}$ such that all the probability masses can be expressed as fractions with denominator n.

6. Let p be any probability distribution over a finite set $\mathcal{X}$ and η be a real number in $(0, 1)$. Prove that for any subset A of $\mathcal{X}^n$ with $p^n(A) \geq \eta$,

   $$|A \cap T^n_{[X]\delta}| \geq 2^{n(H(p) - \delta')},$$

   where $\delta' \to 0$ as $\delta \to 0$ and $n \to \infty$.

   In the following problems, for a sequence $\mathbf{x} \in \mathcal{X}^n$, let $q_\mathbf{x}$ be the empirical distribution of $\mathbf{x}$, i.e., $q_\mathbf{x}(x) = n^{-1}N(x; \mathbf{x})$ for all $x \subset \mathcal{X}$. Similarly, for a pair of sequences $(\mathbf{x}, \mathbf{y}) \in \mathcal{X}^n \times \mathcal{Y}^n$, let $q_{\mathbf{x},\mathbf{y}}$ be the joint empirical distribution of $(\mathbf{x}, \mathbf{y})$, i.e., $q_{\mathbf{x},\mathbf{y}}(x, y) = n^{-1}N(x, y; \mathbf{x}, \mathbf{y})$ for all $(x, y) \in \mathcal{X} \times \mathcal{Y}$.

7. *Alternative definition of strong typicality.* Show that (6.1) is equivalent to

   $$V(q_\mathbf{x}, p) \leq \delta,$$

   where $V(\cdot, \cdot)$ denotes the variational distance. Thus strong typicality can be regarded as requiring the empirical distribution of a sequence to be close to the probability distribution of the generic random variable in variational distance. Also compare the result here with the alternative definition of weak typicality (Problem 5 in Chapter 5).

8. The empirical distribution $q_\mathbf{x}$ of the sequence $\mathbf{x}$ is also called the *type* of $\mathbf{x}$. Assuming that $\mathcal{X}$ is finite, show that there are a total of $\binom{n + |\mathcal{X}| - 1}{n}$ distinct types $q_\mathbf{x}$. Hint: There are $\binom{a+b-1}{a}$ ways to distribute a identical balls in b boxes.

9. *Unified typicality.* Let $\mathbf{X} = (X_1, X_2, \cdots, X_n)$ be an i.i.d. sequence whose generic random variable X is distributed with $p(x)$, where the alphabet $\mathcal{X}$ is countable. For any $\eta > 0$, the unified typical set $U^n_{[X]\eta}$ with respect to $p(x)$ is the set of sequences $\mathbf{x} \in \mathcal{X}^n$ such that

   $$D(q_\mathbf{x} \| p) + |H(q_\mathbf{x}) - H(p)| \leq \eta.$$

   a) Show that for any $\mathbf{x} \in \mathcal{X}^n$, if $\mathbf{x} \in U^n_{[X]\eta}$, then $\mathbf{x} \in W^n_{[X]\eta}$.

b) Show that for any $\mathbf{x} \in \mathcal{X}^n$, if $\mathbf{x} \in U^n_{[X]\eta}$, then $\mathbf{x} \in T^n_{[X]\delta}$, where $\delta = \sqrt{\eta \cdot 2 \ln 2}$.

Therefore, unified typicality implies both weak typicality and strong typicality.

10. *The AEP for unified typicality.* Unified typicality defined in Problem 9, unlike strong typicality, can be applied to random variables whose alphabets are countable . At the same time, it preserves the essential properties of strong typicality. The following outlines the proof of the AEP which has been discussed in Theorem 5.3 and Theorem 6.2 for weak typicality and strong typicality, respectively.

a) Show that

$$2^{-n(H(X)+\eta)} \le p(\mathbf{x}) \le 2^{-n(H(X)-\eta)},$$

i.e., Property 1 of the AEP.

b) Show that for sufficiently large n,

$$\Pr\{H(q_{\mathbf{x}}) - H(p) > \epsilon\} < \epsilon.$$

Hint: Use the results in Problem 9 above and Problem 5 in Chapter 5.

c) It can be proved by means of the result in Problem 9 that

$$\Pr\{H(p) - H(q_{\mathbf{x}}) > \epsilon\} < \epsilon$$

(see Ho and Yeung [167]). By assuming this inequality, prove that

$$\Pr\{|H(q_{\mathbf{x}}) - H(p)| \le \epsilon\} < 1 - 2\epsilon.$$

d) Show that if $|H(q_{\mathbf{x}}) - H(p)| \le \epsilon$ and $|D(q_{\mathbf{x}}\|p) + H(q_{\mathbf{x}}) - H(p)| \le \epsilon$, then

$$D(q_{\mathbf{x}}\|p) + |H(q_{\mathbf{x}}) - H(p)| \le 3\epsilon.$$

e) Use the results in (c) and (d) above and the result in Problem 5, Part (c) in Chapter 5 to show that

$$\Pr\{D(q_{\mathbf{x}}\|p) + |H(q_{\mathbf{x}}) - H(p)| \le \eta\} > 1 - \eta.$$

This proves Property 2 of the AEP. Property 3 of the AEP follows from Property 1 as in the proof of Theorem 5.3.

11. *Consistency of unified typicality* For any $\eta > 0$, the unified jointly typical set $U^n_{[XY]\eta}$ with respect to $p_{XY}(x, y)$ is the set of sequences $(\mathbf{x}, \mathbf{y}) \in \mathcal{X}^n \times \mathcal{Y}^n$ such that

$$D(q_{\mathbf{x},\mathbf{y}}\|p_{XY}) + |H(q_{\mathbf{x},\mathbf{y}}) - H(p_{XY})|$$
$$+ |H(q_{\mathbf{x}}) - H(p_X)| + |H(q_{\mathbf{y}}) - H(p_Y)| \le \eta.$$

Show that if $(\mathbf{x}, \mathbf{y}) \in U^n_{[XY]\eta}$, then $\mathbf{x} \in U^n_{[X]\eta}$ and $\mathbf{y} \in U^n_{[Y]\eta}$.

Historical Notes

Strong typicality was used by Wolfowitz [384] for proving channel coding theorems and by Berger [28] for proving the rate-distortion theorem and various results in multiterminal source coding. The method of types, a refinement of the notion of strong typicality, was systematically developed in the book by Csiszár and Körner [83]. The interpretation of the basic inequalities in Section 6.4 is a preamble to the relation between entropy and groups to be discussed in Chapter 16.

Recently, Ho and Yeung [167] introduced the notion of unified typicality which is stronger than both weak typicality and strong typicality. This notion of typicality can be applied to random variables with countable alphabets, while at the same time can preserve the essential properties of strong typicality. See Problems 9, 10, and 11 for a discussion.

7

Discrete Memoryless Channels

In all practical communication systems, when a signal is transmitted from one point to another point, the signal is inevitably contaminated by random noise, i.e., the signal received is correlated with but possibly different from the signal transmitted. We use a *noisy channel* to model such a situation. A noisy channel is a "system" which has one input terminal and one output terminal,[1] with the input connected to the transmission point and the output connected to the receiving point. When the signal is transmitted through the channel, it is distorted in a random way which depends on the channel characteristics. As a consequence, the signal received may be different from the signal transmitted.

In communication engineering, we are interested in conveying messages reliably through a noisy channel at the maximum possible rate. We first look at a simple channel called the *binary symmetric channel* (BSC), which is represented by the transition diagram in Figure 7.1. In this channel both the input X and the output Y take values in the set $\{0,1\}$. There is a certain probability, denoted by ϵ, that the output is not equal to the input. That is, if the input is 0, then the output is 0 with probability $1 - \epsilon$ and is 1 with probability ϵ. Likewise, if the input is 1, then the output is 1 with probability $1 - \epsilon$ and is 0 with probability ϵ. The parameter ϵ is called the *crossover probability* of the BSC.

Let $\{A, B\}$ be the message set which contains two possible messages to be conveyed through a BSC with $0 \leq \epsilon < 0.5$. We further assume that the two messages A and B are equally likely. If the message is A, we map it to the codeword 0, and if the message is B, we map it to the codeword 1. This is the simplest example of a *channel code*. The codeword is then transmitted through the channel. Our task is to decode the message based on the output of the channel, and an error is said to occur if the message is decoded incorrectly.

---

[1] The discussion on noisy channels here is confined to point-to-point channels.

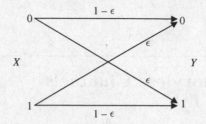

Fig. 7.1. The transition diagram of a binary symmetric channel.

Consider

$$\Pr\{A|Y=0\} = \Pr\{X=0|Y=0\} \tag{7.1}$$

$$= \frac{\Pr\{X=0\}\Pr\{Y=0|X=0\}}{\Pr\{Y=0\}} \tag{7.2}$$

$$= \frac{0.5(1-\epsilon)}{\Pr\{Y=0\}}. \tag{7.3}$$

Since

$$\Pr\{Y=0\} = \Pr\{Y=1\} = 0.5 \tag{7.4}$$

by symmetry,[2] it follows that

$$\Pr\{A|Y=0\} = 1 - \epsilon \tag{7.5}$$

and

$$\Pr\{B|Y=0\} = 1 - \Pr\{A|Y=0\} = \epsilon. \tag{7.6}$$

Since $\epsilon < 0.5$,

$$\Pr\{B|Y=0\} < \Pr\{A|Y=0\}. \tag{7.7}$$

Therefore, in order to minimize the probability of error, we decode a received 0 to the message A. By symmetry, we decode a received 1 to the message B.

An error occurs if a 0 is received and the message is B or if a 1 is received and the message is A. Therefore, the probability of error, denoted by P_e, is given by

---

[2] More explicitly,

$$\Pr\{Y=0\} = \Pr\{A\}\Pr\{Y=0|A\} + \Pr\{B\}\Pr\{Y=0|B\}$$
$$= 0.5\,\Pr\{Y=0|X=0\} + 0.5\,\Pr\{Y=0|X=1\}$$
$$= 0.5(1-\epsilon) + 0.5\epsilon$$
$$= 0.5.$$

$$P_e = \Pr\{Y = 0\}\Pr\{B|Y = 0\} + \Pr\{Y = 1\}\Pr\{A|Y = 1\} \tag{7.8}$$
$$= 0.5\epsilon + 0.5\epsilon \tag{7.9}$$
$$= \epsilon, \tag{7.10}$$

where (7.9) follows from (7.6) because

$$\Pr\{A|Y = 1\} = \Pr\{B|Y = 0\} = \epsilon \tag{7.11}$$

by symmetry.

Let us assume that $\epsilon \neq 0$. Then the above scheme obviously does not provide perfectly reliable communication. If we are allowed to use the channel only once, then this is already the best we can do. However, if we are allowed to use the same channel repeatedly, then we can improve the reliability by generalizing the above scheme.

We now consider the following channel code which we refer to as the binary *repetition code*. Let $n \geq 1$ be an odd positive integer which is called the *block length* of the code. In this code, the message A is mapped to the sequence of n 0s, and the message B is mapped to the sequence of n 1s. The codeword, which consists of a sequence of either n 0s or n 1s, is transmitted through the channel in n uses. Upon receiving a sequence of n bits at the output of the channel, we use the majority vote to decode the message, i.e., if there are more 0s than 1s in the sequence, we decode the sequence to the message A, otherwise we decode the sequence to the message B. Note that the block length is chosen to be odd so that there cannot be a tie. When $n = 1$, this scheme reduces to the previous scheme.

For this more general scheme, we continue to denote the probability of error by P_e. Let N_0 and N_1 be the number of 0s and 1s in the received sequence, respectively. Clearly,

$$N_0 + N_1 = n. \tag{7.12}$$

For large n, if the message is A, the number of 0s received is approximately equal to

$$E[N_0|A] = n(1 - \epsilon) \tag{7.13}$$

and the number of 1s received is approximately equal to

$$E[N_1|A] = n\epsilon \tag{7.14}$$

with high probability by the weak law of large numbers. This implies that the probability of an error, namely the event $\{N_0 < N_1\}$, is small because

$$n(1 - \epsilon) > n\epsilon \tag{7.15}$$

with the assumption that $\epsilon < 0.5$. Specifically,

$$\Pr\{\text{error}|A\} = \Pr\{N_0 < N_1|A\} \tag{7.16}$$
$$= \Pr\{n - N_1 < N_1|A\} \tag{7.17}$$
$$= \Pr\{N_1 > 0.5n|A\} \tag{7.18}$$
$$\leq \Pr\{N_1 > (\epsilon + \phi)n|A\}, \tag{7.19}$$

where

$$0 < \phi < 0.5 - \epsilon, \tag{7.20}$$

so that ϕ is positive and

$$\epsilon + \phi < 0.5. \tag{7.21}$$

Note that such a ϕ exists because $\epsilon < 0.5$. Then by the weak law of large numbers, the upper bound in (7.19) tends to 0 as $n \to \infty$. By symmetry, $\Pr\{\text{error}|B\}$ also tends to 0 as $n \to \infty$. Therefore,

$$P_e = \Pr\{A\}\Pr\{\text{error}|A\} + \Pr\{B\}\Pr\{\text{error}|B\} \tag{7.22}$$

tends to 0 as $n \to \infty$. In other words, by using a long enough repetition code, we can make P_e arbitrarily small. In this sense, we say that reliable communication is achieved asymptotically.

We point out that for a BSC with $\epsilon > 0$, for any given transmitted sequence of length n, the probability of receiving any given sequence of length n is nonzero. It follows that for any two distinct input sequences, there is always a nonzero probability that the same output sequence is produced so that the two input sequences become indistinguishable. Therefore, except for very special channels (e.g., the BSC with $\epsilon = 0$), no matter how the encoding/decoding scheme is devised, a nonzero probability of error is inevitable, and asymptotically reliable communication is the best we can hope for.

Though a rather naive approach, asymptotically reliable communication can be achieved by using the repetition code. The repetition code, however, is not without catch. For a channel code, the *rate* of the code in bit(s) per use is defined as the ratio of the logarithm of the size of the message set in the base 2 to the block length of the code. Roughly speaking, the rate of a channel code is the average number of bits the channel code attempts to convey through the channel per use of the channel. For a binary repetition code with block length n, the rate is $\frac{1}{n}\log 2 = \frac{1}{n}$, which tends to 0 as $n \to \infty$. Thus in order to achieve asymptotic reliability by using the repetition code, we cannot communicate through the noisy channel at any positive rate!

In this chapter, we characterize the maximum rate at which information can be communicated through a *discrete memoryless channel* (DMC) with an arbitrarily small probability of error. This maximum rate, which is generally positive, is known as the *channel capacity*. Then we discuss the use of feedback in communicating through a DMC and show that feedback does not increase the capacity. At the end of the chapter, we discuss transmitting an information source through a DMC, and we show that asymptotic optimality can be achieved by separating source coding and channel coding.

7.1 Definition and Capacity

Definition 7.1. *Let $\mathcal{X}$ and $\mathcal{Y}$ be discrete alphabets, and $p(y|x)$ be a transition matrix from $\mathcal{X}$ to $\mathcal{Y}$. A discrete channel $p(y|x)$ is a single input–single output system with input random variable X taking values in $\mathcal{X}$ and output random variable Y taking values in $\mathcal{Y}$ such that*

$$\Pr\{X = x, Y = y\} = \Pr\{X = x\}p(y|x) \tag{7.23}$$

for all $(x, y) \in \mathcal{X} \times \mathcal{Y}$.

Remark From (7.23), we see that if $\Pr\{X = x\} > 0$, then

$$\Pr\{Y = y|X = x\} = \frac{\Pr\{X = x, Y = y\}}{\Pr\{X = x\}} = p(y|x). \tag{7.24}$$

Note that $\Pr\{Y = y|X = x\}$ is undefined if $\Pr\{X = x\} = 0$. Nevertheless, (7.23) is valid for both cases.

We now present an alternative description of a discrete channel. Let $\mathcal{X}$ and $\mathcal{Y}$ be discrete alphabets. Let X be a random variable taking values in $\mathcal{X}$ and $p(y|x)$ be *any* transition matrix from $\mathcal{X}$ to $\mathcal{Y}$. Define random variables Z_x with $\mathcal{Z}_x = \mathcal{Y}$ for $x \in \mathcal{X}$ such that

$$\Pr\{Z_x = y\} = p(y|x) \tag{7.25}$$

for all $y \in \mathcal{Y}$. We assume that Z_x, $x \in \mathcal{X}$ are mutually independent and also independent of X. Further define the random variable

$$Z = (Z_x : x \in \mathcal{X}), \tag{7.26}$$

called the *noise variable*. Note that Z is independent of X. Now define a random variable taking values in $\mathcal{Y}$ as

$$Y = Z_x \quad \text{if } X = x. \tag{7.27}$$

Evidently, Y is a function of X and Z. Then for $x \in \mathcal{X}$ such that $\Pr\{X = x\} > 0$, we have

$$\Pr\{X = x, Y = y\} = \Pr\{X = x\}\Pr\{Y = y|X = x\} \tag{7.28}$$

$$= \Pr\{X = x\}\Pr\{Z_x = y|X = x\} \tag{7.29}$$

$$= \Pr\{X = x\}\Pr\{Z_x = y\} \tag{7.30}$$

$$= \Pr\{X = x\}p(y|x), \tag{7.31}$$

i.e., (7.23) in Definition 7.1, where (7.30) follows from the assumption that Z_x is independent of X. For $x \in \mathcal{X}$ such that $\Pr\{X = x\} = 0$, since $\Pr\{X = x\} = 0$ implies $\Pr\{X = x, Y = y\} = 0$, (7.23) continues to hold. Then by regarding X and Y as the input and output random variables, we have obtained an alternative description of the discrete channel $p(y|x)$.

Since Y is a function of X and Z, we can write

$$Y = \alpha(X, Z). \tag{7.32}$$

Then we have the following equivalent definition for a discrete channel.

Definition 7.2. *Let $\mathcal{X}$, $\mathcal{Y}$, and $\mathcal{Z}$ be discrete alphabets. Let $\alpha : \mathcal{X} \times \mathcal{Z} \to \mathcal{Y}$, and Z be a random variable taking values in $\mathcal{Z}$, called the noise variable. A discrete channel (α, Z) is a single input–single output system with input alphabet $\mathcal{X}$ and output alphabet $\mathcal{Y}$. For any input random variable X, the noise variable Z is independent of X, and the output random variable Y is given by*

$$Y = \alpha(X, Z). \tag{7.33}$$

Figure 7.2 illustrates a discrete channel $p(y|x)$ and a discrete channel (α, Z). The next definition gives the condition for the equivalence of the two specifications of a discrete channel according to Definitions 7.1 and 7.2, respectively.

Definition 7.3. *Two discrete channels $p(y|x)$ and (α, Z) defined on the same input alphabet $\mathcal{X}$ and output alphabet $\mathcal{Y}$ are equivalent if*

$$\Pr\{\alpha(x, Z) = y\} = p(y|x) \tag{7.34}$$

for all x and y.

We point out that the qualifier "discrete" in a discrete channel refers to the input and output alphabets of the channel being discrete. As part of a discrete-time communication system, a discrete channel can be used repeatedly at every time index $i = 1, 2, \cdots$. As the simplest model, we may assume that

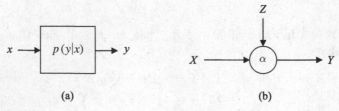

(a)   (b)

Fig. 7.2. Illustrations of (a) a discrete channel $p(y|x)$ and (b) a discrete channel (α, Z).

the noises for the transmission over the channel at different time indices are independent of each other. In the next definition, we will introduce the *discrete memoryless channel* (DMC) as a discrete-time extension of a discrete channel that captures this modeling assumption.

To properly formulate a DMC, we regard it as a subsystem of a discrete-time stochastic system which will be referred to as "the system" in the sequel. In such a system, random variables are generated sequentially in discrete time, and more than one random variable may be generated instantaneously but sequentially at a particular time index.

Definition 7.4. *A discrete memoryless channel (DMC) $p(y|x)$ is a sequence of replicates of a generic discrete channel $p(y|x)$. These discrete channels are indexed by a discrete-time index i, where $i \geq 1$, with the ith channel being available for transmission at time i. Transmission through a channel is assumed to be instantaneous. Let X_i and Y_i be, respectively, the input and the output of the DMC at time i, and let T_{i-} denote all the random variables that are generated in the system before X_i. The equality*

$$\Pr\{Y_i = y, X_i = x, T_{i-} = t\} = \Pr\{X_i = x, T_{i-} = t\}p(y|x) \qquad (7.35)$$

holds for all $(x, y, t) \in \mathcal{X} \times \mathcal{Y} \times T_{i-}$.

Remark Similar to the remark following Definition 7.1, if $\Pr\{X_i = x, T_{i-} = t\} > 0$, then

$$\Pr\{Y_i = y | X_i = x, T_{i-} = t\} = \frac{\Pr\{Y_i = y, X_i = x, T_{i-} = t\}}{\Pr\{X_i = x, T_{i-} = t\}} \qquad (7.36)$$

$$= p(y|x). \qquad (7.37)$$

Note that $\Pr\{Y_i = y | X_i = x, T_{i-} = t\}$ is undefined if $\Pr\{X_i = x, T_{i-} = t\} = 0$. Nevertheless, (7.35) is valid for both cases.

Invoking Proposition 2.5, we see from (7.35) that

$$T_{i-} \to X_i \to Y_i \qquad (7.38)$$

forms a Markov chain, i.e., the output of the DMC at time i is independent of all the random variables that have already been generated in the system conditioning on the input at time i. This captures the memorylessness of a DMC. Figure 7.3 is an illustration of a DMC $p(y|x)$.

Paralleling Definition 7.2 for a discrete channel, we now present an alternative definition of a DMC.

Definition 7.5. *A discrete memoryless channel (α, Z) is a sequence of replicates of a generic discrete channel (α, Z). These discrete channels are indexed by a discrete-time index i, where $i \geq 1$, with the ith channel being available*

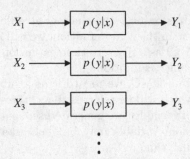

Fig. 7.3. An illustration of a discrete memoryless channel $p(y|x)$.

for transmission at time i. Transmission through a channel is assumed to be instantaneous. Let X_i and Y_i be, respectively, the input and the output of the DMC at time i, and let T_{i-} denote all the random variables that are generated in the system before X_i. The noise variable Z_i for the transmission at time i is a copy of the generic noise variable Z and is independent of (X_i, T_{i-}). The output of the DMC at time i is given by

$$Y_i = \alpha(X_i, Z_i). \tag{7.39}$$

Figure 7.4 is an illustration of a DMC (α, Z). We now show that Definitions 7.4 and 7.5 specify the same DMC provided that the generic discrete channel $p(y|x)$ in Definition 7.4 is equivalent to the generic discrete channel (α, Z) in Definition 7.5, i.e., (7.34) holds. For the DMC (α, Z) in Definition 7.5, consider

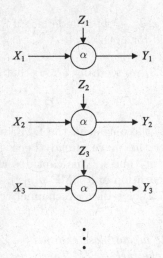

Fig. 7.4. An illustration of a discrete memoryless channel (α, Z).

$$0 \leq I(T_{i-}; Y_i | X_i) \tag{7.40}$$
$$\leq I(T_{i-}; Y_i, X_i, Z_i | X_i) \tag{7.41}$$
$$= I(T_{i-}; X_i, Z_i | X_i) \tag{7.42}$$
$$= I(T_{i-}; Z_i | X_i) \tag{7.43}$$
$$= 0, \tag{7.44}$$

where the first equality follows from (7.39) and the last equality follows from the assumption that Z_i is independent of (X_i, T_{i-}). Therefore,

$$I(T_{i-}; Y_i | X_i) = 0 \tag{7.45}$$

or $T_{i-} \to X_i \to Y_i$ forms a Markov chain. It remains to establish (7.35) for all $(x, y, t) \in \mathcal{X} \times \mathcal{Y} \times T_{i-}$. For $x \in \mathcal{X}$ such that $\Pr\{X_i = x\} = 0$, both $\Pr\{Y_i = y, X_i = x, T_{i-} = t\}$ and $\Pr\{X_i = x, T_{i-} = t\}$ vanish because they are upper bounded by $\Pr\{X_i = x\}$. Therefore (7.35) holds. For $x \in \mathcal{X}$ such that $\Pr\{X_i = x\} > 0$,

$$\Pr\{Y_i = y, X_i = x, T_{i-} = t\}$$
$$\overset{a)}{=} \Pr\{X_i = x, T_{i-} = t\} \Pr\{Y_i = y | X_i = x\} \tag{7.46}$$
$$\overset{b)}{=} \Pr\{X_i = x, T_{i-} = t\} \Pr\{\alpha(X_i, Z_i) = y | X_i = x\} \tag{7.47}$$
$$= \Pr\{X_i = x, T_{i-} = t\} \Pr\{\alpha(x, Z_i) = y | X_i = x\} \tag{7.48}$$
$$\overset{c)}{=} \Pr\{X_i = x, T_{i-} = t\} \Pr\{\alpha(x, Z_i) = y\} \tag{7.49}$$
$$\overset{d)}{=} \Pr\{X_i = x, T_{i-} = t\} \Pr\{\alpha(x, Z) = y\} \tag{7.50}$$
$$\overset{e)}{=} \Pr\{X_i = x, T_{i-} = t\} p(y|x), \tag{7.51}$$

where

(a) follows from the Markov chain $T_{i-} \to X_i \to Y_i$;
(b) follows from (7.39);
(c) follows from Definition 7.5 that Z_i is independent of X_i;
(d) follows from Definition 7.5 that Z_i and the generic noise variable Z have the same distribution;
(e) follows from (7.34).

Hence, (7.35) holds for all $(x, y, t) \in \mathcal{X} \times \mathcal{Y} \times T_{i-}$, proving that the DMC (α, Z) in Definition 7.4 is equivalent to the DMC $p(y|x)$ in Definition 7.5.

Definition 7.5 renders the following physical conceptualization of a DMC. The DMC can be regarded as a "box" which has only two terminals, the input and the output. The box perfectly shields its contents from the rest of the system. At time i, upon the transmission of the input X_i, the noise variable Z_i is generated inside the box according to the distribution of the generic noise variable Z. Since the box is perfectly shielded, the generation of the Z_i is independent of X_i and any other random variable that has already been generated in the system. Then the function α is applied to (X_i, Z_i) to produce the output Y_i.

In the rest of the section, we will define the capacity of a DMC and discuss some of its basic properties. The capacities of two simple DMCs will also be evaluated explicitly. To keep our discussion simple, we will assume that the alphabets $\mathcal{X}$ and $\mathcal{Y}$ are finite.

Definition 7.6. *The capacity of a discrete memoryless channel $p(y|x)$ is defined as*

$$C = \max_{p(x)} I(X;Y), \tag{7.52}$$

where X and Y are, respectively, the input and the output of the generic discrete channel, and the maximum is taken over all input distributions $p(x)$.

From the above definition, we see that

$$C \geq 0 \tag{7.53}$$

because

$$I(X;Y) \geq 0 \tag{7.54}$$

for all input distributions $p(x)$. By Theorem 2.43, we have

$$C = \max_{p(x)} I(X;Y) \leq \max_{p(x)} H(X) = \log|\mathcal{X}|. \tag{7.55}$$

Likewise, we have

$$C \leq \log|\mathcal{Y}|. \tag{7.56}$$

Therefore,

$$C \leq \min(\log|\mathcal{X}|, \log|\mathcal{Y}|). \tag{7.57}$$

Since $I(X;Y)$ is a continuous functional of $p(x)$ and the set of all $p(x)$ is a compact set (i.e., closed and bounded) in $\Re^{|\mathcal{X}|}$, the maximum value of $I(X;Y)$ can be attained.[3] This justifies taking the maximum rather than the supremum in the definition of channel capacity in (7.52).

We will prove subsequently that C is in fact the maximum rate at which information can be communicated reliably through a DMC. We first give some examples of DMCs for which the capacities can be obtained in closed form. In the following, X and Y denote, respectively, the input and the output of the generic discrete channel, and all logarithms are in the base 2.

Example 7.7 (Binary Symmetric Channel). The transition diagram of a BSC has been shown in Figure 7.1. Alternatively, a BSC can be represented by the system in Figure 7.5. Here, Z is a binary random variable representing the noise of the channel, with

$$\Pr\{Z = 0\} = 1 - \epsilon \quad \text{and} \quad \Pr\{Z = 1\} = \epsilon, \tag{7.58}$$

---

[3] The assumption that $\mathcal{X}$ is finite is essential in this argument.

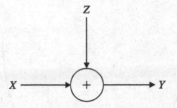

Fig. 7.5. An alternative representation for a binary symmetric channel.

and Z is independent of X. Then

$$Y = X + Z \bmod 2. \tag{7.59}$$

This representation for a BSC is in the form prescribed by Definition 7.2.

In order to determine the capacity of the BSC, we first bound $I(X;Y)$ as follows:

$$I(X;Y) = H(Y) - H(Y|X) \tag{7.60}$$
$$= H(Y) - \sum_x p(x)H(Y|X=x) \tag{7.61}$$
$$= H(Y) - \sum_x p(x)h_b(\epsilon) \tag{7.62}$$
$$= H(Y) - h_b(\epsilon) \tag{7.63}$$
$$\leq 1 - h_b(\epsilon), \tag{7.64}$$

where we have used h_b to denote the binary entropy function in the base 2. In order to achieve this upper bound, we have to make $H(Y) = 1$, i.e., the output distribution of the BSC is uniform. This can be done by letting $p(x)$ be the uniform distribution on $\{0,1\}$. Therefore, the upper bound on $I(X;Y)$ can be achieved, and we conclude that

$$C = 1 - h_b(\epsilon) \quad \text{bit per use.} \tag{7.65}$$

Figure 7.6 is a plot of the capacity C versus the crossover probability ϵ. We see from the plot that C attains the maximum value 1 when $\epsilon = 0$ or $\epsilon = 1$ and attains the minimum value 0 when $\epsilon = 0.5$. When $\epsilon = 0$, it is easy to see that $C = 1$ is the maximum rate at which information can be communicated through the channel reliably. This can be achieved simply by transmitting unencoded bits through the channel, and no decoding is necessary because all the bits are received unchanged. When $\epsilon = 1$, the same can be achieved with the additional decoding step which complements all the received bits. By doing so, the bits transmitted through the channel can be recovered without error. Thus from the communication point of view, for binary channels, a channel which never makes error and a channel which always makes errors are equally good. When $\epsilon = 0.5$, the channel output is independent of the channel input. Therefore, no information can possibly be communicated through the channel.

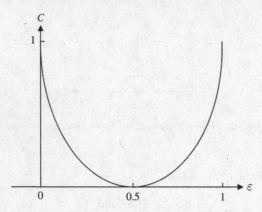

Fig. 7.6. The capacity of a binary symmetric channel.

Example 7.8 (Binary Erasure Channel). The transition diagram of a binary erasure channel is shown in Figure 7.7. In this channel, the input alphabet is $\{0,1\}$, while the output alphabet is $\{0,1,e\}$. With probability γ, the erasure symbol e is produced at the output, which means that the input bit is lost; otherwise the input bit is reproduced at the output without error. The parameter γ is called the erasure probability.

To determine the capacity of this channel, we first consider

$$C = \max_{p(x)} I(X;Y) \tag{7.66}$$

$$= \max_{p(x)} (H(Y) - H(Y|X)) \tag{7.67}$$

$$= \max_{p(x)} H(Y) - h_b(\gamma). \tag{7.68}$$

Thus we only have to maximize $H(Y)$. To this end, let

$$\Pr\{X = 0\} = a \tag{7.69}$$

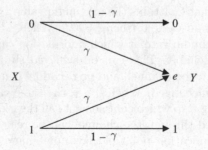

Fig. 7.7. The transition diagram of a binary erasure channel.

and define a binary random variable E by

$$E = \begin{cases} 0 \text{ if } Y \neq e \\ 1 \text{ if } Y = e \end{cases}. \tag{7.70}$$

The random variable E indicates whether an erasure has occurred, and it is a function of Y. Then

$$H(Y) = H(Y, E) \tag{7.71}$$
$$= H(E) + H(Y|E) \tag{7.72}$$
$$= h_b(\gamma) + (1 - \gamma)h_b(a). \tag{7.73}$$

Hence,

$$C = \max_{p(x)} H(Y) - h_b(\gamma) \tag{7.74}$$
$$= \max_a [h_b(\gamma) + (1 - \gamma)h_b(a)] - h_b(\gamma) \tag{7.75}$$
$$= (1 - \gamma) \max_a h_b(a) \tag{7.76}$$
$$= (1 - \gamma) \text{ bit per use}, \tag{7.77}$$

where the capacity is achieved by letting $a = 0.5$, i.e., the input distribution is uniform.

It is in general not possible to obtain the capacity of a DMC in closed form, and we have to resort to numerical computation. In Chapter 9 we will discuss the Blahut–Arimoto algorithm for computing the channel capacity.

7.2 The Channel Coding Theorem

We will justify the definition of the capacity of a DMC by proving the *channel coding theorem*. This theorem, which consists of two parts, will be formally stated at the end of the section. The direct part of the theorem asserts that information can be communicated through a DMC with an arbitrarily small probability of error at any rate less than the channel capacity. Here it is assumed that the decoder knows the transition matrix of the DMC. The converse part of the theorem asserts that if information is communicated through a DMC at a rate higher than the capacity, then the probability of error is bounded away from zero. For better appreciation of the definition of channel capacity, we will first prove the converse part in Section 7.3 and then prove the direct part in Section 7.4.

Definition 7.9. *An (n, M) code for a discrete memoryless channel with input alphabet $\mathcal{X}$ and output alphabet $\mathcal{Y}$ is defined by an encoding function*

$$f : \{1, 2, \cdots, M\} \rightarrow \mathcal{X}^n \tag{7.78}$$

and a decoding function

$$g : \mathcal{Y}^n \rightarrow \{1, 2, \cdots, M\}. \tag{7.79}$$

The set $\{1, 2, \cdots, M\}$, denoted by $\mathcal{W}$, is called the message set. The sequences $f(1), f(2), \cdots, f(M)$ in $\mathcal{X}^n$ are called codewords, and the set of codewords is called the codebook.

In order to distinguish a channel code as defined above from a channel code with feedback which will be discussed in Section 7.6, we will refer to the former as a channel code without feedback.

We assume that a message W is randomly chosen from the message set $\mathcal{W}$ according to the uniform distribution. Therefore,

$$H(W) = \log M. \tag{7.80}$$

With respect to a channel code for a DMC, we let

$$\mathbf{X} = (X_1, X_2, \cdots, X_n) \tag{7.81}$$

and

$$\mathbf{Y} = (Y_1, Y_2, \cdots, Y_n) \tag{7.82}$$

be the input sequence and the output sequence of the channel, respectively. Evidently,

$$\mathbf{X} = f(W). \tag{7.83}$$

We also let

$$\hat{W} = g(\mathbf{Y}) \tag{7.84}$$

be the estimate on the message W by the decoder. Figure 7.8 is the block diagram for a channel code.

Definition 7.10. *For all $1 \leq w \leq M$, let*

$$\lambda_w = \Pr\{\hat{W} \neq w | W = w\} = \sum_{\mathbf{y} \in \mathcal{Y}^n : g(\mathbf{y}) \neq w} \Pr\{\mathbf{Y} = \mathbf{y} | \mathbf{X} = f(w)\} \tag{7.85}$$

be the conditional probability of error given that the message is w.

We now define two performance measures for a channel code.

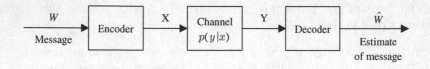

Fig. 7.8. A channel code with block length n.

Definition 7.11. *The maximal probability of error of an* (n, M) *code is defined as*

$$\lambda_{max} = \max_w \lambda_w. \tag{7.86}$$

Definition 7.12. *The average probability of error of an* (n, M) *code is defined as*

$$P_e = \Pr\{\hat{W} \neq W\}. \tag{7.87}$$

From the definition of P_e, we have

$$P_e = \Pr\{\hat{W} \neq W\} \tag{7.88}$$

$$= \sum_w \Pr\{W = w\} \Pr\{\hat{W} \neq W | W = w\} \tag{7.89}$$

$$= \sum_w \frac{1}{M} \Pr\{\hat{W} \neq w | W = w\} \tag{7.90}$$

$$= \frac{1}{M} \sum_w \lambda_w, \tag{7.91}$$

i.e., P_e is the arithmetic mean of λ_w, $1 \leq w \leq M$. It then follows that

$$P_e \leq \lambda_{max}. \tag{7.92}$$

In fact, it can be readily seen that this inequality remains valid even without the assumption that W is distributed uniformly on the message set $\mathcal{W}$.

Definition 7.13. *The rate of an* (n, M) *channel code is* $n^{-1} \log M$ *in bits per use.*

Definition 7.14. *A rate* R *is asymptotically achievable for a discrete memoryless channel if for any* $\epsilon > 0$, *there exists for sufficiently large* n *an* (n, M) *code such that*

$$\frac{1}{n} \log M > R - \epsilon \tag{7.93}$$

and

$$\lambda_{max} < \epsilon. \tag{7.94}$$

For brevity, an asymptotically achievable rate will be referred to as an achievable rate.

In other words, a rate R is achievable if there exists a sequence of codes whose rates approach R and whose probabilities of error approach zero. We end this section by stating the channel coding theorem, which gives a characterization of all achievable rates. This theorem will be proved in the next two sections.

Theorem 7.15 (Channel Coding Theorem). *A rate* R *is achievable for a discrete memoryless channel if and only if* $R \leq C$, *the capacity of the channel.*

7.3 The Converse

Let us consider a channel code with block length n. The random variables involved in this code are W, X_i and Y_i for $1 \leq i \leq n$, and $\hat{W}$. We see from the definition of a channel code in Definition 7.9 that all the random variables are generated sequentially according to some deterministic or probabilistic rules. Specifically, the random variables are generated in the order $W, X_1, Y_1, X_2, Y_2, \cdots, X_n, Y_n, \hat{W}$. The generation of these random variables can be represented by the dependency graph[4] in Figure 7.9. In this graph, a node represents a random variable. If there is a (directed) edge from node X to node Y, then node X is called a *parent* of node Y. We further distinguish a *solid* edge and a *dotted* edge: a solid edge represents functional (deterministic) dependency, while a dotted edge represents the probabilistic dependency induced by the transition matrix $p(y|x)$ of the generic discrete channel. For a node X, its parent nodes represent all the random variables on which random variable X depends when it is generated.

We now explain the specific structure of the dependency graph. First, X_i is a function of W, so each X_i is connected to W by a solid edge. According to Definition 7.4,

$$T_{i-} = (W, X_1, Y_1, \cdots, X_{i-1}, Y_{i-1}). \tag{7.95}$$

By (7.35), the Markov chain

$$(W, X_1, Y_1, \cdots, X_{i-1}, Y_{i-1}) \to X_i \to Y_i \tag{7.96}$$

prevails. Therefore, the generation of Y_i depends only on X_i and not on $W, X_1, Y_1, \cdots, X_{i-1}, Y_{i-1}$. So, Y_i is connected to X_i by a dotted edge representing the discrete channel $p(y|x)$ at time i, and there is no connection

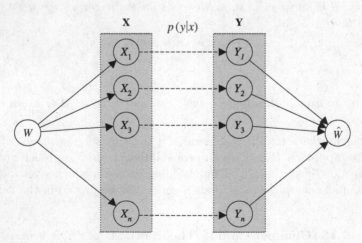

Fig. 7.9. The dependency graph for a channel code without feedback.

---

[4] A dependency graph can be regarded as a Bayesian network [287].

between Y_i and any of the nodes $W, X_1, Y_1, \cdots, X_{i-1}, Y_{i-1}$. Finally, $\hat{W}$ is a function of $Y_1, Y_2, \cdots, Y_n$, so $\hat{W}$ is connected to each Y_i by a solid edge.

We will use q to denote the joint distribution of these random variables as well as all the marginals, and let x_i denote the ith component of a sequence $\mathbf{x}$. From the dependency graph, we see that for all $(w, \mathbf{x}, \mathbf{y}, \hat{w}) \in \mathcal{W} \times \mathcal{X}^n \times \mathcal{Y}^n \times \hat{\mathcal{W}}$ such that $q(\mathbf{x}) > 0$ and $q(\mathbf{y}) > 0$,

$$q(w, \mathbf{x}, \mathbf{y}\, \hat{w}) = q(w) \left(\prod_{i=1}^n q(x_i|w) \right) \left(\prod_{i=1}^n p(y_i|x_i) \right) q(\hat{w}|\mathbf{y}). \qquad (7.97)$$

Note that $q(w) > 0$ for all w so that $q(x_i|w)$ are well defined, and $q(x_i|w)$ and $q(\hat{w}|\mathbf{y})$ are both deterministic. Denote the set of nodes $X_1, X_2, \cdots, X_n$ by $\mathbf{X}$ and the set of nodes $Y_1, Y_2, \cdots, Y_n$ by $\mathbf{Y}$. We notice the following structure in the dependency graph: all the edges from W end in $\mathbf{X}$, all the edges from $\mathbf{X}$ end in $\mathbf{Y}$, and all the edges from $\mathbf{Y}$ end in $\hat{W}$. This suggests that the random variables $W, \mathbf{X}, \mathbf{Y}$, and $\hat{W}$ form the Markov chain

$$W \to \mathbf{X} \to \mathbf{Y} \to \hat{W}. \qquad (7.98)$$

The validity of this Markov chain can be formally justified by applying Proposition 2.9 to (7.97), so that for all $(w, \mathbf{x}, \mathbf{y}, \hat{w}) \in \mathcal{W} \times \mathcal{X}^n \times \mathcal{Y}^n \times \hat{\mathcal{W}}$ such that $q(\mathbf{x}) > 0$ and $q(\mathbf{y}) > 0$, we can write

$$q(w, \mathbf{x}, \mathbf{y}, \hat{w}) = q(w)q(\mathbf{x}|w)q(\mathbf{y}|\mathbf{x})q(\hat{w}|\mathbf{y}). \qquad (7.99)$$

Now $q(\mathbf{x}, \mathbf{y})$ is obtained by summing over all w and $\hat{w}$ in (7.97), and $q(\mathbf{x})$ is obtained by further summing over all $\mathbf{y}$. After some straightforward algebra and using

$$q(\mathbf{y}|\mathbf{x}) = \frac{q(\mathbf{x}, \mathbf{y})}{q(\mathbf{x})} \qquad (7.100)$$

for all $\mathbf{x}$ such that $q(\mathbf{x}) > 0$, we obtain

$$q(\mathbf{y}|\mathbf{x}) = \prod_{i=1}^n p(y_i|x_i). \qquad (7.101)$$

The Markov chain in (7.98) and the relation in (7.101) are apparent from the setup of the problem, and the above justification may seem superfluous. However, the methodology developed here is necessary for handling the more delicate situation which arises when the channel is used with feedback. This will be discussed in Section 7.6.

Consider a channel code whose probability of error is arbitrarily small. Since $W, \mathbf{X}, \mathbf{Y}$, and $\hat{W}$ form the Markov chain in (7.98), the information diagram for these four random variables is as shown in Figure 7.10. Moreover, $\mathbf{X}$ is a function of W, and $\hat{W}$ is a function of $\mathbf{Y}$. These two relations are equivalent to

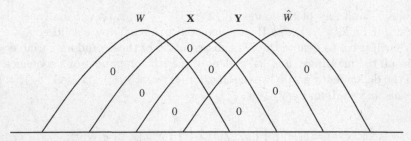

Fig. 7.10. The information diagram for $W \to \mathbf{X} \to \mathbf{Y} \to \hat{W}$.

$$H(\mathbf{X}|W) = 0 \tag{7.102}$$

and

$$H(\hat{W}|\mathbf{Y}) = 0, \tag{7.103}$$

respectively. Since the probability of error is arbitrarily small, W and $\hat{W}$ are essentially identical. To gain insight into the problem, we assume for the time being that W and $\hat{W}$ are equivalent, so that

$$H(\hat{W}|W) = H(W|\hat{W}) = 0. \tag{7.104}$$

Since the I-Measure μ^* for a Markov chain is nonnegative, the constraints in (7.102)–(7.104) imply that μ^* vanishes on all the atoms in Figure 7.10 marked with a "0." Immediately, we see that

$$H(W) = I(\mathbf{X}; \mathbf{Y}). \tag{7.105}$$

That is, the amount of information conveyed through the channel is essentially the mutual information between the input sequence and the output sequence of the channel.

For a single transmission, we see from the definition of channel capacity that the mutual information between the input and the output cannot exceed the capacity of the channel, i.e., for all $1 \le i \le n$,

$$I(X_i; Y_i) \le C. \tag{7.106}$$

Summing i from 1 to n, we have

$$\sum_{i=1}^{n} I(X_i; Y_i) \le nC. \tag{7.107}$$

Upon establishing in the next lemma that

$$I(\mathbf{X}; \mathbf{Y}) \le \sum_{i=1}^{n} I(X_i; Y_i), \tag{7.108}$$

the converse of the channel coding theorem then follows from

$$\frac{1}{n} \log M = \frac{1}{n} H(W) \tag{7.109}$$

$$= \frac{1}{n} I(\mathbf{X}; \mathbf{Y}) \tag{7.110}$$

$$\leq \frac{1}{n} \sum_{i=1}^{n} I(X_i; Y_i) \tag{7.111}$$

$$\leq C. \tag{7.112}$$

Lemma 7.16. *For a discrete memoryless channel used with a channel code without feedback, for any $n \geq 1$,*

$$I(\mathbf{X}; \mathbf{Y}) \leq \sum_{i=1}^{n} I(X_i; Y_i), \tag{7.113}$$

where X_i and Y_i are, respectively, the input and the output of the channel at time i.

Proof. For any $(\mathbf{x}, \mathbf{y}) \in \mathcal{X}^n \times \mathcal{Y}^n$, if $q(\mathbf{x}, \mathbf{y}) > 0$, then $q(\mathbf{x}) > 0$ and (7.101) holds. Therefore,

$$q(\mathbf{Y}|\mathbf{X}) = \prod_{i=1}^{n} p(Y_i|X_i) \tag{7.114}$$

holds for all $(\mathbf{x}, \mathbf{y})$ in the support of $q(\mathbf{x}, \mathbf{y})$. Then

$$-E \log q(\mathbf{Y}|\mathbf{X}) = -E \log \prod_{i=1}^{n} p(Y_i|X_i) = -\sum_{i=1}^{n} E \log p(Y_i|X_i) \tag{7.115}$$

or

$$H(\mathbf{Y}|\mathbf{X}) = \sum_{i=1}^{n} H(Y_i|X_i). \tag{7.116}$$

Hence,

$$I(\mathbf{X}; \mathbf{Y}) = H(\mathbf{Y}) - H(\mathbf{Y}|\mathbf{X}) \tag{7.117}$$

$$\leq \sum_{i=1}^{n} H(Y_i) - \sum_{i=1}^{n} H(Y_i|X_i) \tag{7.118}$$

$$= \sum_{i=1}^{n} I(X_i; Y_i). \tag{7.119}$$

The lemma is proved. $\square$

We now formally prove the converse of the channel coding theorem. Let R be an achievable rate, i.e., for any $\epsilon > 0$, there exists for sufficiently large n an (n, M) code such that

$$\frac{1}{n} \log M > R - \epsilon \tag{7.120}$$

and

$$\lambda_{max} < \epsilon. \tag{7.121}$$

Consider

$$\log M \overset{a)}{=} H(W) \tag{7.122}$$

$$= H(W|\hat{W}) + I(W; \hat{W}) \tag{7.123}$$

$$\overset{b)}{\leq} H(W|\hat{W}) + I(\mathbf{X}; \mathbf{Y}) \tag{7.124}$$

$$\overset{c)}{\leq} H(W|\hat{W}) + \sum_{i=1}^{n} I(X_i; Y_i) \tag{7.125}$$

$$\overset{d)}{\leq} H(W|\hat{W}) + nC, \tag{7.126}$$

where

(a) follows from (7.80);
(b) follows from the data processing theorem since $W \to \mathbf{X} \to \mathbf{Y} \to \hat{W}$;
(c) follows from Lemma 7.16;
(d) follows from (7.107).

From (7.87) and Fano's inequality (cf. Corollary 2.48), we have

$$H(W|\hat{W}) < 1 + P_e \log M. \tag{7.127}$$

Therefore, from (7.126),

$$\log M < 1 + P_e \log M + nC \tag{7.128}$$

$$\leq 1 + \lambda_{max} \log M + nC \tag{7.129}$$

$$< 1 + \epsilon \log M + nC, \tag{7.130}$$

where we have used (7.92) and (7.121), respectively, to obtain the last two inequalities. Dividing by n and rearranging the terms, we have

$$\frac{1}{n} \log M < \frac{\frac{1}{n} + C}{1 - \epsilon}, \tag{7.131}$$

and from (7.120), we obtain

$$R - \epsilon < \frac{\frac{1}{n} + C}{1 - \epsilon}. \tag{7.132}$$

For any $\epsilon > 0$, the above inequality holds for all sufficiently large n. Letting $n \to \infty$ and then $\epsilon \to 0$, we conclude that

$$R \leq C. \tag{7.133}$$

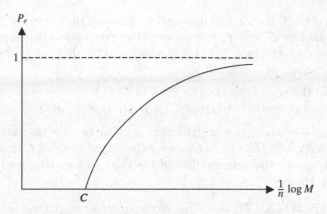

Fig. 7.11. An asymptotic upper bound on P_e.

This completes the proof for the converse of the channel coding theorem.

From the above proof, we can obtain an asymptotic bound on P_e when the rate of the code $\frac{1}{n} \log M$ is greater than C. Consider (7.128) and obtain

$$P_e \geq 1 - \frac{1 + nC}{\log M} = 1 - \frac{\frac{1}{n} + C}{\frac{1}{n} \log M}. \tag{7.134}$$

Then

$$P_e \geq 1 - \frac{\frac{1}{n} + C}{\frac{1}{n} \log M} \approx 1 - \frac{C}{\frac{1}{n} \log M}. \tag{7.135}$$

when n is large. This asymptotic bound on P_e, which is strictly positive if $\frac{1}{n} \log M > C$, is illustrated in Figure 7.11.

In fact, the lower bound in (7.134) implies that $P_e > 0$ for all n if $\frac{1}{n} \log M > C$ because if $P_e^{(n_0)} = 0$ for some n_0, then for all $k \geq 1$, by concatenating k copies of the code, we obtain a code with the same rate and block length equal to kn_0 such that $P_e^{(kn_0)} = 0$, which is a contradiction to our conclusion that $P_e > 0$ when n is large. Therefore, if we use a code whose rate is greater than the channel capacity, the probability of error is nonzero for all block lengths.

The converse of the channel coding theorem we have proved is called the *weak converse*. A stronger version of this result called the *strong converse* can be proved, which says that $P_e \to 1$ as $n \to \infty$ if there exists an $\epsilon > 0$ such that $\frac{1}{n} \log M \geq C + \epsilon$ for all n.

7.4 Achievability

We have shown in the last section that the channel capacity C is an upper bound on all the achievable rates for a DMC. In this section, we show that the rate C is achievable, which implies that any rate $R \leq C$ is achievable.

Consider a DMC $p(y|x)$, and denote the input and the output of the generic discrete channel by X and Y, respectively. For every input distribution $p(x)$, we will prove that the rate $I(X;Y)$ is achievable by showing for large n the existence of a channel code such that

1. the rate of the code is arbitrarily close to $I(X;Y)$;
2. the maximal probability of error λ_{max} is arbitrarily small.

Then by choosing the input distribution $p(x)$ to be one that achieves the channel capacity, i.e., $I(X;Y) = C$, we conclude that the rate C is achievable.

Before we prove the achievability of the channel capacity, we first prove the following lemma.

Lemma 7.17. *Let* $(\mathbf{X}', \mathbf{Y}')$ *be* n *i.i.d. copies of a pair of generic random variables* (X', Y'), *where* X' *and* Y' *are independent and have the same marginal distributions as* X *and* Y, *respectively. Then*

$$\Pr\{(\mathbf{X}', \mathbf{Y}') \in T^n_{[XY]\delta}\} \le 2^{-n(I(X;Y)-\tau)}, \tag{7.136}$$

where $\tau \to 0$ *as* $\delta \to 0$.

Proof. Consider

$$\Pr\{(\mathbf{X}', \mathbf{Y}') \in T^n_{[XY]\delta}\} = \sum_{(\mathbf{x},\mathbf{y}) \in T^n_{[XY]\delta}} p(\mathbf{x})p(\mathbf{y}). \tag{7.137}$$

By the consistency of strong typicality, for $(\mathbf{x}, \mathbf{y}) \in T^n_{[XY]\delta}$, $\mathbf{x} \in T^n_{[X]\delta}$ and $\mathbf{y} \in T^n_{[Y]\delta}$. By the strong AEP, all the $p(\mathbf{x})$ and $p(\mathbf{y})$ in the above summation satisfy

$$p(\mathbf{x}) \le 2^{-n(H(X)-\eta)} \tag{7.138}$$

and

$$p(\mathbf{y}) \le 2^{-n(H(Y)-\zeta)}, \tag{7.139}$$

where $\eta, \zeta \to 0$ as $\delta \to 0$. By the strong JAEP,

$$|T^n_{[XY]\delta}| \le 2^{n(H(X,Y)+\xi)}, \tag{7.140}$$

where $\xi \to 0$ as $\delta \to 0$. Then from (7.137), we have

$$\Pr\{(\mathbf{X}', \mathbf{Y}') \in T^n_{[XY]\delta}\}$$
$$\le 2^{n(H(X,Y)+\xi)} \cdot 2^{-n(H(X)-\eta)} \cdot 2^{-n(H(Y)-\zeta)} \tag{7.141}$$
$$= 2^{-n(H(X)+H(Y)-H(X,Y)-\xi-\eta-\zeta)} \tag{7.142}$$
$$= 2^{-n(I(X;Y)-\xi-\eta-\zeta)} \tag{7.143}$$
$$= 2^{-n(I(X;Y)-\tau)}, \tag{7.144}$$

where

$$\tau = \xi + \eta + \zeta \to 0 \tag{7.145}$$

as $\delta \to 0$. The lemma is proved. $\square$

Fix any $\epsilon > 0$ and let δ be a small positive quantity to be specified later. Toward proving the existence of a desired code, we fix an input distribution $p(x)$ for the generic discrete channel $p(y|x)$, and let M be an *even* integer satisfying

$$I(X;Y) - \frac{\epsilon}{2} < \frac{1}{n} \log M < I(X;Y) - \frac{\epsilon}{4}, \tag{7.146}$$

where n is sufficiently large. We now describe a *random coding scheme* in the following steps:

1. Construct the codebook $\mathcal{C}$ of an (n, M) code randomly by generating M codewords in $\mathcal{X}^n$ independently and identically according to $p(x)^n$. Denote these codewords by $\tilde{\mathbf{X}}(1), \tilde{\mathbf{X}}(2), \cdots, \tilde{\mathbf{X}}(M)$.
2. Reveal the codebook $\mathcal{C}$ to both the encoder and the decoder.
3. A message W is chosen from $\mathcal{W}$ according to the uniform distribution.
4. The sequence $\mathbf{X} = \tilde{\mathbf{X}}(W)$, namely the Wth codeword in the codebook $\mathcal{C}$, is transmitted through the channel.
5. The channel outputs a sequence $\mathbf{Y}$ according to

$$\Pr\{\mathbf{Y} = \mathbf{y} | \tilde{\mathbf{X}}(W) = \mathbf{x}\} = \prod_{i=1}^{n} p(y_i|x_i) \tag{7.147}$$

(cf. (7.101)).
6. The sequence $\mathbf{Y}$ is decoded to the message w if $(\tilde{\mathbf{X}}(w), \mathbf{Y}) \in T^n_{[XY]\delta}$ and there does not exist $w' \neq w$ such that $(\tilde{\mathbf{X}}(w'), \mathbf{Y}) \in T^n_{[XY]\delta}$. Otherwise, $\mathbf{Y}$ is decoded to a constant message in $\mathcal{W}$. Denote by $\hat{W}$ the message to which $\mathbf{Y}$ is decoded.

Remark 1   There are a total of $|\mathcal{X}|^{Mn}$ possible codebooks which can be constructed in Step 1 of the random coding scheme, where we regard two codebooks whose sets of codewords are permutations of each other as two different codebooks.

Remark 2   Strong typicality is used in defining the decoding function in Step 6. This is made possible by the assumption that the alphabets $\mathcal{X}$ and $\mathcal{Y}$ are finite.

We now analyze the performance of this random coding scheme. Let

$$Err = \{\hat{W} \neq W\} \tag{7.148}$$

be the event of a decoding error. In the following, we analyze $\Pr\{Err\}$, the probability of a decoding error for the random code constructed above. For all $1 \leq w \leq M$, define the event

$$E_w = \{(\tilde{\mathbf{X}}(w), \mathbf{Y}) \in T^n_{[XY]\delta}\}. \tag{7.149}$$

Now

$$\Pr\{Err\} = \sum_{w=1}^{M} \Pr\{Err|W = w\}\Pr\{W = w\}. \tag{7.150}$$

Since $\Pr\{Err|W = w\}$ are identical for all w by symmetry in the code construction, we have

$$\Pr\{Err\} = \Pr\{Err|W = 1\} \sum_{w=1}^{M} \Pr\{W = w\} \tag{7.151}$$

$$= \Pr\{Err|W = 1\}, \tag{7.152}$$

i.e., we can assume without loss of generality that the message 1 is chosen. Then decoding is correct if the received sequence $\mathbf{Y}$ is decoded to the message 1. This is the case if E_1 occurs but E_w does not occur for all $2 \leq w \leq M$. It follows that[5]

$$\Pr\{Err^c|W = 1\} \geq \Pr\{E_1 \cap E_2^c \cap E_3^c \cap \cdots \cap E_M^c|W = 1\}, \tag{7.153}$$

which implies

$$\Pr\{Err|W = 1\}$$
$$= 1 - \Pr\{Err^c|W = 1\} \tag{7.154}$$
$$\leq 1 - \Pr\{E_1 \cap E_2^c \cap E_3^c \cap \cdots \cap E_M^c|W = 1\} \tag{7.155}$$
$$= \Pr\{(E_1 \cap E_2^c \cap E_3^c \cap \cdots \cap E_M^c)^c|W = 1\} \tag{7.156}$$
$$= \Pr\{E_1^c \cup E_2 \cup E_3 \cup \cdots \cup E_M|W = 1\}. \tag{7.157}$$

By the union bound, we have

$$\Pr\{Err|W = 1\} \leq \Pr\{E_1^c|W = 1\} + \sum_{w=2}^{M} \Pr\{E_w|W = 1\}. \tag{7.158}$$

First, conditioning on $\{W = 1\}$, $(\tilde{\mathbf{X}}(1), \mathbf{Y})$ are n i.i.d. copies of the pair of generic random variables (X, Y). By the strong JAEP, for any $\nu > 0$,

$$\Pr\{E_1^c|W = 1\} = \Pr\{(\tilde{\mathbf{X}}(1), \mathbf{Y}) \notin T^n_{[XY]\delta}|W = 1\} < \nu \tag{7.159}$$

for sufficiently large n. This gives an upper bound on the first term on the right-hand side of (7.158).

Second, conditioning on $\{W = 1\}$, for $2 \leq w \leq M$, $(\tilde{\mathbf{X}}(w), \mathbf{Y})$ are n i.i.d. copies of the pair of generic random variables (X', Y'), where X'

---

[5] If E_1 does not occur or E_w occurs for some $1 \leq w \leq M$, the received sequence $\mathbf{Y}$ is decoded to the constant message, which may happen to be the message 1. Therefore, the inequality in (7.153) is not an equality in general.

and Y' have the same marginal distributions as X and Y, respectively. Furthermore, from the random coding scheme and the memorylessness of the DMC, it is intuitively correct that X' and Y' are independent because $\tilde{\mathbf{X}}(1)$ and $\tilde{\mathbf{X}}(w)$ are independent and the generation of $\mathbf{Y}$ depends only on $\tilde{\mathbf{X}}(1)$.

A formal proof of this claim requires a more detailed analysis. In our random coding scheme, the random variables are generated in the order $\tilde{\mathbf{X}}(1), \tilde{\mathbf{X}}(2), \cdots, \tilde{\mathbf{X}}(M), W, X_1, Y_1, X_2, Y_2, \cdots, X_n, Y_n, \hat{W}$. By considering the joint distribution of these random variables, similar to the discussion in Section 7.3, the Markov chain

$$(\tilde{\mathbf{X}}(1), \tilde{\mathbf{X}}(2), \cdots, \tilde{\mathbf{X}}(M), W) \to \mathbf{X} \to \mathbf{Y} \to \hat{W} \qquad (7.160)$$

can be established. See Problem 1 for the details. Then for any $2 \le w \le M$, from the above Markov chain, we have

$$I(\mathbf{Y}; \tilde{\mathbf{X}}(w), W | \mathbf{X}) = 0. \qquad (7.161)$$

By the chain rule for mutual information, the left-hand side can be written as

$$I(\mathbf{Y}; W | \mathbf{X}) + I(\mathbf{Y}; \tilde{\mathbf{X}}(w) | \mathbf{X}, W). \qquad (7.162)$$

By the nonnegativity of conditional mutual information, this implies

$$I(\mathbf{Y}; \tilde{\mathbf{X}}(w) | \mathbf{X}, W) = 0 \qquad (7.163)$$

or

$$\sum_{w=1}^{M} \Pr\{W = w\} I(\mathbf{Y}; \tilde{\mathbf{X}}(w) | \mathbf{X}, W = w) = 0. \qquad (7.164)$$

Since $I(\mathbf{Y}; \tilde{\mathbf{X}}(w) | \mathbf{X}, W = w)$ are all nonnegative, we see from the above that they must all vanish. In particular,

$$I(\mathbf{Y}; \tilde{\mathbf{X}}(w) | \mathbf{X}, W = 1) = 0. \qquad (7.165)$$

Then

$$I(\mathbf{Y}; \tilde{\mathbf{X}}(w) | \tilde{\mathbf{X}}(1), W = 1) = I(\mathbf{Y}; \tilde{\mathbf{X}}(w) | \tilde{\mathbf{X}}(W), W = 1) \qquad (7.166)$$
$$= I(\mathbf{Y}; \tilde{\mathbf{X}}(w) | \mathbf{X}, W = 1) \qquad (7.167)$$
$$= 0. \qquad (7.168)$$

On the other hand, since $\tilde{\mathbf{X}}(1), \tilde{\mathbf{X}}(w)$, and W are mutually independent, we have

$$I(\tilde{\mathbf{X}}(1); \tilde{\mathbf{X}}(w) | W = 1) = 0. \qquad (7.169)$$

Hence,

$$I(\mathbf{Y}; \tilde{\mathbf{X}}(w)|W = 1)$$
$$\leq I(\tilde{\mathbf{X}}(1), \mathbf{Y}; \tilde{\mathbf{X}}(w)|W = 1) \tag{7.170}$$
$$= I(\tilde{\mathbf{X}}(1); \tilde{\mathbf{X}}(w)|W = 1) + I(\mathbf{Y}; \tilde{\mathbf{X}}(w)|\tilde{\mathbf{X}}(1), W = 1) \tag{7.171}$$
$$= 0 + 0 \tag{7.172}$$
$$= 0, \tag{7.173}$$

where (7.172) follows from (7.168) and (7.169), proving the claim.

Let us now return to (7.158). For any $2 \leq w \leq M$, it follows from the above claim and Lemma 7.17 that

$$\Pr\{E_w|W = 1\}$$
$$= \Pr\{(\tilde{\mathbf{X}}(w), \mathbf{Y}) \in T^n_{[XY]\delta}|W = 1\} \tag{7.174}$$
$$\leq 2^{-n(I(X;Y)-\tau)}, \tag{7.175}$$

where $\tau \to 0$ as $\delta \to 0$. From the upper bound in (7.146), we have

$$M < 2^{n(I(X;Y)-\frac{\epsilon}{4})}. \tag{7.176}$$

Using (7.159), (7.175), and the above upper bound on M, it follows from (7.152) and (7.158) that

$$\Pr\{Err\} < \nu + 2^{n(I(X;Y)-\frac{\epsilon}{4})} \cdot 2^{-n(I(X;Y)-\tau)} \tag{7.177}$$
$$= \nu + 2^{-n(\frac{\epsilon}{4}-\tau)}. \tag{7.178}$$

Since $\tau \to 0$ as $\delta \to 0$, for sufficiently small δ, we have

$$\frac{\epsilon}{4} - \tau > 0 \tag{7.179}$$

for any $\epsilon > 0$, so that $2^{-n(\frac{\epsilon}{4}-\tau)} \to 0$ as $n \to \infty$. Then by letting $\nu < \frac{\epsilon}{3}$, it follows from (7.178) that

$$\Pr\{Err\} < \frac{\epsilon}{2} \tag{7.180}$$

for sufficiently large n.

The main idea of the above analysis of $\Pr\{Err\}$ is the following. In constructing the codebook, we randomly generate M codewords in $\mathcal{X}^n$ according to $p(x)^n$, and one of the codewords is sent through the channel $p(y|x)$. When n is large, with high probability, the received sequence is jointly typical with the codeword sent with respect to $p(x, y)$. If the number of codewords M grows with n at a rate less than $I(X; Y)$, then the probability that the received sequence is jointly typical with a codeword other than the one sent through the channel is negligible. Accordingly, the message can be decoded correctly with probability arbitrarily close to 1.

In constructing the codebook in Step 1 of the random coding scheme, we choose a codebook $\mathcal{C}$ with a certain probability $\Pr\{\mathcal{C}\}$ from the ensemble of all possible codebooks. By conditioning on the codebook chosen, we have

$$\Pr\{Err\} = \sum_{\mathcal{C}} \Pr\{\mathcal{C}\}\Pr\{Err|\mathcal{C}\}, \tag{7.181}$$

i.e., $\Pr\{Err\}$ is a weighted average of $\Pr\{Err|\mathcal{C}\}$ over all $\mathcal{C}$ in the ensemble of all possible codebooks, where $\Pr\{Err|\mathcal{C}\}$ is the average probability of error of the code, i.e., P_e, when the codebook $\mathcal{C}$ is chosen (cf. Definition 7.12). The reader should compare the two different expansions of $\Pr\{Err\}$ in (7.181) and (7.150).

Therefore, there exists at least one codebook $\mathcal{C}^*$ such that

$$\Pr\{Err|\mathcal{C}^*\} \le \Pr\{Err\} < \frac{\epsilon}{2}. \tag{7.182}$$

Thus we have shown that for any $\epsilon > 0$, there exists for sufficiently large n an (n, M) code such that

$$\frac{1}{n}\log M > I(X;Y) - \frac{\epsilon}{2} \tag{7.183}$$

(cf. (7.146)) and

$$P_e < \frac{\epsilon}{2}. \tag{7.184}$$

We are still one step away from proving that the rate $I(X;Y)$ is achievable because we require that λ_{max} instead of P_e is arbitrarily small. Toward this end, we write (7.184) as

$$\frac{1}{M}\sum_{w=1}^{M}\lambda_w < \frac{\epsilon}{2} \tag{7.185}$$

or

$$\sum_{w=1}^{M}\lambda_w < \left(\frac{M}{2}\right)\epsilon. \tag{7.186}$$

Upon ordering the codewords according to their conditional probabilities of error, we observe that the conditional probabilities of error of the better half of the M codewords are less than ϵ, otherwise the conditional probabilities of error of the worse half of the codewords are at least ϵ, and they contribute at least $(M/2)\epsilon$ to the summation in (7.186), which is a contradiction.

Thus by discarding the worse half of the codewords in $\mathcal{C}^*$, for the resulting codebook, the maximal probability of error λ_{max} is less than ϵ. Using (7.183) and considering

$$\frac{1}{n}\log\frac{M}{2} = \frac{1}{n}\log M - \frac{1}{n} \tag{7.187}$$

$$> \left(I(X;Y) - \frac{\epsilon}{2}\right) - \frac{1}{n} \tag{7.188}$$

$$> I(X;Y) - \epsilon \tag{7.189}$$

when n is sufficiently large, we see that the rate of the resulting code is greater than $I(X;Y) - \epsilon$. Hence, we conclude that the rate $I(X;Y)$ is achievable.

Finally, upon letting the input distribution $p(x)$ be one that achieves the channel capacity, i.e., $I(X;Y) = C$, we have proved that the rate C is achievable. This completes the proof of the direct part of the channel coding theorem.

7.5 A Discussion

In the last two sections, we have proved the channel coding theorem which asserts that reliable communication through a DMC at rate R is possible if and only if $R < C$, the channel capacity. By reliable communication at rate R, we mean that the size of the message set grows exponentially with n at rate R, while the message can be decoded correctly with probability arbitrarily close to 1 as $n \to \infty$. Therefore, the capacity C is a fundamental characterization of a DMC.

The capacity of a noisy channel is analogous to the capacity of a water pipe in the following way. For a water pipe, if we pump water through the pipe at a rate higher than its capacity, the pipe would burst and water would be lost. For a communication channel, if we communicate through the channel at a rate higher than the capacity, the probability of error is bounded away from zero, i.e., information is lost.

In proving the direct part of the channel coding theorem, we showed that there exists a channel code whose rate is arbitrarily close to C and whose probability of error is arbitrarily close to zero. Moreover, the existence of such a code is guaranteed only when the block length n is large. However, the proof does not indicate how we can find such a codebook. For this reason, the proof we gave is called an existence proof (as oppose to a constructive proof).

For a fixed block length n, we in principle can search through the ensemble of all possible codebooks for a good one, but this is quite prohibitive even for small n because the number of all possible codebooks grows doubly exponentially with n. Specifically, the total number of all possible (n, M) codebooks is equal to $|\mathcal{X}|^{Mn}$. When the rate of the code is close to C, M is approximately equal to 2^{nC}. Therefore, the number of codebooks we need to search through is about $|\mathcal{X}|^{n2^{nC}}$.

Nevertheless, the proof of the direct part of the channel coding theorem does indicate that if we generate a codebook randomly as prescribed, the codebook is most likely to be good. More precisely, we now show that the probability of choosing a code $\mathcal{C}$ such that $\Pr\{Err|\mathcal{C}\}$ is greater than any prescribed $\psi > 0$ is arbitrarily small when n is sufficiently large. Consider

$$\Pr\{Err\} = \sum_{\mathcal{C}} \Pr\{\mathcal{C}\}\Pr\{Err|\mathcal{C}\} \tag{7.190}$$

$$= \sum_{\mathcal{C}:\Pr\{Err|\mathcal{C}\}\leq\psi} \Pr\{\mathcal{C}\}\Pr\{Err|\mathcal{C}\}$$

$$+ \sum_{\mathcal{C}:\Pr\{Err|\mathcal{C}\}>\psi} \Pr\{\mathcal{C}\}\Pr\{Err|\mathcal{C}\} \tag{7.191}$$

$$\geq \sum_{\mathcal{C}:\Pr\{Err|\mathcal{C}\}>\psi} \Pr\{\mathcal{C}\}\Pr\{Err|\mathcal{C}\} \tag{7.192}$$

$$> \psi \sum_{\mathcal{C}:\Pr\{Err|\mathcal{C}\}>\psi} \Pr\{\mathcal{C}\}, \tag{7.193}$$

which implies

$$\sum_{\mathcal{C}:\Pr\{Err|\mathcal{C}\}>\psi} \Pr\{\mathcal{C}\} < \frac{\Pr\{Err\}}{\psi}. \qquad (7.194)$$

From (7.182), we have

$$\Pr\{Err\} < \frac{\epsilon}{2} \qquad (7.195)$$

for any $\epsilon > 0$ when n is sufficiently large. Then

$$\sum_{\mathcal{C}:\Pr\{Err|\mathcal{C}\}>\psi} \Pr\{\mathcal{C}\} < \frac{\epsilon}{2\psi}. \qquad (7.196)$$

Since ψ is fixed, this upper bound can be made arbitrarily small by choosing a sufficiently small ϵ.

Although the proof of the direct part of the channel coding theorem does not provide an explicit construction of a good code, it does give much insight into what a good code is like. Figure 7.12 is an illustration of a channel code that achieves the channel capacity. Here we assume that the input distribution $p(x)$ is one that achieves the channel capacity, i.e., $I(X;Y) = C$. The idea is that most of the codewords are typical sequences in $\mathcal{X}^n$ with respect to $p(x)$. (For this reason, the repetition code is not a good code.) When such a codeword is transmitted through the channel, the received sequence is likely to be one of about $2^{nH(Y|X)}$ sequences in $\mathcal{Y}^n$ which are jointly typical with the transmitted codeword with respect to $p(x,y)$. The association between a codeword and the about $2^{nH(Y|X)}$ corresponding sequences in $\mathcal{Y}^n$ is shown as a cone in the figure. As we require that the probability of decoding error is small, the cones essentially do not overlap with each other. Since the number of typical sequences with respect to $p(y)$ is about $2^{nH(Y)}$, the number of codewords cannot exceed about

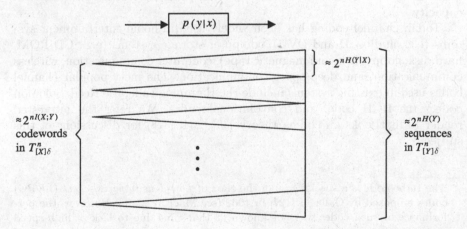

Fig. 7.12. A channel code that achieves capacity.

$$\frac{2^{nH(Y)}}{2^{nH(Y|X)}} = 2^{nI(X;Y)} = 2^{nC}. \tag{7.197}$$

This is consistent with the converse of the channel coding theorem. The direct part of the channel coding theorem says that when n is large, as long as the number of codewords generated randomly is not more than about $2^{n(C-\epsilon)}$, the overlap among the cones is negligible with high probability.

Therefore, instead of searching through the ensemble of all possible code-books for a good one, we can generate a codebook randomly, and it is likely to be good. However, such a code is difficult to use due to the following implementation issues.

A codebook with block length n and rate R consists of $n2^{nR}$ symbols from the input alphabet $\mathcal{X}$. This means that the size of the codebook, i.e., the amount of storage required to store the codebook, grows exponentially with n. This also makes the encoding process inefficient.

Another issue is regarding the computation required for decoding. Based on the sequence received at the output of the channel, the decoder needs to decide which of the about 2^{nR} codewords was the one transmitted. This requires an exponential amount of computation.

In practice, we are satisfied with the reliability of communication once it exceeds a certain level. Therefore, the above implementation issues may eventually be resolved with the advancement of microelectronics. But before then, we still have to deal with these issues. For this reason, the entire field of *coding theory* has been developed since the 1950s. Researchers in this field are devoted to searching for good codes and devising efficient decoding algorithms.

In fact, almost all the codes studied in coding theory are *linear codes*. By taking advantage of the linear structures of these codes, efficient encoding and decoding can be achieved. In particular, Berrou et al. [33] proposed in 1993 a linear code called the *turbo code*[6] that can practically achieve the channel capacity.

Today, channel coding has been widely used in home entertainment systems (e.g., audio CD and DVD), computer storage systems (e.g., CD-ROM, hard disk, floppy disk, and magnetic tape), computer communication, wireless communication, and deep space communication. The most popular channel codes used in existing systems include the Hamming code, the Reed–Solomon code,[7] the BCH code, and convolutional codes. We refer the interested reader to textbooks on coding theory [39] [234] [378] for discussions of this subject.

---

[6] The turbo code is a special case of the class of *Low-density parity-check* (LDPC) codes proposed by Gallager [127] in 1962 (see MacKay [240]). However, the performance of such codes was not known at that time due to lack of high-speed computers for simulation.

[7] The Reed–Solomon code was independently discovered by Arimoto [18].

7.6 Feedback Capacity

Feedback is common in practical communication systems for correcting possible errors which occur during transmission. As an example, during a telephone conversation, we often have to request the speaker to repeat due to poor voice quality of the telephone line. As another example, in data communication, the receiver may request a packet to be retransmitted if the *parity check* bits received are incorrect. In general, when feedback from the receiver is available at the transmitter, the transmitter can at any time decide what to transmit next based on the feedback so far and can potentially transmit information through the channel reliably at a higher rate.

In this section, we study a model in which a DMC is used with complete feedback. The block diagram for the model is shown in Figure 7.13. In this model, the symbol Y_i received at the output of the channel at time i is available instantaneously at the encoder without error. Then depending on the message W and all the previous feedback $Y_1, Y_2, \cdots, Y_i$, the encoder decides the value of X_{i+1}, the next symbol to be transmitted. Such a channel code is formally defined below.

Definition 7.18. *An (n, M) code with complete feedback for a discrete memoryless channel with input alphabet $\mathcal{X}$ and output alphabet $\mathcal{Y}$ is defined by encoding functions*

$$f_i : \{1, 2, \cdots, M\} \times \mathcal{Y}^{i-1} \to \mathcal{X} \tag{7.198}$$

for $1 \le i \le n$ and a decoding function

$$g : \mathcal{Y}^n \to \{1, 2, \cdots, M\}. \tag{7.199}$$

We will use $\mathbf{Y}^i$ to denote $(Y_1, Y_2, \cdots, Y_i)$ and X_i to denote $f_i(W, \mathbf{Y}^{i-1})$. We note that a channel code without feedback is a special case of a channel code with complete feedback because for the latter, the encoder can ignore the feedback.

Definition 7.19. *A rate R is achievable with complete feedback for a discrete memoryless channel $p(y|x)$ if for any $\epsilon > 0$, there exists for sufficiently large n an (n, M) code with complete feedback such that*

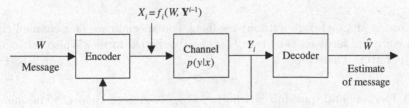

Fig. 7.13. A channel code with feedback.

$$\frac{1}{n} \log M > R - \epsilon \tag{7.200}$$

and

$$\lambda_{max} < \epsilon. \tag{7.201}$$

Definition 7.20. *The feedback capacity, C_{FB}, of a discrete memoryless channel is the supremum of all the rates achievable by codes with complete feedback.*

Proposition 7.21. *The supremum in the definition of C_{FB} in Definition 7.20 is the maximum.*

Proof. Consider rates $R^{(k)}$ which are achievable with complete feedback such that

$$\lim_{k \to \infty} R^{(k)} = R. \tag{7.202}$$

Then for any $\epsilon > 0$, for all k, there exists for sufficiently large n an $(n, M^{(k)})$ code with complete feedback such that

$$\frac{1}{n} \log M^{(k)} > R^{(k)} - \epsilon \tag{7.203}$$

and

$$\lambda_{max}^{(k)} < \epsilon. \tag{7.204}$$

By virtue of (7.202), let $k(\epsilon)$ be an integer such that for all $k > k(\epsilon)$,

$$|R - R^{(k)}| < \epsilon, \tag{7.205}$$

which implies

$$R^{(k)} > R - \epsilon. \tag{7.206}$$

Then for all $k > k(\epsilon)$,

$$\frac{1}{n} \log M^{(k)} > R^{(k)} - \epsilon > R - 2\epsilon. \tag{7.207}$$

Therefore, it follows from (7.207) and (7.204) that R is achievable with complete feedback. This implies that the supremum in Definition 7.20, which can be achieved, is in fact the maximum.   □

Since a channel code without feedback is a special case of a channel code with complete feedback, any rate R achievable by the former is also achievable by the latter. Therefore,

$$C_{FB} \geq C. \tag{7.208}$$

A fundamental question is whether C_{FB} is greater than C. The answer surprisingly turns out to be negative for a DMC, as we now show. From the

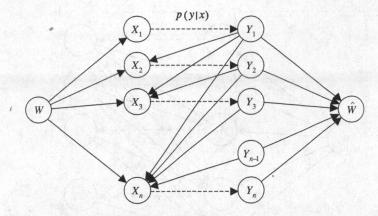

Fig. 7.14. The dependency graph for a channel code with feedback.

description of a channel code with complete feedback, we obtain the dependency graph for the random variables $W, \mathbf{X}, \mathbf{Y}, \hat{W}$ in Figure 7.14. From this dependency graph, we see that

$$q(w, \mathbf{x}, \mathbf{y}, \hat{w}) = q(w) \left(\prod_{i=1}^{n} q(x_i | w, \mathbf{y}^{i-1}) \right) \left(\prod_{i=1}^{n} p(y_i | x_i) \right) q(\hat{w} | \mathbf{y}) \quad (7.209)$$

for all $(w, \mathbf{x}, \mathbf{y}, \hat{w}) \in \mathcal{W} \times \mathcal{X}^n \times \mathcal{Y}^n \times \mathcal{W}$ such that $q(w, \mathbf{y}^{i-1}), q(x_i) > 0$ for $1 \le i \le n$ and $q(\mathbf{y}) > 0$, where $\mathbf{y}^i = (y_1, y_2, \cdots, y_i)$. Note that $q(x_i | w, \mathbf{y}^{i-1})$ and $q(\hat{w} | \mathbf{y})$ are deterministic.

Lemma 7.22. *For all* $1 \le i \le n$,

$$(W, \mathbf{Y}^{i-1}) \to X_i \to Y_i \quad (7.210)$$

forms a Markov chain.

Proof. The dependency graph for the random variables $W, \mathbf{X}^i$, and $\mathbf{Y}^i$ is shown in Figure 7.15. Denote the set of nodes $W, \mathbf{X}^{i-1}$, and $\mathbf{Y}^{i-1}$ by Z. Then we see that all the edges from Z end at X_i, and the only edge from X_i ends at Y_i. This means that Y_i depends on $(W, \mathbf{X}^{i-1}, \mathbf{Y}^{i-1})$ only through X_i, i.e.,

$$(W, \mathbf{X}^{i-1}, \mathbf{Y}^{i-1}) \to X_i \to Y_i \quad (7.211)$$

forms a Markov chain or

$$I(W, \mathbf{X}^{i-1}, \mathbf{Y}^{i-1}; Y_i | X_i) = 0. \quad (7.212)$$

This can be formally justified by Proposition 2.9, and the details are omitted here. Since

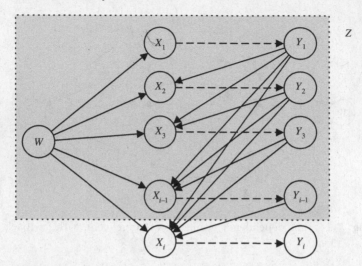

Fig. 7.15. The dependency graph for $W, \mathbf{X}^i$, and $\mathbf{Y}^i$.

$$0 = I(W, \mathbf{X}^{i-1}, \mathbf{Y}^{i-1}; Y_i | X_i) \tag{7.213}$$
$$= I(W, \mathbf{Y}^{i-1}; Y_i | X_i) + I(\mathbf{X}^{i-1}; Y_i | W, X_i, \mathbf{Y}^{i-1}) \tag{7.214}$$

and mutual information is nonnegative, we obtain

$$I(W, \mathbf{Y}^{i-1}; Y_i | X_i) = 0 \tag{7.215}$$

or

$$(W, \mathbf{Y}^{i-1}) \to X_i \to Y_i \tag{7.216}$$

forms a Markov chain. The lemma is proved. $\square$

From the definition of C_{FB} and by virtue of Proposition 7.21, if $R \le C_{FB}$, then R is a rate achievable with complete feedback. We will show that if a rate R is achievable with complete feedback, then $R \le C$. If so, then $R \le C_{FB}$ implies $R \le C$, which can be true if and only if $C_{FB} \le C$. Then from (7.208), we can conclude that $C_{FB} = C$.

Let R be a rate achievable with complete feedback, i.e., for any $\epsilon > 0$, there exists for sufficiently large n an (n, M) code with complete feedback such that

$$n^{-1} \log M > R - \epsilon \tag{7.217}$$

and

$$\lambda_{max} < \epsilon. \tag{7.218}$$

Consider

$$\log M = H(W) = I(W; \mathbf{Y}) + H(W | \mathbf{Y}) \tag{7.219}$$

and bound $I(W; \mathbf{Y})$ and $H(W|\mathbf{Y})$ as follows. First,

$$I(W; \mathbf{Y}) = H(\mathbf{Y}) - H(\mathbf{Y}|W) \tag{7.220}$$

$$= H(\mathbf{Y}) - \sum_{i=1}^{n} H(Y_i|\mathbf{Y}^{i-1}, W) \tag{7.221}$$

$$\overset{a)}{=} H(\mathbf{Y}) - \sum_{i=1}^{n} H(Y_i|\mathbf{Y}^{i-1}, W, X_i) \tag{7.222}$$

$$\overset{b)}{=} H(\mathbf{Y}) - \sum_{i=1}^{n} H(Y_i|X_i) \tag{7.223}$$

$$\leq \sum_{i=1}^{n} H(Y_i) - \sum_{i=1}^{n} H(Y_i|X_i) \tag{7.224}$$

$$= \sum_{i=1}^{n} I(X_i; Y_i) \tag{7.225}$$

$$\leq nC, \tag{7.226}$$

where (a) follows because X_i is a function of W and $\mathbf{Y}^{i-1}$ and (b) follows from Lemma 7.22. Second,

$$H(W|\mathbf{Y}) = H(W|\mathbf{Y}, \hat{W}) \leq H(W|\hat{W}). \tag{7.227}$$

Thus

$$\log M \leq H(W|\hat{W}) + nC, \tag{7.228}$$

which is the same as (7.126). Then by (7.217) and an application of Fano's inequality, we conclude as in the proof for the converse of the channel coding theorem that

$$R \leq C. \tag{7.229}$$

Hence, we have proved that $C_{FB} = C$.

Remark 1 The proof for the converse of the channel coding theorem in Section 7.3 depends critically on the Markov chain

$$W \rightarrow \mathbf{X} \rightarrow \mathbf{Y} \rightarrow \hat{W} \tag{7.230}$$

and the relation in (7.101) (the latter implies Lemma 7.16). Both of them do not hold in general in the presence of feedback.

Remark 2 The proof for $C_{FB} = C$ in this section is also a proof for the converse of the channel coding theorem, so we actually do not need the proof in Section 7.3. However, the proof here and the proof in Section 7.3 have different spirits. Without comparing the two proofs, one cannot possibly understand the subtlety of the result that feedback does not increase the capacity of a DMC.

Remark 3 Although feedback does not increase the capacity of a DMC, the availability of feedback often makes coding much simpler. For some channels, communication through the channel with zero probability of error can be achieved in the presence of feedback by using a *variable-length* channel code. These are discussed in the next example.

Example 7.23. Consider the binary erasure channel in Example 7.8 whose capacity is $1 - \gamma$, where γ is the erasure probability. In the presence of complete feedback, for every information bit to be transmitted, the encoder can transmit the same information bit through the channel until an erasure does not occur, i.e., the information bit is received correctly. Then the number of uses of the channel it takes to transmit an information bit through the channel correctly has a truncated geometrical distribution whose mean is $(1 - \gamma)^{-1}$. Therefore, the effective rate at which information can be transmitted through the channel is $1 - \gamma$. In other words, the channel capacity is achieved by using a very simple variable-length code. Moreover, the channel capacity is achieved with zero probability of error.

In the absence of feedback, the rate $1 - \gamma$ can also be achieved, but with an arbitrarily small probability of error and a much more complicated code.

To conclude this section, we point out that the memoryless assumption of the channel is essential for drawing the conclusion that feedback does not increase the channel capacity not because the proof presented in this section does not go through without this assumption, but because if the channel has memory, feedback actually can increase the channel capacity. For an illustrating example, see Problem 12.

7.7 Separation of Source and Channel Coding

We have so far considered the situation in which we want to convey a message through a DMC, where the message is randomly selected from a finite set according to the uniform distribution. However, in most situations, we want to convey an information source through a DMC. Let $\{U_k, k > -n\}$ be an ergodic stationary information source with entropy rate H. Denote the common alphabet by $\mathcal{U}$ and assume that $\mathcal{U}$ is finite. To convey $\{U_k\}$ through the channel, we can employ a source code with rate R_s and a channel code with rate R_c as shown in Figure 7.16 such that $R_s < R_c$.

Let f^s and g^s be, respectively, the encoding function and the decoding function of the source code, and f^c and g^c be, respectively, the encoding

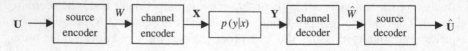

Fig. 7.16. Separation of source coding and channel coding.

function and the decoding function of the channel code. The block of n information symbols $\mathbf{U} = (U_{-(n-1)}, U_{-(n-2)}, \cdots, U_0)$ is first encoded by the source encoder into an index

$$W = f^s(\mathbf{U}), \tag{7.231}$$

called the source codeword. Then W is mapped by the channel encoder to a distinct channel codeword

$$\mathbf{X} = f^c(W), \tag{7.232}$$

where $\mathbf{X} = (X_1, X_2, \cdots, X_n)$. This is possible because there are about 2^{nR_s} source codewords and about 2^{nR_c} channel codewords, and we assume that $R_s < R_c$. Then $\mathbf{X}$ is transmitted through the DMC $p(y|x)$, and the sequence $\mathbf{Y} = (Y_1, Y_2, \cdots, Y_n)$ is received. Based on $\mathbf{Y}$, the channel decoder first estimates W as

$$\hat{W} = g^c(\mathbf{Y}). \tag{7.233}$$

Finally, the source decoder decodes $\hat{W}$ to

$$\hat{\mathbf{U}} = g^s(\hat{W}). \tag{7.234}$$

For this scheme, an error occurs if $\mathbf{U} \neq \hat{\mathbf{U}}$, and we denote the probability of error by P_e.

We now show that if $H < C$, the capacity of the DMC $p(y|x)$, then it is possible to convey $\mathbf{U}$ through the channel with an arbitrarily small probability of error. First, we choose R_s and R_c such that

$$H < R_s < R_c < C. \tag{7.235}$$

Observe that if $\hat{W} = W$ and $g^s(W) = \mathbf{U}$, then from (7.234),

$$\hat{\mathbf{U}} = g^s(\hat{W}) = g^s(W) = \mathbf{U}, \tag{7.236}$$

i.e., an error does not occur. In other words, if an error occurs, either $\hat{W} \neq W$ or $g^s(W) \neq \mathbf{U}$. Then by the union bound, we have

$$P_e \leq \Pr\{\hat{W} \neq W\} + \Pr\{g^s(W) \neq \mathbf{U}\}. \tag{7.237}$$

For any $\epsilon > 0$ and sufficiently large n, by the Shannon–McMillan–Breiman theorem, there exists a source code such that

$$\Pr\{g^s(W) \neq \mathbf{U}\} \leq \epsilon. \tag{7.238}$$

By the channel coding theorem, there exists a channel code such that $\lambda_{max} \leq \epsilon$, where λ_{max} is the maximal probability of error. This implies

$$\Pr\{\hat{W} \neq W\} = \sum_w \Pr\{\hat{W} \neq W | W = w\} \Pr\{W = w\} \tag{7.239}$$

$$\leq \lambda_{max} \sum_w \Pr\{W = w\} \tag{7.240}$$

$$= \lambda_{max} \tag{7.241}$$

$$\leq \epsilon. \tag{7.242}$$

Combining (7.238) and (7.242), we have

$$P_e \leq 2\epsilon. \tag{7.243}$$

Therefore, we conclude that as long as $H < C$, it is possible to convey $\{U_k\}$ through the DMC reliably.

In the scheme we have discussed, source coding and channel coding are separated. In general, source coding and channel coding can be combined. This technique is called *joint source–channel coding*. It is then natural to ask whether it is possible to convey information through the channel reliably at a higher rate by using joint source–channel coding. In the rest of the section, we show that the answer to this question is no to the extent that for asymptotic reliability, we must have $H \leq C$. However, whether asymptotical reliability can be achieved for $H = C$ depends on the specific information source and channel.

We base our discussion on the general assumption that complete feedback is available at the encoder as shown in Figure 7.17. Let f_i^{sc}, $1 \leq i \leq n$, be the encoding functions and g^{sc} be the decoding function of the source–channel code. Then

$$X_i = f_i^{sc}(\mathbf{U}, \mathbf{Y}^{i-1}) \tag{7.244}$$

for $1 \leq i \leq n$, where $\mathbf{Y}^{i-1} = (Y_1, Y_2, \cdots, Y_{i-1})$, and

$$\hat{\mathbf{U}} = g^{sc}(\mathbf{Y}), \tag{7.245}$$

where $\hat{\mathbf{U}} = (\hat{U}_1, \hat{U}_2, \cdots, \hat{U}_n)$. In exactly the same way as we proved (7.226) in the last section, we can prove that

$$I(\mathbf{U}; \mathbf{Y}) \leq nC. \tag{7.246}$$

Since $\hat{\mathbf{U}}$ is a function of $\mathbf{Y}$,

$$I(\mathbf{U}; \hat{\mathbf{U}}) \leq I(\mathbf{U}; \hat{\mathbf{U}}, \mathbf{Y}) \tag{7.247}$$
$$= I(\mathbf{U}; \mathbf{Y}) \tag{7.248}$$
$$\leq nC. \tag{7.249}$$

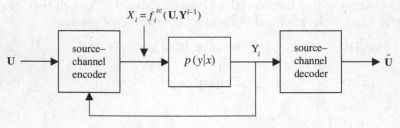

Fig. 7.17. Joint source–channel coding.

For any $\epsilon > 0$,

$$H(\mathbf{U}) \geq n(H - \epsilon) \tag{7.250}$$

for sufficiently large n. Then

$$n(H - \epsilon) \leq H(\mathbf{U}) = H(\mathbf{U}|\hat{\mathbf{U}}) + I(\mathbf{U}; \hat{\mathbf{U}}) \leq H(\mathbf{U}|\hat{\mathbf{U}}) + nC. \tag{7.251}$$

Applying Fano's inequality (Corollary 2.48), we obtain

$$n(H - \epsilon) \leq 1 + nP_e \log |\mathcal{U}| + nC \tag{7.252}$$

or

$$H - \epsilon \leq \frac{1}{n} + P_e \log |\mathcal{U}| + C. \tag{7.253}$$

For asymptotic reliability, $P_e \to 0$ as $n \to \infty$. Therefore, by letting $n \to \infty$ and then $\epsilon \to 0$, we conclude that

$$H \leq C. \tag{7.254}$$

This result, sometimes called the *separation theorem for source and channel coding*, says that asymptotic optimality can be achieved by separating source coding and channel coding. This theorem has significant engineering implication because the source code and the channel code can be designed separately without losing asymptotic optimality. Specifically, we only need to design the best source code for the information source and design the best channel code for the channel. Moreover, separation of source coding and channel coding facilitates the transmission of different information sources on the same channel because we need only change the source code for different information sources. Likewise, separation of source coding and channel coding also facilitates the transmission of an information source on different channels because we need only change the channel code for different channels.

We remark that although asymptotic optimality can be achieved by separating source coding and channel coding, for finite block length, the probability of error generally can be reduced by using joint source–channel coding.

Chapter Summary

Capacity of Discrete Memoryless Channel:

$$C = \max_{p(x)} I(X; Y),$$

where $p(x)$ is the input distribution of the channel.

1. $C \leq \min(\log |\mathcal{X}|, \log |\mathcal{Y}|)$.
2. For a binary symmetric channel with crossover probability ϵ, $C = 1 - h_b(\epsilon)$.
3. For a binary erasure channel with erasure probability γ, $C = 1 - \gamma$.

Lemma: Let X and Y be a pair of random variables and $(\mathbf{X}', \mathbf{Y}')$ be n i.i.d. copies of a pair of generic random variables (X', Y'), where X' and Y' are independent and have the same marginal distributions as X and Y, respectively. Then

$$\Pr\{(\mathbf{X}', \mathbf{Y}') \in T_{[XY]\delta}^n\} \leq 2^{-n(I(X;Y)-\tau)},$$

where $\tau \to 0$ as $\delta \to 0$.

Channel Coding Theorem: A message drawn uniformly from the set $\{1, 2, \cdots, 2^{n(R-\epsilon)}\}$ can be transmitted through a discrete memoryless channel with negligible probability of error as $n \to \infty$ if and only if $R \leq C$.

Feedback: The capacity of a discrete memoryless channel is not increased by feedback.

Separation of Source and Channel Coding: An information source with entropy rate H can be transmitted through a discrete memoryless channel with capacity C reliably if $H < C$ (only if $H \leq C$), and asymptotic optimality can be achieved by separating source coding and channel coding.

Problems

In the following, $\mathbf{X} = (X_1, X_2, \cdots, X_n)$, $\mathbf{x} = (x_1, x_2, \cdots, x_n)$, and so on.

1. Refer to the discussion in Section 7.4.
   a) Construct the dependency graph for the random variables involved in the random coding scheme.
   b) By considering the joint distribution of these random variables, prove the Markov chain in (7.160).
2. Show that the capacity of a DMC with complete feedback cannot be increased by using probabilistic encoding and/or decoding schemes.
3. *Memory increases capacity.* Consider a BSC with crossover probability $0 < \epsilon < 1$ represented by $X_i = Y_i + Z_i$ mod 2, where X_i, Y_i, and Z_i are, respectively, the input, the output, and the noise variable at time i. Then

$$\Pr\{Z_i = 0\} = 1 - \epsilon \quad \text{and} \quad \Pr\{Z_i = 1\} = \epsilon$$

for all i. We assume that $\{X_i\}$ and $\{Z_i\}$ are independent, but we make no assumption that Z_i are i.i.d. so that the channel may have memory.
   a) Prove that
$$I(\mathbf{X}; \mathbf{Y}) \leq n - h_b(\epsilon).$$
   b) Show that the upper bound in (a) can be achieved by letting X_i be i.i.d. bits taking the values 0 and 1 with equal probability and $Z_1 = Z_2 = \cdots = Z_n$.

c) Show that with the assumptions in (b), $I(\mathbf{X}; \mathbf{Y}) > nC$, where $C = 1 - h_b(\epsilon)$ is the capacity of the BSC if it is memoryless.

4. Consider the channel in Problem 3, Part (b).

   a) Show that the channel capacity is not increased by feedback.

   b) Devise a coding scheme without feedback that achieves the channel capacity.

5. In Remark 1 toward the end of Section 7.6, it was mentioned that in the presence of feedback, both the Markov chain $W \to \mathbf{X} \to \mathbf{Y} \to \hat{W}$ and Lemma 7.16 do not hold in general. Give examples to substantiate this remark.

6. Prove that when a DMC is used with complete feedback,

$$\Pr\{Y_i = y_i | \mathbf{X}^i = \mathbf{x}^i, \mathbf{Y}^{i-1} = \mathbf{y}^{i-1}\} = \Pr\{Y_i = y_i | X_i = x_i\}$$

for all $i \geq 1$. This relation, which is a consequence of the causality of the code, says that given the current input, the current output does not depend on all the past inputs and outputs of the DMC.

7. Let

$$P(\epsilon) = \begin{bmatrix} 1 - \epsilon & \epsilon \\ \epsilon & 1 - \epsilon \end{bmatrix}$$

be the transition matrix for a BSC with crossover probability ϵ. Define $a * b = (1 - a)b + a(1 - b)$ for $0 \leq a, b \leq 1$.

   a) Prove that a DMC with transition matrix $P(\epsilon_1)P(\epsilon_2)$ is equivalent to a BSC with crossover probability $\epsilon_1 * \epsilon_2$. Such a channel is the cascade of two BSCs with crossover probabilities ϵ_1 and ϵ_2, respectively.

   b) Repeat (a) for a DMC with transition matrix $P(\epsilon_2)P(\epsilon_1)$.

   c) Prove that

$$1 - h_b(\epsilon_1 * \epsilon_2) \leq \min(1 - h_b(\epsilon_1), 1 - h_b(\epsilon_2)).$$

   This means that the capacity of the cascade of two BSCs is upper bounded by the capacity of either of the two BSCs.

   d) Prove that a DMC with transition matrix $P(\epsilon)^n$ is equivalent to a BSC with crossover probabilities $\frac{1}{2}(1 - (1 - 2\epsilon)^n)$.

8. *Symmetric channel.* A DMC is *symmetric* if the rows of the transition matrix $p(y|x)$ are permutations of each other and so are the columns. Determine the capacity of such a channel.
   See Section 4.5 in Gallager [129] for a more general discussion.

9. Let C_1 and C_2 be the capacities of two DMCs with transition matrices P_1 and P_2, respectively, and let C be the capacity of the DMC with transition matrix $P_1 P_2$. Prove that $C \leq \min(C_1, C_2)$.

10. *Two parallel channels.* Let C_1 and C_2 be the capacities of two DMCs $p_1(y_1|x_1)$ and $p_2(y_2|x_2)$, respectively. Determine the capacity of the DMC

$$p(y_1, y_2 | x_1, x_2) = p_1(y_1|x_1)p_2(y_2|x_2).$$

Hint: Prove that

$$I(X_1, X_2; Y_1, Y_2) \leq I(X_1; Y_1) + I(X_2; Y_2)$$

if $p(y_1, y_2|x_1, x_2) = p_1(y_1|x_1)p_2(y_2|x_2)$.

11. In the system below, there are two channels with transition matrices $p_1(y_1|x)$ and $p_2(y_2|x)$. These two channels have a common input alphabet $\mathcal{X}$ and output alphabets $\mathcal{Y}_1$ and $\mathcal{Y}_2$, respectively, where $\mathcal{Y}_1$ and $\mathcal{Y}_2$ are disjoint. The position of the switch is determined by a random variable Z which is independent of X, where $\Pr\{Z = 1\} = \lambda$.

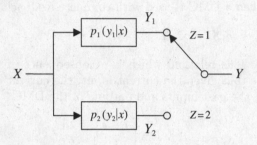

a) Show that

$$I(X; Y) = \lambda I(X; Y_1) + (1 - \lambda)I(X; Y_2).$$

b) The capacity of the system is given by $C = \max_{p(x)} I(X; Y)$. Show that $C \leq \lambda C_1 + (1 - \lambda)C_2$, where $C_i = \max_{p(x)} I(X; Y_i)$ is the capacity of the channel with transition matrix $p_i(y_i|x)$, $i = 1, 2$.

c) If both C_1 and C_2 can be achieved by a common input distribution, show that $C = \lambda C_1 + (1 - \lambda)C_2$.

12. *Feedback increases capacity.* Consider a ternary channel with memory with input/output alphabet $\{0, 1, 2\}$ as follows. At time 1, the output of the channel Y_1 has a uniform distribution on $\{0, 1, 2\}$ and is independent of the input X_1 (i.e., the channel outputs each of the values 0, 1, and 2 with probability $\frac{1}{3}$ regardless of the input). At time 2, the transition from X_2 to Y_2 which depends on the value of Y_1 is depicted below:

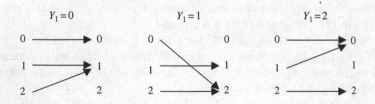

For every two subsequent transmissions, the channel replicates itself independently. So we only need to consider the first two transmissions. In the sequel, we regard this channel as described by a generic discrete channel (with transmission duration equals 2) with two input symbols X_1 and X_2 and two output symbols Y_1 and Y_2, and we will refer to this channel as the *block* channel.

a) Determine the capacity of this block channel when it is used without feedback. Hint: Use the results in Problems 8 and 11.

b) Consider the following coding scheme when the block channel is used with feedback. Let the message $W = (W_1, W_2)$ with $\mathcal{W}_1 = \{0, 1, 2\}$ and $\mathcal{W}_2 = \{0, 1\}$. Let W_1 and W_2 be independent, and each of them is distributed uniformly on its alphabet. First, Let $X_1 = W_1$ and transmit X_1 through the channel to obtain Y_1, which is independent of X_1. Then based on the value of Y_1, we determine X_2 as follows:

   i) If $Y_1 = 0$, let $X_2 = 0$ if $W_2 = 0$, and let $X_2 = 1$ if $W_2 = 1$.
   ii) If $Y_1 = 1$, let $X_2 = 1$ if $W_2 = 0$, and let $X_2 = 2$ if $W_2 = 1$.
   iii) If $Y_1 = 2$, let $X_2 = 0$ if $W_2 = 0$, and let $X_2 = 2$ if $W_2 = 1$.

   Then transmit X_2 through the channel to obtain Y_2. Based on this coding scheme, show that the capacity of this block channel can be increased by feedback.

13. *Channel with memory and directed information.* The memorylessness of a DMC is characterized by the Markov chain $T_{i-} \to X_i \to Y_i$ according to the discussion following Definition 7.4. In general, a channel with memory satisfies the Markov chain $T'_{i-} \to (\mathbf{X}^i, \mathbf{Y}^{i-1}) \to Y_i$, where T'_{i-} denotes all the random variables generated in the system before X_i (i.e., the random variables denoted by T_{i-}) except for $\mathbf{X}^{i-1}$ and $\mathbf{Y}^{i-1}$. Consider the use of such a channel in the presence of complete feedback.

a) Give the dependency graph for all the random variables involved in the coding scheme. Note that the memory of the channel is manifested by the dependence of Y_i on $\mathbf{X}^{i-1}$ and $\mathbf{Y}^{i-1}$ (in addition to its dependence on X_i) for $1 \le i \le n$.

b) Verify the correctness of the following derivation:

$$I(W; \mathbf{Y}) = H(\mathbf{Y}) - H(\mathbf{Y}|W)$$

$$= \sum_{i=1}^{n} [H(Y_i|\mathbf{Y}^{i-1}) - H(Y_i|W, \mathbf{Y}^{i-1})]$$

$$\le \sum_{i=1}^{n} [H(Y_i|\mathbf{Y}^{i-1}) - H(Y_i|W, \mathbf{X}^i, \mathbf{Y}^{i-1})]$$

$$= \sum_{i=1}^{n} [H(Y_i|\mathbf{Y}^{i-1}) - H(Y_i|\mathbf{X}^i, \mathbf{Y}^{i-1})]$$

$$= \sum_{i=1}^{n} I(Y_i; \mathbf{X}^i|\mathbf{Y}^{i-1}).$$

The above upper bound on $I(W; \mathbf{Y})$, denoted by $I(\mathbf{X} \to \mathbf{Y})$, is called the *directed information* from $\mathbf{X}$ to $\mathbf{Y}$.

c) Show that the inequality in the derivation in (b) is in fact an equality. Hint: Use Definition 7.18.

d) In the spirit of the informal discussion in Section 7.3, we impose the constraint $H(W|\mathbf{Y}) = 0$. Show that

$$H(W) = I(\mathbf{X} \to \mathbf{Y}).$$

This is the generalization of (7.105) for a channel with memory in the presence of complete feedback.

e) Show that $I(\mathbf{X} \to \mathbf{Y}) = I(\mathbf{X}; \mathbf{Y})$ if the channel code does not make use of the feedback. Hint: First show that

$$H(Y_i|\mathbf{X}^i, \mathbf{Y}^{i-1}) = H(Y_i|W, \mathbf{X}^i, \mathbf{Y}^{i-1}) = H(Y_i|W, \mathbf{X}, \mathbf{Y}^{i-1}).$$

(Marko [245] and Massey [250].)

14. *Maximum likelihood decoding.* In maximum likelihood decoding for a given channel and a given codebook, if a received sequence $\mathbf{y}$ is decoded to a codeword $\mathbf{x}$, then $\mathbf{x}$ maximizes $\Pr\{\mathbf{y}|\mathbf{x}'\}$ among all codewords $\mathbf{x}'$ in the codebook.

a) Prove that maximum likelihood decoding minimizes the average probability of error.

b) Does maximum likelihood decoding also minimize the maximal probability of error? Give an example if your answer is no.

15. *Minimum distance decoding.* The Hamming distance between two binary sequences $\mathbf{x}$ and $\mathbf{y}$, denoted by $d(\mathbf{x}, \mathbf{y})$, is the number of places where $\mathbf{x}$ and $\mathbf{y}$ differ. In minimum distance decoding for a memoryless BSC, if a received sequence $\mathbf{y}$ is decoded to a codeword $\mathbf{x}$, then $\mathbf{x}$ minimizes $d(\mathbf{x}', \mathbf{y})$ over all codewords $\mathbf{x}'$ in the codebook. Prove that minimum distance decoding is equivalent to maximum likelihood decoding if the crossover probability of the BSC is less than 0.5.

16. The following figure shows a communication system with two DMCs with complete feedback. The capacities of the two channels are, respectively, C_1 and C_2.

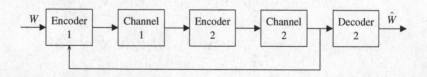

a) Give the dependency graph for all the random variables involved in the coding scheme.

b) Prove that the capacity of the system is $\min(C_1, C_2)$.

17. *Binary arbitrarily varying channel.* Consider a memoryless BSC whose crossover probability is time-varying. Specifically, the crossover probability $\epsilon(i)$ at time i is an arbitrary value in $[\epsilon_1, \epsilon_2]$, where $0 \le \epsilon_1 < \epsilon_2 < 0.5$.

Prove that the capacity of this channel is $1 - h_b(\epsilon_2)$.   (Ahlswede and Wolfowitz [12].)

18. Consider a BSC with crossover probability $\epsilon \in [\epsilon_1, \epsilon_2]$, where $0 < \epsilon_1 < \epsilon_2 < 0.5$, but the exact value of ϵ is unknown. Prove that the capacity of this channel is $1 - h_b(\epsilon_2)$.

Historical Notes

The concept of channel capacity was introduced in Shannon's original paper [322], where he stated the channel coding theorem and outlined a proof. The first rigorous proof was due to Feinstein [110]. The random coding error exponent was developed by Gallager [128] in a simplified proof.

The converse of the channel coding theorem was proved by Fano [107], where he used an inequality now bearing his name. The strong converse was first proved by Wolfowitz [383]. An iterative algorithm for calculating the channel capacity developed independently by Arimoto [19] and Blahut [37] will be discussed in Chapter 9. Shannon [326] proved that the capacity of a discrete memoryless channel cannot be increased by feedback.

The definition of a discrete memoryless channel in this chapter is new. With this definition, coding over such a channel with or without feedback can be rigorously formulated.

8

Rate-Distortion Theory

Consider an information source with entropy rate H. By the source coding theorem, it is possible to design a source code with rate R which reconstructs the source sequence $\mathbf{X} = (X_1, X_2, \cdots, X_n)$ with an arbitrarily small probability of error provided $R > H$ and the block length n is sufficiently large. However, there are situations in which we want to convey an information source by a source code with rate less than H. Then we are motivated to ask: what is the best we can do when $R < H$?

A natural approach is to design a source code such that for part of the time the source sequence is reconstructed correctly, while for the other part of the time the source sequence is reconstructed incorrectly, i.e., an error occurs. In designing such a code, we try to minimize the probability of error. However, this approach is not viable asymptotically because the converse of the source coding theorem says that if $R < H$, then the probability of error inevitably tends to 1 as $n \to \infty$.

Therefore, if $R < H$, no matter how the source code is designed, the source sequence is almost always reconstructed incorrectly when n is large. An alternative approach is to design a source code called a *rate-distortion code* which reproduces the source sequence with distortion. In order to formulate the problem properly, we need a *distortion measure* between each source sequence and each reproduction sequence. Then we try to design a rate-distortion code which with high probability reproduces the source sequence with a distortion within a tolerance level.

Clearly, a smaller distortion can potentially be achieved if we are allowed to use a higher coding rate. *Rate-distortion theory*, the subject matter of this chapter, gives a characterization of the asymptotic optimal tradeoff between the coding rate of a rate-distortion code for a given information source and the allowed distortion in the reproduction sequence with respect to a distortion measure.

8.1 Single-Letter Distortion Measures

Let $\{X_k, k \geq 1\}$ be an i.i.d. information source with generic random variable X. We assume that the source alphabet $\mathcal{X}$ is finite. Let $p(x)$ be the probability distribution of X, and we assume without loss of generality that the support of X is equal to $\mathcal{X}$. Consider a source sequence

$$\mathbf{x} = (x_1, x_2, \cdots, x_n) \tag{8.1}$$

and a reproduction sequence

$$\hat{\mathbf{x}} = (\hat{x}_1, \hat{x}_2, \cdots, \hat{x}_n). \tag{8.2}$$

The components of $\hat{\mathbf{x}}$ can take values in $\mathcal{X}$, but more generally, they can take values in any finite set $\hat{\mathcal{X}}$ which may be different from $\mathcal{X}$. The set $\hat{\mathcal{X}}$, which is also assumed to be finite, is called the reproduction alphabet. To measure the distortion between $\mathbf{x}$ and $\hat{\mathbf{x}}$, we introduce the single-letter distortion measure and the average distortion measure.

Definition 8.1. *A single-letter distortion measure is a mapping*

$$d : \mathcal{X} \times \hat{\mathcal{X}} \to \Re^+, \tag{8.3}$$

where $\Re^+$ is the set of nonnegative real numbers.[1] The value $d(x, \hat{x})$ denotes the distortion incurred when a source symbol x is reproduced as $\hat{x}$.

Definition 8.2. *The average distortion between a source sequence $\mathbf{x} \in \mathcal{X}^n$ and a reproduction sequence $\hat{\mathbf{x}} \in \hat{\mathcal{X}}^n$ induced by a single-letter distortion measure d is defined by*

$$d(\mathbf{x}, \hat{\mathbf{x}}) = \frac{1}{n} \sum_{k=1}^{n} d(x_k, \hat{x}_k). \tag{8.4}$$

In Definition 8.2, we have used d to denote both the single-letter distortion measure and the average distortion measure, but this abuse of notation should cause no ambiguity. Henceforth, we will refer to a single-letter distortion measure simply as a distortion measure.

Very often, the source sequence $\mathbf{x}$ represents quantized samples of a continuous signal, and the user attempts to recognize certain objects and derive meaning from the reproduction sequence $\hat{\mathbf{x}}$. For example, $\mathbf{x}$ may represent a video signal, an audio signal, or an image. The ultimate purpose of a distortion measure is to reflect the distortion between $\mathbf{x}$ and $\hat{\mathbf{x}}$ as *perceived* by the user. This goal is difficult to achieve in general because measurements of the distortion between $\mathbf{x}$ and $\hat{\mathbf{x}}$ must be made within contextunless the symbols

---

[1] Note that $d(x, \hat{x})$ is finite for all $(x, \hat{x}) \in \mathcal{X} \times \hat{\mathcal{X}}$.

in $\mathcal{X}$ carry no physical meaning. Specifically, when the user derives meaning from $\hat{x}$, the distortion in $\hat{x}$ as perceived by the user depends on the context. For example, the perceived distortion is small for a portrait contaminated by a fairly large noise, while the perceived distortion is large for the image of a book page contaminated by the same noise. Hence, a good distortion measure should be context dependent.

Although the average distortion is not necessarily the best way to measure the distortion between a source sequence and a reproduction sequence, it has the merit of being simple and easy to use. Moreover, rate-distortion theory, which is based on the average distortion measure, provides a framework for data compression when distortion is inevitable.

Example 8.3. When the symbols in $\mathcal{X}$ and $\hat{\mathcal{X}}$ represent real values, a popular distortion measure is the square-error distortion measure which is defined by

$$d(x,\hat{x}) = (x - \hat{x})^2. \tag{8.5}$$

The average distortion measure so induced is often referred to as the mean-square error.

Example 8.4. When $\mathcal{X}$ and $\hat{\mathcal{X}}$ are identical and the symbols in $\mathcal{X}$ do not carry any particular meaning, a frequently used distortion measure is the Hamming distortion measure, which is defined by

$$d(x,\hat{x}) = \begin{cases} 0 \text{ if } x = \hat{x} \\ 1 \text{ if } x \neq \hat{x} \end{cases}. \tag{8.6}$$

The Hamming distortion measure indicates the occurrence of an error. In particular, for an estimate $\hat{X}$ of X, we have

$$Ed(X,\hat{X}) = \Pr\{X = \hat{X}\} \cdot 0 + \Pr\{X \neq \hat{X}\} \cdot 1 = \Pr\{X \neq \hat{X}\}, \tag{8.7}$$

i.e., the expectation of the Hamming distortion measure between X and $\hat{X}$ is the probability of error.

For $\mathbf{x} \in \mathcal{X}^n$ and $\hat{\mathbf{x}} \in \hat{\mathcal{X}}^n$, the average distortion $d(\mathbf{x},\hat{\mathbf{x}})$ induced by the Hamming distortion measure gives the frequency of error in the reproduction sequence $\hat{\mathbf{x}}$.

Definition 8.5. *For a distortion measure d, for each $x \in \mathcal{X}$, let $\hat{x}^*(x) \in \hat{\mathcal{X}}$ minimize $d(x,\hat{x})$ over all $\hat{x} \in \hat{\mathcal{X}}$. A distortion measure d is said to be normal if*

$$c_x \stackrel{\text{def}}{=} d(x,\hat{x}^*(x)) = 0 \tag{8.8}$$

for all $x \in \mathcal{X}$.

The square-error distortion measure and the Hamming distortion measure are examples of normal distortion measures. Basically, a normal distortion measure is one which allows X to be reproduced with zero distortion. Although a distortion measure d is not normal in general, a normalization of d can always be obtained by defining the distortion measure

$$\tilde{d}(x,\hat{x}) = d(x,\hat{x}) - c_x \tag{8.9}$$

for all $(x,\hat{x}) \in \mathcal{X} \times \hat{\mathcal{X}}$. Evidently, $\tilde{d}$ is a normal distortion measure, and it is referred to as the normalization of d.

Example 8.6. Let d be a distortion measure defined by

$d(x,\hat{x})$	a b c
1	2 7 5 .
2	4 3 8

Then $\tilde{d}$, the normalization of d, is given by

$\tilde{d}(x,\hat{x})$	a b c
1	0 5 3 .
2	1 0 5

Note that for every $x \in \mathcal{X}$, there exists an $\hat{x} \in \hat{\mathcal{X}}$ such that $\tilde{d}(x,\hat{x}) = 0$.

Let $\hat{X}$ be any estimate of X which takes values in $\hat{\mathcal{X}}$, and denote the joint distribution for X and $\hat{X}$ by $p(x,\hat{x})$. Then

$$Ed(X,\hat{X}) = \sum_x \sum_{\hat{x}} p(x,\hat{x}) d(x,\hat{x}) \tag{8.10}$$

$$= \sum_x \sum_{\hat{x}} p(x,\hat{x}) \left[\tilde{d}(x,\hat{x}) + c_x \right] \tag{8.11}$$

$$= E\tilde{d}(X,\hat{X}) + \sum_x p(x) \sum_{\hat{x}} p(\hat{x}|x) c_x \tag{8.12}$$

$$= E\tilde{d}(X,\hat{X}) + \sum_x p(x) c_x \left(\sum_{\hat{x}} p(\hat{x}|x) \right) \tag{8.13}$$

$$= E\tilde{d}(X,\hat{X}) + \sum_x p(x) c_x \tag{8.14}$$

$$= E\tilde{d}(X,\hat{X}) + \Delta, \tag{8.15}$$

where

$$\Delta = \sum_x p(x) c_x \tag{8.16}$$

is a constant which depends only on $p(x)$ and d but not on the conditional distribution $p(\hat{x}|x)$. In other words, for a given X and a distortion measure d, the expected distortion between X and an estimate $\hat{X}$ of X is always reduced by a constant upon using $\tilde{d}$ instead of d as the distortion measure. For reasons which will be explained in Section 8.3, it is sufficient for us to assume that a distortion measure is normal.

Definition 8.7. *Let $\hat{x}^*$ minimizes $Ed(X, \hat{x})$ over all $\hat{x} \in \hat{\mathcal{X}}$, and define*

$$D_{max} = Ed(X, \hat{x}^*). \tag{8.17}$$

If we know nothing about X, then $\hat{x}^*$ is the best estimate of X, and D_{max} is the minimum expected distortion between X and a constant estimate of X. The significance of D_{max} can be seen by taking the reproduction sequence $\hat{\mathbf{X}}$ to be $(\hat{x}^*, \hat{x}^*, \cdots, \hat{x}^*)$. Since $d(X_k, \hat{x}^*)$ are i.i.d., by the weak law of large numbers

$$d(\mathbf{X}, \hat{\mathbf{X}}) = \frac{1}{n} \sum_{k=1}^{n} d(X_k, \hat{x}^*) \to Ed(X, \hat{x}^*) = D_{max} \tag{8.18}$$

in probability, i.e., for any $\epsilon > 0$,

$$\Pr\{d(\mathbf{X}, \hat{\mathbf{X}}) > D_{max} + \epsilon\} \le \epsilon \tag{8.19}$$

for sufficiently large n. Note that $\hat{\mathbf{X}}$ is a constant sequence which does not depend on $\mathbf{X}$. In other words, even when no description of $\mathbf{X}$ is available, we can still achieve an average distortion no more than $D_{max} + \epsilon$ with probability arbitrarily close to 1 when n is sufficiently large.

The notation D_{max} may seem confusing because the quantity stands for the minimum rather than the maximum expected distortion between X and a constant estimate of X. But we see from the above discussion that this notation is in fact appropriate because D_{max} is the maximum distortion we have to be concerned about. Specifically, it is not meanful to impose a constraint $D \ge D_{max}$ on the reproduction sequence because it can be achieved even without receiving any information about the sequence produced by the source.

8.2 The Rate-Distortion Function $R(D)$

Throughout this chapter, all the discussions are with respect to an i.i.d. information source $\{X_k, k \ge 1\}$ with generic random variable X and a distortion measure d. All logarithms are in the base 2 unless otherwise specified.

Definition 8.8. *An (n, M) rate-distortion code is defined by an encoding function*

$$f : \mathcal{X}^n \to \{1, 2, \cdots, M\} \tag{8.20}$$

and a decoding function

$$g : \{1, 2, \cdots, M\} \to \hat{\mathcal{X}}^n. \tag{8.21}$$

The set $\{1, 2, \cdots, M\}$, denoted by $\mathcal{I}$, is called the index set. The reproduction sequences $g(f(1)), g(f(2)), \cdots, g(f(M))$ in $\hat{\mathcal{X}}^n$ are called codewords, and the set of codewords is called the codebook.

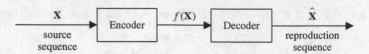

Fig. 8.1. A rate-distortion code with block length n.

Figure 8.1 is an illustration of a rate-distortion code.

Definition 8.9. *The rate of an (n, M) rate-distortion code is $n^{-1} \log M$ in bits per symbol.*

Definition 8.10. *A rate-distortion pair (R, D) is asymptotically achievable if for any $\epsilon > 0$, there exists for sufficiently large n an (n, M) rate-distortion code such that*

$$\frac{1}{n} \log M \leq R + \epsilon \tag{8.22}$$

and

$$\Pr\{d(\mathbf{X}, \hat{\mathbf{X}}) > D + \epsilon\} \leq \epsilon, \tag{8.23}$$

where $\hat{\mathbf{X}} = g(f(\mathbf{X}))$. For brevity, an asymptotically achievable pair will be referred to as an achievable pair.

Remark It is clear from the definition that if (R, D) is achievable, then (R', D) and (R, D') are also achievable for all $R' \geq R$ and $D' \geq D$.

Definition 8.11. *The rate-distortion region is the subset of $\Re^2$ containing all achievable pairs (R, D).*

Theorem 8.12. *The rate-distortion region is closed and convex.*

Proof. We first show that the rate-distortion region is closed. Consider achievable rate-distortion pairs $(R^{(k)}, D^{(k)})$ such that

$$\lim_{k \to \infty} (R^{(k)}, D^{(k)}) = (R, D) \tag{8.24}$$

componentwise. Then for any $\epsilon > 0$, for all k, there exists for sufficiently large n an $(n, M^{(k)})$ code such that

$$\frac{1}{n} \log M^{(k)} \leq R^{(k)} + \epsilon \tag{8.25}$$

and

$$\Pr\{d(\mathbf{X}^{(k)}, \hat{\mathbf{X}}^{(k)}) > D^{(k)} + \epsilon\} \leq \epsilon, \tag{8.26}$$

where $f^{(k)}$ and $g^{(k)}$ are, respectively, the encoding function and the decoding function of the $(n, M^{(k)})$ code, and $\hat{\mathbf{X}}^{(k)} = g^{(k)}(f^{(k)}(\mathbf{X}))$. By virtue of (8.24), let $k(\epsilon)$ be an integer such that for all $k > k(\epsilon)$,

$$|R - R^{(k)}| < \epsilon \tag{8.27}$$

and

$$|D - D^{(k)}| < \epsilon, \tag{8.28}$$

which imply

$$R^{(k)} < R + \epsilon \tag{8.29}$$

and

$$D^{(k)} < D + \epsilon, \tag{8.30}$$

respectively. Then for all $k > k(\epsilon)$,

$$\frac{1}{n} \log M^{(k)} \leq R^{(k)} + \epsilon < R + 2\epsilon \tag{8.31}$$

and

$$\Pr\{d(\mathbf{X}^{(k)}, \hat{\mathbf{X}}^{(k)}) > D + 2\epsilon\} \leq \Pr\{d(\mathbf{X}^{(k)}, \hat{\mathbf{X}}^{(k)}) > D^{(k)} + \epsilon\} \tag{8.32}$$

$$\leq \epsilon. \tag{8.33}$$

Note that (8.32) follows because

$$D + 2\epsilon > D^{(k)} + \epsilon \tag{8.34}$$

by (8.30). From (8.31) and (8.33), we see that (R, D) is also achievable. Thus we have proved that the rate-distortion region is closed.

We will prove the convexity of the rate-distortion region by a time-sharing argument whose idea is the following. Roughly speaking, if we can use a code $\mathcal{C}_1$ to achieve $(R^{(1)}, D^{(1)})$ and a code $\mathcal{C}_2$ to achieve $(R^{(2)}, D^{(2)})$, then for any rational number λ between 0 and 1, we can use $\mathcal{C}_1$ for a fraction λ of the time and $\mathcal{C}_2$ for a fraction $\bar{\lambda}$ of the time to achieve $(R^{(\lambda)}, D^{(\lambda)})$, where

$$R^{(\lambda)} = \lambda R^{(1)} + \bar{\lambda} R^{(2)}, \tag{8.35}$$
$$D^{(\lambda)} = \lambda D^{(1)} + \bar{\lambda} D^{(2)}, \tag{8.36}$$

and $\bar{\lambda} = 1 - \lambda$. Since the rate-distortion region is closed as we have proved, λ can be taken as any real number between 0 and 1, and the convexity of the region follows.

We now give a formal proof for the convexity of the rate-distortion region. Let

$$\lambda = \frac{r}{r+s}, \tag{8.37}$$

where r and s are positive integers. Then λ is a rational number between 0 and 1. We now prove that if $(R^{(1)}, D^{(1)})$ and $(R^{(2)}, D^{(2)})$ are achievable, then $(R^{(\lambda)}, D^{(\lambda)})$ is also achievable. Assume $(R^{(1)}, D^{(1)})$ and $(R^{(2)}, D^{(2)})$ are achievable. Then for any $\epsilon > 0$ and sufficiently large n, there exist an $(n, M^{(1)})$ code and an $(n, M^{(2)})$ code such that

$$\frac{1}{n} \log M^{(i)} \le R^{(i)} + \epsilon \tag{8.38}$$

and

$$\Pr\{d(\mathbf{X}, \hat{\mathbf{X}}^{(i)}) > D^{(i)} + \epsilon\} \le \epsilon, \tag{8.39}$$

$i = 1, 2$. Let

$$M(\lambda) = (M^{(1)})^r (M^{(2)})^s \tag{8.40}$$

and

$$n(\lambda) = (r + s)n. \tag{8.41}$$

We now construct an $(n(\lambda), M(\lambda))$ code by concatenating r copies of the $(n, M^{(1)})$ code followed by s copies of the $(n, M^{(2)})$ code. We call these $r + s$ codes subcodes of the $(n(\lambda), M(\lambda))$ code. For this code, let

$$\mathbf{Y} = (\mathbf{X}(1), \mathbf{X}(2), \cdots, \mathbf{X}(r + s)) \tag{8.42}$$

and

$$\hat{\mathbf{Y}} = (\hat{\mathbf{X}}(1), \hat{\mathbf{X}}(2), \cdots, \hat{\mathbf{X}}(r + s)), \tag{8.43}$$

where $\mathbf{X}(j)$ and $\hat{\mathbf{X}}(j)$ are the source sequence and the reproduction sequence of the jth subcode, respectively. Then for this $(n(\lambda), M(\lambda))$ code,

$$\frac{1}{n(\lambda)} \log M(\lambda) = \frac{1}{(r + s)n} \log[(M^{(1)})^r (M^{(2)})^s] \tag{8.44}$$

$$= \frac{1}{(r + s)n} (r \log M^{(1)} + s \log M^{(2)}) \tag{8.45}$$

$$= \lambda \left(\frac{1}{n} \log M^{(1)}\right) + \bar{\lambda} \left(\frac{1}{n} \log M^{(2)}\right) \tag{8.46}$$

$$\le \lambda(R^{(1)} + \epsilon) + \bar{\lambda}(R^{(2)} + \epsilon) \tag{8.47}$$

$$= (\lambda R^{(1)} + \bar{\lambda} R^{(2)}) + \epsilon \tag{8.48}$$

$$= R^{(\lambda)} + \epsilon, \tag{8.49}$$

where (8.47) follows from (8.38), and

$$\Pr\{d(\mathbf{Y}, \hat{\mathbf{Y}}) > D^{(\lambda)} + \epsilon\}$$

$$= \Pr\left\{\frac{1}{r + s} \sum_{j=1}^{r+s} d(\mathbf{X}(j), \hat{\mathbf{X}}(j)) > D^{(\lambda)} + \epsilon\right\} \tag{8.50}$$

$$\le \Pr\left\{d(\mathbf{X}(j), \hat{\mathbf{X}}(j)) > D^{(1)} + \epsilon \text{ for some } 1 \le j \le r \text{ or}\right.$$

$$\left. d(\mathbf{X}(j), \hat{\mathbf{X}}(j)) > D^{(2)} + \epsilon \text{ for some } r + 1 \le j \le r + s\right\} \tag{8.51}$$

$$\le \sum_{j=1}^{r} \Pr\{d(\mathbf{X}(j), \hat{\mathbf{X}}(j)) > D^{(1)} + \epsilon\}$$

$$+ \sum_{j=r+1}^{r+s} \Pr\{d(\mathbf{X}(j), \hat{\mathbf{X}}(j)) > D^{(2)} + \epsilon\} \tag{8.52}$$

$$\leq (r+s)\epsilon, \tag{8.53}$$

where (8.52) follows from the union bound and (8.53) follows from (8.39). Hence, we conclude that the rate-distortion pair $(R^{(\lambda)}, D^{(\lambda)})$ is achievable. This completes the proof of the theorem. □

Definition 8.13. *The rate-distortion function $R(D)$ is the minimum of all rates R for a given distortion D such that (R, D) is achievable.*

Definition 8.14. *The distortion-rate function $D(R)$ is the minimum of all distortions D for a given rate R such that (R, D) is achievable.*

Both the functions $R(D)$ and $D(R)$ are equivalent descriptions of the boundary of the rate-distortion region. They are sufficient to describe the rate-distortion region because the region is closed. Note that in defining $R(D)$, the minimum instead of the infimum is taken because for a fixed D, the set of all R such that (R, D) is achievable is closed and lower bounded by zero. Similarly, the minimum instead of the infimum is taken in defining $D(R)$. In the subsequent discussions, only $R(D)$ will be used.

Theorem 8.15. *The following properties hold for the rate-distortion function $R(D)$:*

1. *$R(D)$ is non-increasing in D.*
2. *$R(D)$ is convex.*
3. *$R(D) = 0$ for $D \geq D_{max}$.*
4. *$R(0) \leq H(X)$.*

Proof. From the remark following Definition 8.10, since $(R(D), D)$ is achievable, $(R(D), D')$ is also achievable for all $D' \geq D$. Therefore, $R(D) \geq R(D')$ because $R(D')$ is the minimum of all R such that (R, D') is achievable. This proves Property 1.

Property 2 follows immediately from the convexity of the rate-distortion region which was proved in Theorem 8.12. From the discussion toward the end of the last section, we see for any $\epsilon > 0$, it is possible to achieve

$$\Pr\{d(\mathbf{X}, \hat{\mathbf{X}}) > D_{max} + \epsilon\} \leq \epsilon \tag{8.54}$$

for sufficiently large n with no description of $\mathbf{X}$ available. Therefore, $(0, D)$ is achievable for all $D \geq D_{max}$, proving Property 3.

Property 4 is a consequence of the assumption that the distortion measure d is normalized, which can be seen as follows. By the source coding theorem, for any $\epsilon > 0$, by using a rate no more than $H(X) + \epsilon$, we can describe the source sequence $\mathbf{X}$ of length n with probability of error less than ϵ when n is sufficiently large. Since d is normalized, for each $k \geq 1$, let

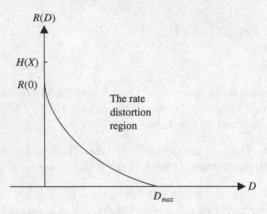

Fig. 8.2. A rate-distortion function $R(D)$.

$$\hat{X}_k = \hat{x}^*(X_k) \tag{8.55}$$

(cf. Definition 8.5), so that whenever an error does not occur,

$$d(X_k, \hat{X}_k) = d(X_k, \hat{x}^*(X_k)) = 0 \tag{8.56}$$

by (8.8) for each k, and

$$d(\mathbf{X}, \hat{\mathbf{X}}) = \frac{1}{n} \sum_{k=1}^{n} d(X_k, \hat{X}_k) = \frac{1}{n} \sum_{k=1}^{n} d(X_k, \hat{x}^*(X_k)) = 0. \tag{8.57}$$

Therefore,

$$\Pr\{d(\mathbf{X}, \hat{\mathbf{X}}) > \epsilon\} \leq \epsilon, \tag{8.58}$$

which shows that the pair $(H(X), 0)$ is achievable. This in turn implies that $R(0) \leq H(X)$ because $R(0)$ is the minimum of all R such that $(R, 0)$ is achievable. $\square$

Figure 8.2 is an illustration of a rate-distortion function $R(D)$. The reader should note the four properties of $R(D)$ in Theorem 8.15. The rate-distortion theorem, which will be stated in the next section, gives a characterization of $R(D)$.

8.3 The Rate-Distortion Theorem

Definition 8.16. *For $D \geq 0$, the information rate-distortion function is defined by*

$$R_I(D) = \min_{\hat{X}: Ed(X,\hat{X}) \leq D} I(X; \hat{X}). \tag{8.59}$$

In defining $R_I(D)$, the minimization is taken over all random variables $\hat{X}$ jointly distributed with X such that

$$Ed(X, \hat{X}) \leq D. \tag{8.60}$$

Since $p(x)$ is given, the minimization is taken over the set of all $p(\hat{x}|x)$ such that (8.60) is satisfied, namely the set

$$\left\{ p(\hat{x}|x) : \sum_{x,\hat{x}} p(x)p(\hat{x}|x)d(x,\hat{x}) \leq D \right\}. \tag{8.61}$$

Since this set is compact in $\Re^{|\mathcal{X}||\hat{\mathcal{X}}|}$ and $I(X; \hat{X})$ is a continuous functional of $p(\hat{x}|x)$, the minimum value of $I(X; \hat{X})$ can be attained.[2] This justifies taking the minimum instead of the infimum in the definition of $R_I(D)$.

We have seen in Section 8.1 that we can obtain a normalization $\tilde{d}$ for any distortion measure d with

$$E\tilde{d}(X, \hat{X}) = Ed(X, \hat{X}) - \Delta \tag{8.62}$$

for any $\hat{X}$, where Δ is a constant which depends only on $p(x)$ and d. Thus if d is not normal, we can always replace d by $\tilde{d}$ and D by $D - \Delta$ in the definition of $R_I(D)$ without changing the minimization problem. Therefore, we do not lose any generality by assuming that a distortion measure d is normal.

Theorem 8.17 (The Rate-Distortion Theorem). $R(D) = R_I(D)$.

The rate-distortion theorem, which is the main result in rate-distortion theory, says that the minimum coding rate for achieving a distortion D is $R_I(D)$. This theorem will be proved in the next two sections. In the next section, we will prove the converse of this theorem, i.e., $R(D) \geq R_I(D)$, and in Section 8.5, we will prove the achievability of $R_I(D)$, i.e., $R(D) \leq R_I(D)$.

In order for $R_I(D)$ to be a characterization of $R(D)$, it has to satisfy the same properties as $R(D)$. In particular, the four properties of $R(D)$ in Theorem 8.15 should also be satisfied by $R_I(D)$.

Theorem 8.18. *The following properties hold for the information rate-distortion function* $R_I(D)$:

1. $R_I(D)$ *is non-increasing in* D.
2. $R_I(D)$ *is convex.*
3. $R_I(D) = 0$ *for* $D \geq D_{max}$.
4. $R_I(0) \leq H(X)$.

---

[2] The assumption that both $\mathcal{X}$ and $\hat{\mathcal{X}}$ are finite is essential in this argument.

Proof. Referring to the definition of $R_I(D)$ in (8.59), for a larger D, the minimization is taken over a larger set. Therefore, $R_I(D)$ is non-increasing in D, proving Property 1.

To prove Property 2, consider any $D^{(1)}, D^{(2)} \geq 0$ and let λ be any number between 0 and 1. Let $\hat{X}^{(i)}$ achieve $R_I(D^{(i)})$ for $i = 1, 2$, i.e.,

$$R_I(D^{(i)}) = I(X; \hat{X}^{(i)}), \tag{8.63}$$

where

$$Ed(X, \hat{X}^{(i)}) \leq D^{(i)}, \tag{8.64}$$

and let $\hat{X}^{(i)}$ be defined by the transition matrix $p_i(\hat{x}|x)$. Let $\hat{X}^{(\lambda)}$ be jointly distributed with X which is defined by

$$p_\lambda(\hat{x}|x) = \lambda p_1(\hat{x}|x) + \bar{\lambda} p_2(\hat{x}|x), \tag{8.65}$$

where $\bar{\lambda} = 1 - \lambda$. Then

$$Ed(X, \hat{X}^{(\lambda)})$$
$$= \sum_{x, \hat{x}} p(x) p_\lambda(\hat{x}|x) d(x, \hat{x}) \tag{8.66}$$
$$= \sum_{x, \hat{x}} p(x) (\lambda p_1(\hat{x}|x) + \bar{\lambda} p_2(\hat{x}|x)) d(x, \hat{x}) \tag{8.67}$$
$$= \lambda \left(\sum_{x, \hat{x}} p(x) p_1(\hat{x}|x) d(x, \hat{x}) \right) + \bar{\lambda} \left(\sum_{x, \hat{x}} p(x) p_2(\hat{x}|x) d(x, \hat{x}) \right) \tag{8.68}$$
$$= \lambda Ed(X, \hat{X}^{(1)}) + \bar{\lambda} Ed(X, \hat{X}^{(2)}) \tag{8.69}$$
$$\leq \lambda D^{(1)} + \bar{\lambda} D^{(2)} \tag{8.70}$$
$$= D^{(\lambda)}, \tag{8.71}$$

where

$$D^{(\lambda)} = \lambda D^{(1)} + \bar{\lambda} D^{(2)}, \tag{8.72}$$

and (8.70) follows from (8.64). Now consider

$$\lambda R_I(D^{(1)}) + \bar{\lambda} R_I(D^{(2)}) = \lambda I(X; \hat{X}^{(1)}) + \bar{\lambda} I(X; \hat{X}^{(2)}) \tag{8.73}$$
$$\geq I(X; \hat{X}^{(\lambda)}) \tag{8.74}$$
$$\geq R_I(D^{(\lambda)}), \tag{8.75}$$

where the inequality in (8.74) follows from the convexity of mutual information with respect to the transition matrix $p(\hat{x}|x)$ (see Example 3.13), and the inequality in (8.75) follows from (8.71) and the definition of $R_I(D)$. Therefore, we have proved Property 2.

To prove Property 3, let $\hat{X}$ take the value $\hat{x}^*$ as defined in Definition 8.7 with probability 1. Then

$$I(X; \hat{X}) = 0 \tag{8.76}$$

and

$$Ed(X; \hat{X}) = Ed(X; \hat{x}^*) = D_{max}. \tag{8.77}$$

Then for $D \geq D_{max}$,

$$R_I(D) \leq I(X; \hat{X}) = 0. \tag{8.78}$$

On the other hand, since $R_I(D)$ is nonnegative, we conclude that

$$R_I(D) = 0. \tag{8.79}$$

This proves Property 3.

Finally, to prove Property 4, we let

$$\hat{X} = \hat{x}^*(X), \tag{8.80}$$

where $\hat{x}^*(x)$ is defined in Definition 8.5. Then

$$Ed(X, \hat{X}) = Ed(X, \hat{x}^*(X)) \tag{8.81}$$

$$= \sum_x p(x) d(x, \hat{x}^*(x)) \tag{8.82}$$

$$= 0 \tag{8.83}$$

by (8.8) since we assume that d is a normal distortion measure. Moreover,

$$R_I(0) \leq I(X; \hat{X}) \leq H(X). \tag{8.84}$$

Then Property 4 and hence the theorem is proved. $\square$

Corollary 8.19. *If $R_I(0) > 0$, then $R_I(D)$ is strictly decreasing for $0 \leq D \leq D_{max}$, and the inequality constraint in Definition 8.16 for $R_I(D)$ can be replaced by an equality constraint.*

Proof. Assume that $R_I(0) > 0$. We first show that $R_I(D) > 0$ for $0 \leq D < D_{max}$ by contradiction. Suppose $R_I(D') = 0$ for some $0 \leq D' < D_{max}$, and let $R_I(D')$ be achieved by some $\hat{X}$. Then

$$R_I(D') = I(X; \hat{X}) = 0 \tag{8.85}$$

implies that X and $\hat{X}$ are independent or

$$p(x, \hat{x}) = p(x) p(\hat{x}) \tag{8.86}$$

for all x and $\hat{x}$. It follows that

$$D' \geq Ed(X, \hat{X}) \tag{8.87}$$

$$= \sum_x \sum_{\hat{x}} p(x, \hat{x}) d(x, \hat{x}) \tag{8.88}$$

$$= \sum_x \sum_{\hat{x}} p(x) p(\hat{x}) d(x, \hat{x}) \tag{8.89}$$

$$= \sum_{\hat{x}} p(\hat{x}) \sum_x p(x) d(x, \hat{x}) \tag{8.90}$$

$$= \sum_{\hat{x}} p(\hat{x}) Ed(X, \hat{x}) \tag{8.91}$$

$$\geq \sum_{\hat{x}} p(\hat{x}) Ed(X, \hat{x}^*) \tag{8.92}$$

$$= \sum_{\hat{x}} p(\hat{x}) D_{max} \tag{8.93}$$

$$= D_{max}, \tag{8.94}$$

where $\hat{x}^*$ and D_{max} are defined in Definition 8.7. This leads to a contradiction because we have assumed that $0 \leq D' < D_{max}$. Therefore, we conclude that $R_I(D) > 0$ for $0 \leq D < D_{max}$.

Since $R_I(0) > 0$ and $R_I(D_{max}) = 0$ and $R_I(D)$ is non-increasing and convex from the above theorem, $R_I(D)$ must be strictly decreasing for $0 \leq D \leq D_{max}$. We now prove by contradiction that the inequality constraint in Definition 8.16 for $R_I(D)$ can be replaced by an equality constraint. Assume that $R_I(D)$ is achieved by some $\hat{X}^*$ such that

$$Ed(X, \hat{X}^*) = D'' < D. \tag{8.95}$$

Then

$$R_I(D'') = \min_{\hat{X}:Ed(X,\hat{X}) \leq D''} I(X; \hat{X}) \leq I(X; \hat{X}^*) = R_I(D). \tag{8.96}$$

This is a contradiction because $R_I(D)$ is strictly decreasing for $0 \leq D \leq D_{max}$. Hence,

$$Ed(X, \hat{X}^*) = D. \tag{8.97}$$

This implies that the inequality constraint in Definition 8.16 for $R_I(D)$ can be replaced by an equality constraint. □

Remark In all problems of interest, $R(0) = R_I(0) > 0$. Otherwise, $R(D) = 0$ for all $D \geq 0$ because $R(D)$ is nonnegative and non-increasing.

Example 8.20 (Binary Source). Let X be a binary random variable with

$$\Pr\{X = 0\} = 1 - \gamma \quad \text{and} \quad \Pr\{X = 1\} = \gamma. \tag{8.98}$$

Let $\hat{\mathcal{X}} = \{0, 1\}$ be the reproduction alphabet for X, and let d be the Hamming distortion measure. We first consider the case that $0 \leq \gamma \leq 1/2$. Then if we make a guess on the value of X, we should guess 0 in order to minimize the expected distortion. Therefore, $\hat{x}^* = 0$ and

$$
\begin{align}
D_{max} &= Ed(X, 0) \tag{8.99} \\
&= \Pr\{X = 1\} \tag{8.100} \\
&= \gamma. \tag{8.101}
\end{align}
$$

We will show that for $0 \leq \gamma \leq \frac{1}{2}$,

$$
R_I(D) = \begin{cases} h_b(\gamma) - h_b(D) & \text{if } 0 \leq D < \gamma \\ 0 & \text{if } D \geq \gamma \end{cases}. \tag{8.102}
$$

Let $\hat{X}$ be an estimate of X taking values in $\hat{\mathcal{X}}$, and let Y be the Hamming distortion measure between X and $\hat{X}$, i.e.,

$$
Y = d(X, \hat{X}). \tag{8.103}
$$

Observe that conditioning on $\hat{X}$, X, and Y determine each other. Therefore,

$$
H(X|\hat{X}) = H(Y|\hat{X}). \tag{8.104}
$$

Then for $D < \gamma = D_{max}$ and any $\hat{X}$ such that

$$
Ed(X, \hat{X}) \leq D, \tag{8.105}
$$

we have

$$
\begin{align}
I(X; \hat{X}) &= H(X) - H(X|\hat{X}) \tag{8.106} \\
&= h_b(\gamma) - H(Y|\hat{X}) \tag{8.107} \\
&\geq h_b(\gamma) - H(Y) \tag{8.108} \\
&= h_b(\gamma) - h_b(\Pr\{X \neq \hat{X}\}) \tag{8.109} \\
&\geq h_b(\gamma) - h_b(D), \tag{8.110}
\end{align}
$$

where the last inequality is justified because

$$
\Pr\{X \neq \hat{X}\} = Ed(X, \hat{X}) \leq D \tag{8.111}
$$

and $h_b(a)$ is increasing for $0 \leq a \leq 1/2$. Minimizing over all $\hat{X}$ satisfying (8.105) in (8.110), we obtain the lower bound

$$
R_I(D) \geq h_b(\gamma) - h_b(D). \tag{8.112}
$$

To show that this lower bound is achievable, we need to construct an $\hat{X}$ such that the inequalities in both (8.108) and (8.110) are tight. The tightness of the inequality in (8.110) simply says that

$$
\Pr\{X \neq \hat{X}\} = D, \tag{8.113}
$$

while the tightness of the inequality in (8.108) says that Y should be independent of $\hat{X}$.

It would be more difficult to make Y independent of $\hat{X}$ if we specify $\hat{X}$ by $p(\hat{x}|x)$. Instead, we specify the joint distribution of X and $\hat{X}$ by means of a reverse binary symmetric channel (BSC) with crossover probability D as shown in Figure 8.3. Here, we regard $\hat{X}$ as the input and X as the output of the BSC. Then Y is independent of the input $\hat{X}$ because the error event is independent of the input for a BSC, and (8.113) is satisfied by setting the crossover probability to D. However, we need to ensure that the marginal distribution of X so specified is equal to $p(x)$. Toward this end, we let

$$\Pr\{\hat{X} = 1\} = \alpha, \tag{8.114}$$

and consider

$$\begin{aligned}\Pr\{X = 1\} = {}&\Pr\{\hat{X} = 0\}\Pr\{X = 1|\hat{X} = 0\} \\ &+\Pr\{\hat{X} = 1\}\Pr\{X = 1|\hat{X} = 1\}\end{aligned} \tag{8.115}$$

or

$$\gamma = (1 - \alpha)D + \alpha(1 - D), \tag{8.116}$$

which gives

$$\alpha = \frac{\gamma - D}{1 - 2D}. \tag{8.117}$$

Since

$$D < D_{max} = \gamma \le \frac{1}{2}, \tag{8.118}$$

we have $\alpha \ge 0$. On the other hand,

$$\gamma, D \le \frac{1}{2} \tag{8.119}$$

gives

$$\gamma + D \le 1. \tag{8.120}$$

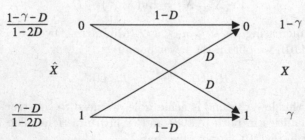

Fig. 8.3. Achieving $R_I(D)$ for a binary source via a reverse binary symmetric channel.

This implies

$$\gamma - D \leq 1 - 2D,$$
(8.121)

or $\alpha \leq 1$. Therefore,

$$0 \leq \alpha = \Pr\{\hat{X} = 1\} \leq 1$$
(8.122)

and

$$0 \leq 1 - \alpha = \Pr\{\hat{X} = 0\} \leq 1.$$
(8.123)

Hence, we have shown that the lower bound on $R_I(D)$ in (8.110) can be achieved, and $R_I(D)$ is as given in (8.102).

For $1/2 \leq \gamma \leq 1$, by exchanging the roles of the symbols 0 and 1 in the above argument, we obtain $R_I(D)$ as in (8.102) except that γ is replaced by $1 - \gamma$. Combining the two cases, we have

$$R_I(D) = \begin{cases} h_b(\gamma) - h_b(D) & \text{if } 0 \leq D < \min(\gamma, 1 - \gamma) \\ 0 & \text{if } D \geq \min(\gamma, 1 - \gamma) \end{cases}$$
(8.124)

for $0 \leq \gamma \leq 1$. The function $R_I(D)$ for $\gamma = 1/2$ is illustrated in Figure 8.4.

Remark In the above example, we see that $R_I(0) = h_b(\gamma) = H(X)$. Then by the rate-distortion theorem, $H(X)$ is the minimum rate of a rate-distortion code which achieves an arbitrarily small average Hamming distortion. It is tempting to regard this special case of the rate-distortion theorem as a version of the source coding theorem and conclude that the rate-distortion theorem is a generalization of the source coding theorem. However, this is incorrect because the rate-distortion theorem only guarantees that the *average* Hamming distortion between $\mathbf{X}$ and $\hat{\mathbf{X}}$ is small with probability arbitrarily close to 1, but the source coding theorem guarantees that $\mathbf{X} = \hat{\mathbf{X}}$ with probability arbitrarily close to 1, which is much stronger.

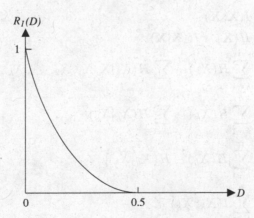

Fig. 8.4. The function $R_I(D)$ for the uniform binary source with the Hamming distortion measure.

It is in general not possible to obtain the rate-distortion function in closed form, and we have to resort to numerical computation. In Chapter 9, we will discuss the Blahut–Arimoto algorithm for computing the rate-distortion function.

8.4 The Converse

In this section, we prove that the rate-distortion function $R(D)$ is lower bounded by the information rate-distortion function $R_I(D)$, i.e., $R(D) \geq R_I(D)$. Specifically, we will prove that for any achievable rate-distortion pair (R, D), $R \geq R_I(D)$. Then by fixing D and minimizing R over all achievable pairs (R, D), we conclude that $R(D) \geq R_I(D)$.

Let (R, D) be any achievable rate-distortion pair. Then for any $\epsilon > 0$, there exists for sufficiently large n an (n, M) code such that

$$\frac{1}{n} \log M \leq R + \epsilon \tag{8.125}$$

and

$$\Pr\{d(\mathbf{X}, \hat{\mathbf{X}}) > D + \epsilon\} \leq \epsilon, \tag{8.126}$$

where $\hat{\mathbf{X}} = g(f(\mathbf{X}))$. Then

$$n(R + \epsilon) \overset{a)}{\geq} \log M \tag{8.127}$$

$$\geq H(f(\mathbf{X})) \tag{8.128}$$

$$\geq H(g(f(\mathbf{X}))) \tag{8.129}$$

$$= H(\hat{\mathbf{X}}) \tag{8.130}$$

$$= H(\hat{\mathbf{X}}) - H(\hat{\mathbf{X}}|\mathbf{X}) \tag{8.131}$$

$$= I(\hat{\mathbf{X}}; \mathbf{X}) \tag{8.132}$$

$$= H(\mathbf{X}) - H(\mathbf{X}|\hat{\mathbf{X}}) \tag{8.133}$$

$$= \sum_{k=1}^{n} H(X_k) - \sum_{k=1}^{n} H(X_k|\hat{\mathbf{X}}, X_1, X_2, \cdots, X_{k-1}) \tag{8.134}$$

$$\overset{b)}{\geq} \sum_{k=1}^{n} H(X_k) - \sum_{k=1}^{n} H(X_k|\hat{X}_k) \tag{8.135}$$

$$= \sum_{k=1}^{n} [H(X_k) - H(X_k|\hat{X}_k)] \tag{8.136}$$

$$= \sum_{k=1}^{n} I(X_k; \hat{X}_k) \tag{8.137}$$

$$\overset{c)}{\geq} \sum_{k=1}^{n} R_I(Ed(X_k, \hat{X}_k)) \tag{8.138}$$

$$= n \left[\frac{1}{n} \sum_{k=1}^{n} R_I(Ed(X_k, \hat{X}_k)) \right] \tag{8.139}$$

$$\overset{d)}{\geq} nR_I \left(\frac{1}{n} \sum_{k=1}^{n} Ed(X_k, \hat{X}_k) \right) \tag{8.140}$$

$$= nR_I(Ed(\mathbf{X}, \hat{\mathbf{X}})). \tag{8.141}$$

In the above,

(a) follows from (8.125);
(b) follows because conditioning does not increase entropy;
(c) follows from the definition of $R_I(D)$ in Definition 8.16;
(d) follows from the convexity of $R_I(D)$ proved in Theorem 8.18 and Jensen's inequality.

Now let

$$d_{max} = \max_{x, \hat{x}} d(x, \hat{x}) \tag{8.142}$$

be the maximum value which can be taken by the distortion measure d. The reader should not confuse d_{max} with D_{max} in Definition 8.7. Then from (8.126), we have

$$Ed(\mathbf{X}, \hat{\mathbf{X}})$$
$$= E[d(\mathbf{X}, \hat{\mathbf{X}})|d(\mathbf{X}, \hat{\mathbf{X}}) > D + \epsilon]\mathrm{Pr}\{d(\mathbf{X}, \hat{\mathbf{X}}) > D + \epsilon\}$$
$$+ E[d(\mathbf{X}, \hat{\mathbf{X}})|d(\mathbf{X}, \hat{\mathbf{X}}) \leq D + \epsilon]\mathrm{Pr}\{d(\mathbf{X}, \hat{\mathbf{X}}) \leq D + \epsilon\} \tag{8.143}$$
$$\leq d_{max} \cdot \epsilon + (D + \epsilon) \cdot 1 \tag{8.144}$$
$$= D + (d_{max} + 1)\epsilon. \tag{8.145}$$

This shows that if the probability that the average distortion between $\mathbf{X}$ and $\hat{\mathbf{X}}$ exceeds $D + \epsilon$ is small, then the expected average distortion between $\mathbf{X}$ and $\hat{\mathbf{X}}$ can exceed D only by a small amount.[3] Following (8.141), we have

$$R + \epsilon \geq R_I(Ed(\mathbf{X}, \hat{\mathbf{X}})) \tag{8.146}$$
$$\geq R_I(D + (d_{max} + 1)\epsilon), \tag{8.147}$$

where the last inequality follows from (8.145) because $R_I(D)$ is non-increasing in D. We note that the convexity of $R_I(D)$ implies that it is a continuous function of D. Then taking the limit as $\epsilon \to 0$, we obtain

$$R \geq \lim_{\epsilon \to 0} R_I(D + (d_{max} + 1)\epsilon) \tag{8.148}$$
$$= R_I \left(D + (d_{max} + 1) \lim_{\epsilon \to 0} \epsilon \right) \tag{8.149}$$
$$= R_I(D), \tag{8.150}$$

---

[3] The converse is not true.

where we have invoked the continuity of $R_I(D)$ in obtaining (8.149). Upon minimizing R over all achievable pairs (R, D) for a fixed D in (8.150), we have proved that

$$R(D) \geq R_I(D). \qquad (8.151)$$

This completes the proof for the converse of the rate-distortion theorem.

8.5 Achievability of $R_I(D)$

In this section, we prove that the rate-distortion function $R(D)$ is upper bounded by the information rate-distortion function $R_I(D)$, i.e., $R(D) \leq R_I(D)$. Then by combining with the result that $R(D) \geq R_I(D)$ from the last section, we conclude that $R(D) = R_I(D)$, and the rate-distortion theorem is proved.

For any $0 \leq D \leq D_{max}$, we will prove that for every random variable $\hat{X}$ taking values in $\hat{\mathcal{X}}$ such that

$$Ed(X, \hat{X}) \leq D, \qquad (8.152)$$

the rate-distortion pair $(I(X; \hat{X}), D)$ is achievable. This will be proved by showing for sufficiently large n the existence of a rate-distortion code such that

1. the rate of the code is not more than $I(X; \hat{X}) + \epsilon$;
2. $d(\mathbf{X}, \hat{\mathbf{X}}) \leq D + \epsilon$ with probability almost 1.

Then by minimizing $I(X; \hat{X})$ over all $\hat{X}$ satisfying (8.152), we conclude that the rate-distortion pair $(R_I(D), D)$ is achievable, which implies $R_I(D) \geq R(D)$ because $R(D)$ is the minimum of all R such that (R, D) is achievable.

Fix any $0 \leq D \leq D_{max}$ and any $\epsilon > 0$, and let δ be a small positive quantity to be specified later. Toward proving the existence of a desired code, we fix a random variable $\hat{X}$ which satisfies (8.152) and let M be an integer satisfying

$$I(X; \hat{X}) + \frac{\epsilon}{2} \leq \frac{1}{n} \log M \leq I(X; \hat{X}) + \epsilon, \qquad (8.153)$$

where n is sufficiently large.

We now describe a random coding scheme in the following steps:

1. Construct a codebook $\mathcal{C}$ of an (n, M) code by randomly generating M codewords in $\hat{\mathcal{X}}^n$ independently and identically according to $p(\hat{x})^n$. Denote these codewords by $\hat{\mathbf{X}}(1), \hat{\mathbf{X}}(2), \cdots, \hat{\mathbf{X}}(M)$.
2. Reveal the codebook $\mathcal{C}$ to both the encoder and the decoder.
3. The source sequence $\mathbf{X}$ is generated according to $p(x)^n$.
4. The encoder encodes the source sequence $\mathbf{X}$ into an index K in the set $\mathcal{I} = \{1, 2, \cdots, M\}$. The index K takes the value i if

   a) $(\mathbf{X}, \hat{\mathbf{X}}(i)) \in T^n_{[X\hat{X}]\delta}$,

   b) for all $i' \in \mathcal{I}$, if $(\mathbf{X}, \hat{\mathbf{X}}(i')) \in T^n_{[X\hat{X}]\delta}$, then $i' \leq i$;

otherwise, K takes the constant value 1.

5. The index K is delivered to the decoder.
6. The decoder outputs $\hat{\mathbf{X}}(K)$ as the reproduction sequence $\hat{\mathbf{X}}$.

Remark Strong typicality is used in defining the encoding function in Step 4. This is made possible by the assumption that both the source alphabet $\mathcal{X}$ and the reproduction alphabet $\hat{\mathcal{X}}$ are finite.

Let us further explain the encoding scheme described in Step 4. After the source sequence $\mathbf{X}$ is generated, we search through all the codewords in the codebook $\mathcal{C}$ for those which are jointly typical with $\mathbf{X}$ with respect to $p(x, \hat{x})$. If there is at least one such codeword, we let i be the largest index of such codewords and let $K = i$. If such a codeword does not exist, we let $K = 1$.

The event $\{K = 1\}$ occurs in one of the following two scenarios:

1. $\hat{X}(1)$ is the only codeword in $\mathcal{C}$ which is jointly typical with $\mathbf{X}$.
2. No codeword in $\mathcal{C}$ is jointly typical with $\mathbf{X}$.

In either scenario, $\mathbf{X}$ is not jointly typical with the codewords $\hat{X}(2), \hat{X}(3), \cdots,$ $\hat{X}(M)$. In other words, if $K = 1$, then $\mathbf{X}$ is jointly typical with none of the codewords $\hat{X}(2), \hat{X}(3), \cdots, \hat{X}(M)$.

Define

$$E_i = \left\{ (\mathbf{X}, \hat{\mathbf{X}}(i)) \in T^n_{[X\hat{X}]\delta} \right\} \tag{8.154}$$

to be the event that $\mathbf{X}$ is jointly typical with the codeword $\hat{X}(i)$. We see from the above discussion that

$$\{K = 1\} \subset E_2^c \cap E_3^c \cap \cdots \cap E_M^c. \tag{8.155}$$

Since the codewords are generated i.i.d., conditioning on $\{\mathbf{X} = \mathbf{x}\}$ for any $\mathbf{x} \in \mathcal{X}^n$, the events E_i are mutually independent,[4] and they all have the same probability. Then for any $\mathbf{x} \in \mathcal{X}^n$,

$$\Pr\{K = 1 | \mathbf{X} = \mathbf{x}\} \leq \Pr\{E_2^c \cap E_3^c \cap \cdots \cap E_M^c | \mathbf{X} = \mathbf{x}\} \tag{8.156}$$

$$= \prod_{i=2}^{M} \Pr\{E_i^c | \mathbf{X} = \mathbf{x}\} \tag{8.157}$$

$$= (\Pr\{E_1^c | \mathbf{X} = \mathbf{x}\})^{M-1} \tag{8.158}$$

$$= (1 - \Pr\{E_1 | \mathbf{X} = \mathbf{x}\})^{M-1}. \tag{8.159}$$

---

[4] Without conditioning on $\{\mathbf{X} = \mathbf{x}\}$, the events E_i are not mutually independent because they depend on each other through $\mathbf{X}$.

We now obtain a lower bound on $\Pr\{E_1|\mathbf{X} = \mathbf{x}\}$ for $\mathbf{x} \in S^n_{[X]\delta}$, where

$$S^n_{[X]\delta} = \{\mathbf{x} \in T^n_{[X]\delta} : |T^n_{[\hat{X}|X]\delta}(\mathbf{x})| \geq 1\} \tag{8.160}$$

(cf. Section 6.3). Consider

$$\Pr\{E_1|\mathbf{X} = \mathbf{x}\} = \Pr\left\{(\mathbf{x}, \hat{\mathbf{X}}(1)) \in T^n_{[X\hat{X}]\delta}\right\} \tag{8.161}$$

$$= \sum_{\hat{\mathbf{x}}:(\mathbf{x},\hat{\mathbf{x}})\in T^n_{[X\hat{X}]\delta}} p(\hat{\mathbf{x}}). \tag{8.162}$$

The summation above is over all $\hat{\mathbf{x}}$ such that $(\mathbf{x}, \hat{\mathbf{x}}) \in T^n_{[X\hat{X}]\delta}$. From the consistency of strong typicality (Theorem 6.7), if $(\mathbf{x}, \hat{\mathbf{x}}) \in T^n_{[X\hat{X}]\delta}$, then $\hat{\mathbf{x}} \in T^n_{[\hat{X}]\delta}$. By the strong AEP (Theorem 6.2), all $p(\hat{\mathbf{x}})$ in the above summation satisfy

$$p(\hat{\mathbf{x}}) \geq 2^{-n(H(\hat{X})+\eta)}, \tag{8.163}$$

where $\eta \to 0$ as $\delta \to 0$. By the conditional strong AEP (Theorem 6.10),

$$|T^n_{[\hat{X}|X]\delta}(\mathbf{x})| \geq 2^{n(H(\hat{X}|X)-\xi)}, \tag{8.164}$$

where $\xi \to 0$ as $\delta \to 0$. Then from (8.162), we have

$$\Pr\{E_1|\mathbf{X} = \mathbf{x}\} \geq 2^{n(H(\hat{X}|X)-\xi)}2^{-n(H(\hat{X})+\eta)} \tag{8.165}$$

$$= 2^{-n(H(\hat{X})-H(\hat{X}|X)+\xi+\eta)} \tag{8.166}$$

$$= 2^{-n(I(X;\hat{X})+\zeta)}, \tag{8.167}$$

where

$$\zeta = \xi + \eta \to 0 \tag{8.168}$$

as $\delta \to 0$. Following (8.159), we have

$$\Pr\{K = 1|\mathbf{X} = \mathbf{x}\} \leq \left[1 - 2^{-n(I(X;\hat{X})+\zeta)}\right]^{M-1}. \tag{8.169}$$

The lower bound in (8.153) implies

$$M \geq 2^{n(I(X;\hat{X})+\frac{\epsilon}{2})}. \tag{8.170}$$

Then upon taking natural logarithm in (8.169), we obtain

$$\ln \Pr\{K = 1|\mathbf{X} = \mathbf{x}\}$$
$$\leq (M-1)\ln\left[1 - 2^{-n(I(X;\hat{X})+\zeta)}\right] \tag{8.171}$$

$$\overset{a)}{\leq} \left(2^{n(I(X;\hat{X})+\frac{\epsilon}{2})} - 1\right)\ln\left[1 - 2^{-n(I(X;\hat{X})+\zeta)}\right] \tag{8.172}$$

$$\overset{b)}{\leq} -\left(2^{n(I(X;\hat{X})+\frac{\epsilon}{2})} - 1\right)2^{-n(I(X;\hat{X})+\zeta)} \tag{8.173}$$

$$= -\left[2^{n(\frac{\epsilon}{2}-\zeta)} - 2^{-n(I(X;\hat{X})+\zeta)}\right]. \tag{8.174}$$

In the above, (a) follows from (8.170) by noting that the logarithm in (8.171) is negative, and (b) follows from the fundamental inequality $\ln a \leq a - 1$. By letting δ be sufficiently small so that

$$\frac{\epsilon}{2} - \zeta > 0, \tag{8.175}$$

the above upper bound on $\ln \Pr\{K = 1 | \mathbf{X} = \mathbf{x}\}$ tends to $-\infty$ as $n \to \infty$, i.e., $\Pr\{K = 1 | \mathbf{X} = \mathbf{x}\} \to 0$ as $n \to \infty$. This implies

$$\Pr\{K = 1 | \mathbf{X} = \mathbf{x}\} \leq \frac{\epsilon}{2} \tag{8.176}$$

for sufficiently large n. It then follows that

$$\Pr\{K = 1\} \tag{8.177}$$

$$= \sum_{\mathbf{x} \in S^n_{[X]\delta}} \Pr\{K = 1 | \mathbf{X} = \mathbf{x}\} \Pr\{\mathbf{X} = \mathbf{x}\}$$

$$+ \sum_{\mathbf{x} \notin S^n_{[X]\delta}} \Pr\{K = 1 | \mathbf{X} = \mathbf{x}\} \Pr\{\mathbf{X} = \mathbf{x}\} \tag{8.178}$$

$$\leq \sum_{\mathbf{x} \in S^n_{[X]\delta}} \frac{\epsilon}{2} \cdot \Pr\{\mathbf{X} = \mathbf{x}\} + \sum_{\mathbf{x} \notin S^n_{[X]\delta}} 1 \cdot \Pr\{\mathbf{X} = \mathbf{x}\} \tag{8.179}$$

$$= \frac{\epsilon}{2} \cdot \Pr\{\mathbf{X} \in S^n_{[X]\delta}\} + \Pr\{\mathbf{X} \notin S^n_{[X]\delta}\} \tag{8.180}$$

$$\leq \frac{\epsilon}{2} \cdot 1 + (1 - \Pr\{\mathbf{X} \in S^n_{[X]\delta}\}) \tag{8.181}$$

$$< \frac{\epsilon}{2} + \delta, \tag{8.182}$$

where we have invoked Proposition 6.13 in the last step. By letting δ be sufficiently small so that

$$\delta < \frac{\epsilon}{2} \tag{8.183}$$

and (8.175) is satisfied, we obtain

$$\Pr\{K = 1\} < \epsilon. \tag{8.184}$$

The main idea of the above upper bound on $\Pr\{K = 1\}$ for sufficiently large n is the following. In constructing the codebook, we randomly generate M codewords in $\hat{\mathcal{X}}^n$ according to $p(\hat{x})^n$. If M grows with n at a rate higher than $I(X; \hat{X})$, then the probability that there exists at least one codeword which is jointly typical with the source sequence $\mathbf{X}$ with respect to $p(x, \hat{x})$ is very high when n is large. Further, the average distortion between $\mathbf{X}$ and such a codeword is close to $Ed(X, \hat{X})$ because the empirical joint distribution of the symbol pairs in $\mathbf{X}$ and such a codeword is close to $p(x, \hat{x})$. Then by letting the reproduction sequence $\hat{\mathbf{X}}$ be such a codeword, the average distortion between $\mathbf{X}$ and $\hat{\mathbf{X}}$ is less than $D + \epsilon$ with probability arbitrarily close to 1 since $Ed(X, \hat{X}) \leq D$. These will be formally shown in the rest of the proof.

Now for sufficiently large n, consider

$$\Pr\{d(\mathbf{X}, \hat{\mathbf{X}}) > D + \epsilon\}$$

$$= \Pr\{d(\mathbf{X}, \hat{\mathbf{X}}) > D + \epsilon | K = 1\}\Pr\{K = 1\}$$

$$+ \Pr\{d(\mathbf{X}, \hat{\mathbf{X}}) > D + \epsilon | K \neq 1\}\Pr\{K \neq 1\} \tag{8.185}$$

$$\leq 1 \cdot \epsilon + \Pr\{d(\mathbf{X}, \hat{\mathbf{X}}) > D + \epsilon | K \neq 1\} \cdot 1 \tag{8.186}$$

$$= \epsilon + \Pr\{d(\mathbf{X}, \hat{\mathbf{X}}) > D + \epsilon | K \neq 1\}. \tag{8.187}$$

We will show that by choosing the value of δ carefully, it is possible to make $d(\mathbf{X}, \hat{\mathbf{X}})$ always less than or equal to $D + \epsilon$ provided $K \neq 1$. Since $(\mathbf{X}, \hat{\mathbf{X}}) \in T^n_{[X\hat{X}]\delta}$ conditioning on $\{K \neq 1\}$, we have

$$d(\mathbf{X}, \hat{\mathbf{X}})$$

$$= \frac{1}{n} \sum_{k=1}^{n} d(X_k, \hat{X}_k) \tag{8.188}$$

$$= \frac{1}{n} \sum_{x,\hat{x}} d(x, \hat{x}) N(x, \hat{x} | \mathbf{X}, \hat{\mathbf{X}}) \tag{8.189}$$

$$= \frac{1}{n} \sum_{x,\hat{x}} d(x, \hat{x})(np(x, \hat{x}) + N(x, \hat{x} | \mathbf{X}, \hat{\mathbf{X}}) - np(x, \hat{x})) \tag{8.190}$$

$$= \left[\sum_{x,\hat{x}} p(x, \hat{x}) d(x, \hat{x}) \right] + \left[\sum_{x,\hat{x}} d(x, \hat{x}) \left(\frac{1}{n} N(x, \hat{x} | \mathbf{X}, \hat{\mathbf{X}}) - p(x, \hat{x}) \right) \right] \tag{8.191}$$

$$= Ed(X, \hat{X}) + \sum_{x,\hat{x}} d(x, \hat{x}) \left(\frac{1}{n} N(x, \hat{x} | \mathbf{X}, \hat{\mathbf{X}}) - p(x, \hat{x}) \right) \tag{8.192}$$

$$\leq Ed(X, \hat{X}) + \sum_{x,\hat{x}} d(x, \hat{x}) \left| \frac{1}{n} N(x, \hat{x} | \mathbf{X}, \hat{\mathbf{X}}) - p(x, \hat{x}) \right| \tag{8.193}$$

$$\overset{a)}{\leq} Ed(X, \hat{X}) + d_{max} \sum_{x,\hat{x}} \left| \frac{1}{n} N(x, \hat{x} | \mathbf{X}, \hat{\mathbf{X}}) - p(x, \hat{x}) \right| \tag{8.194}$$

$$\overset{b)}{\leq} Ed(X, \hat{X}) + d_{max}\delta \tag{8.195}$$

$$\overset{c)}{\leq} D + d_{max}\delta, \tag{8.196}$$

where

(a) follows from the definition of d_{max} in (8.142);
(b) follows because $(\mathbf{X}, \hat{\mathbf{X}}) \in T^n_{[X\hat{X}]\delta}$;
(c) follows from (8.152).

By taking

$$\delta \leq \frac{\epsilon}{d_{max}}, \tag{8.197}$$

we obtain

$$d(\mathbf{X}, \hat{\mathbf{X}}) \leq D + d_{max} \left(\frac{\epsilon}{d_{max}} \right) = D + \epsilon \tag{8.198}$$

if $K \neq 1$. Therefore,

$$\Pr\{d(\mathbf{X}, \hat{\mathbf{X}}) > D + \epsilon | K \neq 1\} = 0, \tag{8.199}$$

and it follows from (8.187) that

$$\Pr\{d(\mathbf{X}, \hat{\mathbf{X}}) > D + \epsilon\} \leq \epsilon. \tag{8.200}$$

Thus we have shown that for sufficiently large n, there exists an (n, M) random code which satisfies

$$\frac{1}{n} \log M \leq I(X; \hat{X}) + \epsilon \tag{8.201}$$

(this follows from the upper bound in (8.153)) and (8.200). This implies the existence of an (n, M) rate-distortion code which satisfies (8.201) and (8.200). Therefore, the rate-distortion pair $(I(X; \hat{X}), D)$ is achievable. Then upon minimizing over all $\hat{X}$ which satisfy (8.152), we conclude that the rate-distortion pair $(R_I(D), D)$ is achievable, which implies $R_I(D) \geq R(D)$. The proof is completed.

Chapter Summary

Rate-Distortion Function: For an information source X and a single-letter distortion measure $d : \mathcal{X} \times \hat{\mathcal{X}} \rightarrow \Re$, the rate-distortion function is defined as

$$R(D) = \min_{\hat{X} : Ed(X, \hat{X}) \leq D} I(X; \hat{X}).$$

Rate-Distortion Theorem: An i.i.d. random sequence $X_1, X_2, \cdots, X_n$ with generic random variable X can be compressed at rate $R + \epsilon$ such that $\Pr\{d(\mathbf{X}, \hat{\mathbf{X}}) > D + \epsilon\} \rightarrow 0$ as $n \rightarrow \infty$ if and only if $R \geq R(D)$.

Binary Source: Let X be binary with distribution $\{\gamma, 1 - \gamma\}$ and let d be the Hamming distortion measure. Then

$$R(D) = \begin{cases} h_b(\gamma) - h_b(D) & \text{if } 0 \leq D < \min(\gamma, 1 - \gamma) \\ 0 & \text{if } D \geq \min(\gamma, 1 - \gamma) \end{cases}.$$

Problems

1. Obtain the forward channel description of $R(D)$ for the binary source with the Hamming distortion measure.

2. *Binary covering radius.* The Hamming ball with center $\mathbf{c} = (c_1, c_2, \cdots, c_n) \in \{0,1\}^n$ and radius r is the set

$$S_r(\mathbf{c}) = \left\{ \mathbf{x} \in \{0,1\}^n : \sum_{i=1}^{n} |x_i - c_i| \leq r \right\}.$$

Let $M_{r,n}$ be the minimum number M such that there exists Hamming balls $S_r(\mathbf{c}_j)$, $j = 1, 2, \cdots, M$ such that for all $\mathbf{x} \in \{0,1\}^n$, $\mathbf{x} \in S_r(\mathbf{c}_j)$ for some j.

   a) Show that

$$M_{r,n} \geq \frac{2^n}{\sum_{k=0}^{r} \binom{n}{k}}.$$

   b) What is the relation between $M_{r,n}$ and the rate-distortion function for the binary source with the Hamming distortion measure?

3. Consider a source random variable X with the Hamming distortion measure.

   a) Prove that

$$R(D) \geq H(X) - D \log(|\mathcal{X}| - 1) - h_b(D)$$

   for $0 \leq D \leq D_{max}$.

   b) Show that the above lower bound on $R(D)$ is tight if X is distributed uniformly on $\mathcal{X}$.

   See Jerohin [190] (also see [83], p. 133) for the tightness of this lower bound for a general source. This bound is a special case of the Shannon lower bound for the rate-distortion function [327] (also see [79], p. 369).

4. *Product source.* Let X and Y be two independent source random variables with reproduction alphabets $\hat{\mathcal{X}}$ and $\hat{\mathcal{Y}}$ and distortion measures d_x and d_y, and the rate-distortion functions for X and Y are denoted by $R_x(D_x)$ and $R_y(D_y)$, respectively. Now for the product source (X, Y), define a distortion measure $d : \mathcal{X} \times \mathcal{Y} \to \hat{\mathcal{X}} \times \hat{\mathcal{Y}}$ by

$$d((x, y), (\hat{x}, \hat{y})) = d_x(x, \hat{x}) + d_y(y, \hat{y}).$$

Prove that the rate-distortion function $R(D)$ for (X, Y) with distortion measure d is given by

$$R(D) = \min_{D_x + D_y = D} (R_x(D_x) + R_y(D_y)).$$

Hint: Prove that $I(X, Y; \hat{X}, \hat{Y}) \geq I(X; \hat{X}) + I(Y; \hat{Y})$ if X and Y are independent. (Shannon [327].)

5. *Compound source.* Let Θ be an index set and $\mathcal{Z}_\Theta = \{X_\theta : \theta \in \Theta\}$ be a collection of source random variables. The random variables in $\mathcal{Z}_\Theta$ have a common source alphabet $\mathcal{X}$, a common reproduction alphabet $\hat{\mathcal{X}}$, and a common distortion measure d. A compound source is an i.i.d. information source whose generic random variable is X_Φ, where Φ is equal to some $\theta \in \Theta$ but we do not know which one it is. The rate-distortion function $R_\Phi(D)$ for X_Φ has the same definition as the rate-distortion function defined in this chapter except that (8.23) is replaced by

$$\Pr\{d(\mathbf{X}_\theta, \hat{\mathbf{X}}) > D + \epsilon\} \le \epsilon \quad \text{for all } \theta \in \Theta.$$

Show that

$$R_\Phi(D) = \sup_{\theta \in \Theta} R_\theta(D),$$

where $R_\theta(D)$ is the rate-distortion function for X_θ.

6. Show that asymptotic optimality can always be achieved by separating rate-distortion coding and channel coding when the information source is i.i.d. (with a single-letter distortion measure) and the channel is memoryless.

7. *Slepian–Wolf coding.* Let ϵ, γ, and δ be small positive quantities. For $1 \le i \le 2^{n(H(Y|X)+\epsilon)}$, randomly and independently select with replacement $2^{n(I(X;Y)-\gamma)}$ sequences from $T_{[Y]\delta}^n$ according to the uniform distribution to form a bin B_i. Let $(\mathbf{x}, \mathbf{y})$ be a fixed pair of sequences in $T_{[XY]\delta}^n$. Prove the following by choosing ϵ, γ, and δ appropriately:

   a) The probability that $\mathbf{y}$ is in some B_i tends to 1 as $n \to \infty$.
   b) Given that $\mathbf{y} \in B_i$, the probability that there exists another $\mathbf{y}' \in B_i$ such that $(\mathbf{x}, \mathbf{y}') \in T_{[XY]\delta}^n$ tends to 0 as $n \to \infty$.

Let $(\mathbf{X}, \mathbf{Y}) \sim p^n(x, y)$. The results in (a) and (b) say that if $(\mathbf{X}, \mathbf{Y})$ is jointly typical, which happens with probability close to 1 for large n, then it is very likely that $\mathbf{Y}$ is in some bin B_i, and that $\mathbf{Y}$ is the unique vector in B_i which is jointly typical with $\mathbf{X}$. If $\mathbf{X}$ is available as side-information, then by specifying the index of the bin containing $\mathbf{Y}$, which takes about $2^{nH(Y|X)}$ bits, $\mathbf{Y}$ can be uniquely specified. Note that no knowledge about $\mathbf{X}$ is involved in specifying the index of the bin containing $\mathbf{Y}$. This is the basis of the Slepian–Wolf coding [339] which launched the whole area of multiterminal source coding (see Berger [28]).

Historical Notes

Transmission of an information source with distortion was first conceived by Shannon in his 1948 paper [322]. He returned to the problem in 1959 and proved the rate-distortion theorem [327]. The normalization of the rate-distortion function is due to Pinkston [290]. The rate-distortion theorem proved here is a stronger version of the original theorem. Extensions of the theorem to more general sources were proved in the book by Berger [27]. An iterative algorithm for computing the rate-distortion function developed by Blahut [37] will be discussed in Chapter 9. Rose [312] has developed an algorithm for the same purpose based on a mapping approach.

The Blahut–Arimoto Algorithms

For a discrete memoryless channel $p(y|x)$, the capacity

$$C = \max_{r(x)} I(X;Y), \tag{9.1}$$

where X and Y are, respectively, the input and the output of the generic channel and $r(x)$ is the input distribution, characterizes the maximum asymptotically achievable rate at which information can be transmitted through the channel reliably. The expression for C in (9.1) is called a *single-letter characterization* in the sense that it depends only on the transition matrix of the generic channel but not on the block length n of a code for the channel. When both the input alphabet $\mathcal{X}$ and the output alphabet $\mathcal{Y}$ are finite, the computation of C becomes a finite-dimensional maximization problem.

For an i.i.d. information source $\{X_k, k \geq 1\}$ with generic random variable X, the rate-distortion function

$$R(D) = \min_{Q(\hat{x}|x):Ed(X,\hat{X}) \leq D} I(X;\hat{X}) \tag{9.2}$$

characterizes the minimum asymptotically achievable rate of a rate-distortion code which reproduces the information source with an average distortion no more than D with respect to a single-letter distortion measure d. Again, the expression for $R(D)$ in (9.2) is a single-letter characterization because it depends only on the generic random variable X but not on the block length n of a rate-distortion code. When both the source alphabet $\mathcal{X}$ and the reproduction alphabet $\hat{\mathcal{X}}$ are finite, the computation of $R(D)$ becomes a finite-dimensional minimization problem.

Unless for very special cases, it is not possible to obtain an expression for C or $R(D)$ in closed form, and we have to resort to numerical computation. However, computing these quantities is not straightforward because the associated optimization problem is nonlinear. In this chapter, we discuss the *Blahut–Arimoto algorithms* (henceforth the BA algorithms), which is an iterative algorithm devised for this purpose.

In order to better understand how and why the BA algorithm works, we will first describe the algorithm in a general setting in the next section. Specializations of the algorithm for the computation of C and $R(D)$ will be discussed in Section 9.2, and convergence of the algorithm will be proved in Section 9.3.

9.1 Alternating Optimization

In this section, we describe an alternating optimization algorithm. This algorithm will be specialized in the next section for computing the channel capacity and the rate-distortion function.

Consider the double supremum

$$\sup_{\mathbf{u}_1 \in A_1} \sup_{\mathbf{u}_2 \in A_2} f(\mathbf{u}_1, \mathbf{u}_2), \tag{9.3}$$

where A_i is a convex subset of $\Re^{n_i}$ for $i = 1, 2$, and f is a real function defined on $A_1 \times A_2$. The function f is bounded from above, and is continuous and has continuous partial derivatives on $A_1 \times A_2$. Further assume that for all $\mathbf{u}_2 \in A_2$, there exists a unique $c_1(\mathbf{u}_2) \in A_1$ such that

$$f(c_1(\mathbf{u}_2), \mathbf{u}_2) = \max_{\mathbf{u}_1' \in A_1} f(\mathbf{u}_1', \mathbf{u}_2), \tag{9.4}$$

and for all $\mathbf{u}_1 \in A_1$, there exists a unique $c_2(\mathbf{u}_1) \in A_2$ such that

$$f(\mathbf{u}_1, c_2(\mathbf{u}_1)) = \max_{\mathbf{u}_2' \in A_2} f(\mathbf{u}_1, \mathbf{u}_2'). \tag{9.5}$$

Let $\mathbf{u} = (\mathbf{u}_1, \mathbf{u}_2)$ and $A = A_1 \times A_2$. Then (9.3) can be written as

$$\sup_{\mathbf{u} \in A} f(\mathbf{u}). \tag{9.6}$$

In other words, the supremum of f is taken over a subset of $\Re^{n_1 + n_2}$ which is equal to the Cartesian product of two convex subsets of $\Re^{n_1}$ and $\Re^{n_2}$, respectively.

We now describe an alternating optimization algorithm for computing f^*, the value of the double supremum in (9.3). Let $\mathbf{u}^{(k)} = (\mathbf{u}_1^{(k)}, \mathbf{u}_2^{(k)})$ for $k \geq 0$ which are defined as follows. Let $\mathbf{u}_1^{(0)}$ be an arbitrarily chosen vector in A_1, and let $\mathbf{u}_2^{(0)} = c_2(\mathbf{u}_1^{(0)})$. For $k \geq 1$, $\mathbf{u}^{(k)}$ is defined by

$$\mathbf{u}_1^{(k)} = c_1(\mathbf{u}_2^{(k-1)}) \tag{9.7}$$

and

$$\mathbf{u}_2^{(k)} = c_2(\mathbf{u}_1^{(k)}). \tag{9.8}$$

In other words, $\mathbf{u}_1^{(k)}$ and $\mathbf{u}_2^{(k)}$ are generated in the order $\mathbf{u}_1^{(0)}$, $\mathbf{u}_2^{(0)}$, $\mathbf{u}_1^{(1)}$, $\mathbf{u}_2^{(1)}$, $\mathbf{u}_1^{(2)}$, $\mathbf{u}_2^{(2)}$, $\cdots$, where each vector in the sequence is a function of the previous vector except that $\mathbf{u}_1^{(0)}$ is arbitrarily chosen in A_1. Let

$$f^{(k)} = f(\mathbf{u}^{(k)}). \tag{9.9}$$

Then from (9.4) and (9.5),

$$f^{(k)} = f(\mathbf{u}_1^{(k)}, \mathbf{u}_2^{(k)}) \tag{9.10}$$
$$\geq f(\mathbf{u}_1^{(k)}, \mathbf{u}_2^{(k-1)}) \tag{9.11}$$
$$\geq f(\mathbf{u}_1^{(k-1)}, \mathbf{u}_2^{(k-1)}) \tag{9.12}$$
$$= f^{(k-1)} \tag{9.13}$$

for $k \geq 1$. Since the sequence $f^{(k)}$ is non-decreasing, it must converge because f is bounded from above. We will show in Section 9.3 that $f^{(k)} \to f^*$ if f is concave. Figure 9.1 is an illustration of the alternating maximization algorithm, where in this case both n_1 and n_2 are equal to 1, and $f^{(k)} \to f^*$.

The alternating optimization algorithm can be explained by the following analogy. Suppose a hiker wants to reach the summit of a mountain. Starting from a certain point in the mountain, the hiker moves north–south and east–west alternately. (In our problem, the north–south and east–west directions can be multi-dimensional.) In each move, the hiker moves to the highest possible point. The question is whether the hiker can eventually approach the summit starting from any point in the mountain.

Replacing f by $-f$ in (9.3), the double supremum becomes the double infimum

$$\inf_{\mathbf{u}_1 \in A_1} \inf_{\mathbf{u}_2 \in A_2} f(\mathbf{u}_1, \mathbf{u}_2). \tag{9.14}$$

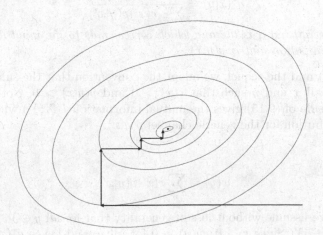

Fig. 9.1. Alternating optimization.

All the previous assumptions on A_1, A_2, and f remain valid except that f is now assumed to be bounded from below instead of bounded from above. The double infimum in (9.14) can be computed by the same alternating optimization algorithm. Note that with f replaced by $-f$, the maximums in (9.4) and (9.5) become minimums, and the inequalities in (9.11) and (9.12) are reversed.

9.2 The Algorithms

In this section, we specialize the alternating optimization algorithm described in the last section to compute the channel capacity and the rate-distortion function. The corresponding algorithms are known as the BA algorithms.

9.2.1 Channel Capacity

We will use $\mathbf{r}$ to denote an input distribution $r(x)$, and we write $\mathbf{r} > 0$ if $\mathbf{r}$ is strictly positive, i.e., $r(x) > 0$ for all $x \in \mathcal{X}$. If $\mathbf{r}$ is not strictly positive, we write $\mathbf{r} \geq 0$. Similar notations will be introduced as appropriate.

Lemma 9.1. *Let $r(x)p(y|x)$ be a given joint distribution on $\mathcal{X} \times \mathcal{Y}$ such that $\mathbf{r} > 0$, and let $\mathbf{q}$ be a transition matrix from $\mathcal{Y}$ to $\mathcal{X}$. Then*

$$\max_{\mathbf{q}} \sum_{x} \sum_{y} r(x)p(y|x) \log \frac{q(x|y)}{r(x)} = \sum_{x} \sum_{y} r(x)p(y|x) \log \frac{q^*(x|y)}{r(x)}, \quad (9.15)$$

where the maximization is taken over all $\mathbf{q}$ such that

$$q(x|y) = 0 \quad \textit{if and only if} \quad p(y|x) = 0, \quad\quad (9.16)$$

and

$$q^*(x|y) = \frac{r(x)p(y|x)}{\sum_{x'} r(x')p(y|x')}, \quad\quad (9.17)$$

i.e., the maximizing $\mathbf{q}$ is the one which corresponds to the input distribution $\mathbf{r}$ and the transition matrix $p(y|x)$.

In (9.15) and the sequel, we adopt the convention that the summation is taken over all x and y such that $r(x) > 0$ and $p(y|x) > 0$. Note that the right-hand side of (9.15) gives the mutual information $I(X;Y)$ when $\mathbf{r}$ is the input distribution for the generic channel $p(y|x)$.

Proof. Let

$$w(y) = \sum_{x'} r(x')p(y|x') \quad\quad (9.18)$$

in (9.17). We assume without loss of generality that for all $y \in \mathcal{Y}$, $p(y|x) > 0$ for some $x \in \mathcal{X}$. Since $\mathbf{r} > 0$, $w(y) > 0$ for all y, and hence $q^*(x|y)$ is well defined. Rearranging (9.17), we have

$$r(x)p(y|x) = w(y)q^*(x|y). \tag{9.19}$$

Consider

$$\sum_x \sum_y r(x)p(y|x) \log \frac{q^*(x|y)}{r(x)} - \sum_x \sum_y r(x)p(y|x) \log \frac{q(x|y)}{r(x)}$$

$$= \sum_x \sum_y r(x)p(y|x) \log \frac{q^*(x|y)}{q(x|y)} \tag{9.20}$$

$$= \sum_y \sum_x w(y)q^*(x|y) \log \frac{q^*(x|y)}{q(x|y)} \tag{9.21}$$

$$= \sum_y w(y) \sum_x q^*(x|y) \log \frac{q^*(x|y)}{q(x|y)} \tag{9.22}$$

$$= \sum_y w(y)D(q^*(x|y)\|q(x|y)) \tag{9.23}$$

$$\geq 0, \tag{9.24}$$

where (9.21) follows from (9.19), and the last step is an application of the divergence inequality. Then the proof is completed by noting in (9.17) that $\mathbf{q}^*$ satisfies (9.16) because $\mathbf{r} > 0$. $\square$

Theorem 9.2. *For a discrete memoryless channel $p(y|x)$,*

$$C = \sup_{\mathbf{r}>0} \max_{\mathbf{q}} \sum_x \sum_y r(x)p(y|x) \log \frac{q(x|y)}{r(x)}, \tag{9.25}$$

where the maximization is taken over all $\mathbf{q}$ that satisfies (9.16).

Proof. Let $I(\mathbf{r}, \mathbf{p})$ denote the mutual information $I(X;Y)$ when $\mathbf{r}$ is the input distribution for the generic channel $p(y|x)$. Then we can write

$$C = \max_{\mathbf{r}\geq 0} I(\mathbf{r}, \mathbf{p}). \tag{9.26}$$

Let $\mathbf{r}^*$ achieves C. If $\mathbf{r}^* > 0$, then

$$C = \max_{\mathbf{r}\geq 0} I(\mathbf{r}, \mathbf{p}) \tag{9.27}$$

$$= \max_{\mathbf{r}>0} I(\mathbf{r}, \mathbf{p}) \tag{9.28}$$

$$= \max_{\mathbf{r}>0} \max_{\mathbf{q}} \sum_x \sum_y r(x)p(y|x) \log \frac{q(x|y)}{r(x)} \tag{9.29}$$

$$= \sup_{\mathbf{r}>0} \max_{\mathbf{q}} \sum_x \sum_y r(x)p(y|x) \log \frac{q(x|y)}{r(x)}, \tag{9.30}$$

where (9.29) follows from Lemma 9.1 (and the maximization is over all $\mathbf{q}$ that satisfies (9.16)).

Next, we consider the case when $\mathbf{r}^* \geq 0$. Since $I(\mathbf{r}, \mathbf{p})$ is continuous in $\mathbf{r}$, for any $\epsilon > 0$, there exists $\delta > 0$ such that if

$$\|\mathbf{r} - \mathbf{r}^*\| < \delta, \tag{9.31}$$

then

$$C - I(\mathbf{r}, \mathbf{p}) < \epsilon, \tag{9.32}$$

where $\|\mathbf{r} - \mathbf{r}^*\|$ denotes the Euclidean distance between $\mathbf{r}$ and $\mathbf{r}^*$. In particular, there exists $\tilde{\mathbf{r}} > 0$ which satisfies (9.31) and (9.32). Then

$$C = \max_{\mathbf{r} \geq 0} I(\mathbf{r}, \mathbf{p}) \tag{9.33}$$

$$\geq \sup_{\mathbf{r} > 0} I(\mathbf{r}, \mathbf{p}) \tag{9.34}$$

$$\geq I(\tilde{\mathbf{r}}, \mathbf{p}) \tag{9.35}$$

$$> C - \epsilon, \tag{9.36}$$

where the last step follows because $\tilde{\mathbf{r}}$ satisfies (9.32). Thus we have

$$C - \epsilon < \sup_{\mathbf{r} > 0} I(\mathbf{r}, \mathbf{p}) \leq C. \tag{9.37}$$

Finally, by letting $\epsilon \to 0$, we conclude that

$$C = \sup_{\mathbf{r} > 0} I(\mathbf{r}, \mathbf{p}) = \sup_{\mathbf{r} > 0} \max_{\mathbf{q}} \sum_x \sum_y r(x) p(y|x) \log \frac{q(x|y)}{r(x)}. \tag{9.38}$$

This accomplishes the proof. $\square$

Now for the double supremum in (9.3), let

$$f(\mathbf{r}, \mathbf{q}) = \sum_x \sum_y r(x) p(y|x) \log \frac{q(x|y)}{r(x)}, \tag{9.39}$$

with $\mathbf{r}$ and $\mathbf{q}$ playing the roles of $\mathbf{u}_1$ and $\mathbf{u}_2$, respectively. Let

$$A_1 = \{(r(x), x \in \mathcal{X}) : \ r(x) > 0 \text{ and } \textstyle\sum_x r(x) = 1\} \tag{9.40}$$

and

$$A_2 = \{(q(x|y), (x,y) \in \mathcal{X} \times \mathcal{Y}) : \ q(x|y) > 0$$
$$\text{if } p(x|y) > 0, \ q(x|y) = 0 \text{ if } p(y|x) = 0,$$
$$\text{and } \textstyle\sum_x q(x|y) = 1 \text{ for all } y \in \mathcal{Y}\}. \tag{9.41}$$

Then A_1 is a subset of $\Re^{|\mathcal{X}|}$ and A_2 is a subset of $\Re^{|\mathcal{X}||\mathcal{Y}|}$, and it can readily be checked that both A_1 and A_2 are convex. For all $\mathbf{r} \in A_1$ and $\mathbf{q} \in A_2$, by Lemma 9.1,

$$f(\mathbf{r}, \mathbf{q}) = \sum_x \sum_y r(x) p(y|x) \log \frac{q(x|y)}{r(x)} \tag{9.42}$$

$$\leq \sum_x \sum_y r(x) p(y|x) \log \frac{q^*(x|y)}{r(x)} \tag{9.43}$$

$$= I(X;Y) \tag{9.44}$$

$$\leq H(X) \tag{9.45}$$

$$\leq \log |\mathcal{X}|. \tag{9.46}$$

Thus f is bounded from above. Since for all $\mathbf{q} \in A_2$, $q(x|y) = 0$ for all x and y such that $p(x|y) = 0$, these components of $\mathbf{q}$ are degenerated. In fact, these components of $\mathbf{q}$ do not appear in the definition of $f(\mathbf{r}, \mathbf{q})$ in (9.39), which can be seen as follows. Recall the convention that the double summation in (9.39) is over all x and y such that $r(x) > 0$ and $p(y|x) > 0$. If $q(x|y) = 0$, then $p(y|x) = 0$, and hence the corresponding term is not included in the double summation. Therefore, it is readily seen that f is continuous and has continuous partial derivatives on A because all the probabilities involved in the double summation in (9.39) are strictly positive. Moreover, for any given $\mathbf{r} \in A_1$, by Lemma 9.1, there exists a unique $\mathbf{q} \in A_2$ that maximizes f. It will be shown shortly that for any given $\mathbf{q} \in A_2$, there also exists a unique $\mathbf{r} \in A_1$ that maximizes f.

The double supremum in (9.3) now becomes

$$\sup_{\mathbf{r} \in A_1} \sup_{\mathbf{q} \in A_2} \sum_x \sum_y r(x) p(y|x) \log \frac{q(x|y)}{r(x)}, \tag{9.47}$$

which by Theorem 9.2 is equal to C, where the supremum over all $\mathbf{q} \in A_2$ is in fact a maximum. We then apply the alternating optimization algorithm in the last section to compute C. First, we arbitrarily choose a *strictly positive* input distribution in A_1 and let it be $\mathbf{r}^{(0)}$. Then we define $\mathbf{q}^{(0)}$ and in general $\mathbf{q}^{(k)}$ for $k \geq 0$ by

$$q^{(k)}(x|y) = \frac{r^{(k)}(x) p(y|x)}{\sum_{x'} r^{(k)}(x') p(y|x')} \tag{9.48}$$

in view of Lemma 9.1. In order to define $\mathbf{r}^{(1)}$ and in general $\mathbf{r}^{(k)}$ for $k \geq 1$, we need to find the $\mathbf{r} \in A_1$ that maximizes f for a given $\mathbf{q} \in A_2$, where the constraints on $\mathbf{r}$ are

$$\sum_x r(x) = 1 \tag{9.49}$$

and

$$r(x) > 0 \quad \text{for all } x \in \mathcal{X}. \tag{9.50}$$

We now use the method of Lagrange multipliers to find the best $\mathbf{r}$ by ignoring temporarily the positivity constraints in (9.50). Let

$$J = \sum_x \sum_y r(x)p(y|x) \log \frac{q(x|y)}{r(x)} - \lambda \sum_x r(x). \tag{9.51}$$

For convenience sake, we assume that the logarithm is the natural logarithm. Differentiating with respect to $r(x)$ gives

$$\frac{\partial J}{\partial r(x)} = \sum_y p(y|x) \log q(x|y) - \log r(x) - 1 - \lambda. \tag{9.52}$$

Upon setting $\frac{\partial J}{\partial r(x)} = 0$, we have

$$\log r(x) = \sum_y p(y|x) \log q(x|y) - 1 - \lambda \tag{9.53}$$

or

$$r(x) = e^{-(\lambda+1)} \prod_y q(x|y)^{p(y|x)}. \tag{9.54}$$

By considering the normalization constraint in (9.49), we can eliminate λ and obtain

$$r(x) = \frac{\prod_y q(x|y)^{p(y|x)}}{\sum_{x'} \prod_y q(x'|y)^{p(y|x')}}. \tag{9.55}$$

The above product is over all y such that $p(y|x) > 0$ and $q(x|y) > 0$ for all such y. This implies that both the numerator and the denominator on the right-hand side above are positive, and therefore $r(x) > 0$. In other words, the $\mathbf{r}$ thus obtained happen to satisfy the positivity constraints in (9.50) although these constraints were ignored when we set up the Lagrange multipliers. We will show in Section 9.3.2 that f is concave. Then $\mathbf{r}$ as given in (9.55), which is unique, indeed achieves the maximum of f for a given $\mathbf{q} \in A_2$ because $\mathbf{r}$ is in the interior of A_1. In view of (9.55), we define $\mathbf{r}^{(k)}$ for $k \geq 1$ by

$$r^{(k)}(x) = \frac{\prod_y q^{(k-1)}(x|y)^{p(y|x)}}{\sum_{x'} \prod_y q^{(k-1)}(x'|y)^{p(y|x')}}. \tag{9.56}$$

The vectors $\mathbf{r}^{(k)}$ and $\mathbf{q}^{(k)}$ are defined in the order $\mathbf{r}^{(0)}, \mathbf{q}^{(0)}, \mathbf{r}^{(1)}, \mathbf{q}^{(1)}, \mathbf{r}^{(2)}, \mathbf{q}^{(2)}, \cdots$, where each vector in the sequence is a function of the previous vector except that $\mathbf{r}^{(0)}$ is arbitrarily chosen in A_1. It remains to show by induction that $\mathbf{r}^{(k)} \in A_1$ for $k \geq 1$ and $\mathbf{q}^{(k)} \in A_2$ for $k \geq 0$. If $\mathbf{r}^{(k)} \in A_1$, i.e., $\mathbf{r}^{(k)} > 0$, then we see from (9.48) that $q^{(k)}(x|y) = 0$ if and only if $p(x|y) = 0$, i.e., $\mathbf{q}^{(k)} \in A_2$. On the other hand, if $\mathbf{q}^{(k)} \in A_2$, then we see from (9.56) that $\mathbf{r}^{(k+1)} > 0$, i.e., $\mathbf{r}^{(k+1)} \in A_2$. Therefore, $\mathbf{r}^{(k)} \in A_1$ and $\mathbf{q}^{(k)} \in A_2$ for all $k \geq 0$. Upon determining $(\mathbf{r}^{(k)}, \mathbf{q}^{(k)})$, we can compute $f^{(k)} = f(\mathbf{r}^{(k)}, \mathbf{q}^{(k)})$ for all k. It will be shown in Section 9.3 that $f^{(k)} \to C$.

9.2.2 The Rate-Distortion Function

The discussion in this section is analogous to the discussion in Section 9.2.1. Some of the details will be omitted for brevity.

For all problems of interest, $R(0) > 0$. Otherwise, $R(D) = 0$ for all $D \geq 0$ since $R(D)$ is nonnegative and non-increasing. Therefore, we assume without loss of generality that $R(0) > 0$.

We have shown in Corollary 8.19 that if $R(0) > 0$, then $R(D)$ is strictly decreasing for $0 \leq D \leq D_{\max}$. Since $R(D)$ is convex, for any $s \leq 0$, there exists a point on the $R(D)$ curve for $0 \leq D \leq D_{\max}$ such that the slope of a tangent[1] to the $R(D)$ curve at that point is equal to s. Denote such a point on the $R(D)$ curve by $(D_s, R(D_s))$, which is not necessarily unique. Then this tangent intersects with the ordinate at $R(D_s) - sD_s$. This is illustrated in Figure 9.2.

Let $I(\mathbf{p}, \mathbf{Q})$ denote the mutual information $I(X, \hat{X})$ and $D(\mathbf{p}, \mathbf{Q})$ denote the expected distortion $Ed(X, \hat{X})$ when $\mathbf{p}$ is the distribution for X and $\mathbf{Q}$ is the transition matrix from $\mathcal{X}$ to $\hat{\mathcal{X}}$ defining $\hat{X}$. Then for any $\mathbf{Q}$, $(I(\mathbf{p}, \mathbf{Q}), D(\mathbf{p}, \mathbf{Q}))$ is a point in the rate-distortion region, and the line with slope s passing through $(I(\mathbf{p}, \mathbf{Q}), D(\mathbf{p}, \mathbf{Q}))$ intersects the ordinate at $I(\mathbf{p}, \mathbf{Q}) - sD(\mathbf{p}, \mathbf{Q})$. Since the $R(D)$ curve defines the boundary of the rate-distortion region and it is above the tangent in Figure 9.2, we see that

$$R(D_s) - sD_s = \min_{\mathbf{Q}}[I(\mathbf{p}, \mathbf{Q}) - sD(\mathbf{p}, \mathbf{Q})]. \tag{9.57}$$

For each $s \leq 0$, if we can find a $\mathbf{Q}_s$ that achieves the above minimum, then the line passing through $(0, I(\mathbf{p}, \mathbf{Q}_s) - sD(\mathbf{p}, \mathbf{Q}_s))$, i.e., the tangent in Figure 9.2,

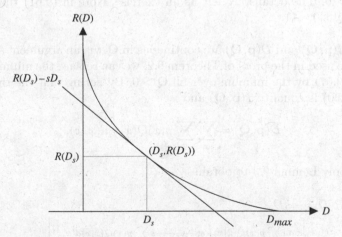

Fig. 9.2. A tangent to the $R(D)$ curve with slope equal to s.

---

[1] We say that a line is a tangent to the $R(D)$ curve if it touches the $R(D)$ curve from below.

gives a tight lower bound on the $R(D)$ curve. In particular, if $(R(D_s), D_s)$ is unique,

$$D_s = D(\mathbf{p}, \mathbf{Q}_s) \tag{9.58}$$

and

$$R(D_s) = I(\mathbf{p}, \mathbf{Q}_s). \tag{9.59}$$

By varying over all $s \leq 0$, we can then trace out the whole $R(D)$ curve. In the rest of the section, we will devise an iterative algorithm for the minimization problem in (9.57).

Lemma 9.3. *Let $p(x)Q(\hat{x}|x)$ be a given joint distribution on $\mathcal{X} \times \hat{\mathcal{X}}$ such that $\mathbf{Q} > 0$, and let $\mathbf{t}$ be any distribution on $\hat{\mathcal{X}}$ such that $\mathbf{t} > 0$. Then*

$$\min_{\mathbf{t}>0} \sum_x \sum_{\hat{x}} p(x)Q(\hat{x}|x) \log \frac{Q(\hat{x}|x)}{t(\hat{x})} = \sum_x \sum_{\hat{x}} p(x)Q(\hat{x}|x) \log \frac{Q(\hat{x}|x)}{t^*(\hat{x})}, \tag{9.60}$$

where

$$t^*(\hat{x}) = \sum_x p(x)Q(\hat{x}|x), \tag{9.61}$$

i.e., the minimizing $\mathbf{t}$ is the one which corresponds to the input distribution $\mathbf{p}$ and the transition matrix $\mathbf{Q}$.

Proof. It suffices to prove that

$$\sum_x \sum_{\hat{x}} p(x)Q(\hat{x}|x) \log \frac{Q(\hat{x}|x)}{t(\hat{x})} \geq \sum_x \sum_{\hat{x}} p(x)Q(\hat{x}|x) \log \frac{Q(\hat{x}|x)}{t^*(\hat{x})} \tag{9.62}$$

for all $\mathbf{t} > 0$. The details are left as an exercise. Note in (9.61) that $\mathbf{t}^* > 0$ because $\mathbf{Q} > 0$. $\square$

Since $I(\mathbf{p}, \mathbf{Q})$ and $D(\mathbf{p}, \mathbf{Q})$ are continuous in $\mathbf{Q}$, via an argument similar to the one we used in the proof of Theorem 9.2, we can replace the minimum over all $\mathbf{Q}$ in (9.57) by the infimum over all $\mathbf{Q} > 0$. By noting that the right-hand side of (9.60) is equal to $I(\mathbf{p}, \mathbf{Q})$ and

$$D(\mathbf{p}, \mathbf{Q}) = \sum_x \sum_{\hat{x}} p(x)Q(\hat{x}|x)d(x, \hat{x}), \tag{9.63}$$

we can apply Lemma 9.3 to obtain

$$R(D_s) - sD_s$$

$$= \inf_{\mathbf{Q}>0} \left[\min_{\mathbf{t}>0} \sum_{x,\hat{x}} p(x)Q(\hat{x}|x) \log \frac{Q(\hat{x}|x)}{t(\hat{x})} - s \sum_{x,\hat{x}} p(x)Q(\hat{x}|x)d(x,\hat{x}) \right] \tag{9.64}$$

$$= \inf_{\mathbf{Q}>0} \min_{\mathbf{t}>0} \left[\sum_{x,\hat{x}} p(x)Q(\hat{x}|x) \log \frac{Q(\hat{x}|x)}{t(\hat{x})} - s \sum_{x,\hat{x}} p(x)Q(\hat{x}|x)d(x,\hat{x}) \right]. \tag{9.65}$$

Now in the double infimum in (9.14), let

$$f(\mathbf{Q}, \mathbf{t}) = \sum_x \sum_{\hat{x}} p(x) Q(\hat{x}|x) \log \frac{Q(\hat{x}|x)}{t(\hat{x})}$$

$$- s \sum_x \sum_{\hat{x}} p(x) Q(\hat{x}|x) d(x, \hat{x}), \tag{9.66}$$

$$A_1 = \left\{ (Q(\hat{x}|x), (x, \hat{x}) \in \mathcal{X} \times \hat{\mathcal{X}}) : Q(\hat{x}|x) > 0, \right.$$

$$\left. \sum_{\hat{x}} Q(\hat{x}|x) = 1 \text{ for all } x \in \mathcal{X} \right\}, \tag{9.67}$$

and

$$A_2 = \{ (t(\hat{x}), \hat{x} \in \hat{\mathcal{X}}) : t(\hat{x}) > 0 \text{ and } \sum_{\hat{x}} t(\hat{x}) = 1 \}, \tag{9.68}$$

with $\mathbf{Q}$ and $\mathbf{t}$ playing the roles of $\mathbf{u}_1$ and $\mathbf{u}_2$, respectively. Then A_1 is a subset of $\Re^{|\mathcal{X}||\hat{\mathcal{X}}|}$ and A_2 is a subset of $\Re^{|\hat{\mathcal{X}}|}$, and it can readily be checked that both A_1 and A_2 are convex. Since $s \leq 0$,

$$f(\mathbf{Q}, \mathbf{t})$$

$$= \sum_x \sum_{\hat{x}} p(x) Q(\hat{x}|x) \log \frac{Q(\hat{x}|x)}{t(\hat{x})} - s \sum_x \sum_{\hat{x}} p(x) Q(\hat{x}|x) d(x, \hat{x})$$

$$\tag{9.69}$$

$$\geq \sum_x \sum_{\hat{x}} p(x) Q(\hat{x}|x) \log \frac{Q(\hat{x}|x)}{t^*(\hat{x})} + 0 \tag{9.70}$$

$$= I(X; \hat{X}) \tag{9.71}$$

$$\geq 0. \tag{9.72}$$

Therefore, f is bounded from below.

The double infimum in (9.14) now becomes

$$\inf_{\mathbf{Q} \in A_1} \inf_{\mathbf{t} \in A_2} \left[\sum_x \sum_{\hat{x}} p(x) Q(\hat{x}|x) \log \frac{Q(\hat{x}|x)}{t(\hat{x})} - s \sum_x \sum_{\hat{x}} p(x) Q(\hat{x}|x) d(x, \hat{x}) \right],$$

$$\tag{9.73}$$

where the infimum over all $\mathbf{t} \in A_2$ is in fact a minimum. We then apply the alternating optimization algorithm described in Section 9.2 to compute f^*, the value of (9.73). First, we arbitrarily choose a *strictly positive* transition matrix in A_1 and let it be $\mathbf{Q}^{(0)}$. Then we define $\mathbf{t}^{(0)}$ and in general $\mathbf{t}^{(k)}$ for $k \geq 1$ by

$$t^{(k)}(\hat{x}) = \sum_x p(x) Q^{(k)}(\hat{x}|x) \tag{9.74}$$

in view of Lemma 9.3. In order to define $\mathbf{Q}^{(1)}$ and in general $\mathbf{Q}^{(k)}$ for $k \geq 1$, we need to find the $\mathbf{Q} \in A_1$ that minimizes f for a given $\mathbf{t} \in A_2$, where the constraints on $\mathbf{Q}$ are

$$Q(\hat{x}|x) > 0 \quad \text{for all } (x, \hat{x}) \in \mathcal{X} \times \hat{\mathcal{X}} \tag{9.75}$$

and

$$\sum_{\hat{x}} Q(\hat{x}|x) = 1 \quad \text{for all } x \in \mathcal{X}. \tag{9.76}$$

As we did for the computation of the channel capacity, we first ignore the positivity constraints in (9.75) when setting up the Lagrange multipliers. Then we obtain

$$Q(\hat{x}|x) = \frac{t(\hat{x})e^{sd(x,\hat{x})}}{\sum_{\hat{x}'} t(\hat{x}')e^{sd(x,\hat{x}')}} > 0. \tag{9.77}$$

The details are left as an exercise. We then define $\mathbf{Q}^{(k)}$ for $k \geq 1$ by

$$Q^{(k)}(\hat{x}|x) = \frac{t^{(k-1)}(\hat{x})e^{sd(x,\hat{x})}}{\sum_{\hat{x}'} t^{(k-1)}(\hat{x}')e^{sd(x,\hat{x}')}}. \tag{9.78}$$

It will be shown in the next section that $f^{(k)} = f(\mathbf{Q}^{(k)}, \mathbf{t}^{(k)}) \to f^*$ as $k \to \infty$. If there exists a unique point $(R(D_s), D_s)$ on the $R(D)$ curve such that the slope of a tangent at that point is equal to s, then

$$(I(\mathbf{p}, \mathbf{Q}^{(k)}), D(\mathbf{p}, \mathbf{Q}^{(k)})) \to (R(D_s), D_s). \tag{9.79}$$

Otherwise, $(I(\mathbf{p}, \mathbf{Q}^{(k)}), D(\mathbf{p}, \mathbf{Q}^{(k)}))$ is arbitrarily close to the segment of the $R(D)$ curve at which the slope is equal to s when k is sufficiently large. These facts are easily shown to be true.

9.3 Convergence

In this section, we first prove that if f is concave, then $f^{(k)} \to f^*$. We then apply this sufficient condition to prove the convergence of the BA algorithm for computing the channel capacity. The convergence of the BA algorithm for computing the rate-distortion function can be proved likewise. The details are omitted.

9.3.1 A Sufficient Condition

In the alternating optimization algorithm in Section 9.1, we see from (9.7) and (9.8) that

$$\mathbf{u}^{(k+1)} = (\mathbf{u}_1^{(k+1)}, \mathbf{u}_2^{(k+1)}) = (c_1(\mathbf{u}_2^{(k)}), c_2(c_1(\mathbf{u}_2^{(k)}))) \tag{9.80}$$

for $k \geq 0$. Define

$$\Delta f(\mathbf{u}) = f(c_1(\mathbf{u}_2), c_2(c_1(\mathbf{u}_2))) - f(\mathbf{u}_1, \mathbf{u}_2). \tag{9.81}$$

Then

$$f^{(k+1)} - f^{(k)} = f(\mathbf{u}^{(k+1)}) - f(\mathbf{u}^{(k)}) \tag{9.82}$$

$$= f(c_1(\mathbf{u}_2^{(k)}), c_2(c_1(\mathbf{u}_2^{(k)}))) - f(\mathbf{u}_1^{(k)}, \mathbf{u}_2^{(k)}) \tag{9.83}$$

$$= \Delta f(\mathbf{u}^{(k)}). \tag{9.84}$$

We will prove that f being concave is sufficient for $f^{(k)} \to f^*$. To this end, we first prove that if f is concave, then the algorithm cannot be trapped at $\mathbf{u}$ if $f(\mathbf{u}) < f^*$.

Lemma 9.4. *Let f be concave. If $f^{(k)} < f^*$, then $f^{(k+1)} > f^{(k)}$.*

Proof. We will prove that $\Delta f(\mathbf{u}) > 0$ for any $\mathbf{u} \in A$ such that $f(\mathbf{u}) < f^*$. Then if $f^{(k)} = f(\mathbf{u}^{(k)}) < f^*$, we see from (9.84) that

$$f^{(k+1)} - f^{(k)} = \Delta f(\mathbf{u}^{(k)}) > 0, \tag{9.85}$$

and the lemma is proved.

Consider any $\mathbf{u} \in A$ such that $f(\mathbf{u}) < f^*$. We will prove by contradiction that $\Delta f(\mathbf{u}) > 0$. Assume $\Delta f(\mathbf{u}) = 0$. Then it follows from (9.81) that

$$f(c_1(\mathbf{u}_2), c_2(c_1(\mathbf{u}_2))) = f(\mathbf{u}_1, \mathbf{u}_2). \tag{9.86}$$

Now we see from (9.5) that

$$f(c_1(\mathbf{u}_2), c_2(c_1(\mathbf{u}_2))) \geq f(c_1(\mathbf{u}_2), \mathbf{u}_2). \tag{9.87}$$

If $c_1(\mathbf{u}_2) \neq \mathbf{u}_1$, then

$$f(c_1(\mathbf{u}_2), \mathbf{u}_2) > f(\mathbf{u}_1, \mathbf{u}_2) \tag{9.88}$$

because $c_1(\mathbf{u}_2)$ is unique. Combining (9.87) and (9.88), we have

$$f(c_1(\mathbf{u}_2), c_2(c_1(\mathbf{u}_2))) > f(\mathbf{u}_1, \mathbf{u}_2), \tag{9.89}$$

which is a contradiction to (9.86). Therefore,

$$\mathbf{u}_1 = c_1(\mathbf{u}_2). \tag{9.90}$$

Using this, we see from (9.86) that

$$f(\mathbf{u}_1, c_2(\mathbf{u}_1)) = f(\mathbf{u}_1, \mathbf{u}_2), \tag{9.91}$$

which implies

$$\mathbf{u}_2 = c_2(\mathbf{u}_1). \tag{9.92}$$

because $c_2(c_1(\mathbf{u}_2))$ is unique.

Since $f(\mathbf{u}) < f^*$, there exists $\mathbf{v} \in A$ such that

$$f(\mathbf{u}) < f(\mathbf{v}). \tag{9.93}$$

Consider

$$\mathbf{v} - \mathbf{u} = (\mathbf{v}_1 - \mathbf{u}_1, 0) + (0, \mathbf{v}_2 - \mathbf{u}_2). \tag{9.94}$$

Let $\tilde{\mathbf{z}}$ be the unit vector in the direction of $\mathbf{v} - \mathbf{u}$, $\mathbf{z}_1$ be the unit vector in the direction of $(\mathbf{v}_1 - \mathbf{u}_1, 0)$, and $\mathbf{z}_2$ be the unit vector in the direction of $(\mathbf{v}_2 - \mathbf{u}_2, 0)$. Then

$$\|\mathbf{v} - \mathbf{u}\|\tilde{\mathbf{z}} = \|\mathbf{v}_1 - \mathbf{u}_1\|\mathbf{z}_1 + \|\mathbf{v}_2 - \mathbf{u}_2\|\mathbf{z}_2 \tag{9.95}$$

or

$$\tilde{\mathbf{z}} = \alpha_1 \mathbf{z}_1 + \alpha_2 \mathbf{z}_2, \tag{9.96}$$

where

$$\alpha_i = \frac{\|\mathbf{v}_i - \mathbf{u}_i\|}{\|\mathbf{v} - \mathbf{u}\|}, \tag{9.97}$$

$i = 1, 2$. Figure 9.3 is an illustration of the vectors $\mathbf{u}$, $\mathbf{v}$, $\tilde{\mathbf{z}}, \mathbf{z}_1$, and $\mathbf{z}_2$.

We see from (9.90) that f attains its maximum value at $\mathbf{u} = (\mathbf{u}_1, \mathbf{u}_2)$ when $\mathbf{u}_2$ is fixed. In particular, f attains its maximum value at $\mathbf{u}$ along the line passing through $(\mathbf{u}_1, \mathbf{u}_2)$ and $(\mathbf{v}_1, \mathbf{u}_2)$. Let $\bigtriangledown f$ denotes the gradient of f. Since f is continuous and has continuous partial derivatives, the directional derivative of f at $\mathbf{u}$ in the direction of $\mathbf{z}_1$ exists and is given by $\bigtriangledown f \cdot \mathbf{z}_1$. It follows from the concavity of f that f is concave along the line passing through $(\mathbf{u}_1, \mathbf{u}_2)$ and $(\mathbf{v}_1, \mathbf{u}_2)$. Since f attains its maximum value at $\mathbf{u}$, the derivative of f along the line passing through $(\mathbf{u}_1, \mathbf{u}_2)$ and $(\mathbf{v}_1, \mathbf{u}_2)$ vanishes. Then we see that

$$\bigtriangledown f \cdot \mathbf{z}_1 = 0. \tag{9.98}$$

Similarly, we see from (9.92) that

$$\bigtriangledown f \cdot \mathbf{z}_2 = 0. \tag{9.99}$$

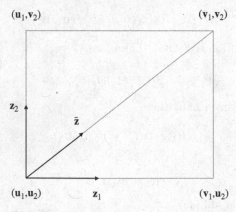

Fig. 9.3. The vectors $\mathbf{u}$, $\mathbf{v}$, $\tilde{\mathbf{z}}, \mathbf{z}_1$, and $\mathbf{z}_2$.

Then from (9.96), the directional derivative of f at $\mathbf{u}$ in the direction of $\tilde{\mathbf{z}}$ is given by

$$\triangledown f \cdot \tilde{\mathbf{z}} = \alpha_1(\triangledown f \cdot \mathbf{z}_1) + \alpha_2(\triangledown f \cdot \mathbf{z}_2) = 0. \tag{9.100}$$

Since f is concave along the line passing through $\mathbf{u}$ and $\mathbf{v}$, this implies

$$f(\mathbf{u}) \geq f(\mathbf{v}), \tag{9.101}$$

which is a contradiction to (9.93). Hence, we conclude that $\Delta f(\mathbf{u}) > 0$. $\square$

Although we have proved that the algorithm cannot be trapped at $\mathbf{u}$ if $f(\mathbf{u}) < f^*$, $f^{(k)}$ does not necessarily converge to f^* because the increment in $f^{(k)}$ in each step may be arbitrarily small. In order to prove the desired convergence, we will show in next theorem that this cannot be the case.

Theorem 9.5. *If f is concave, then $f^{(k)} \to f^*$.*

Proof. We have already shown in Section 9.1 that $f^{(k)}$ necessarily converges, say to f'. Hence, for any $\epsilon > 0$ and all sufficiently large k,

$$f' - \epsilon \leq f^{(k)} \leq f'. \tag{9.102}$$

Let

$$\gamma = \min_{\mathbf{u} \in A'} \Delta f(\mathbf{u}), \tag{9.103}$$

where

$$A' = \{\mathbf{u} \in A : f' - \epsilon \leq f(\mathbf{u}) \leq f'\}. \tag{9.104}$$

Since f has continuous partial derivatives, $\Delta f(\mathbf{u})$ is a continuous function of $\mathbf{u}$. Then the minimum in (9.103) exists because A' is compact.[2]

We now show that $f' < f^*$ will lead to a contradiction if f is concave. If $f' < f^*$, then from Lemma 9.4, we see that $\Delta f(\mathbf{u}) > 0$ for all $\mathbf{u} \in A'$ and hence $\gamma > 0$. Since $f^{(k)} = f(\mathbf{u}^{(k)})$ satisfies (9.102), $\mathbf{u}^{(k)} \in A'$, and

$$f^{(k+1)} - f^{(k)} = \Delta f(\mathbf{u}^{(k)}) \geq \gamma \tag{9.105}$$

for all sufficiently large k. Therefore, no matter how smaller γ is, $f^{(k)}$ will eventually be greater than f', which is a contradiction to $f^{(k)} \to f'$. Hence, we conclude that $f^{(k)} \to f^*$. $\square$

9.3.2 Convergence to the Channel Capacity

In order to show that the BA algorithm for computing the channel capacity converges as intended, i.e., $f^{(k)} \to C$, we only need to show that the function f defined in (9.39) is concave. Toward this end, for

---

[2] A' is compact because it is the inverse image of a closed interval under a continuous function and A is bounded.

$$f(\mathbf{r}, \mathbf{q}) = \sum_x \sum_y r(x) p(y|x) \log \frac{q(x|y)}{r(x)} \tag{9.106}$$

defined in (9.39), we consider two ordered pairs $(\mathbf{r}_1, \mathbf{q}_1)$ and $(\mathbf{r}_2, \mathbf{q}_2)$ in A, where A_1 and A_2 are defined in (9.40) and (9.41), respectively. For any $0 \le \lambda \le 1$ and $\bar\lambda = 1 - \lambda$, an application of the log-sum inequality (Theorem 2.32) gives

$$(\lambda r_1(x) + \bar\lambda r_2(x)) \log \frac{\lambda r_1(x) + \bar\lambda r_2(x)}{\lambda q_1(x|y) + \bar\lambda q_2(x|y)}$$
$$\le \lambda r_1(x) \log \frac{r_1(x)}{q_1(x|y)} + \bar\lambda r_2(x) \log \frac{r_2(x)}{q_2(x|y)}. \tag{9.107}$$

Taking reciprocal in the logarithms yields

$$(\lambda r_1(x) + \bar\lambda r_2(x)) \log \frac{\lambda q_1(x|y) + \bar\lambda q_2(x|y)}{\lambda r_1(x) + \bar\lambda r_2(x)}$$
$$\ge \lambda r_1(x) \log \frac{q_1(x|y)}{r_1(x)} + \bar\lambda r_2(x) \log \frac{q_2(x|y)}{r_2(x)}, \tag{9.108}$$

and upon multiplying by $p(y|x)$ and summing over all x and y, we obtain

$$f(\lambda \mathbf{r}_1 + \bar\lambda \mathbf{r}_2, \lambda \mathbf{q}_1 + \bar\lambda \mathbf{q}_2) \ge \lambda f(\mathbf{r}_1, \mathbf{q}_1) + \bar\lambda f(\mathbf{r}_2, \mathbf{q}_2). \tag{9.109}$$

Therefore, f is concave. Hence, we have shown that $f^{(k)} \to C$.

Chapter Summary

Channel Capacity: For a discrete memoryless channel $p(y|x)$,

$$C = \sup_{r>0} \max_{\mathbf{q}} \sum_x \sum_y r(x) p(y|x) \log \frac{q(x|y)}{r(x)},$$

where the maximization is taken over all $\mathbf{q}$ that satisfies $q(x|y) = 0$ if and only if $p(y|x) = 0$.

Computation of Channel Capacity: Start with any strictly positive input distribution $\mathbf{r}^{(0)}$. Compute $\mathbf{q}^{(0)}, \mathbf{r}^{(1)}, \mathbf{q}^{(1)}, \mathbf{r}^{(2)}, \cdots$ alternately by

$$q^{(k)}(x|y) = \frac{r^{(k)}(x) p(y|x)}{\sum_{x'} r^{(k)}(x') p(y|x')}$$

and

$$r^{(k)}(x) = \frac{\prod_y q^{(k-1)}(x|y)^{p(y|x)}}{\sum_{x'} \prod_y q^{(k-1)}(x'|y)^{p(y|x')}}.$$

Then $\mathbf{r}^{(k)}$ tends to the capacity-achieving input distribution as $k \to \infty$.

Rate-Distortion Function: For $s \leq 0$, the tangent to the rate-distortion function $R(D)$ at $(D_s, R(D_s))$ has slope s and intersects with the ordinate at $R(D_s) - sD_s$, which is given by

$$\inf_{Q > 0} \min_{t > 0} \left[\sum_{x, \hat{x}} p(x) Q(\hat{x}|x) \log \frac{Q(\hat{x}|x)}{t(\hat{x})} - s \sum_{x, \hat{x}} p(x) Q(\hat{x}|x) d(x, \hat{x}) \right].$$

The curve $R(D)$, $0 \leq D \leq D_{\max}$ is traced out by the collection of all such tangents.

Computation of Rate-Distortion Function: Start with any strictly positive transition matrix $\mathbf{Q}^{(0)}$. Compute $\mathbf{t}^{(0)}, \mathbf{Q}^{(1)}, \mathbf{t}^{(1)}, \mathbf{Q}^{(2)}, \cdots$ alternately by

$$t^{(k)}(\hat{x}) = \sum_{x} p(x) Q^{(k)}(\hat{x}|x)$$

and

$$Q^{(k)}(\hat{x}|x) = \frac{t^{(k-1)}(\hat{x}) e^{sd(x, \hat{x})}}{\sum_{\hat{x}'} t^{(k-1)}(\hat{x}') e^{sd(x, \hat{x}')}}.$$

Let

$$f(\mathbf{Q}, \mathbf{t}) = \sum_{x} \sum_{\hat{x}} p(x) Q(\hat{x}|x) \log \frac{Q(\hat{x}|x)}{t(\hat{x})} - s \sum_{x} \sum_{\hat{x}} p(x) Q(\hat{x}|x) d(x, \hat{x}).$$

Then $f(\mathbf{Q}^{(k)}, \mathbf{t}^{(k)}) \to R(D_s) - sD_s$ as $k \to \infty$.

Problems

1. Implement the BA algorithm for computing channel capacity.
2. Implement the BA algorithm for computing the rate-distortion function.
3. Explain why in the BA algorithm for computing channel capacity, we should not choose an initial input distribution which contains zero probability masses.
4. Prove Lemma 9.3.
5. Consider $f(\mathbf{Q}, \mathbf{t})$ in the BA algorithm for computing the rate-distortion function.

    a) Show that for fixed s and $\mathbf{t}$, $f(\mathbf{Q}, \mathbf{t})$ is minimized by

    $$Q(\hat{x}|x) = \frac{t(\hat{x}) e^{sd(x, \hat{x})}}{\sum_{\hat{x}'} t(\hat{x}') e^{sd(x, \hat{x}')}}.$$

    b) Show that $f(\mathbf{Q}, \mathbf{t})$ is convex.

Historical Notes

An iterative algorithm for computing the channel capacity was developed by Arimoto [19], where the convergence of the algorithm was proved. Blahut [37] independently developed two similar algorithms, the first for computing the channel capacity and the second for computing the rate-distortion function. The convergence of Blahut's second algorithm was proved by Csiszár [81]. These two algorithms are now commonly referred to as the Blahut–Arimoto algorithms. The simplified proof of convergence in this chapter is based on Yeung and Berger [404].

The Blahut–Arimoto algorithms are special cases of a general iterative algorithm due to Csiszár and Tusnády [88] which also include the expectation–maximization (EM) algorithm for fitting models from incomplete data [93] and the algorithm for finding the log-optimal portfolio for a stock market due to Cover [74].

10

Differential Entropy

Our discussion in the previous chapters involved only discrete random variables. The actual values taken by these random variables did not play any role in establishing the results. In this chapter and the next, our discussion will involve random variables taking real values. The values taken by these random variables do play a crucial role in the discussion.

Let X be a real random variable with *cumulative distribution function* (CDF) $F_X(x) = \Pr\{X \leq x\}$, which by definition is right-continuous. The random variable X is said to be

- *discrete* if $F_X(x)$ increases only at a countable number of values of x;
- *continuous* if $F_X(x)$ is continuous or, equivalently, $\Pr\{X = x\} = 0$ for every value of x;
- *mixed* if it is neither discrete nor continuous.

The support of X, denoted by $\mathcal{S}_X$, is the set of all x such that $F_X(x) > F_X(x - \epsilon)$ for all $\epsilon > 0$. For a function g defined on $\mathcal{S}_X$, we write

$$Eg(X) = \int_{\mathcal{S}_X} g(x) dF_X(x), \tag{10.1}$$

where the right-hand side is a *Lebesgue–Stieltjes integration* which covers all cases (i.e., discrete, continuous, and mixed) for the CDF $F_X(x)$. It may be regarded as a notation for the expectation of $g(X)$ with respect to $F_X(x)$ if the reader is not familiar with measure theory.

A nonnegative function $f_X(x)$ is called a *probability density function* (pdf) of X if

$$F_X(x) = \int_{-\infty}^{x} f_X(u) du \tag{10.2}$$

for all x. Since

$$\int f_X(x) dx = F_X(\infty) = 1 < \infty, \tag{10.3}$$

a pdf $f_X(x)$ can possibly take infinite value only on a set with zero Lebesgue measure. Therefore, we can assume without loss of generality that $f_X(x)$ instead takes any finite values on this set. If X has a pdf, then X is continuous, but not vice versa.

Let X and Y be two real random variables with joint CDF $F_{XY}(x, y) = \Pr\{X \le x, Y \le y\}$. The marginal CDF of X is given by $F_X(x) = F_{XY}(x, \infty)$ (likewise for Y). A nonnegative function $f_{XY}(x, y)$ is called a joint pdf of X and Y if

$$F_{XY}(x, y) = \int_{-\infty}^{x} \int_{-\infty}^{y} f_{XY}(u, v)\, dv du \qquad (10.4)$$

for all x and y. As for the case of a single random variable, we can assume without loss of generality that a joint pdf $f_{XY}(x, y)$ is finite for all x and y. For $x \in S_X$, the conditional CDF of Y given $\{X = x\}$ is defined as

$$F_{Y|X}(y|x) = \int_{-\infty}^{y} f_{Y|X}(v|x) dv, \qquad (10.5)$$

where

$$f_{Y|X}(y|x) = \frac{f_{XY}(x, y)}{f_X(x)} \qquad (10.6)$$

is the conditional pdf of Y given $\{X = x\}$.

All the above definitions and notations naturally extend to more than two real random variables. When there is no ambiguity, the subscripts specifying the random variables will be omitted.

All the random variables in this chapter are assumed to be real.[1] The *variance* of a random variable X is defined as

$$\mathrm{var}X = E(X - EX)^2 = EX^2 - (EX)^2. \qquad (10.7)$$

The *covariance* between two random variables X and Y is defined as

$$\mathrm{cov}(X, Y) = E(X - EX)(Y - EY) = E(XY) - (EX)(EY). \qquad (10.8)$$

For a random vector $\mathbf{X} = [X_1 X_2 \cdots X_n]^{\top}$, the covariance matrix is defined as

$$K_{\mathbf{X}} = E(\mathbf{X} - E\mathbf{X})(\mathbf{X} - E\mathbf{X})^{\top} = [\mathrm{cov}(X_i, X_j)], \qquad (10.9)$$

and the correlation matrix is defined as

$$\tilde{K}_{\mathbf{X}} = E\mathbf{X}\mathbf{X}^{\top} = [EX_i X_j]. \qquad (10.10)$$

---

[1] For a discrete random variable X with a countable alphabet $\mathcal{X}$, by replacing $\mathcal{X}$ by any countable subset of $\Re$, all information measures involving X (and possibly other random variables) are unchanged. Therefore, we assume without loss of generality that a discrete random variable is real.

Then

$$K_{\mathbf{X}} = E(\mathbf{X} - E\mathbf{X})(\mathbf{X} - E\mathbf{X})^{\top} \tag{10.11}$$

$$= E[\mathbf{X}\mathbf{X}^{\top} - \mathbf{X}(E\mathbf{X}^{\top}) - (E\mathbf{X})\mathbf{X}^{\top} + (E\mathbf{X})(E\mathbf{X}^{\top})] \tag{10.12}$$

$$= E\mathbf{X}\mathbf{X}^{\top} - (E\mathbf{X})(E\mathbf{X}^{\top}) - (E\mathbf{X})(E\mathbf{X}^{\top}) + (E\mathbf{X})(E\mathbf{X}^{\top}) \tag{10.13}$$

$$= E\mathbf{X}\mathbf{X}^{\top} - (E\mathbf{X})(E\mathbf{X}^{\top}) \tag{10.14}$$

$$= \tilde{K}_{\mathbf{X}} - (E\mathbf{X})(E\mathbf{X})^{\top}. \tag{10.15}$$

This implies that if $E\mathbf{X} = 0$, then $\tilde{K}_{\mathbf{X}} = K_{\mathbf{X}}$. On the other hand, it can readily be verified that

$$K_{\mathbf{X}} = \tilde{K}_{\mathbf{X}-E\mathbf{X}}. \tag{10.16}$$

Therefore, a covariance matrix is a correlation matrix. When there is no ambiguity, the subscripts in $K_{\mathbf{X}}$ and $\tilde{K}_{\mathbf{X}}$ will be omitted.

Let $\mathcal{N}(\mu, \sigma^2)$ denote the Gaussian distribution with mean μ and variance σ^2, i.e., the pdf of the distribution is given by

$$f(x) = \frac{1}{\sqrt{2\pi\sigma^2}} e^{-\frac{(x-\mu)^2}{2\sigma^2}} \tag{10.17}$$

for $-\infty < x < \infty$. More generally, let $\mathcal{N}(\boldsymbol{\mu}, K)$ denote the multivariate Gaussian distribution with mean $\boldsymbol{\mu}$ and covariance matrix K, i.e., the joint pdf of the distribution is given by

$$f(\mathbf{x}) = \frac{1}{\left(\sqrt{2\pi}\right)^n |K|^{1/2}} e^{-\frac{1}{2}(\mathbf{x}-\boldsymbol{\mu})^{\top} K^{-1}(\mathbf{x}-\boldsymbol{\mu})} \tag{10.18}$$

for all $\mathbf{x} \in \Re^n$, where K is a symmetric positive definite matrix[2] and $|K|$ is the determinant of K.

In the rest of the chapter, we will define various information measures under suitable conditions. Whenever these information measures are subsequently used, they are assumed to be defined.

10.1 Preliminaries

In this section, we present some preliminary results on matrices and linear transformation of random variables. All vectors and matrices are assumed to be real.

Definition 10.1. *A square matrix K is symmetric if $K^{\top} = K$.*

---

[2] See Definitions 10.1 and 10.2.

Definition 10.2. *An $n \times n$ matrix K is positive definite if*

$$\mathbf{x}^\top K \mathbf{x} > 0 \tag{10.19}$$

for all nonzero column n-vector $\mathbf{x}$ and is positive semidefinite if

$$\mathbf{x}^\top K \mathbf{x} \geq 0 \tag{10.20}$$

for all column n-vector $\mathbf{x}$.

Proposition 10.3. *A covariance matrix is both symmetric and positive semidefinite.*

Proof. Omitted. □

If a matrix K is symmetric, it can be diagonalized as

$$K = Q \Lambda Q^\top, \tag{10.21}$$

where Λ is a diagonal matrix and Q (also $Q^\top$) is an orthogonal matrix, i.e.,

$$Q^{-1} = Q^\top \tag{10.22}$$

or

$$QQ^\top = Q^\top Q = I. \tag{10.23}$$

The latter says that the rows (columns) of Q form an orthonormal system. Since

$$|Q|^2 = |Q||Q^\top| = |QQ^\top| = |I| = 1, \tag{10.24}$$

we have

$$|Q| = |Q^\top| = 1. \tag{10.25}$$

If (10.21) holds, we also say that $Q \Lambda Q^\top$ is a diagonalization of K.

From (10.21) and (10.23), we have

$$KQ = (Q \Lambda Q^\top)Q = Q \Lambda (Q^\top Q) = Q \Lambda. \tag{10.26}$$

Let λ_i and $\mathbf{q}_i \neq 0$ denote the ith diagonal element of Λ and the ith column of Q, respectively. Then (10.26) can be written as

$$K \mathbf{q}_i = \lambda_i \mathbf{q}_i \tag{10.27}$$

for all i, i.e., $\mathbf{q}_i$ is an eigenvector of K with eigenvalue λ_i. The next proposition further shows that these eigenvalues are nonnegative if K is positive semidefinite.

Proposition 10.4. *The eigenvalues of a positive semidefinite matrix are non-negative.*

Proof. Let K be a positive semidefinite matrix, and let $\mathbf{q} \neq 0$ be an eigenvector of K with eigenvalue λ, i.e.,

$$K\mathbf{q} = \lambda\mathbf{q}. \tag{10.28}$$

Since K is positive semidefinite,

$$0 \leq \mathbf{q}^\top K\mathbf{q} = \mathbf{q}^\top(\lambda\mathbf{q}) = \lambda(\mathbf{q}^\top\mathbf{q}). \tag{10.29}$$

Then we conclude that $\lambda \geq 0$ because $\mathbf{q}^\top\mathbf{q} > 0$. $\square$

The above discussions on diagonalization apply to a covariance matrix because a covariance matrix is both symmetric and positive semidefinite. As we will see, by diagonalizing the covariance matrix, a set of correlated random variables can be decorrelated by an orthogonal transformation. On the other hand, a set of correlated random variables can be regarded as an orthogonal transformation of a set of uncorrelated random variables. This is particularly important in the context of Gaussian random variables because a set of jointly distributed Gaussian random variables are mutually independent if and only if they are uncorrelated.

Proposition 10.5. *Let* $\mathbf{Y} = A\mathbf{X}$, *where* $\mathbf{X}$ *and* $\mathbf{Y}$ *are column vectors of* n *random variables and* A *is an* $n \times n$ *matrix. Then*

$$K_\mathbf{Y} = AK_\mathbf{X}A^\top \tag{10.30}$$

and

$$\tilde{K}_\mathbf{Y} = A\tilde{K}_\mathbf{X}A^\top. \tag{10.31}$$

Proof. To prove (10.30), consider

$$K_\mathbf{Y} = E(\mathbf{Y} - E\mathbf{Y})(\mathbf{Y} - E\mathbf{Y})^\top \tag{10.32}$$

$$= E[A(\mathbf{X} - E\mathbf{X})][A(\mathbf{X} - E\mathbf{X})]^\top \tag{10.33}$$

$$= E[A(\mathbf{X} - E\mathbf{X})(\mathbf{X} - E\mathbf{X})^\top A^\top] \tag{10.34}$$

$$= A[E(\mathbf{X} - E\mathbf{X})(\mathbf{X} - E\mathbf{X})^\top]A^\top \tag{10.35}$$

$$= AK_\mathbf{X}A^\top. \tag{10.36}$$

The proof of (10.31) is similar. $\square$

Proposition 10.6. *Let* $\mathbf{X}$ *and* $\mathbf{Y}$ *be column vectors of* n *random variables such that*

$$\mathbf{Y} = Q^\top\mathbf{X}, \tag{10.37}$$

where $Q\Lambda Q^\top$ *is a diagonalization of* $K_\mathbf{X}$. *Then* $K_\mathbf{Y} = \Lambda$, *i.e., the random variables in* $\mathbf{Y}$ *are uncorrelated and* $\mathrm{var}\,Y_i = \lambda_i$, *the* i*th diagonal element of* Λ.

Remark The matrix $K_{\mathbf{X}}$ is positive semidefinite, so that λ_i, being an eigenvalue of $K_{\mathbf{X}}$, is nonnegative by Proposition 10.4, as required for being the variance of a random variable.

Proof of Proposition 10.6. By Proposition 10.5,

$$K_{\mathbf{Y}} = Q^{\top} K_{\mathbf{X}} Q \tag{10.38}$$

$$= Q^{\top}(Q\Lambda Q^{\top})Q \tag{10.39}$$

$$= (Q^{\top}Q)\Lambda(Q^{\top}Q) \tag{10.40}$$

$$= \Lambda. \tag{10.41}$$

Since $K_{\mathbf{Y}} = \Lambda$ is a diagonal matrix, the random variables in $\mathbf{Y}$ are uncorrelated. Furthermore, the variance of Y_i is given by the ith diagonal element of $K_{\mathbf{Y}} = \Lambda$, i.e., λ_i. The proposition is proved. $\square$

Corollary 10.7. *Let $\mathbf{X}$ be a column vector of n random variables such that $Q\Lambda Q^{\top}$ is a diagonalization of $K_{\mathbf{X}}$. Then*

$$\mathbf{X} = Q\mathbf{Y}, \tag{10.42}$$

where $\mathbf{Y}$ is the column vector of n uncorrelated random variables prescribed in Proposition 10.6.

Proposition 10.8. *Let $\mathbf{X}$, $\mathbf{Y}$, and $\mathbf{Z}$ be vectors of n random variables such that $\mathbf{X}$ and $\mathbf{Z}$ are independent and $\mathbf{Y} = \mathbf{X} + \mathbf{Z}$. Then*

$$K_{\mathbf{Y}} = K_{\mathbf{X}} + K_{\mathbf{Z}}. \tag{10.43}$$

Proof. Omitted. $\square$

In communication engineering, the second moment of a random variable X is very often referred to as the *energy* of X. The total energy of a random vector $\mathbf{X}$ is then equal to $E \sum_i X_i^2$. The following proposition shows that the total energy of a random vector is preserved by an orthogonal transformation.

Proposition 10.9. *Let $\mathbf{Y} = Q\mathbf{X}$, where $\mathbf{X}$ and $\mathbf{Y}$ are column vectors of n random variables and Q is an orthogonal matrix. Then*

$$E \sum_{i=1}^{n} Y_i^2 = E \sum_{i=1}^{n} X_i^2. \tag{10.44}$$

Proof. Consider

$$\sum_{i=1}^{n} Y_i^2 = \mathbf{Y}^\top \mathbf{Y} \tag{10.45}$$

$$= (Q\mathbf{X})^\top (Q\mathbf{X}) \tag{10.46}$$

$$= \mathbf{X}^\top (Q^\top Q)\mathbf{X} \tag{10.47}$$

$$= \mathbf{X}^\top \mathbf{X} \tag{10.48}$$

$$= \sum_{i=1}^{n} X_i^2. \tag{10.49}$$

The proposition is proved upon taking expectation on both sides.   □

10.2 Definition

We now introduce the *differential entropy* for continuous random variables as the analog of the entropy for discrete random variables.

Definition 10.10. *The differential entropy $h(X)$ of a continuous random variable X with pdf $f(x)$ is defined as*

$$h(X) = -\int_S f(x) \log f(x) dx = -E \log f(X). \tag{10.50}$$

The entropy of a discrete random variable X is a measure of the average amount of information contained in X or, equivalently, the average amount of uncertainty removed upon revealing the outcome of X. This was justified by the asymptotic achievability of the entropy bound for zero-error data compression discussed in Chapter 4 as well as the source coding theorem discussed in Chapter 5.

However, although entropy and differential entropy have similar mathematical forms, the latter does not serve as a measure of the average amount of information contained in a continuous random variable. In fact, a continuous random variable generally contains an infinite amount of information, as explained in the following example.

Example 10.11. Let X be uniformly distributed on $[0, 1)$. Then we can write

$$X = .X_1 X_2 X_3 \cdots, \tag{10.51}$$

the dyadic expansion of X, where $X_1, X_2, X_3, \cdots$ is a sequence of fair bits.[3] Then

---

[3] Fair bits refer to i.i.d. bits, each distributed uniformly on $\{0, 1\}$.

$$H(X) = H(X_1, X_2, X_3, \cdots) \tag{10.52}$$

$$= \sum_{i=1}^{\infty} H(X_i) \tag{10.53}$$

$$= \sum_{i=1}^{\infty} 1 \tag{10.54}$$

$$= \infty. \tag{10.55}$$

In the following, we give two examples in which the differential entropy can be evaluated explicitly.

Example 10.12 (Uniform Distribution). Let X be uniformly distributed on $[0, a)$. Then

$$h(X) = -\int_0^a \frac{1}{a} \log \frac{1}{a} dx = \log a. \tag{10.56}$$

From this example, we see immediately that $h(X) < 0$ if $a < 1$. This poses no contradiction because as we have mentioned, the differential entropy does not serve as a measure of the average amount of information contained in X. The physical meaning of differential entropy will be understood through the AEP for continuous random variables to be discussed in Section 10.4.

Example 10.13 (Gaussian Distribution). Let $X \sim \mathcal{N}(0, \sigma^2)$ and let e be the base of the logarithm. Then

$$h(X) = -\int f(x) \ln f(x) dx \tag{10.57}$$

$$= -\int f(x) \left(-\frac{x^2}{2\sigma^2} - \ln \sqrt{2\pi\sigma^2} \right) dx \tag{10.58}$$

$$= \frac{1}{2\sigma^2} \int x^2 f(x) dx + \ln \sqrt{2\pi\sigma^2} \int f(x) dx \tag{10.59}$$

$$= \frac{EX^2}{2\sigma^2} + \frac{1}{2} \ln(2\pi\sigma^2) \tag{10.60}$$

$$= \frac{\text{var} X + (EX)^2}{2\sigma^2} + \frac{1}{2} \ln(2\pi\sigma^2) \tag{10.61}$$

$$= \frac{\sigma^2 + 0}{2\sigma^2} + \frac{1}{2} \ln(2\pi\sigma^2) \tag{10.62}$$

$$= \frac{1}{2} + \frac{1}{2} \ln(2\pi\sigma^2) \tag{10.63}$$

$$= \frac{1}{2} \ln e + \frac{1}{2} \ln(2\pi\sigma^2) \tag{10.64}$$

$$= \frac{1}{2} \ln(2\pi e \sigma^2) \tag{10.65}$$

in nats. Changing the base of the logarithm to any chosen positive value, we obtain

$$h(X) = \frac{1}{2} \log(2\pi e \sigma^2). \tag{10.66}$$

The following two basic properties of differential entropy can readily be proved from the definition.

Theorem 10.14 (Translation).

$$h(X + c) = h(X). \tag{10.67}$$

Proof. Let $Y = X + c$. Then $f_Y(y) = f_X(y - c)$ and $S_Y = \{x + c : x \in S_X\}$. Letting $x = y - c$ in (10.50), we have

$$h(X) = -\int_{S_X} f_X(x) \log f_X(x) dx \tag{10.68}$$

$$= -\int_{S_Y} f_X(y - c) \log f_X(y - c) dy \tag{10.69}$$

$$= -\int_{S_Y} f_Y(y) \log f_Y(y) dy \tag{10.70}$$

$$= h(Y) \tag{10.71}$$

$$= h(X + c), \tag{10.72}$$

accomplishing the proof. □

Theorem 10.15 (Scaling). *For $a \neq 0$,*

$$h(aX) = h(X) + \log |a|. \tag{10.73}$$

Proof. Let $Y = aX$. Then $f_Y(y) = \frac{1}{|a|} f_X(\frac{y}{a})$ and $S_Y = \{ax : x \in S_X\}$. Letting $x = \frac{y}{a}$ in (10.50), we have

$$h(X) = -\int_{S_X} f_X(x) \log f_X(x) dx \tag{10.74}$$

$$= -\int_{S_Y} f_X\left(\frac{y}{a}\right) \log f_X\left(\frac{y}{a}\right) \frac{dy}{|a|} \tag{10.75}$$

$$= -\int_{S_Y} \frac{1}{|a|} f_X\left(\frac{y}{a}\right) \left[\log\left(\frac{1}{|a|} f_X\left(\frac{y}{a}\right)\right) + \log |a|\right] dy \tag{10.76}$$

$$= -\int_{S_Y} f_Y(y) \log f_Y(y) dy - \log |a| \int_{S_Y} f_Y(y) dy \tag{10.77}$$

$$= h(Y) - \log |a| \tag{10.78}$$

$$= h(aX) - \log |a|. \tag{10.79}$$

Hence,

$$h(aX) = h(X) + \log|a|, \tag{10.80}$$

accomplishing the proof. □

Example 10.16. We illustrate Theorems 10.14 and 10.15 by means of the Gaussian distribution. Let $X \sim \mathcal{N}(\mu_X, \sigma_X^2)$. By Theorem 10.14 (and Example 10.13),

$$h(X) = \frac{1}{2}\log(2\pi e \sigma_X^2). \tag{10.81}$$

Let $Y = aX$. Then $Y \sim \mathcal{N}(\mu_Y, \sigma_Y^2)$, where $\mu_Y = a\mu_X$ and $\sigma_Y^2 = a^2\sigma_X^2$. By (10.81),

$$h(Y) = \frac{1}{2}\log(2\pi e \sigma_Y^2) = \frac{1}{2}\log(2\pi e a^2 \sigma_X^2) = \frac{1}{2}\log(2\pi e \sigma_X^2) + \log|a|, \tag{10.82}$$

verifying Theorem 10.15.

Theorem 10.14 says that the differential entropy of a random variable is unchanged by translation. Theorem 10.15 says that the differential entropy of a random variable is generally changed by scaling. Specifically, if $|a| > 1$, the differential entropy is increased by $\log|a|$. If $|a| < 1$, the differential entropy is decreased by $-\log|a|$ (note that $-\log|a| > 0$). If $a = -1$, the differential entropy is unchanged.

These properties suggest that the differential entropy of a random variable depends only on the "spread" of the pdf. More specifically, the differential entropy increases with the "spread" of the pdf. This point will be further elaborated in Section 10.6.

10.3 Joint Differential Entropy, Conditional (Differential) Entropy, and Mutual Information

The definition for differential entropy is readily extended to multiple continuous random variables. In the rest of the chapter, we let $\mathbf{X} = [X_1\, X_2 \cdots X_n]$.

Definition 10.17. *The joint differential entropy $h(\mathbf{X})$ of a random vector $\mathbf{X}$ with joint pdf $f(\mathbf{x})$ is defined as*

$$h(\mathbf{X}) = -\int_{\mathcal{S}} f(\mathbf{x})\log f(\mathbf{x})d\mathbf{x} = -E\log f(\mathbf{X}). \tag{10.83}$$

It follows immediately from the above definition that if $X_1, X_2, \cdots, X_n$ are mutually independent, then

$$h(\mathbf{X}) = \sum_{i=1}^{n} h(X_i). \tag{10.84}$$

The following two theorems are straightforward generalizations of Theorems 10.14 and 10.15, respectively. The proofs are omitted.

Theorem 10.18 (Translation). *Let* **c** *be a column vector in* $\Re^n$. *Then*

$$h(\mathbf{X} + \mathbf{c}) = h(\mathbf{X}).\tag{10.85}$$

Theorem 10.19 (Scaling). *Let* A *be a nonsingular* $n \times n$ *matrix. Then*

$$h(A\mathbf{X}) = h(\mathbf{X}) + \log|\det(A)|.\tag{10.86}$$

Theorem 10.20 (Multivariate Gaussian Distribution). *Let* $\mathbf{X} \sim \mathcal{N}(\boldsymbol{\mu}, K)$. *Then*

$$h(\mathbf{X}) = \frac{1}{2}\log\left[(2\pi e)^n|K|\right].\tag{10.87}$$

Proof. Let K be diagonalizable as $Q\Lambda Q^\top$. Write $\mathbf{X} = Q\mathbf{Y}$ as in Corollary 10.7, where the random variables in $\mathbf{Y}$ are uncorrelated with $\operatorname{var} Y_i = \lambda_i$, the ith diagonal element of Λ. Since $\mathbf{X}$ is Gaussian, so is $\mathbf{Y}$. Then the random variables in $\mathbf{Y}$ are mutually independent because they are uncorrelated. Now consider

$$h(\mathbf{X}) = h(Q\mathbf{Y})\tag{10.88}$$

$$\overset{a)}{=} h(\mathbf{Y}) + \log|\det(Q)|\tag{10.89}$$

$$\overset{b)}{=} h(\mathbf{Y}) + 0\tag{10.90}$$

$$\overset{c)}{=} \sum_{i=1}^{n} h(Y_i)\tag{10.91}$$

$$\overset{d)}{=} \sum_{i=1}^{n} \frac{1}{2}\log(2\pi e\lambda_i)\tag{10.92}$$

$$= \frac{1}{2}\log\left[(2\pi e)^n \prod_{i=1}^{n} \lambda_i\right]\tag{10.93}$$

$$\overset{e)}{=} \frac{1}{2}\log[(2\pi e)^n|\Lambda|]\tag{10.94}$$

$$\overset{f)}{=} \frac{1}{2}\log[(2\pi e)^n|K|].\tag{10.95}$$

In the above

(a) follows from Theorem 10.19;
(b) follows from (10.25);

(c) follows from (10.84) since $Y_1, Y_2, \cdots, Y_n$ are mutually independent;
(d) follows from Example 10.16;
(e) follows because Λ is a diagonal matrix;
(f) follows because

$$|\Lambda| = |Q||\Lambda||Q^\top| = |Q\Lambda Q^\top| = |K|. \qquad (10.96)$$

The theorem is proved. □

In describing a communication system, we very often specify the relation between two random variables X and Y through a conditional distribution $p(y|x)$ (if Y is discrete) or a conditional pdf $f(y|x)$ (if Y is continuous) defined for all x, even though certain x may not be in $\mathcal{S}_X$. This is made precise by the following two definitions.

Definition 10.21. *Let X and Y be two jointly distributed random variables with Y being discrete. The random variable Y is related to the random variable X through a conditional distribution $p(y|x)$ defined for all x means that for all x and y,*

$$\Pr\{X \leq x, Y = y\} = \int_{-\infty}^{x} p_{Y|X}(y|u)dF_X(u). \qquad (10.97)$$

Definition 10.22. *Let X and Y be two jointly distributed random variables with Y being continuous. The random variable Y is related to the random variable X through a conditional pdf $f(y|x)$ defined for all x means that for all x and y,*

$$F_{XY}(x,y) = \int_{-\infty}^{x} F_{Y|X}(y|u)dF_X(u), \qquad (10.98)$$

where

$$F_{Y|X}(y|x) = \int_{-\infty}^{y} f_{Y|X}(v|x)dv. \qquad (10.99)$$

Definition 10.23. *Let X and Y be jointly distributed random variables where Y is continuous and is related to X through a conditional pdf $f(y|x)$ defined for all x. The conditional differential entropy of Y given $\{X = x\}$ is defined as*

$$h(Y|X = x) = -\int_{\mathcal{S}_Y(x)} f(y|x) \log f(y|x)dy, \qquad (10.100)$$

where $\mathcal{S}_Y(x) = \{y : f(y|x) > 0\}$, and the conditional differential entropy of Y given X is defined as

$$h(Y|X) = -\int_{S_X} h(Y|X = x)dF(x) = -E\log f(Y|X).$$ (10.101)

Proposition 10.24. *Let X and Y be jointly distributed random variables where Y is continuous and is related to X through a conditional pdf $f(y|x)$ defined for all x. Then $f(y)$ exists and is given by*

$$f(y) = \int f(y|x)dF(x).$$ (10.102)

Proof. From (10.98) and (10.99), we have

$$F_Y(y) = F_{XY}(\infty, y) = \int\int_{-\infty}^{y} f_{Y|X}(v|x)\,dv\,dF(x).$$ (10.103)

Since $f_{Y|X}(v|x)$ is nonnegative and

$$\int\int_{-\infty}^{y} f_{Y|X}(v|x)\,dv\,dF(x) \le \int\int f_{Y|X}(v|x)\,dv\,dF(x)$$ (10.104)

$$= \int dF(x)$$ (10.105)

$$= 1,$$ (10.106)

$f_{Y|X}(v|x)$ is absolutely integrable. By Fubini's theorem,[4] the order of integration in the iterated integral in (10.103) can be exchanged. Therefore,

$$F_Y(y) = \int_{-\infty}^{y}\left[\int f_{Y|X}(v|x)dF(x)\right]dv,$$ (10.107)

implying (10.102) (cf. (10.2)). The proposition is proved.   □

The above proposition says that if Y is related to X through a conditional pdf $f(y|x)$, then the pdf of Y exists regardless of the distribution of X. The next proposition is a generalization to random vectors, and the proof is omitted. The theory in the rest of this chapter and in the next chapter will be developed around this important fact.

Proposition 10.25. *Let $\mathbf{X}$ and $\mathbf{Y}$ be jointly distributed random vectors where $\mathbf{Y}$ is continuous and is related to $\mathbf{X}$ through a conditional pdf $f(\mathbf{y}|\mathbf{x})$ defined for all $\mathbf{x}$. Then $f(\mathbf{y})$ exists and is given by*

$$f(\mathbf{y}) = \int f(\mathbf{y}|\mathbf{x})dF(\mathbf{x}).$$ (10.108)

---

[4] See for example [314].

Definition 10.26. *Let X and Y be jointly distributed random variables where Y is continuous and is related to X through a conditional pdf $f(y|x)$ defined for all x. The mutual information between X and Y is defined as*

$$I(X;Y) = \int_{S_X} \int_{S_Y(x)} f(y|x) \log \frac{f(y|x)}{f(y)} dy \, dF(x) \qquad (10.109)$$

$$= E \log \frac{f(Y|X)}{f(Y)}, \qquad (10.110)$$

where $f(y)$ exists and is given in (10.102) by Proposition 10.24. When both X and Y are continuous and $f(x,y)$ exists,

$$I(X;Y) = E \log \frac{f(Y|X)}{f(Y)} = E \log \frac{f(X,Y)}{f(X)f(Y)}. \qquad (10.111)$$

Together with our discussion on discrete random variables in Chapter 2, the mutual information $I(X;Y)$ is defined when each of the random variables involved can be either discrete or continuous. In the same way, we can define the conditional mutual information $I(X;Y|T)$.

Definition 10.27. *Let X, Y, and T be jointly distributed random variables where Y is continuous and is related to (X,T) through a conditional pdf $f(y|x,t)$ defined for all x and t. The mutual information between X and Y given T is defined as*

$$I(X;Y|T) = \int_{S_T} I(X;Y|T = t) dF(t) = E \log \frac{f(Y|X,T)}{f(Y|T)}, \qquad (10.112)$$

where

$$I(X;Y|T = t) = \int_{S_X(t)} \int_{S_Y(x,t)} f(y|x,t) \log \frac{f(y|x,t)}{f(y|t)} dy \, dF(x|t). \qquad (10.113)$$

We now give a physical interpretation of $I(X;Y)$ when X and Y have a joint pdf $f(x,y)$. For simplicity, we assume that $f(x,y) > 0$ for all x and y. Let Δ be a small positive quantity. For all integer i, define the interval

$$A_x^i = [i\Delta, (i+1)\Delta) \qquad (10.114)$$

in $\Re$, and for all integer j, define the interval

$$A_y^j = [j\Delta, (j+1)\Delta). \qquad (10.115)$$

For all integers i and j, define the set

$$A_{xy}^{i,j} = A_x^i \times A_y^j, \qquad (10.116)$$

which corresponds to a rectangle in $\Re^2$.

We now introduce a pair of discrete random variables $\hat{X}_\Delta$ and $\hat{Y}_\Delta$ defined by

$$\begin{cases} \hat{X}_\Delta = i \text{ if } X \in A_x^i \\ \hat{Y}_\Delta = j \text{ if } Y \in A_y^j \end{cases}. \tag{10.117}$$

The random variables $\hat{X}_\Delta$ and $\hat{Y}_\Delta$ are quantizations of the continuous random variables X and Y, respectively. For all i and j, let $(x_i, y_j) \in A_{xy}^{i,j}$. Then

$$I(\hat{X}_\Delta; \hat{Y}_\Delta)$$

$$= \sum_i \sum_j \Pr\{(\hat{X}_\Delta, \hat{Y}_\Delta) = (i,j)\} \log \frac{\Pr\{(\hat{X}_\Delta, \hat{Y}_\Delta) = (i,j)\}}{\Pr\{\hat{X}_\Delta = i\}\Pr\{\hat{Y}_\Delta = j\}} \tag{10.118}$$

$$\approx \sum_i \sum_j f(x_i, y_j)\Delta^2 \log \frac{f(x_i, y_j)\Delta^2}{(f(x_i)\Delta)(f(y_j)\Delta)} \tag{10.119}$$

$$= \sum_i \sum_j f(x_i, y_j)\Delta^2 \log \frac{f(x_i, y_j)}{f(x_i)f(y_j)} \tag{10.120}$$

$$\approx \int \int f(x, y) \log \frac{f(x, y)}{f(x)f(y)} dx dy \tag{10.121}$$

$$= I(X; Y). \tag{10.122}$$

Therefore, $I(X; Y)$ can be interpreted as the limit of $I(\hat{X}_\Delta; \hat{Y}_\Delta)$ as $\Delta \to 0$. This interpretation carries over to the case when X and Y have a general joint distribution[5] (see Dobrushin [95]). As $I(\hat{X}_\Delta; \hat{Y}_\Delta)$ is always nonnegative, this suggests that $I(X; Y)$ is also always nonnegative, which will be established in Theorem 10.31.

Definition 10.28. *Let Y be a continuous random variable and X be a discrete random variable, where Y is related to X through a conditional pdf $f(y|x)$. The conditional entropy of X given Y is defined as*

$$H(X|Y) = H(X) - I(X; Y), \tag{10.123}$$

where $I(X; Y)$ is defined as in Definition 10.26.

Proposition 10.29. *For two random variables X and Y,*

---

[5] In the general setting, the mutual information between X and Y is defined as

$$I(X; Y) = \int_{S_{XY}} \left(\log \frac{dP_{XY}}{d(P_X \times P_Y)} \right) dP_{XY},$$

where P_{XY}, P_X, and P_Y are the probability measures of (X, Y), X, and Y, respectively, and $\frac{dP_{XY}}{d(P_X \times P_Y)}$ denotes the Radon–Nikodym derivative of P_{XY} with respect to the product measure $P_X \times P_Y$.

$$h(Y) = h(Y|X) + I(X;Y) \qquad (10.124)$$

if Y is continuous, and

$$H(Y) = H(Y|X) + I(X;Y) \qquad (10.125)$$

if Y is discrete.

Proposition 10.30 (Chain Rule for Differential Entropy).

$$h(X_1, X_2, \cdots, X_n) = \sum_{i=1}^{n} h(X_i|X_1, \cdots, X_{i-1}). \qquad (10.126)$$

The proofs of these propositions are left as an exercise.

Theorem 10.31.

$$I(X;Y) \geq 0, \qquad (10.127)$$

with equality if and only if X is independent of Y.

Proof. Consider

$$I(X;Y)$$

$$= \int_{S_X} \int_{S_Y(x)} f(y|x) \log \frac{f(y|x)}{f(y)} dy \, dF_X(x) \qquad (10.128)$$

$$\geq (\log e) \int_{S_X} \int_{S_Y(x)} f(y|x) \left(1 - \frac{f(y)}{f(y|x)}\right) dy \, dF_X(x) \qquad (10.129)$$

$$= (\log e) \int_{S_X} \left[\int_{S_Y(x)} f(y|x) dy - \int_{S_Y(x)} f(y) dy \right] dF_X(x) \qquad (10.130)$$

$$\geq 0, \qquad (10.131)$$

where (10.129) results from an application of the fundamental inequality (Corollary 2.30), and (10.131) follows from

$$\int_{S_Y(x)} f(y) dy \leq 1 = \int_{S_Y(x)} f(y|x) dy. \qquad (10.132)$$

This proves (10.127).

For equality to hold in (10.127), equality must hold in (10.129) for all $x \in S_X$ and all $y \in S_Y(x)$, and equality must hold in (10.131) for all $x \in S_X$. For the former, this is the case if and only if

$$f(y|x) = f(y) \quad \text{for all } x \in S_X \text{ and } y \in S_Y(x), \qquad (10.133)$$

which implies

$$\int_{S_Y(x)} f(y) dy = \int_{S_Y(x)} f(y|x) dy = 1, \qquad (10.134)$$

i.e., equality holds in (10.131). Thus (10.133) is a necessary and sufficient condition for equality to hold in (10.127).

It is immediate that if X and Y are independent, then (10.133) holds. It remains to prove the converse. To this end, observe that (10.134), implied by (10.133), is equivalent to that $f(y) = 0$ on $\mathcal{S}_Y \setminus \mathcal{S}_Y(x)$ a.e. (almost everywhere). By the definition of $\mathcal{S}_Y$, this means that $\mathcal{S}_Y \setminus \mathcal{S}_Y(x) \subset \mathcal{S}_Y^c$ or $\mathcal{S}_Y = \mathcal{S}_Y(x)$. Since this holds for all $x \in \mathcal{S}_X$, we conclude that $f(y|x) = f(y)$ for all $(x, y) \in \mathcal{S}_X \times \mathcal{S}_Y$, i.e., X and Y are independent. The theorem is proved. $\square$

Corollary 10.32.
$$I(X; Y|T) \geq 0, \tag{10.135}$$

with equality if and only if X is independent of Y conditioning on T.

Proof. This follows directly from (10.112). $\square$

Corollary 10.33 (Conditioning Does Not Increase Differential Entropy).
$$h(X|Y) \leq h(X) \tag{10.136}$$

with equality if and only if X and Y are independent.

Corollary 10.34 (Independence Bound for Differential Entropy).

$$h(X_1, X_2, \cdots, X_n) \leq \sum_{i=1}^{n} h(X_i) \tag{10.137}$$

with equality if and only if $i = 1, 2, \cdots, n$ are mutually independent.

10.4 The AEP for Continuous Random Variables

The weak AEP for discrete random variables discussed in Chapter 5 states that for n i.i.d. random variables $X_1, X_2, \cdots, X_n$ with generic discrete random variable X, $p(X_1, X_2, \cdots, X_n)$ is close to $2^{-nH(X)}$ with high probability when n is large (Theorem 5.1, Weak AEP I). This fundamental property of entropy leads to the definition of weak typicality, and as a consequence, the total number of weakly typical sequences is approximately equal to $2^{nH(X)}$ (Theorem 5.3, Weak AEP II).

In the following, we develop the AEP for continuous random variables in the same way we developed the weak AEP for discrete random variables. Some of the proofs are exactly the same as their discrete analogs, and they are omitted. We note that for continuous random variables, the notion of strong typicality does not apply because the probability that a continuous random variable takes a particular value is equal to zero.

Theorem 10.35 (AEP I for Continuous Random Variables).

$$-\frac{1}{n}\log f(\mathbf{X}) \to h(X) \tag{10.138}$$

in probability as $n \to \infty$, *i.e., for any* $\epsilon > 0$, *for* n *sufficiently large,*

$$\Pr\left\{\left|-\frac{1}{n}\log f(\mathbf{X}) - h(X)\right| < \epsilon\right\} > 1 - \epsilon. \tag{10.139}$$

Definition 10.36. *The typical set* $W_{[X]\epsilon}^n$ *with respect to* $f(x)$ *is the set of sequences* $\mathbf{x} = (x_1, x_2, \cdots, x_n) \in \mathcal{X}^n$ *such that*

$$\left|-\frac{1}{n}\log f(\mathbf{x}) - h(X)\right| < \epsilon \tag{10.140}$$

or, equivalently,

$$h(X) - \epsilon < -\frac{1}{n}\log f(\mathbf{x}) < h(X) + \epsilon, \tag{10.141}$$

where ϵ *is an arbitrarily small positive real number. The sequences in* $W_{[X]\epsilon}^n$ *are called* ϵ-*typical sequences.*

The quantity

$$-\frac{1}{n}\log f(\mathbf{x}) = -\frac{1}{n}\sum_{k=1}^n \log f(x_k) \tag{10.142}$$

is called the *empirical differential entropy* of the sequence $\mathbf{x}$. The empirical differential entropy of a typical sequence is close to the true differential entropy $h(X)$.

If the pdf $f(x)$ is continuous, we see from (10.142) that the empirical differential entropy is continuous in $\mathbf{x}$. Therefore, if $\mathbf{x}$ is ϵ-typical, then all the sequences in the neighborhood of $\mathbf{x}$ are also ϵ-typical. As a consequence, the number of ϵ-typical sequences is uncountable, and it is not meaningful to discuss the cardinality of a typical set as in the discrete case. Instead, the "size" of a typical set is measured by its volume.

Definition 10.37. *The volume of a set* A *in* $\Re^n$ *is defined as*

$$\text{Vol}(A) = \int_A d\mathbf{x}. \tag{10.143}$$

Theorem 10.38 (AEP II for Continuous Random Variables). *The following hold for any* $\epsilon > 0$:

1) If $\mathbf{x} \in W_{[X]\epsilon}^n$, *then*

$$2^{-n(h(X)+\epsilon)} < f(\mathbf{x}) < 2^{-n(h(X)-\epsilon)}. \tag{10.144}$$

2) For n sufficiently large,

$$\Pr\{\mathbf{X} \in W_{[X]\epsilon}^n\} > 1 - \epsilon. \tag{10.145}$$

3) For n sufficiently large,

$$(1 - \epsilon)2^{n(h(X)-\epsilon)} < \mathrm{Vol}(W_{[X]\epsilon}^n) < 2^{n(h(X)+\epsilon)}. \tag{10.146}$$

Proof. Property 1 follows immediately from the definition of $W_{[X]\epsilon}^n$ in (10.141). Property 2 is equivalent to Theorem 10.35. To prove Property 3, we use the lower bound in (10.144) and consider

$$1 \geq \Pr\{W_{[X]\epsilon}^n\} \tag{10.147}$$

$$= \int_{W_{[X]\epsilon}^n} f(\mathbf{x}) \, d\mathbf{x} \tag{10.148}$$

$$> \int_{W_{[X]\epsilon}^n} 2^{-n(h(X)+\epsilon)} \, d\mathbf{x} \tag{10.149}$$

$$> 2^{-n(h(X)+\epsilon)} \int_{W_{[X]\epsilon}^n} d\mathbf{x} \tag{10.150}$$

$$= 2^{-n(h(X)+\epsilon)} \mathrm{Vol}(W_{[X]\epsilon}^n), \tag{10.151}$$

which implies

$$\mathrm{Vol}(W_{[X]\epsilon}^n) < 2^{n(h(X)+\epsilon)}. \tag{10.152}$$

Note that this upper bound holds for any $n \geq 1$. On the other hand, using the upper bound in (10.144) and Theorem 10.35, for n sufficiently large, we have

$$1 - \epsilon < \Pr\{W_{[X]\epsilon}^n\} \tag{10.153}$$

$$= \int_{W_{[X]\epsilon}^n} f(\mathbf{x}) \, d\mathbf{x} \tag{10.154}$$

$$< \int_{W_{[X]\epsilon}^n} 2^{-n(h(X)-\epsilon)} \, d\mathbf{x} \tag{10.155}$$

$$= 2^{-n(h(X)-\epsilon)} \mathrm{Vol}(W_{[X]\epsilon}^n). \tag{10.156}$$

Then

$$\mathrm{Vol}(W_{[X]\epsilon}^n) > (1 - \epsilon)2^{n(h(X)-\epsilon)}. \tag{10.157}$$

Combining (10.152) and (10.157) gives Property 3. The theorem is proved. □

From the AEP for continuous random variables, we see that the volume of the typical set is approximately equal to $2^{nh(X)}$ when n is large. This gives the following physical interpretations of differential entropy. First, the fact that

$h(X)$ can be negative does not incur any difficulty because $2^{nh(X)}$ is always positive. Second, if the differential entropy is large, then the volume of the typical set is large; if the differential entropy is small (not in magnitude but in value), then the volume of the typical set is small.

10.5 Informational Divergence

We first extend the definition of informational divergence introduced in Section 2.5 to pdfs.

Definition 10.39. *Let f and g be two pdfs defined on $\Re^n$ with supports S_f and S_g, respectively. The informational divergence between f and g is defined as*

$$D(f\|g) = \int_{S_f} f(x) \log \frac{f(x)}{g(x)} dx = E_f \log \frac{f(X)}{g(X)}, \tag{10.158}$$

where E_f denotes expectation with respect to f.

Remark In the above definition, we adopt the convention $c \log \frac{c}{0} = \infty$ for $c > 0$. Therefore, if $D(f\|g) < \infty$, then

$$S_f \setminus S_g = \{x : f(x) > 0 \text{ and } g(x) = 0\} \tag{10.159}$$

has zero Lebesgue measure, i.e., S_f is essentially a subset of S_g.

Theorem 10.40 (Divergence Inequality). *Let f and g be two pdfs defined on $\Re^n$. Then*

$$D(f\|g) \geq 0, \tag{10.160}$$

with equality if and only if $f = g$ a.e.

Proof. Consider

$$D(f\|g) = \int_{S_f} f(x) \log \frac{f(x)}{g(x)} dx \tag{10.161}$$

$$= (\log e) \int_{S_f} f(x) \ln \frac{f(x)}{g(x)} dx \tag{10.162}$$

$$\geq (\log e) \int_{S_f} f(x) \left(1 - \frac{g(x)}{f(x)}\right) dx \tag{10.163}$$

$$= (\log e) \left[\int_{S_f} f(x) dx - \int_{S_f} g(x) dx\right] \tag{10.164}$$

$$\geq 0, \tag{10.165}$$

where (10.163) follows from the fundamental inequality (Corollary 2.30) and (10.165) follows from

$$\int_{\mathcal{S}_f} g(x)dx \leq 1 = \int_{\mathcal{S}_f} f(x)dx. \tag{10.166}$$

Equality holds in (10.163) if and only if $f(x) = g(x)$ on $\mathcal{S}_f$ a.e., which implies

$$\int_{\mathcal{S}_f} g(x)dx = \int_{\mathcal{S}_f} f(x)dx = 1, \tag{10.167}$$

i.e., equality holds in (10.165). Then we see from (10.167) that $g(x) = 0$ on $\mathcal{S}_f^c$ a.e. Hence, we conclude that equality holds in (10.160) if and only if $f = g$ a.e. The theorem is proved. $\square$

10.6 Maximum Differential Entropy Distributions

In Section 2.9, we have discussed maximum entropy distributions for a discrete random variable. We now extend this theme to multiple continuous random variables. Specifically, we are interested in the following problem:

Maximize $h(f)$ over all pdf f defined on a subset $\mathcal{S}$ of $\Re^n$, subject to

$$\int_{\mathcal{S}_f} r_i(\mathbf{x})f(\mathbf{x})d\mathbf{x} = a_i \quad \text{for } 1 \leq i \leq m, \tag{10.168}$$

where $\mathcal{S}_f \subset \mathcal{S}$ and $r_i(\mathbf{x})$ is defined for all $\mathbf{x} \in \mathcal{S}$.

Theorem 10.41. *Let*

$$f^*(\mathbf{x}) = e^{-\lambda_0 - \sum_{i=1}^m \lambda_i r_i(\mathbf{x})} \tag{10.169}$$

for all $\mathbf{x} \in \mathcal{S}$, where $\lambda_0, \lambda_1, \cdots, \lambda_m$ are chosen such that the constraints in (10.168) are satisfied. Then f^* maximizes $h(f)$ over all pdf f defined on $\mathcal{S}$, subject to the constraints in (10.168).

Proof. The proof is analogous to that of Theorem 2.50. The details are omitted. $\square$

Corollary 10.42. *Let f^* be a pdf defined on $\mathcal{S}$ with*

$$f^*(\mathbf{x}) = e^{-\lambda_0 - \sum_{i=1}^m \lambda_i r_i(\mathbf{x})} \tag{10.170}$$

for all $\mathbf{x} \in \mathcal{S}$. Then f^* maximizes $h(f)$ over all pdf f defined on $\mathcal{S}$, subject to the constraints

$$\int_{\mathcal{S}_f} r_i(\mathbf{x})f(\mathbf{x})d\mathbf{x} = \int_{\mathcal{S}} r_i(\mathbf{x})f^*(\mathbf{x})d\mathbf{x} \quad \text{for } 1 \leq i \leq m. \tag{10.171}$$

Theorem 10.43. *Let X be a continuous random variable with $EX^2 = \kappa$. Then*

$$h(X) \leq \frac{1}{2}\log(2\pi e\kappa), \tag{10.172}$$

with equality if and only if $X \sim \mathcal{N}(0, \kappa)$.

Proof. The problem here is to maximize $h(f)$ subject to the constraint

$$\int x^2 f(x)dx = \kappa. \tag{10.173}$$

An application of Theorem 10.41 yields

$$f^*(x) = ae^{-bx^2} \tag{10.174}$$

which is identified as a Gaussian distribution with zero mean. In order that the constraint (10.173) is satisfied, we must have

$$a = \frac{1}{\sqrt{2\pi\kappa}} \quad \text{and} \quad b = \frac{1}{2\kappa}. \tag{10.175}$$

Hence, in light of (10.66) in Example 10.13, we have proved (10.172) with equality if and only if $X \sim \mathcal{N}(0, \kappa)$. $\square$

Theorem 10.44. *Let X be a continuous random variable with mean μ and variance σ^2. Then*

$$h(X) \leq \frac{1}{2}\log(2\pi e\sigma^2), \tag{10.176}$$

with equality if and only if $X \sim \mathcal{N}(\mu, \sigma^2)$.

Proof. Let $X' = X - \mu$. Then

$$EX' = E(X - \mu) = EX - \mu = 0 \tag{10.177}$$

and

$$E(X')^2 = E(X - \mu)^2 = \text{var}X = \sigma^2. \tag{10.178}$$

Applying Theorem 10.14 and Theorem 10.43, we have

$$h(X) = h(X') \leq \frac{1}{2}\log(2\pi e\sigma^2), \tag{10.179}$$

and equality holds if and only if $X' \sim \mathcal{N}(0, \sigma^2)$ or $X \sim \mathcal{N}(\mu, \sigma^2)$. The theorem is proved. $\square$

Remark Theorem 10.43 says that with the constraint $EX^2 = \kappa$, the differential entropy is maximized by the distribution $\mathcal{N}(0, \kappa)$. If we impose the additional constraint that $EX = 0$, then $\text{var}X = EX^2 = \kappa$. By Theorem 10.44, the differential entropy is still maximized by $\mathcal{N}(0, \kappa)$.

We have mentioned at the end of Section 10.2 that the differential entropy of a random variable increases with the "spread" of the pdf. Though a simple consequence of Theorem 10.43, the above theorem makes this important interpretation precise. By rewriting the upper bound in (10.179), we obtain

$$h(X) \leq \log \sigma + \frac{1}{2} \log(2\pi e). \qquad (10.180)$$

That is, the differential entropy is at most equal to the logarithm of the standard deviation plus a constant. In particular, the differential entropy tends to $-\infty$ as the standard deviation tends to 0.

The next two theorems are the vector generalizations of Theorems 10.43 and 10.44.

Theorem 10.45. *Let* $\mathbf{X}$ *be a vector of* n *continuous random variables with correlation matrix* $\tilde{K}$. *Then*

$$h(\mathbf{X}) \leq \frac{1}{2} \log \left[(2\pi e)^n |\tilde{K}| \right], \qquad (10.181)$$

with equality if and only if $\mathbf{X} \sim \mathcal{N}(0, \tilde{K})$.

Proof. By Theorem 10.41, the joint pdf that maximizes $h(\mathbf{X})$ has the form

$$f^*(\mathbf{x}) = e^{-\lambda_0 - \sum_{i,j} \lambda_{ij} x_i x_j} = e^{-\lambda_0 - \mathbf{x}^\top L \mathbf{x}}, \qquad (10.182)$$

where $L = [\lambda_{ij}]$. Thus f^* is a multivariate Gaussian distribution with zero mean. Therefore,

$$\mathrm{cov}(X_i, X_j) = EX_i X_j - (EX_i)(EX_j) = EX_i X_j \qquad (10.183)$$

for all i and j. Since f^* is constrained by $\tilde{K}$, λ_0 and L have the unique solution given by

$$e^{-\lambda_0} = \frac{1}{\left(\sqrt{2\pi}\right)^n |\tilde{K}|^{1/2}} \qquad (10.184)$$

and

$$L = \frac{1}{2} \tilde{K}^{-1}, \qquad (10.185)$$

so that

$$f^*(\mathbf{x}) = \frac{1}{\left(\sqrt{2\pi}\right)^n |\tilde{K}|^{1/2}} e^{-\frac{1}{2} \mathbf{x}^\top \tilde{K}^{-1} \mathbf{x}}, \qquad (10.186)$$

the joint pdf of $\mathbf{X} \sim \mathcal{N}(0, \tilde{K})$. Hence, by Theorem 10.20, we have proved (10.181) with equality if and only if $\mathbf{X} \sim \mathcal{N}(0, \tilde{K})$. $\square$

Theorem 10.46. *Let* $\mathbf{X}$ *be a vector of* n *continuous random variables with mean* $\boldsymbol{\mu}$ *and covariance matrix* K. *Then*

$$h(\mathbf{X}) \leq \frac{1}{2} \log \left[(2\pi e)^n |K| \right], \qquad (10.187)$$

with equality if and only if $\mathbf{X} \sim \mathcal{N}(\boldsymbol{\mu}, K)$.

Proof. Similar to the proof of Theorem 10.44. $\square$

Chapter Summary

In the following, $\mathbf{X} = [X_1\,X_2\,\cdots\,X_n]^\top$.

Covariance Matrix: $K_\mathbf{X} = E(\mathbf{X} - E\mathbf{X})(\mathbf{X} - E\mathbf{X})^\top = [\mathrm{cov}(X_i, X_j)]$.

Correlation Matrix: $\tilde{K}_\mathbf{X} = E\mathbf{X}\mathbf{X}^\top = [EX_iX_j]$.

Diagonalization of a Covariance Matrix: A covariance matrix can be diagonalized as $Q\Lambda Q^\top$. The diagonal elements of Λ, which are nonnegative, are the eigenvalues of the covariance matrix.

Linear Transformation of a Random Vector: Let $\mathbf{Y} = A\mathbf{X}$. Then $K_\mathbf{Y} = AK_\mathbf{X}A^\top$ and $\tilde{K}_\mathbf{Y} = A\tilde{K}_\mathbf{X}A^\top$.

Decorrelation of a Random Vector: Let $\mathbf{Y} = Q^\top\mathbf{X}$, where $Q\Lambda Q^\top$ is a diagonalization of $K_\mathbf{X}$. Then $K_\mathbf{Y} = \Lambda$, i.e., the random variables in $\mathbf{Y}$ are uncorrelated and $\mathrm{var}\,Y_i = \lambda_i$, the ith diagonal element of Λ.

Differential Entropy:

$$h(X) = -\int_S f(x)\log f(x)dx = -E\log f(X) = -E\log f(X).$$

1. Translation: $h(X + c) = h(X)$.
2. Scaling: $h(aX) = h(X) + \log|a|$.
3. Uniform Distribution on $[0, a)$: $h(X) = \log a$.
4. Gaussian Distribution $\mathcal{N}(\mu, \sigma^2)$: $h(X) = \frac{1}{2}\log(2\pi e\sigma^2)$.

Joint Differential Entropy:

$$h(\mathbf{X}) = -\int_S f(\mathbf{x})\log f(\mathbf{x})d\mathbf{x} = -E\log f(\mathbf{X}).$$

1. Translation: $h(\mathbf{X} + \mathbf{c}) = h(\mathbf{X})$.
2. Scaling: $h(A\mathbf{X}) = h(\mathbf{X}) + \log|\det(A)|$.
3. Multivariate Gaussian Distribution $\mathcal{N}(\boldsymbol{\mu}, K)$: $h(\mathbf{X}) = \frac{1}{2}\log[(2\pi e)^n|K|]$.

Proposition: For fixed $f(y|x)$, $f(y)$ exists for any $F(x)$ and is given by

$$f(y) = \int f(y|x)dF(x).$$

Conditional (Differential) Entropy and Mutual Information:

1. If Y is continuous,

$$h(Y|X = x) = -\int_{\mathcal{S}_Y(x)} f(y|x) \log f(y|x) dy,$$

$$h(Y|X) = -\int_{\mathcal{S}_X} h(Y|X = x) dF(x) = -E \log f(Y|X),$$

$$I(X;Y) = \int_{\mathcal{S}_X} \int_{\mathcal{S}_Y(x)} f(y|x) \log \frac{f(y|x)}{f(y)} dy\, dF(x) = E \log \frac{f(Y|X)}{f(Y)},$$

$$I(X;Y|T) = \int_{\mathcal{S}_T} I(X;Y|T = t) dF(t) = E \log \frac{f(Y|X,T)}{f(Y|T)},$$

$$h(Y) = h(Y|X) + I(X;Y).$$

2. If Y is discrete,

$$H(Y|X) = H(Y) - I(X;Y),$$
$$H(Y) = H(Y|X) + I(X;Y).$$

Chain Rule for Differential Entropy:

$$h(X_1, X_2, \cdots, X_n) = \sum_{i=1}^{n} h(X_i|X_1, \cdots, X_{i-1}).$$

Some Useful Inequalities:

$$I(X;Y) \geq 0,$$
$$I(X;Y|T) \geq 0,$$
$$h(Y|X) \leq h(Y),$$
$$h(X_1, X_2, \cdots, X_n) \leq \sum_{i=1}^{n} h(X_i).$$

AEP I for Continuous Random Variables:

$$-\frac{1}{n} \log f(\mathbf{X}) \to h(X) \text{ in probability.}$$

Typical Set:

$$W_{[X]\epsilon}^n = \{\mathbf{x} \in \mathcal{X}^n : \left| -n^{-1} \log f(\mathbf{x}) - h(X) \right| < \epsilon\}.$$

AEP II for Continuous Random Variables:

1. $2^{-n(h(X)+\epsilon)} < f(\mathbf{x}) < 2^{-n(h(X)-\epsilon)}$ for $\mathbf{x} \in W_{[X]\epsilon}^n$;
2. $\Pr\{\mathbf{X} \in W_{[X]\epsilon}^n\} > 1 - \epsilon$ for sufficiently large n;
3. $(1-\epsilon)2^{n(h(X)-\epsilon)} < \mathrm{Vol}(W_{[X]\epsilon}^n) < 2^{n(h(X)+\epsilon)}$ for sufficiently large n.

Informational Divergence: For two probability density functions f and g defined on $\Re^n$,

$$D(f\|g) = \int_{\mathcal{S}_f} f(x) \log \frac{f(x)}{g(x)} dx = E_f \log \frac{f(X)}{g(X)}.$$

Divergence Inequality: $D(f\|g) \geq 0$, with equality if and only if $f = g$ a.e.

Maximum Differential Entropy Distributions: Let

$$f^*(\mathbf{x}) = e^{-\lambda_0 - \sum_{i=1}^m \lambda_i r_i(\mathbf{x})}$$

for all $\mathbf{x} \in \mathcal{S}$, where $\lambda_0, \lambda_1, \cdots, \lambda_m$ are chosen such that the constraints

$$\int_{\mathcal{S}_f} r_i(\mathbf{x}) f(\mathbf{x}) d\mathbf{x} = a_i \quad \text{for } 1 \leq i \leq m$$

are satisfied. Then f^* maximizes $h(f)$ over all pdf f defined on $\mathcal{S}$ subject to the above constraints.

Maximum Differential Entropy for a Given Correlation Matrix:

$$h(\mathbf{X}) \leq \frac{1}{2} \log \left[(2\pi e)^n |\tilde{K}| \right],$$

with equality if and only if $\mathbf{X} \sim \mathcal{N}(0, \tilde{K})$, where $\tilde{K}$ is the correlation matrix of $\mathbf{X}$.

Maximum Differential Entropy for a Given Covariance Matrix:

$$h(\mathbf{X}) \leq \frac{1}{2} \log \left[(2\pi e)^n |K| \right],$$

with equality if and only if $\mathbf{X} \sim \mathcal{N}(\boldsymbol{\mu}, K)$, where $\boldsymbol{\mu}$ and K are the mean and the covariance matrix of $\mathbf{X}$, respectively.

Problems

1. Prove Propositions 10.3 and 10.8.
2. Show that the joint pdf of a multivariate Gaussian distribution integrates to 1.
3. Show that $|K| > 0$ in (10.18), the formula for the joint pdf of a multivariate Gaussian distribution.
4. Show that a symmetric positive definite matrix is a covariance matrix.
5. Let

$$K = \begin{bmatrix} 7/4 & \sqrt{2}/4 & -3/4 \\ \sqrt{2}/4 & 5/2 & -\sqrt{2}/4 \\ -3/4 & -\sqrt{2}/4 & 7/4 \end{bmatrix}.$$

   a) Find the eigenvalues and eigenvectors of K.
   b) Show that K is positive definite.
   c) Suppose K is the covariance matrix of a random vector $\mathbf{X} = [X_1\ X_2\ X_3]^\top$.
      i) Find the coefficient of correlation between X_i and X_j for $1 \le i < j \le 3$.
      ii) Find an uncorrelated random vector $\mathbf{Y} = [Y_1\ Y_2\ Y_3]$ such that $\mathbf{X}$ is a linear transformation of $\mathbf{Y}$.
      iii) Determine the covariance matrix of $\mathbf{Y}$.
6. Prove Theorem 10.19.
7. For continuous random variables X and Y, discuss why $I(X; X)$ is not equal to $h(X)$.
8. Each of the following continuous distributions can be obtained as the distribution that maximizes the differential entropy subject to a suitable set of constraints:
   a) the exponential distribution,

$$f(x) = \lambda e^{-\lambda x}$$

   for $x \ge 0$, where $\lambda > 0$;
   b) the Laplace distribution,

$$f(x) = \frac{1}{2}\lambda e^{-\lambda|x|}$$

   for $-\infty < x < \infty$, where $\lambda > 0$;
   c) the gamma distribution,

$$f(x) = \frac{\lambda}{\Gamma(\alpha)}(\lambda x)^{\alpha-1} e^{-\lambda x}$$

   for $x \ge 0$, where $\lambda, \alpha > 0$ and $\Gamma(z) = \int_0^\infty t^{z-1} e^{-t} dt$;

d) the beta distribution,

$$f(x) = \frac{\Gamma(p+q)}{\Gamma(p)\Gamma(q)} x^{p-1}(1-x)^{q-1}$$

for $0 \le x \le 1$, where $p, q > 0$;
e) the Cauchy distribution,

$$f(x) = \frac{1}{\pi(1+x^2)}$$

for $-\infty < x < \infty$.
Identify the corresponding set of constraints for each of these distributions.

9. Let μ be the mean of a continuous random variable X defined on $\Re^+$. Obtain an upper bound on $h(X)$ in terms of μ.

10. The inequality in (10.180) gives an upper bound on the differential entropy in terms of the variance. Can you give an upper bound on the variance in terms of the differential entropy?

11. For $i = 1, 2$, suppose f_i maximizes $h(f)$ over all the pdfs defined on $S_i \subset \Re^n$ subject to the constraints in (10.168), where $S_1 \subset S_2$. Show that $h(f_1) \le h(f_2)$.

12. *Hadamard's inequality.* Show that for a positive semidefinite matrix K, $|K| \le \prod_{i=1}^{n} K_{ii}$, with equality if and only if K is diagonal. Hint: Consider the differential entropy of a multivariate Gaussian distribution.

13. Let $K_{\mathbf{X}}$ and $\tilde{K}_{\mathbf{X}}$ be the covariance matrix and the correlation matrix of a random vector $\mathbf{X}$, respectively. Show that $|K_{\mathbf{X}}| \le |\tilde{K}_{\mathbf{X}}|$. This is a generalization of $\mathrm{var}X \le EX^2$ for a random variable X. Hint: Consider a multivariate Gaussian distribution with another multivariate Gaussian distribution with zero mean and the same correlation matrix.

Historical Notes

The concept of differential entropy was introduced by Shannon [322]. Informational divergence and mutual information were subsequently defined in Kolmogorov [204] and Pinsker [292] in the general setting of measure theory. A measure-theoretic treatment of information theory for continuous systems can be found in the book by Ihara [180].

The treatment in this chapter and the next chapter aims to keep the generality of the results without resorting to heavy use of measure theory. The bounds in Section 10.6 for differential entropy subject to constraints are developed in the spirit of maximum entropy expounded in Jayes [186].

Continuous-Valued Channels

In Chapter 7, we have studied the discrete memoryless channel. For such a channel, transmission is in discrete time, and the input and output are discrete. In a physical communication system, the input and output of a channel often take continuous real values. If transmission is in continuous time, the channel is called a *waveform channel*.

In this chapter, we first discuss discrete-time channels with real input and output. We will then extend our discussion to waveform channels. All the logarithms in this chapter are in the base 2.

11.1 Discrete-Time Channels

Definition 11.1. *Let* $f(y|x)$ *be a conditional pdf defined for all* x, *where*

$$-\int_{\mathcal{S}_Y(x)} f(y|x) \log f(y|x) dy < \infty \qquad (11.1)$$

for all x. *A discrete-time continuous channel* $f(y|x)$ *is a system with input random variable* X *and output random variable* Y *such that* Y *is related to* X *through* $f(y|x)$ *(cf. Definition 10.22).*

Remark The integral in (11.1) is precisely the conditional differential entropy $h(Y|X=x)$ defined in (10.100), which is required to be finite in this definition of a discrete-time continuous channel.

Definition 11.2. *Let* $\alpha : \Re \times \Re \to \Re$, *and* Z *be a real random variable, called the noise variable. A discrete-time continuous channel* (α, Z) *is a system with a real input and a real output. For any input random variable* X, *the noise random variable* Z *is independent of* X, *and the output random variable* Y *is given by*

$$Y = \alpha(X, Z). \qquad (11.2)$$

For brevity, a discrete-time continuous channel will be referred to as a continuous channel.

Definition 11.3. *Two continuous channels* $f(y|x)$ *and* (α, Z) *are equivalent if for every input distribution* $F(x)$,

$$\Pr\{\alpha(X, Z) \leq y, X \leq x\} = \int_{-\infty}^{x} \int_{-\infty}^{y} f_{Y|X}(v|u)dv \, dF_X(u) \qquad (11.3)$$

for all x *and* y.

Remark In the above definitions, the input random variable X is not necessarily continuous.

Definitions 11.1 and 11.2 are two definitions for a continuous channel which are analogous to Definitions 7.1 and 7.2 for a discrete channel. While Definitions 7.1 and 7.2 are equivalent, Definition 11.2 is more general than Definition 11.1. For a continuous channel defined in Definition 11.2, the noise random variable Z may not have a pdf, and the function $\alpha(x, \cdot)$ may be many-to-one. As a result, the corresponding conditional pdf $f(y|x)$ as required in Definition 11.1 may not exist. In this chapter, we confine our discussion to continuous channels that can be defined by Definition 11.1 (and hence also by Definition 11.2).

Definition 11.4. *A continuous memoryless channel (CMC)* $f(y|x)$ *is a sequence of replicates of a generic continuous channel* $f(y|x)$. *These continuous channels are indexed by a discrete-time index* i, *where* $i \geq 1$, *with the* ith *channel being available for transmission at time* i. *Transmission through a channel is assumed to be instantaneous. Let* X_i *and* Y_i *be, respectively, the input and the output of the CMC at time* i, *and let* T_{i-} *denote all the random variables that are generated in the system before* X_i. *The Markov chain* $T_{i-} \to X_i \to Y_i$ *holds, and*

$$\Pr\{Y_i \leq y, X_i \leq x\} = \int_{-\infty}^{x} \int_{-\infty}^{y} f_{Y|X}(v|u)dv \, dF_{X_i}(u). \qquad (11.4)$$

Definition 11.5. *A continuous memoryless channel* (α, Z) *is a sequence of replicates of a generic continuous channel* (α, Z). *These continuous channels are indexed by a discrete-time index* i, *where* $i \geq 1$, *with the* ith *channel being available for transmission at time* i. *Transmission through a channel is assumed to be instantaneous. Let* X_i *and* Y_i *be, respectively, the input and the output of the CMC at time* i, *and let* T_{i-} *denote all the random variables that are generated in the system before* X_i. *The noise variable* Z_i *for the transmission at time* i *is a copy of the generic noise variable* Z *and is independent of* (X_i, T_{i-}). *The output of the CMC at time* i *is given by*

$$Y_i = \alpha(X_i, Z_i). \qquad (11.5)$$

Definition 11.6. *Let κ be a real function. An average input constraint (κ, P) for a CMC is the requirement that for any codeword $(x_1, x_2, \cdots, x_n)$ transmitted over the channel,*

$$\frac{1}{n}\sum_{i=1}^{n}\kappa(x_i) \le P.$$ (11.6)

For brevity, an average input constraint is referred to as an input constraint.

Definition 11.7. *The capacity of a continuous memoryless channel $f(y|x)$ with input constraint (κ, P) is defined as*

$$C(P) = \sup_{F(x):E\kappa(X)\le P} I(X;Y),$$ (11.7)

where X and Y are, respectively, the input and output of the generic continuous channel, and $F(x)$ is the distribution of X.

Theorem 11.8. *$C(P)$ is non-decreasing, concave, and left-continuous.*

Proof. In the definition of $C(P)$, the supremum is taken over a larger set for a larger P. Therefore, $C(P)$ is non-decreasing in P.

We now show that $C(P)$ is concave. Let $j = 1, 2$. For an input distribution $F_j(x)$, denote the corresponding input and output random variables by X_j and Y_j, respectively. Then for any P_j, for all $\epsilon > 0$, there exists $F_j(x)$ such that

$$E\kappa(X_j) \le P_j$$ (11.8)

and

$$I(X_j; Y_j) \ge C(P_j) - \epsilon.$$ (11.9)

For $0 \le \lambda \le 1$, let $\bar{\lambda} = 1 - \lambda$, and define the random variable

$$X^{(\lambda)} \sim \lambda F_1(x) + \bar{\lambda} F_2(x).$$ (11.10)

Then

$$E\kappa(X^{(\lambda)}) = \lambda E\kappa(X_1) + \bar{\lambda} E\kappa(X_2) \le \lambda P_1 + \bar{\lambda} P_2.$$ (11.11)

By the concavity of mutual information with respect to the input distribution,[1] we have

$$I(X^{(\lambda)}; Y^{(\lambda)}) \ge \lambda I(X_1; Y_1) + \bar{\lambda} I(X_2; Y_2)$$ (11.12)

$$\ge \lambda(C(P_1) - \epsilon) + \bar{\lambda}(C(P_2) - \epsilon)$$ (11.13)

$$= \lambda C(P_1) + \bar{\lambda} C(P_2) - \epsilon.$$ (11.14)

Then

---

[1] Specifically, we refer to the inequality (3.124) in Example 3.14 with X and Y being real random variables related by a conditional pdf $f(y|x)$. The proof of this inequality is left as an exercise.

$$C(\lambda P_1 + \bar{\lambda} P_2) \geq I(X^{(\lambda)}; Y^{(\lambda)}) \geq \lambda C(P_1) + \bar{\lambda} C(P_2) - \epsilon. \qquad (11.15)$$

Letting $\epsilon \to 0$, we have

$$C(\lambda P_1 + \bar{\lambda} P_2) \geq \lambda C(P_1) + \bar{\lambda} C(P_2), \qquad (11.16)$$

proving that $C(P)$ is concave.

Finally, we prove that $C(P)$ is left-continuous. Let $P_1 < P_2$ in (11.16). Since $C(P)$ is non-decreasing, we have

$$C(P_2) \geq C(\lambda P_1 + \bar{\lambda} P_2) \geq \lambda C(P_1) + \bar{\lambda} C(P_2). \qquad (11.17)$$

Letting $\lambda \to 0$, we have

$$C(P_2) \geq \lim_{\lambda \to 0} C(\lambda P_1 + \bar{\lambda} P_2) \geq C(P_2), \qquad (11.18)$$

which implies

$$\lim_{\lambda \to 0} C(\lambda P_1 + \bar{\lambda} P_2) = C(P_2). \qquad (11.19)$$

Hence, we conclude that

$$\lim_{P \uparrow P_2} C(P) = C(P_2), \qquad (11.20)$$

i.e., $C(P)$ is left-continuous. The theorem is proved. $\square$

11.2 The Channel Coding Theorem

Definition 11.9. *An (n, M) code for a continuous memoryless channel with input constraint (κ, P) is defined by an encoding function*

$$e : \{1, 2, \cdots, M\} \to \Re^n \qquad (11.21)$$

and a decoding function

$$g : \Re^n \to \{1, 2, \cdots, M\}. \qquad (11.22)$$

The set $\{1, 2, \cdots, M\}$, denoted by $\mathcal{W}$, is called the message set. The sequences $e(1), e(2), \cdots, e(M)$ in $\mathcal{X}^n$ are called codewords, and the set of codewords is called the codebook. Moreover,

$$\frac{1}{n} \sum_{i=1}^{n} \kappa(x_i(w)) \leq P \quad \text{for } 1 \leq w \leq M, \qquad (11.23)$$

where $e(w) = (x_1(w), x_2(w), \cdots, x_n(w))$.

We assume that a message W is randomly chosen from the message set $\mathcal{W}$ according to the uniform distribution. Therefore,

$$H(W) = \log M. \tag{11.24}$$

With respect to a channel code for a given CMC, we let

$$\mathbf{X} = (X_1, X_2, \cdots, X_n) \tag{11.25}$$

and

$$\mathbf{Y} = (Y_1, Y_2, \cdots, Y_n) \tag{11.26}$$

be the input sequence and the output sequence of the channel, respectively. Evidently,

$$\mathbf{X} = e(W). \tag{11.27}$$

We also let

$$\hat{W} = g(\mathbf{Y}) \tag{11.28}$$

be the estimate of the message W by the decoder.

Definition 11.10. *For all* $1 \leq w \leq M$, *let*

$$\lambda_w = \Pr\{\hat{W} \neq w | W = w\} = \sum_{\mathbf{y} \in \mathcal{Y}^n : g(\mathbf{y}) \neq w} \Pr\{\mathbf{Y} = \mathbf{y} | \mathbf{X} = e(w)\} \tag{11.29}$$

be the conditional probability of error given that the message is w.

We now define two performance measures for a channel code.

Definition 11.11. *The maximal probability of error of an* (n, M) *code is defined as*

$$\lambda_{max} = \max_w \lambda_w. \tag{11.30}$$

Definition 11.12. *The average probability of error of an* (n, M) *code is defined as*

$$P_e = \Pr\{\hat{W} \neq W\}. \tag{11.31}$$

Evidently, $P_e \leq \lambda_{max}$.

Definition 11.13. *A rate* R *is asymptotically achievable for a continuous memoryless channel if for any* $\epsilon > 0$, *there exists for sufficiently large* n *an* (n, M) *code such that*

$$\frac{1}{n} \log M > R - \epsilon \tag{11.32}$$

and

$$\lambda_{max} < \epsilon. \tag{11.33}$$

For brevity, an asymptotically achievable rate will be referred to as an achievable rate.

Theorem 11.14 (Channel Coding Theorem). *A rate* R *is achievable for a continuous memoryless channel if and only if* $R \leq C$, *the capacity of the channel.*

11.3 Proof of the Channel Coding Theorem

11.3.1 The Converse

We can establish the Markov chain

$$W \to \mathbf{X} \to \mathbf{Y} \to \hat{W} \tag{11.34}$$

very much like the discrete case as discussed in Section 7.3. Here, although $\mathbf{X}$ is a real random vector, it takes only discrete values as it is a function of the message W which is discrete. The only continuous random variable in the above Markov chain is the random vector $\mathbf{Y}$, which needs to be handled with caution. The following lemma is essentially the data processing theorem we proved in Theorem 2.42 except that $\mathbf{Y}$ is continuous. The reader may skip the proof at the first reading.

Lemma 11.15.

$$I(W; \hat{W}) \leq I(\mathbf{X}; \mathbf{Y}). \tag{11.35}$$

Proof. We first consider

$$I(W; \hat{W}) \leq I(W, \mathbf{X}; \hat{W}) \tag{11.36}$$

$$= I(\mathbf{X}; \hat{W}) + I(W; \hat{W}|\mathbf{X}) \tag{11.37}$$

$$= I(\mathbf{X}; \hat{W}). \tag{11.38}$$

Note that all the random variables above are discrete. Continuing from the above, we have

$$I(W; \hat{W}) \leq I(\mathbf{X}; \hat{W}) \tag{11.39}$$

$$\leq I(\mathbf{X}; \hat{W}) + I(\mathbf{X}; \mathbf{Y}|\hat{W}) \tag{11.40}$$

$$= E \log \frac{p(\mathbf{X}, \hat{W})}{p(\mathbf{X})p(\hat{W})} + E \log \frac{f(\mathbf{Y}|\mathbf{X}, \hat{W})}{f(\mathbf{Y}|\hat{W})} \tag{11.41}$$

$$= E \log \frac{p(\mathbf{X}, \hat{W})f(\mathbf{Y}|\mathbf{X}, \hat{W})}{p(\mathbf{X})[p(\hat{W})f(\mathbf{Y}|\hat{W})]} \tag{11.42}$$

$$= E \log \frac{f(\mathbf{Y})p(\mathbf{X}, \hat{W}|\mathbf{Y})}{p(\mathbf{X})[f(\mathbf{Y})p(\hat{W}|\mathbf{Y})]} \tag{11.43}$$

$$= E \log \frac{p(\mathbf{X}, \hat{W}|\mathbf{Y})}{p(\mathbf{X})p(\hat{W}|\mathbf{Y})} \tag{11.44}$$

$$= E \log \frac{p(\mathbf{X}|\mathbf{Y})p(\hat{W}|\mathbf{X}, \mathbf{Y})}{p(\mathbf{X})p(\hat{W}|\mathbf{Y})} \tag{11.45}$$

$$= E \log \frac{p(\mathbf{X}|\mathbf{Y})}{p(\mathbf{X})} + E \log \frac{p(\hat{W}|\mathbf{X}, \mathbf{Y})}{p(\hat{W}|\mathbf{Y})} \tag{11.46}$$

$$= E \log \frac{f(\mathbf{Y}|\mathbf{X})}{f(\mathbf{Y})} + E \log \frac{p(\mathbf{X}|\mathbf{Y}, \hat{W})}{p(\mathbf{X}|\mathbf{Y})} \tag{11.47}$$

$$= I(\mathbf{X}; \mathbf{Y}) + E \log \frac{p(\mathbf{X}|\mathbf{Y})}{p(\mathbf{X}|\mathbf{Y})} \tag{11.48}$$

$$= I(\mathbf{X}; \mathbf{Y}) + E \log 1 \tag{11.49}$$

$$= I(\mathbf{X}; \mathbf{Y}) + 0 \tag{11.50}$$

$$= I(\mathbf{X}; \mathbf{Y}). \tag{11.51}$$

The above steps are justified as follows:

- The relation

$$f(\mathbf{y}|\mathbf{x}) = \prod_{i=1}^{n} f(y_i|x_i) \tag{11.52}$$

  can be established in exactly the same way as we established (7.101) for the discrete case (when the channel is used without feedback). Then

$$f(\mathbf{y}|\mathbf{x}, \hat{w}) = \frac{p(\mathbf{x})f(\mathbf{y}|\mathbf{x})p(\hat{w}|\mathbf{y})}{p(\mathbf{x}, \hat{w})}, \tag{11.53}$$

  and $f(\mathbf{y}|\hat{w})$ exists by Proposition 10.24. Therefore, $I(\mathbf{X}; \mathbf{Y}|\hat{W})$ in (11.40) can be defined according to Definition 10.27.
- (11.40) follows from Corollary 10.32.
- In (11.43), given $f(\mathbf{y}|\mathbf{x})$ as in (11.52), it follows from Proposition 10.24 that $f(\mathbf{y})$ exists.
- (11.47) follows from

$$p(\mathbf{x})f(\mathbf{y}|\mathbf{x}) = f(\mathbf{y})p(\mathbf{x}|\mathbf{y}) \tag{11.54}$$

  and

$$p(\mathbf{x}|\mathbf{y})p(\hat{w}|\mathbf{x}, \mathbf{y}) = p(\hat{w}|\mathbf{y})p(\mathbf{x}|\mathbf{y}, \hat{w}). \tag{11.55}$$

- (11.48) follows from the Markov chain $\mathbf{X} \to \mathbf{Y} \to \hat{W}$.

The proof is accomplished. $\square$

We now proceed to prove the converse. Let R be an achievable rate, i.e., for any $\epsilon > 0$, there exists for sufficiently large n and (n, M) code such that

$$\frac{1}{n} \log M > R - \epsilon \tag{11.56}$$

and

$$\lambda_{max} < \epsilon. \tag{11.57}$$

Consider

$$\log M = H(W) \tag{11.58}$$
$$= H(W|\hat{W}) + I(W; \hat{W}) \tag{11.59}$$
$$\leq H(W|\hat{W}) + I(\mathbf{X}; \mathbf{Y}) \tag{11.60}$$
$$= H(W|\hat{W}) + h(\mathbf{Y}) - h(\mathbf{Y}|\mathbf{X}) \tag{11.61}$$
$$\leq H(W|\hat{W}) + \sum_{i=1}^{n} h(Y_i) - h(\mathbf{Y}|\mathbf{X}) \tag{11.62}$$
$$= H(W|\hat{W}) + \sum_{i=1}^{n} h(Y_i) - \sum_{i=1}^{n} h(Y_i|X_i) \tag{11.63}$$
$$= H(W|\hat{W}) + \sum_{i=1}^{n} I(X_i; Y_i). \tag{11.64}$$

The above steps are justified as follows:

- (11.60) follows from Lemma 11.15.
- It follows from (11.52) that

$$h(\mathbf{Y}|\mathbf{X}) = \sum_{i=1}^{n} h(Y_i|X_i). \tag{11.65}$$

  Then (11.1) in Definition 11.1 implies that $h(Y_i|X_i)$ is finite for all i, and hence $h(\mathbf{Y}|\mathbf{X})$ is also finite.
- From the foregoing, $f(\mathbf{y})$ exists. Therefore, $h(\mathbf{Y})$ can be defined according to Definition 10.10 (but $h(\mathbf{Y})$ may be infinite), and (11.61) follows from Proposition 10.29 because $h(\mathbf{Y}|\mathbf{X})$ is finite. Note that it is necessary to require $h(\mathbf{Y}|\mathbf{X})$ to be finite because otherwise $h(\mathbf{Y})$ is also infinite and Proposition 10.29 cannot be applied.
- (11.62) follows from Corollary 10.34, the independence bound for differential entropy.
- (11.63) follows from (11.65) above.
- (11.64) follows from Proposition 10.29.

Let V be a mixing random variable distributed uniformly on $\{1, 2, \cdots, n\}$ which is independent of X_i, $1 \leq i \leq n$. Let $X = X_V$ and Y be the output of the channel with X being the input. Then

$$E\kappa(X) = EE[\kappa(X)|V] \tag{11.66}$$
$$= \sum_{i=1}^{n} \Pr\{V = i\}E[\kappa(X)|V = i] \tag{11.67}$$
$$= \sum_{i=1}^{n} \Pr\{V = i\}E[\kappa(X_i)|V = i] \tag{11.68}$$
$$= \sum_{i=1}^{n} \frac{1}{n}E\kappa(X_i) \tag{11.69}$$

$$= E\left[\frac{1}{n}\sum_{i=1}^{n}\kappa(X_i)\right] \tag{11.70}$$

$$\leq P, \tag{11.71}$$

where the above inequality follows from (11.23) in the definition of the code. By the concavity of mutual information with respect to the input distribution, we have

$$\frac{1}{n}\sum_{i=1}^{n}I(X_i;Y_i) \leq I(X;Y) \leq C, \tag{11.72}$$

where the last inequality holds in light of the definition of C and (11.71). Then it follows from (11.64) that

$$\log M \leq H(W|\hat{W}) + nC, \tag{11.73}$$

which is precisely (7.126) in the proof of the converse of the channel coding theorem for the DMC. Following exactly the same steps therein, we conclude that

$$R \leq C. \tag{11.74}$$

11.3.2 Achievability

The proof of the achievability of the channel capacity, which involves the construction of a random code, is somewhat different from the construction for the discrete case in Section 7.4. On the one hand, we need to take into account the input constraint. On the other hand, since the input distribution $F(x)$ we use for constructing the random code may not have a pdf, it is difficult to formulate the notion of joint typicality as in the discrete case. Instead, we will introduce a different notion of typicality based on mutual information.

Consider a bivariate information source $\{(X_k, Y_k), k \geq 1\}$, where (X_k, Y_k) are i.i.d. with (X,Y) being the pair of generic real random variables. The conditional pdf $f(y|x)$ exists in the sense prescribed in Definition 10.22. By Proposition 10.24, $f(y)$ exists so that the mutual information $I(X;Y)$ can be defined according to Definition 10.26.

Definition 11.16. *The mutually typical set $\Psi_{[XY]\delta}^n$ with respect to $F(x,y)$ is the set of $(\mathbf{x}, \mathbf{y}) \in \mathcal{X}^n \times \mathcal{Y}^n$ such that*

$$\left|\frac{1}{n}\log\frac{f(\mathbf{y}|\mathbf{x})}{f(\mathbf{y})} - I(X;Y)\right| \leq \delta, \tag{11.75}$$

where

$$f(\mathbf{y}|\mathbf{x}) = \prod_{i=1}^{n}f(y_i|x_i) \tag{11.76}$$

and

$$f(\mathbf{y}) = \prod_{i=1}^{n} f(y_i), \tag{11.77}$$

and δ is an arbitrarily small positive number. A pair of sequences $(\mathbf{x}, \mathbf{y})$ is called mutually δ-typical if it is in $\Psi_{[XY]\delta}^n$.

Lemma 11.17. For any $\delta > 0$, for sufficiently large n,

$$\Pr\{(\mathbf{X}, \mathbf{Y}) \in \Psi_{[XY]\delta}^n)\} \geq 1 - \delta. \tag{11.78}$$

Proof. By (11.76) and (11.77), we write

$$\frac{1}{n} \log \frac{f(\mathbf{Y}|\mathbf{X})}{f(\mathbf{Y})} = \frac{1}{n} \log \prod_{i=1}^{n} \frac{f(Y_i|X_i)}{f(Y_i)} = \frac{1}{n} \sum_{i=1}^{n} \log \frac{f(Y_i|X_i)}{f(Y_i)}. \tag{11.79}$$

Since (X_i, Y_i) are i.i.d., so are the random variables $\log \frac{f(Y_i|X_i)}{f(Y_i)}$. Thus we conclude by the weak law of large numbers that

$$\frac{1}{n} \sum_{i=1}^{n} \log \frac{f(Y_i|X_i)}{f(Y_i)} \to E \log \frac{f(Y|X)}{f(Y)} = I(X;Y) \tag{11.80}$$

in probability, i.e., (11.78) holds for all sufficiently large n, proving the lemma. $\square$

The following lemma is analogous to Lemma 7.17 for the discrete case.

Lemma 11.18. Let $(\mathbf{X}', \mathbf{Y}')$ be n i.i.d. copies of a pair of generic random variables (X', Y'), where X' and Y' are independent and have the same marginal distributions as X and Y, respectively. Then

$$\Pr\{(\mathbf{X}', \mathbf{Y}') \in \Psi_{[XY]\delta}^n\} \leq 2^{-n(I(X;Y)-\delta)}. \tag{11.81}$$

Proof. For $(\mathbf{x}, \mathbf{y}) \in \Psi_{[XY]\delta}^n$, from (11.75), we obtain

$$\frac{1}{n} \log \frac{f(\mathbf{y}|\mathbf{x})}{f(\mathbf{y})} \geq I(X;Y) - \delta \tag{11.82}$$

or

$$f(\mathbf{y}|\mathbf{x}) \geq f(\mathbf{y}) 2^{n(I(X;Y)-\delta)}. \tag{11.83}$$

Then

$$1 \geq \Pr\{(\mathbf{X}, \mathbf{Y}) \in \Psi_{[XY]\delta}^n)\} \tag{11.84}$$

$$= \int \int_{\Psi_{[XY]\delta}^n} f(\mathbf{y}|\mathbf{x}) dF(\mathbf{x}) \, d\mathbf{y} \tag{11.85}$$

$$\geq 2^{n(I(X;Y)-\delta)} \int \int_{\Psi_{[XY]\delta}^n} f(\mathbf{y}) dF(\mathbf{x}) \, d\mathbf{y} \tag{11.86}$$

$$= 2^{n(I(X;Y)-\delta)} \Pr\{(\mathbf{X}', \mathbf{Y}') \in \Psi_{[XY]\delta}^n\}, \tag{11.87}$$

where the last inequality follows from (11.83). Therefore,

$$\Pr\{(\mathbf{X}', \mathbf{Y}') \in \Psi^n_{[XY]\delta}\} \leq 2^{-n(I(X;Y)-\delta)}, \tag{11.88}$$

proving the lemma. □

Fix any $\epsilon > 0$ and let δ be a small quantity to be specified later. Since $C(P)$ is left-continuous, there exists a sufficiently small $\gamma > 0$ such that

$$C(P - \gamma) > C(P) - \frac{\epsilon}{6}. \tag{11.89}$$

By the definition of $C(P - \gamma)$, there exists an input random variable X such that

$$E\kappa(X) \leq P - \gamma \tag{11.90}$$

and

$$I(X;Y) \geq C(P - \gamma) - \frac{\epsilon}{6}. \tag{11.91}$$

Then choose for a sufficiently large n an even integer M satisfying

$$I(X;Y) - \frac{\epsilon}{6} < \frac{1}{n}\log M < I(X;Y) - \frac{\epsilon}{8}, \tag{11.92}$$

from which we obtain

$$\frac{1}{n}\log M > I(X;Y) - \frac{\epsilon}{6} \tag{11.93}$$

$$\geq C(P - \gamma) - \frac{\epsilon}{3} \tag{11.94}$$

$$> C(P) - \frac{\epsilon}{2}. \tag{11.95}$$

We now describe a random coding scheme:

1. Construct the codebook $\mathcal{C}$ of an (n, M) code randomly by generating M codewords in $\Re^n$ independently and identically according to $F(x)^n$. Denote these codewords by $\tilde{\mathbf{X}}(1), \tilde{\mathbf{X}}(2), \cdots, \tilde{\mathbf{X}}(M)$.
2. Reveal the codebook $\mathcal{C}$ to both the encoder and the decoder.
3. A message W is chosen from $\mathcal{W}$ according to the uniform distribution.
4. The sequence $\mathbf{X} = \tilde{\mathbf{X}}(W)$, namely the Wth codeword in the codebook $\mathcal{C}$, is transmitted through the channel.
5. The channel outputs a sequence $\mathbf{Y}$ according to

$$\Pr\{Y_i \leq y_i, 1 \leq i \leq n | \mathbf{X}(W) = \mathbf{x}\} = \prod_{i=1}^{n} \int_{-\infty}^{y_i} f(y|x_i) dy. \tag{11.96}$$

This is the continuous analog of (7.101) and can be established similarly.

6. The sequence $\mathbf{Y}$ is decoded to the message w if $(\mathbf{X}(w), \mathbf{Y}) \in \Psi^n_{[XY]\delta}$ and there does not exist $w' \neq w$ such that $(\mathbf{X}(w'), \mathbf{Y}) \in \Psi^n_{[XY]\delta}$. Otherwise, $\mathbf{Y}$ is decoded to a constant message in $\mathcal{W}$. Denote by $\hat{W}$ the message to which $\mathbf{Y}$ is decoded.

We now analyze the performance of this random coding scheme. Let

$$\tilde{\mathbf{X}}(w) = (\tilde{X}_1(w), \tilde{X}_2(w), \cdots, \tilde{X}_n(w)) \tag{11.97}$$

and define the error event

$$Err = E_e \cup E_d, \tag{11.98}$$

where

$$E_e = \left\{ \frac{1}{n} \sum_{i=1}^{n} \kappa(\tilde{X}_i(W)) > P \right\} \tag{11.99}$$

is the event that the input constraint is violated, and

$$E_d = \{\hat{W} \neq W\} \tag{11.100}$$

is the event that a decoding error occurs. By symmetry in the code construction,

$$\Pr\{Err\} = \Pr\{Err|W = 1\} \tag{11.101}$$
$$\leq \Pr\{E_e|W = 1\} + \Pr\{E_d|W = 1\}. \tag{11.102}$$

With Lemma 11.18 in place of Lemma 7.17, the analysis of $\Pr\{E_d|W = 1\}$ is exactly the same as the analysis of the decoding error in the discrete case. The details are omitted, and we conclude that by choosing δ to be a sufficiently small positive quantity,

$$\Pr\{E_d|W = 1\} \leq \frac{\epsilon}{4} \tag{11.103}$$

for sufficiently large n.

We now analyze $\Pr\{E_e|W = 1\}$. By the weak law of large numbers,

$$\Pr\{E_e|W = 1\} = \Pr\left\{ \frac{1}{n} \sum_{i=1}^{n} \kappa(\tilde{X}_i(1)) > P \,\middle|\, W = 1 \right\} \tag{11.104}$$

$$= \Pr\left\{ \frac{1}{n} \sum_{i=1}^{n} \kappa(\tilde{X}_i(1)) > P \right\} \tag{11.105}$$

$$= \Pr\left\{ \frac{1}{n} \sum_{i=1}^{n} \kappa(\tilde{X}_i(1)) > (P - \gamma) + \gamma \right\} \tag{11.106}$$

$$\leq \Pr\left\{ \frac{1}{n} \sum_{i=1}^{n} \kappa(\tilde{X}_i(1)) > E\kappa(X) + \gamma \right\} \tag{11.107}$$

$$\leq \frac{\epsilon}{4} \tag{11.108}$$

for sufficiently large n. It then follows from (11.102) and (11.103) that

$$\Pr\{Err\} \leq \frac{\epsilon}{2} \qquad (11.109)$$

for sufficiently large n.

It remains to show the existence of a codebook such that $\lambda_{max} < \epsilon$ and the input constraint (11.23) is satisfied by every codeword. Consider

$$\Pr\{Err\} = \sum_{\mathcal{C}} \Pr\{\mathcal{C}\}\Pr\{Err|\mathcal{C}\}, \qquad (11.110)$$

where $\Pr\{\mathcal{C}\}$ is the probability of choosing a codebook $\mathcal{C}$ from the ensemble of all possible codebooks in Step 1 of the random coding scheme. In light of (11.109), there exists at least one codebook $\mathcal{C}^*$ such that

$$\Pr\{Err|\mathcal{C}^*\} \leq \frac{\epsilon}{2}. \qquad (11.111)$$

Furthermore,

$$\Pr\{Err|\mathcal{C}^*\} = \sum_{w=1}^{M} \Pr\{W = w|\mathcal{C}^*\}\Pr\{Err|\mathcal{C}^*, W = w\} \qquad (11.112)$$

$$= \sum_{w=1}^{M} \Pr\{W = w\}\Pr\{Err|\mathcal{C}^*, W = w\} \qquad (11.113)$$

$$= \frac{1}{M} \sum_{w=1}^{M} \Pr\{Err|\mathcal{C}^*, W = w\}. \qquad (11.114)$$

By discarding the worst half of the codewords in $\mathcal{C}^*$, if a codeword $\tilde{\mathbf{X}}(w)$ remains in $\mathcal{C}^*$, then

$$\Pr\{Err|\mathcal{C}^*, W = w\} \leq \epsilon. \qquad (11.115)$$

Since $Err = E_e \cup E_d$, this implies

$$\Pr\{E_e|\mathcal{C}^*, W = w\} \leq \epsilon \qquad (11.116)$$

and

$$\Pr\{E_d|\mathcal{C}^*, W = w\} \leq \epsilon, \qquad (11.117)$$

where the latter implies $\lambda_{max} \leq \epsilon$ for the codebook $\mathcal{C}^*$. Finally, observe that conditioning on $\{\mathcal{C}^*, W = w\}$, the codeword $\tilde{\mathbf{X}}(w)$ is deterministic. Therefore, $\Pr\{E_e|\mathcal{C}^*, W = w\}$ is equal to 1 if the codeword $\tilde{\mathbf{X}}(w)$ violates the input constraint (11.23), otherwise it is equal to 0. Then (11.116) implies that for every codeword $\tilde{\mathbf{X}}(w)$ that remains in $\mathcal{C}^*$, $\Pr\{E_e|\mathcal{C}^*, W = w\} = 0$, i.e., the input constraint is satisfied. This completes the proof.

11.4 Memoryless Gaussian Channels

In communication engineering, the Gaussian channel is the most commonly used model for a noisy channel with real input and output. The reasons are two-fold. First, the Gaussian channel is highly analytically tractable. Second, the Gaussian noise can be regarded as the worst kind of additive noise subject to a constraint on the noise power. This will be discussed in Section 11.9.

We first give two equivalent definitions of a Gaussian channel.

Definition 11.19 (Gaussian Channel). *A Gaussian channel with noise energy N is a continuous channel with the following two equivalent specifications:*

1. $f(y|x) = \frac{1}{\sqrt{2\pi N}} e^{-\frac{(y-x)^2}{2N}}$.

2. $Z \sim \mathcal{N}(0, N)$ *and* $\alpha(X, Z) = X + Z$.

Definition 11.20 (Memoryless Gaussian Channel). *A memoryless Gaussian channel with noise power N and input power constraint P is a memoryless continuous channel with the generic continuous channel being the Gaussian channel with noise energy N. The input power constraint P refers to the input constraint (κ, P) with $\kappa(x) = x^2$.*

Using the formula in Definition 11.7 for the capacity of a CMC, the capacity of a Gaussian channel can be evaluated.

Theorem 11.21 (Capacity of a Memoryless Gaussian Channel). *The capacity of a memoryless Gaussian channel with noise power N and input power constraint P is*

$$\frac{1}{2} \log \left(1 + \frac{P}{N} \right). \tag{11.118}$$

The capacity is achieved by the input distribution $\mathcal{N}(0, P)$.

We first prove the following lemma.

Lemma 11.22. *Let $Y = X + Z$. Then $h(Y|X) = h(Z|X)$ provided that $f_{Z|X}(z|x)$ exists for all $x \in \mathcal{S}_X$.*

Proof. For all $x \in \mathcal{S}_X$, since $f_{Z|X}(z|x)$ exists, $f_{Y|X}(y|x)$ also exists and is given by

$$f_{Y|X}(y|x) = f_{Z|X}(y - x|x). \tag{11.119}$$

Then $h(Y|X = x)$ is defined as in (10.100), and

$$h(Y|X) = \int h(Y|X = x)dF_X(x) \tag{11.120}$$

$$= \int h(X + Z|X = x)dF_X(x) \tag{11.121}$$

$$= \int h(x + Z|X = x)dF_X(x) \tag{11.122}$$

$$= \int h(Z|X = x)dF_X(x) \tag{11.123}$$

$$= h(Z|X). \tag{11.124}$$

In the above, (11.120) and (11.124) follow from (10.101), while (11.123) follows from the translation property of differential entropy (Theorem 10.14). □

Remark Since Y and Z uniquely determine each other given X, it is tempting to write $h(Y|X) = h(Z|X)$ immediately. However, this interpretation is incorrect because differential entropy is not the same as entropy.

Proof of Theorem 11.21. Let $F(x)$ be the CDF of the input random variable X such that $EX^2 \leq P$, where X is not necessarily continuous. Since $Z \sim \mathcal{N}(0, N)$, $f(y|x)$ is given by (11.119). Then by Proposition 10.24, $f(y)$ exists and hence $h(Y)$ is defined. Since Z is independent of X, by Lemma 11.22 and Corollary 10.33,

$$h(Y|X) = h(Z|X) = h(Z). \tag{11.125}$$

Then

$$I(X;Y) = h(Y) - h(Y|X) \tag{11.126}$$

$$= h(Y) - h(Z), \tag{11.127}$$

where (11.126) follows from Proposition 10.29 and (11.127) follows from (11.125).

Since $Y = X + Z$ and Z is independent of X, we have

$$EY^2 = E(X + Z)^2 \tag{11.128}$$
$$= EX^2 + 2(EXZ) + EZ^2 \tag{11.129}$$
$$= EX^2 + 2(EX)(EZ) + EZ^2 \tag{11.130}$$
$$= EX^2 + 2(EX)(0) + EZ^2 \tag{11.131}$$
$$= EX^2 + EZ^2 \tag{11.132}$$
$$\leq P + N. \tag{11.133}$$

Given the above constraint on Y, by Theorem 10.43, we have

$$h(Y) \leq \frac{1}{2}\log[2\pi e(P + N)], \tag{11.134}$$

with equality if $Y \sim \mathcal{N}(0, P + N)$.

Recall from Example 10.13 that

$$h(Z) = \frac{1}{2} \log(2\pi e N). \tag{11.135}$$

It then follows from (11.127), (11.134), and (11.135) that

$$I(X;Y) = h(Y) - h(Z) \tag{11.136}$$

$$\leq \frac{1}{2} \log[2\pi e(P+N)] - \frac{1}{2} \log(2\pi e N) \tag{11.137}$$

$$= \frac{1}{2} \log\left(1 + \frac{P}{N}\right). \tag{11.138}$$

Evidently, this upper bound is tight if $X \sim \mathcal{N}(0, P)$, because then

$$Y = X + Z \sim \mathcal{N}(0, P + N). \tag{11.139}$$

Therefore,

$$C = \sup_{F(x):EX^2 \leq P} I(X;Y) \tag{11.140}$$

$$= \max_{F(x):EX^2 \leq P} I(X;Y) \tag{11.141}$$

$$= \frac{1}{2} \log\left(1 + \frac{P}{N}\right). \tag{11.142}$$

The theorem is proved.   □

Theorem 11.21 says that the capacity of a memoryless Gaussian channel depends only on the ratio of the input power constraint P to the noise power N. This important quantity is called the *signal-to-noise ratio*. Note that no matter how small the signal-to-noise ratio is, the capacity is still strictly positive. In other words, reliable communication can still be achieved, though at a low rate, when the noise power is much higher than the signal power. We also see that the capacity is infinite if there is no constraint on the input power.

11.5 Parallel Gaussian Channels

In Section 11.4, we have discussed the capacity of a memoryless Gaussian channel. Now suppose k such channels are available for communication, where $k \geq 1$. This is illustrated in Figure 11.1, with X_i, Y_i, and Z_i being the input, the output, and the noise variable of the ith channel, respectively, where $Z_i \sim \mathcal{N}(0, N_i)$ and Z_i, $1 \leq i \leq k$ are independent.

We are interested in the capacity of such a system of parallel Gaussian channels, with the total input power constraint

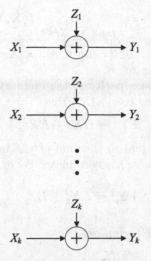

Fig. 11.1. A system of parallel Gaussian channels.

$$E \sum_{i=1}^{k} X_i^2 \leq P. \tag{11.143}$$

Let $\mathbf{X} = [X_1\, X_2\, \cdots\, X_k]$, $\mathbf{Y} = [Y_1\, Y_2\, \cdots\, Y_k]$, and $\mathbf{Z} = [Z_1\, Z_2\, \cdots\, Z_k]$. Then

$$f_{\mathbf{Y}|\mathbf{X}}(\mathbf{y}|\mathbf{x}) = \prod_{i=1}^{k} f_{Y_i|X_i}(y_i|x_i) \tag{11.144}$$

$$= \prod_{i=1}^{k} f_{Z_i|X_i}(y_i - x_i|x_i) \tag{11.145}$$

$$= \prod_{i=1}^{k} f_{Z_i}(y_i - x_i). \tag{11.146}$$

With the existence of $f(\mathbf{y}|\mathbf{x})$, by extending Definition 10.23, we have

$$h(\mathbf{Y}|\mathbf{X}) = - \int_{\mathcal{S}_{\mathbf{X}}} \int_{\mathcal{S}_{\mathbf{Y}}(\mathbf{x})} f(\mathbf{y}|\mathbf{x}) \log f(\mathbf{y}|\mathbf{x}) d\mathbf{y}\, dF(\mathbf{x}). \tag{11.147}$$

Then by Proposition 10.25, $f(\mathbf{y})$ exists and therefore $h(\mathbf{Y})$ is defined. By extending Definition 10.26, we have

$$I(\mathbf{X}; \mathbf{Y}) = \int_{\mathcal{S}_{\mathbf{X}}} \int_{\mathcal{S}_{\mathbf{Y}}(\mathbf{x})} f(\mathbf{y}|\mathbf{x}) \log \frac{f(\mathbf{y}|\mathbf{x})}{f(\mathbf{y})} d\mathbf{y}\, dF(\mathbf{x}). \tag{11.148}$$

It then follows from Definition 11.7 that the capacity of the system is given by

$$C(P) = \sup_{F(\mathbf{x}):E\sum_i X_i^2 \le P} I(\mathbf{X}; \mathbf{Y}), \tag{11.149}$$

where $F(\mathbf{x})$ is the joint CDF of the input vector $\mathbf{X}$. As we will see, the supremum above is indeed a maximum.

When we calculated the capacity of the memoryless Gaussian channel in Theorem 11.21, we obtained in (11.132) that

$$EY^2 = EX^2 + EZ^2, \tag{11.150}$$

i.e., the output power is equal to the sum of the input power and the noise power, provided that the noise has zero mean. By exactly the same argument, we see that

$$EY_i^2 = EX_i^2 + EZ_i^2 \tag{11.151}$$

for all i.

Toward calculating $C(P)$, consider

$$I(\mathbf{X}; \mathbf{Y}) = h(\mathbf{Y}) - h(\mathbf{Y}|\mathbf{X}) \tag{11.152}$$

$$= h(\mathbf{Y}) - h(\mathbf{Z}|\mathbf{X}) \tag{11.153}$$

$$= h(\mathbf{Y}) - h(\mathbf{Z}) \tag{11.154}$$

$$= h(\mathbf{Y}) - \sum_{i=1}^{k} h(Z_i) \tag{11.155}$$

$$= h(\mathbf{Y}) - \frac{1}{2}\sum_{i=1}^{k} \log(2\pi e N_i) \tag{11.156}$$

$$\le \sum_{i=1}^{k} h(Y_i) - \frac{1}{2}\sum_{i=1}^{k} \log(2\pi e N_i) \tag{11.157}$$

$$\le \frac{1}{2}\sum_{i=1}^{k} \log[2\pi e(EY_i^2)] - \frac{1}{2}\sum_{i=1}^{k} \log(2\pi e N_i) \tag{11.158}$$

$$= \frac{1}{2}\sum_{i=1}^{k} \log(EY_i^2) - \frac{1}{2}\sum_{i=1}^{k} \log N_i \tag{11.159}$$

$$= \frac{1}{2}\sum_{i=1}^{k} \log(EX_i^2 + EZ_i^2) - \frac{1}{2}\sum_{i=1}^{k} \log N_i \tag{11.160}$$

$$= \frac{1}{2}\sum_{i=1}^{k} \log(P_i + N_i) - \frac{1}{2}\sum_{i=1}^{k} \log N_i \tag{11.161}$$

$$= \frac{1}{2}\sum_{i=1}^{k} \log\left(1 + \frac{P_i}{N_i}\right), \tag{11.162}$$

where $P_i = EX_i^2$ is the input power of the ith channel. In the above, (11.153) is the vector generalization of Lemma 11.22, (11.155) follows because Z_i are independent, and (11.160) follows from (11.151).

Equality holds in (11.157) and (11.158) if Y_i, $1 \le i \le k$ are independent and $Y_i \sim \mathcal{N}(0, P_i + N_i)$. This happens when X_i are independent of each other and $X_i \sim \mathcal{N}(0, P_i)$. Therefore, maximizing $I(\mathbf{X}; \mathbf{Y})$ becomes maximizing $\sum_i \log(P_i + N_i)$ in (11.161) with the constraint $\sum_i P_i \le P$ and $P_i \ge 0$ for all i. In other words, we are to find the optimal input power allocation among the channels. Comparing (11.162) with (11.142), we see that the capacity of the system of parallel Gaussian channels is equal to the sum of the capacities of the individual Gaussian channels with the input power optimally allocated.

Toward this end, we first apply the method of Lagrange multipliers by temporarily ignoring the nonnegativity constraints on P_i. Observe that in order for the summation $\sum_i \log(P_i + N_i)$ in (11.161) to be maximized, $\sum_i P_i = P$ must hold because $\log(P_i + N_i)$ is increasing in P_i. Therefore, the inequality constraint $\sum_i P_i \le P$ can be replaced by the equality constraint $\sum_i P_i = P$. Let

$$J = \sum_{i=1}^{k} \log(P_i + N_i) - \mu \sum_{i=1}^{k} P_i. \tag{11.163}$$

Differentiating with respect to P_i gives

$$0 = \frac{\partial J}{\partial P_i} = \frac{\log e}{P_i + N_i} - \mu, \tag{11.164}$$

which implies

$$P_i = \frac{\log e}{\mu} - N_i. \tag{11.165}$$

Upon letting $\nu = \frac{\log e}{\mu}$, we have

$$P_i = \nu - N_i, \tag{11.166}$$

where ν is chosen such that

$$\sum_{i=1}^{k} P_i = \sum_{i=1}^{k} (\nu - N_i) = P. \tag{11.167}$$

However, P_i as given in (11.166) is not guaranteed to be nonnegative, so it may not be a valid solution. Nevertheless, (11.166) suggests the general solution to be proved in the next proposition.

Proposition 11.23. *The problem*

For given $\lambda_i \ge 0$, maximize $\sum_{i=1}^{k} \log(a_i + \lambda_i)$ subject to $\sum_i a_i \le P$ and $a_i \ge 0$

has the solution

$$a_i^* = (\nu - \lambda_i)^+, \quad 1 \le i \le k, \tag{11.168}$$

where

$$(x)^+ = \begin{cases} x \ if \ x \geq 0 \\ 0 \ if \ x < 0 \end{cases} \tag{11.169}$$

and ν satisfies

$$\sum_{i=1}^{k}(\nu - \lambda_i)^+ = P. \tag{11.170}$$

Proof. Rewrite the maximization problem as

For given $\lambda_i \geq 0$, maximize $\sum_i \log(a_i + \lambda_i)$ subject to

$$\sum_{i=1}^{k} a_i \leq P \tag{11.171}$$

$$-a_i \leq 0, \quad 1 \leq i \leq k. \tag{11.172}$$

We will prove the proposition by verifying that the proposed solution in (11.168) satisfies the Karush–Kuhn–Tucker (KKT) condition. This is done by finding nonnegative μ and μ_i satisfying the equations

$$\frac{\log e}{a_i^* + \lambda_i} - \mu + \mu_i = 0 \tag{11.173}$$

$$\mu \left(P - \sum_{i=1}^{k} a_i^* \right) = 0 \tag{11.174}$$

$$\mu_i a_i^* = 0, \quad 1 \leq i \leq k, \tag{11.175}$$

where μ and μ_i are the multipliers associated with the constraints in (11.171) and (11.172), respectively.

To avoid triviality, assume $P > 0$ so that $\nu > 0$, and observe that there exists at least one i such that $a_i^* > 0$. For those i, (11.175) implies

$$\mu_i = 0. \tag{11.176}$$

On the other hand,

$$a_i^* = (\nu - \lambda_i)^+ = \nu - \lambda_i. \tag{11.177}$$

Substituting (11.176) and (11.177) into (11.173), we obtain

$$\mu = \frac{\log e}{\nu} > 0. \tag{11.178}$$

For those i such that $a_i^* = 0$, it follows from (11.168) that $\nu \leq \lambda_i$. From (11.178) and (11.173), we obtain

$$\mu_i = (\log e)\left(\frac{1}{\nu} - \frac{1}{\lambda_i}\right) \geq 0. \tag{11.179}$$

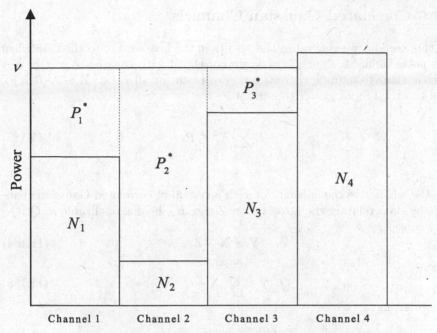

Fig. 11.2. Water-filling for parallel Gaussian channels.

Thus we have found nonnegative μ and μ_i that satisfy (11.173)–(11.175), verifying the KKT condition. The proposition is proved. $\square$

Hence, following (11.162) and applying the above proposition with $a_i = P_i$ and $\lambda_i = N_i$, we see that the capacity of the system of parallel Gaussian channels is equal to

$$\frac{1}{2} \sum_{i=1}^{k} \log \left(1 + \frac{P_i^*}{N_i} \right), \tag{11.180}$$

where $\{P_i^*, \ 1 \leq i \leq k\}$ is the optimal input power allocation among the channels given by

$$P_i^* = (\nu - N_i)^+, \quad 1 \leq i \leq k \tag{11.181}$$

with ν satisfying

$$\sum_{i=1}^{k} (\nu - N_i)^+ = P. \tag{11.182}$$

The process for obtaining $\{P_i^*\}$, called *water-filling*, is illustrated in Figure 11.2. One can image that an amount P of water is poured into a reservoir with an uneven bottom, and ν is the level the water rises to. Under this scheme, high input power is allocated to a channel with low noise power. For a channel with noise power higher than ν, no input power is allocated, i.e., the channel is not used.

11.6 Correlated Gaussian Channels

In this section, we generalize the results in the last section to the case when the noise variables Z_i, $1 \leq i \leq k$ are correlated with covariance matrix $K_{\mathbf{Z}}$. We continue to assume that Z_i has zero mean for all i, i.e., $\mathbf{Z} \sim \mathcal{N}(0, K_{\mathbf{Z}})$, and the total input power constraint

$$E \sum_{i=1}^{k} X_i^2 \leq P \tag{11.183}$$

prevails.

We will derive the capacity of such a system of correlated Gaussian channels by decorrelating the noise vector $\mathbf{Z}$. Let $K_{\mathbf{Z}}$ be diagonalizable as $Q \Lambda Q^{\top}$ and consider

$$\mathbf{Y} = \mathbf{X} + \mathbf{Z}. \tag{11.184}$$

Then

$$Q^{\top} \mathbf{Y} = Q^{\top} \mathbf{X} + Q^{\top} \mathbf{Z}. \tag{11.185}$$

Upon letting

$$\mathbf{X}' = Q^{\top} \mathbf{X}, \tag{11.186}$$
$$\mathbf{Y}' = Q^{\top} \mathbf{Y}, \tag{11.187}$$

and

$$\mathbf{Z}' = Q^{\top} \mathbf{Z}, \tag{11.188}$$

we obtain

$$\mathbf{Y}' = \mathbf{X}' + \mathbf{Z}'. \tag{11.189}$$

Note that

$$E\mathbf{Z}' = E(Q\mathbf{Z}) = Q(E\mathbf{Z}) = Q \cdot 0 = 0, \tag{11.190}$$

and $\mathbf{Z}'$ is jointly Gaussian because it is a linear transformation of $\mathbf{Z}$. By Proposition 10.6, the random variables in $\mathbf{Z}'$ are uncorrelated, and

$$K_{\mathbf{Z}'} = \Lambda. \tag{11.191}$$

Hence, $Z_i' \sim \mathcal{N}(0, \lambda_i)$, where λ_i is the ith diagonal element of Λ, and Z_i', $1 \leq i \leq k$ are mutually independent.

We are then motivated to convert the given system of correlated Gaussian channels into the system shown in Figure 11.3, with $\mathbf{X}'$ and $\mathbf{Y}'$ being the input and output, respectively. Note that in this system, $\mathbf{X}'$ and $\mathbf{Y}'$ are related to $\mathbf{X}$ and $\mathbf{Y}$ as prescribed in (11.186) and (11.187), respectively. We then see from (11.189) that $\mathbf{Z}'$ is the equivalent noise vector of the system with Z_i' being the noise variable of the ith channel. Hence, the system in Figure 11.3 is a system of parallel Gaussian channels. By Proposition 10.9, the total input

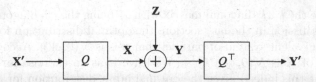

Fig. 11.3. An equivalent system of parallel Gaussian channels.

power constraint in (11.183) for the original system translates to the total input power constraint

$$E \sum_{i=1}^{k} (X_i')^2 \leq P \tag{11.192}$$

for the system in Figure 11.3.

The question is whether the capacity of the system in Figure 11.3 is the same as the capacity of the original system. Let us call these two capacities C' and C, respectively. Intuitively, C' and C should be the same because the matrix Q is invertible. A formal proof goes as follows. We remind the reader that the capacity of a channel is the highest possible asymptotic rate at which information can be transmitted reliably through the channel by means of any encoding/decoding process. In Figure 11.3, by regarding the transformation Q on $\mathbf{X}'$ as part of the encoding process and the transformation $Q^\top$ on $\mathbf{Y}$ as part of the decoding process, we see that $C' \leq C$. Now further convert the system in Figure 11.3 into the system in Figure 11.4 with input $\mathbf{X}''$ and output $\mathbf{Y}''$, and call the capacity of this system C''. By repeating the same argument, we see that $C'' \leq C'$. Thus $C'' \leq C' \leq C$. However, the system in Figure 11.4 is equivalent to the original system because $Q^\top Q = I$. Therefore, $C'' = C$, which implies $C' = C$.

Upon converting the given system of correlated Gaussian channels into an equivalent system of parallel Gaussian channels, we see that the capacity of the system is equal to

$$\frac{1}{2} \sum_{i=1}^{k} \log \left(1 + \frac{a_i^*}{\lambda_i} \right), \tag{11.193}$$

where a_i^* is the optimal power allocated to the ith channel in the equivalent system, and its value can be obtained by water-filling as prescribed in Proposition 11.23. The reader should compare (11.193) with the formula in (11.180) for the capacity of parallel Gaussian channels.

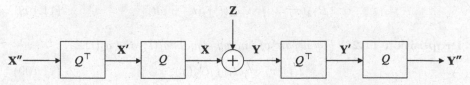

Fig. 11.4. A system identical to the system of correlated Gaussian channels.

Let A^* be the $k \times k$ diagonal matrix with a_i^* being the ith diagonal element. From the discussion in the last section, the optimal distribution for the input $\mathbf{X}'$ in the equivalent system of parallel channels is $\mathcal{N}(0, A^*)$. Accordingly, the distribution of $\mathbf{X}$ is $\mathcal{N}(0, QA^*Q^\top)$. We leave it as an exercise for the reader to verify that this indeed gives the optimal input distribution for the original system of correlated Gaussian channels.

11.7 The Bandlimited White Gaussian Channel

In this section, we discuss a bandlimited waveform channel with zero-mean *additive white Gaussian noise* (AWGN). In the rest of this chapter, the letters j and f are reserved for $\sqrt{-1}$ and "frequency," respectively. We begin with a few definitions from signal analysis. All the signals are assumed to be real.

Definition 11.24. *The Fourier transform of a signal $g(t)$ is defined as*

$$G(f) = \int_{-\infty}^{\infty} g(t)e^{-j2\pi ft}dt. \tag{11.194}$$

The signal $g(t)$ can be recovered from $G(f)$ as

$$g(t) = \int_{-\infty}^{\infty} G(f)e^{j2\pi ft}df, \tag{11.195}$$

and $g(t)$ is called the inverse Fourier transform of $G(f)$. The functions $g(t)$ and $G(f)$ are said to form a transform pair, denoted by

$$g(t) \rightleftharpoons G(f). \tag{11.196}$$

The variables t and f are referred to as time and frequency, respectively.

In general, the Fourier transform of a signal $g(t)$ may not exist. A sufficient condition for the Fourier transform of $g(t)$ to exist is that $g(t)$ has finite energy, i.e.,

$$\int_{-\infty}^{\infty} |g(t)|^2 dt < \infty. \tag{11.197}$$

A signal with finite energy is called an *energy signal*.

Definition 11.25. *Let $g_1(t)$ and $g_2(t)$ be a pair of energy signals. The cross-correlation function for $g_1(t)$ and $g_2(t)$ is defined as*

$$R_{12}(\tau) = \int_{-\infty}^{\infty} g_1(t)g_2(t-\tau)dt. \tag{11.198}$$

Proposition 11.26. *For a pair of energy signals $g_1(t)$ and $g_2(t)$,*

$$R_{12}(\tau) \rightleftharpoons G_1(f)G_2^*(f), \tag{11.199}$$

where $G_2^*(f)$ denotes the complex conjugate of $G_2(f)$.

Definition 11.27. *For a wide-sense stationary process* $\{X(t), -\infty < t < \infty\}$, *the autocorrelation function is defined as*

$$R_X(\tau) = E[X(t+\tau)X(t)], \tag{11.200}$$

which does not depend on t, and the power spectral density is defined as

$$S_X(f) = \int_{-\infty}^{\infty} R_X(\tau)e^{-j2\pi f\tau}d\tau, \tag{11.201}$$

i.e.,

$$R_X(\tau) \rightleftharpoons S_X(f). \tag{11.202}$$

Definition 11.28. *Let* $\{(X(t), Y(t)), -\infty < t < \infty\}$ *be a bivariate wide-sense stationary process. Their cross-correlation functions are defined as*

$$R_{XY}(\tau) = E[X(t+\tau)Y(t)] \tag{11.203}$$

and

$$R_{YX}(\tau) = E[Y(t+\tau)X(t)], \tag{11.204}$$

which do not depend on t. The cross-spectral densities are defined as

$$S_{XY}(f) = \int_{-\infty}^{\infty} R_{XY}(\tau)e^{-j2\pi f\tau}d\tau \tag{11.205}$$

and

$$S_{YX}(f) = \int_{-\infty}^{\infty} R_{YX}(\tau)e^{-j2\pi f\tau}d\tau, \tag{11.206}$$

i.e.,

$$R_{XY}(\tau) \rightleftharpoons S_{XY}(f) \tag{11.207}$$

and

$$R_{YX}(\tau) \rightleftharpoons S_{YX}(f). \tag{11.208}$$

We now describe the simplest nontrivial model for a waveform channel. In wired-line and wireless communication, the frequency spectrum of the medium is often partitioned into a number of communication channels, where each channel occupies a certain frequency band. Consider such a channel that occupies the frequency band $[f_l, f_h]$ with $0 \le f_l < f_h$, where

$$W = f_h - f_l \tag{11.209}$$

is called the *bandwidth*. The input process $X(t)$ is contaminated by a zero-mean additive white Gaussian noise process $Z(t)$ with power $\frac{N_0}{2}$, i.e.,

$$S_Z(f) = \frac{N_0}{2}, \quad -\infty < f < \infty. \tag{11.210}$$

In reality, such a noise process cannot exist because its total power is infinite. For practical purposes, one can regard the power spectral density to be constant within the range of interest of the problem.

Let $h(t)$ be the impulse response of an ideal bandpass filter for the frequency band $[f_l, f_h]$, i.e.,

$$H(f) = \begin{cases} 1 \text{ if } f_l \leq |f| \leq f_h \\ 0 \text{ otherwise} \end{cases}. \tag{11.211}$$

At the receiver for this channel, the ideal bandpass filter $h(t)$ is applied to the received signal in order to filter out the frequency components due to other channels. Effectively, we can regard this filtered version of the received signal given by

$$Y(t) = [X(t) + Z(t)] * h(t) = X(t) * h(t) + Z(t) * h(t) \tag{11.212}$$

as the output of the channel, where $*$ denotes convolution in the time domain. Letting

$$X'(t) = X(t) * h(t) \tag{11.213}$$

and

$$Z'(t) = Z(t) * h(t), \tag{11.214}$$

(11.212) can be written as

$$Y(t) = X'(t) + Z'(t). \tag{11.215}$$

The only difference between $X(t)$ and $X'(t)$ is that all the frequency components of $X'(t)$ are in $[f_l, f_h]$, while $X(t)$ can have frequency components outside this range. However, even if such frequency components exist in $X(t)$, they are filtered out by the ideal bandpass filter $h(t)$ and do not appear in the output process $Y(t)$. Therefore, we can regard $X'(t)$ instead of $X(t)$ as the input process of the channel. By the same token, we regard $Z'(t)$ instead of $Z(t)$ as the noise process of the channel. As for the memoryless Gaussian channel discussed in the last section, we impose an average power constraint P on the input process $X'(t)$.

For simplicity, we consider in this section the case that the channel we have described occupies the frequency band $[0, W]$. This channel, called the *bandlimited white Gaussian channel*, is the special case of the general model with $f_l = 0$.

While a rigorous formulation of the bandlimited white Gaussian channel involves mathematical tools beyond the scope of this book, we will nevertheless give a heuristic argument that suggests the formula for the channel capacity. The sampling theorem in signal analysis will allow us to "convert" this waveform channel into a memoryless Gaussian channel discussed in the last section.

Theorem 11.29 (Sampling Theorem). *Let $g(t)$ be a signal with Fourier transform $G(f)$ that vanishes for $f \notin [-W, W]$. Then*

$$g(t) = \sum_{i=-\infty}^{\infty} g\left(\frac{i}{2W}\right) \text{sinc}(2Wt - i) \tag{11.216}$$

for $-\infty < t < \infty$, where

$$\text{sinc}(t) = \frac{sin(\pi t)}{\pi t}, \tag{11.217}$$

called the sinc function, is defined to be 1 at $t = 0$ by continuity.

Letting

$$g_i = \frac{1}{\sqrt{2W}} g\left(\frac{i}{2W}\right) \tag{11.218}$$

and

$$\psi_i(t) = \sqrt{2W}\,\text{sinc}(2Wt - i), \tag{11.219}$$

the formula in (11.216) can be rewritten as

$$g(t) = \sum_{i=-\infty}^{\infty} g_i \psi_i(t). \tag{11.220}$$

Proposition 11.30. $\psi_i(t)$, $-\infty < i < \infty$ *form an orthonormal basis for signals which are bandlimited to $[0, W]$.*

Proof. Since

$$\psi_i(t) = \psi_0\left(t - \frac{i}{2W}\right), \tag{11.221}$$

$\psi_i(t)$ and $\psi_0(t)$ have the same energy. We first show that

$$\int_{-\infty}^{\infty} \text{sinc}^2(2Wt)dt = \frac{1}{2W}. \tag{11.222}$$

This integral is difficult to evaluate directly. Instead we consider

$$\text{sinc}(2Wt) \rightleftharpoons \frac{1}{2W}\text{rect}\left(\frac{f}{2W}\right), \tag{11.223}$$

where

$$\text{rect}(f) = \begin{cases} 1 & -\frac{1}{2} \leq f \leq \frac{1}{2} \\ 0 & \text{otherwise} \end{cases}. \tag{11.224}$$

Then by *Rayleigh's energy theorem*, we have

$$\int_{-\infty}^{\infty} \text{sinc}^2(2Wt)dt = \int_{-\infty}^{\infty} \left(\frac{1}{2W}\right)^2 \text{rect}^2\left(\frac{f}{2W}\right) df = \frac{1}{2W}. \tag{11.225}$$

It then follows that

$$\int_{-\infty}^{\infty} \psi_i^2(t)dt = \int_{-\infty}^{\infty} \psi_0^2(t)dt \tag{11.226}$$

$$= (\sqrt{2W})^2 \int_{-\infty}^{\infty} \operatorname{sinc}^2(2Wt)dt \tag{11.227}$$

$$= 2W\left(\frac{1}{2W}\right) \tag{11.228}$$

$$= 1. \tag{11.229}$$

Next, we show that

$$\int_{-\infty}^{\infty} \operatorname{sinc}(2Wt - i)\operatorname{sinc}(2Wt - i')dt \tag{11.230}$$

vanishes whenever $i \neq i'$. Again, this integral is difficult to evaluate directly. Since (11.225) implies that both $\operatorname{sinc}(2Wt-i)$ and $\operatorname{sinc}(2Wt-i')$ have finite energy, we can consider their cross-correlation function, denoted by $R_{ii'}(\tau)$. Now

$$\operatorname{sinc}(2Wt - i) \rightleftharpoons e^{-j2\pi f\left(\frac{i}{2W}\right)}\left(\frac{1}{2W}\right)\operatorname{rect}\left(\frac{f}{2W}\right) := G_i(f) \tag{11.231}$$

and

$$\operatorname{sinc}(2Wt - i') \rightleftharpoons e^{-j2\pi f\left(\frac{i'}{2W}\right)}\left(\frac{1}{2W}\right)\operatorname{rect}\left(\frac{f}{2W}\right) := G_{i'}(f). \tag{11.232}$$

Then we have

$$R_{ii'}(\tau) \rightleftharpoons G_i(f)G_{i'}^*(f), \tag{11.233}$$

and the integral in (11.230) is given by

$$R_{ii'}(0) = \int_{-\infty}^{\infty} G_i(f)G_{i'}^*(f)df \tag{11.234}$$

(cf. (11.195)), which vanishes whenever $i \neq i'$. Therefore,

$$\int_{-\infty}^{\infty} \psi_i(t)\psi_{i'}(t)dt = 2W\int_{-\infty}^{\infty} \operatorname{sinc}(2Wt - i)\operatorname{sinc}(2Wt - i')dt \tag{11.235}$$

$$= 0. \tag{11.236}$$

Together with (11.229), this shows that $\psi_i(t)$, $-\infty < i < \infty$ form an orthonormal set. Finally, since $g(t)$ in (11.220) is any signal bandlimited to $[0, W]$, we conclude that $\psi_i(t)$, $-\infty < i < \infty$ form an orthonormal basis for such signals. The theorem is proved. $\square$

Let us return to our discussion of the waveform channel. The sampling theorem implies that the input process $X'(t)$, assuming the existence of the Fourier transform, can be written as

$$X'(t) = \sum_{i=-\infty}^{\infty} X_i' \psi_i(t), \tag{11.237}$$

where

$$X_i' = \frac{1}{\sqrt{2W}} X'\left(\frac{i}{2W}\right), \tag{11.238}$$

and there is a one-to-one correspondence between $X'(t)$ and $\{X_i', -\infty < i < \infty\}$. The same applies to (a realization of) the output process $Y(t)$, which we assume can be written as

$$Y(t) = \sum_{i=-\infty}^{\infty} Y_i' \psi_i(t), \tag{11.239}$$

where

$$Y_i = \frac{1}{\sqrt{2W}} Y\left(\frac{i}{2W}\right). \tag{11.240}$$

With these assumptions on $X'(t)$ and $Y(t)$, the waveform channel is equivalent to a discrete-time channel defined at $t = i/2W$, with the ith input and output of the channel being X_i' and Y_i, respectively.

Toward determining the capacity of this equivalent discrete-time channel, we prove in the following a characterization of the effect of the noise process $Z'(t)$ at the sampling times.

Proposition 11.31. $Z'\left(\frac{i}{2W}\right)$, $-\infty < i < \infty$ are i.i.d. Gaussian random variables with zero mean and variance $N_0 W$.

Proof. First of all, $Z(t)$ is a zero-mean Gaussian process and $Z'(t)$ is a filtered version of $Z(t)$, so $Z'(t)$ is also a zero-mean Gaussian process. Consequently, $Z'\left(\frac{i}{2W}\right)$, $-\infty < i < \infty$ are zero-mean Gaussian random variables. The power spectral density of $Z'(t)$ is given by

$$S_{Z'}(f) = \begin{cases} \frac{N_0}{2} & -W \le f \le W \\ 0 & \text{otherwise} \end{cases}. \tag{11.241}$$

Then the autocorrelation function of $Z'(t)$, which is the inverse Fourier transform of $S_{Z'}(f)$, is given by

$$R_{Z'}(\tau) = N_0 W \operatorname{sinc}(2W\tau). \tag{11.242}$$

It is seen that the value of $R_{Z'}(\tau)$ is equal to 0 when $\tau = \frac{i}{2W}$ for all $i \ne 0$, because the sinc function in (11.217) vanishes at all nonzero integer values of t. This shows that $Z'\left(\frac{i}{2W}\right)$, $-\infty < i < \infty$ are uncorrelated and hence independent because they are jointly Gaussian. Finally, since $Z'\left(\frac{i}{2W}\right)$ has zero mean, in light of (11.200), its variance is given by $R_{Z'}(0) = N_0 W$. $\square$

Recall from (11.215) that

$$Y(t) = X'(t) + Z'(t). \tag{11.243}$$

Then

$$Y\left(\frac{i}{2W}\right) = X'\left(\frac{i}{2W}\right) + Z'\left(\frac{i}{2W}\right). \tag{11.244}$$

Upon dividing by $\sqrt{2W}$ and letting

$$Z_i' = \frac{1}{\sqrt{2W}} Z'\left(\frac{i}{2W}\right), \tag{11.245}$$

it follows from (11.238) and (11.240) that

$$Y_i = X_i' + Z_i'. \tag{11.246}$$

Since $Z'\left(\frac{i}{2W}\right)$, $-\infty < i < \infty$ are i.i.d. with distribution $\mathcal{N}(0, N_0 W)$, Z_i', $-\infty < i < \infty$ are i.i.d. with distribution $\mathcal{N}(0, \frac{N_0}{2})$.

Thus we have shown that the bandlimited white Gaussian channel is equivalent to a memoryless Gaussian channel with noise power equal to $N_0/2$. As we are converting the waveform channel into a discrete-time channel, we need to relate the input power constraint of the waveform channel to the input power constraint of the discrete-time channel. Let P' be the average energy (i.e., the second moment) of the X_is. We now calculate the average power of $X'(t)$ in terms of P'. Since $\psi_i(t)$ has unit energy, the average contribution to the energy of $X'(t)$ by each sample is P'. As there are $2W$ samples per unit time and $\psi_i(t)$, $-\infty < i < \infty$ are orthonormal, $X'(t)$ accumulates energy from the samples at a rate equal to $2WP'$. Upon considering

$$2WP' \le P, \tag{11.247}$$

where P is the average power constraint on the input process $X'(t)$, we obtain

$$P' \le \frac{P}{2W}, \tag{11.248}$$

i.e., an input power constraint P for the bandlimited Gaussian channel translates to an input power constraint $P/2W$ for the discrete-time channel. By Theorem 11.21, the capacity of the memoryless Gaussian channel is

$$\frac{1}{2} \log\left(1 + \frac{P/2W}{N_0/2}\right) = \frac{1}{2} \log\left(1 + \frac{P}{N_0 W}\right) \text{ bits per sample.} \tag{11.249}$$

Since there are $2W$ samples per unit time, we conclude that the capacity of the bandlimited Gaussian channel is

$$W \log\left(1 + \frac{P}{N_0 W}\right) \text{ bits per unit time.} \tag{11.250}$$

The argument we have given above is evidently not rigorous because if there is no additional constraint on the X_is other than their average energy not exceeding $P/2W$, then $X'(t)$ may not have finite energy. This induces a gap in the argument because the Fourier transform of $X'(t)$ may not exist and hence the sampling theorem cannot be applied.

A rigorous formulation of the bandlimited white Gaussian channel involves the consideration of an input signal of finite duration, which is analogous to a code for the DMC with a finite block length. Since a signal with finite duration cannot be bandlimited, this immediately leads to a contradiction. Overcoming this technical difficulty requires the use of *prolate spheroidal wave functions* [338, 218, 219] which are bandlimited functions with most of the energy on a finite interval. The main idea is that there are approximately $2WT$ orthonormal basis functions for the set of signals which are bandlimited to W and have most of the energy on $[0, T)$ in time. We refer the reader to Gallager [129] for a rigorous treatment of the bandlimited white Gaussian channel.

11.8 The Bandlimited Colored Gaussian Channel

In the last section, we have discussed the bandlimited white Gaussian channel occupying the frequency band $[0, W]$. We presented a heuristic argument that led to the formula in (11.250) for the channel capacity. Suppose the channel instead occupies the frequency band $[f_l, f_h]$, with f_l being a multiple of $W = f_h - f_l$. Then the noise process $Z'(t)$ has power spectral density

$$S_{Z'}(f) = \begin{cases} \frac{N_0}{2} & \text{if } f_l \leq |f| \leq f_h \\ 0 & \text{otherwise} \end{cases}. \tag{11.251}$$

We refer to such a channel as the *bandpass white Gaussian channel*. By an extension of the heuristic argument for the bandlimited white Gaussian channel, which would involve the bandpass version of the sampling theorem, the same formula for the channel capacity can be obtained. The details are omitted here.

We now consider a waveform channel occupying the frequency band $[0, W]$ with input power constraint P and zero-mean additive *colored* Gaussian noise $Z(t)$. We refer to this channel as the *bandlimited colored Gaussian channel*. To analyze the capacity of this channel, divide the interval $[0, W]$ into subintervals $[f_l^i, f_h^i]$ for $1 \leq i \leq k$, where

$$f_l^i = (i - 1)\Delta_k, \tag{11.252}$$
$$f_h^i = i\Delta_k, \tag{11.253}$$

and

$$\Delta_k = \frac{W}{k} \tag{11.254}$$

is the width of each subinterval. As an approximation, assume $S_Z(f)$ is equal to a constant $S_{Z,i}$ over the subinterval $[f_l^i, f_h^i]$. Then the channel consists of k sub-channels, with the ith sub-channel being a bandpass (bandlimited if

$i = 1$) white Gaussian channel occupying the frequency band $[f_l^i, f_h^i]$. Thus by letting $N_0 = 2S_{Z,i}$ in (11.251), we obtain from (11.250) that the capacity of the ith sub-channel is equal to

$$\Delta_k \log\left(1 + \frac{P_i}{2S_{Z,i}\Delta_k}\right) \tag{11.255}$$

if P_i is the input power allocated to that sub-channel.

The noise process of the ith sub-channel, denoted by $Z_i'(t)$, is obtained by passing $Z(t)$ through the ideal bandpass filter with frequency response

$$H_i(f) = \begin{cases} 1 \text{ if } f_l^i \leq |f| \leq f_h^i \\ 0 \text{ otherwise} \end{cases}. \tag{11.256}$$

It can be shown (see Problem 9) that the noise processes $Z_i(t)$, $1 \leq i \leq k$ are independent. By converting each sub-channel into an equivalent memoryless Gaussian channel as discussed in the last section, we see that the k sub-channels can be regarded as a system of parallel Gaussian channels. Thus the channel capacity is equal to the sum of the capacities of the individual sub-channels when the power allocation among the k sub-channels is optimal.

Let P_i^* be the optimal power allocation for the ith sub-channel. Then it follows from (11.255) that the channel capacity is equal to

$$\sum_{i=1}^{k} \Delta_k \log\left(1 + \frac{P_i^*}{2S_{Z,i}\Delta_k}\right) = \sum_{i=1}^{k} \Delta_k \log\left(1 + \frac{\frac{P_i^*}{2\Delta_k}}{S_{Z,i}}\right), \tag{11.257}$$

where by Proposition 11.23,

$$\frac{P_i^*}{2\Delta_k} = (\nu - S_{Z,i})^+ \tag{11.258}$$

or

$$P_i^* = 2\Delta_k(\nu - S_{Z,i})^+ \tag{11.259}$$

with

$$\sum_{i=1}^{k} P_i^* = P. \tag{11.260}$$

Then from (11.259) and (11.260), we obtain

$$\sum_{i=1}^{k} (\nu - S_{Z,i})^+ \Delta_k = \frac{P}{2}. \tag{11.261}$$

As $k \to \infty$, following (11.257) and (11.258),

$$\sum_{i=1}^{k} \Delta_k \log \left(1 + \frac{\frac{P_i^*}{2\Delta_k}}{S_{Z,i}} \right) = \sum_{i=1}^{k} \Delta_k \log \left(1 + \frac{(\nu - S_{Z,i})^+}{S_{Z,i}} \right) \tag{11.262}$$

$$\rightarrow \int_0^W \log \left(1 + \frac{(\nu - S_Z(f))^+}{S_Z(f)} \right) df \tag{11.263}$$

$$= \frac{1}{2} \int_{-W}^{W} \log \left(1 + \frac{(\nu - S_Z(f))^+}{S_Z(f)} \right) df, \tag{11.264}$$

and following (11.261),

$$\sum_{i=1}^{k} (\nu - S_{Z,i})^+ \Delta_k \rightarrow \int_0^W (\nu - S_Z(f))^+ df \tag{11.265}$$

$$= \frac{1}{2} \int_{-W}^{W} (\nu - S_Z(f))^+ df, \tag{11.266}$$

where (11.264) and (11.266) are obtained by noting that

$$S_{Z'}(-f) = S_{Z'}(f) \tag{11.267}$$

for $-\infty < f < \infty$ (see Problem 8). Hence, we conclude that the capacity of the bandlimited colored Gaussian channel is equal to

$$\frac{1}{2} \int_{-W}^{W} \log \left(1 + \frac{(\nu - S_Z(f))^+}{S_Z(f)} \right) df \quad \text{bits per unit time,} \tag{11.268}$$

where ν satisfies

$$\int_{-W}^{W} (\nu - S_Z(f))^+ df = P \tag{11.269}$$

in view of (11.261) and (11.266). Figure 11.5 is an illustration of the water-filling process for determining ν, where the amount of water to be poured into the reservoir is equal to P.

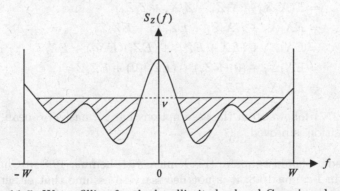

Fig. 11.5. Water-filling for the bandlimited colored Gaussian channel.

11.9 Zero-Mean Gaussian Noise Is the Worst Additive Noise

In the last section, we derived the capacity for a system of correlated Gaussian channels, where the noise vector is a zero-mean Gaussian random vector. In this section, we show that in terms of the capacity of the system, the zero-mean Gaussian noise is the worst additive noise given that the noise vector has a fixed correlation matrix. Note that the diagonal elements of this correlation matrix specify the power of the individual noise variables, while the other elements in the matrix give a characterization of the correlation between the noise variables.

Theorem 11.32. *For a fixed zero-mean Gaussian random vector* $\mathbf{X}^*$, *let*

$$\mathbf{Y} = \mathbf{X}^* + \mathbf{Z}, \tag{11.270}$$

where the joint pdf of $\mathbf{Z}$ *exists and* $\mathbf{Z}$ *is independent of* $\mathbf{X}^*$. *Under the constraint that the correlation matrix of* $\mathbf{Z}$ *is equal to* K, *where* K *is any symmetric positive definite matrix,* $I(\mathbf{X}^*; \mathbf{Y})$ *is minimized if and only if* $\mathbf{Z} \sim \mathcal{N}(0, K)$.

Before proving this theorem, we first prove the following two lemmas.

Lemma 11.33. *Let* $\mathbf{X}$ *be a zero-mean random vector and*

$$\mathbf{Y} = \mathbf{X} + \mathbf{Z}, \tag{11.271}$$

where $\mathbf{Z}$ *is independent of* $\mathbf{X}$. *Then*

$$\tilde{K}_{\mathbf{Y}} = \tilde{K}_{\mathbf{X}} + \tilde{K}_{\mathbf{Z}}. \tag{11.272}$$

Proof. For any i and j, consider

$$
\begin{align}
EY_i Y_j &= E(X_i + Z_i)(X_j + Z_j) \tag{11.273} \\
&= E(X_i X_j + X_i Z_j + Z_i X_j + Z_i Z_j) \tag{11.274} \\
&= EX_i X_j + EX_i Z_j + EZ_i X_j + EZ_i Z_j \tag{11.275} \\
&= EX_i X_j + (EX_i)(EZ_j) + (EZ_i)(EX_j) + EZ_i Z_j \tag{11.276} \\
&= EX_i X_j + (0)(EZ_j) + (EZ_i)(0) + EZ_i Z_j \tag{11.277} \\
&= EX_i X_j + EZ_i Z_j, \tag{11.278}
\end{align}
$$

where (11.277) follows from the assumption that X_i has zero mean for all i. The proposition is proved.   □

The reader should compare this lemma with Lemma 10.8 in Chapter 10. Note that in Lemma 10.8, it is not necessary to assume that either $\mathbf{X}$ or $\mathbf{Z}$ has zero mean.

Lemma 11.34. *Let* $\mathbf{Y}^* \sim \mathcal{N}(0, K)$ *and* $\mathbf{Y}$ *be any random vector with correlation matrix K. Then*

$$\int f_{\mathbf{Y}^*}(\mathbf{y}) \log f_{\mathbf{Y}^*}(\mathbf{y})d\mathbf{y} = \int_{\mathcal{S}_{\mathbf{Y}}} f_{\mathbf{Y}}(\mathbf{y}) \log f_{\mathbf{Y}^*}(\mathbf{y})d\mathbf{y}. \tag{11.279}$$

Proof. The random vector $\mathbf{Y}^*$ has joint pdf

$$f_{\mathbf{Y}^*}(\mathbf{y}) = \frac{1}{\left(\sqrt{2\pi}\right)^k |K|^{1/2}} e^{-\frac{1}{2}(\mathbf{y}^\top K^{-1}\mathbf{y})} \tag{11.280}$$

for all $\mathbf{y} \in \Re^k$. Since $E\mathbf{Y}^* = 0$, $\tilde{K}_{\mathbf{Y}^*} = K_{\mathbf{Y}^*} = K$. Therefore, $\mathbf{Y}^*$ and $\mathbf{Y}$ have the same correlation matrix. Consider

$$\int [\ln f_{\mathbf{Y}^*}(\mathbf{y})] f_{\mathbf{Y}^*}(\mathbf{y})d\mathbf{y}$$

$$= \int \left[-\frac{1}{2}(\mathbf{y}^\top K^{-1}\mathbf{y}) - \ln\left[(\sqrt{2\pi})^k |K|^{1/2}\right] \right] f_{\mathbf{Y}^*}(\mathbf{y})d\mathbf{y} \tag{11.281}$$

$$= -\frac{1}{2} \int (\mathbf{y}^\top K^{-1}\mathbf{y}) f_{\mathbf{Y}^*}(\mathbf{y})d\mathbf{y} - \ln\left[(\sqrt{2\pi})^k |K|^{1/2}\right] \tag{11.282}$$

$$= -\frac{1}{2} \int \left[\sum_{i,j} (K^{-1})_{ij} y_i y_j \right] f_{\mathbf{Y}^*}(\mathbf{y})d\mathbf{y} - \ln\left[(\sqrt{2\pi})^k |K|^{1/2}\right] \tag{11.283}$$

$$= -\frac{1}{2} \sum_{i,j} (K^{-1})_{ij} \int (y_i y_j) f_{\mathbf{Y}^*}(\mathbf{y})d\mathbf{y} - \ln\left[(\sqrt{2\pi})^k |K|^{1/2}\right] \tag{11.284}$$

$$= -\frac{1}{2} \sum_{i,j} (K^{-1})_{ij} \int_{\mathcal{S}_{\mathbf{Y}}} (y_i y_j) f_{\mathbf{Y}}(\mathbf{y})d\mathbf{y} - \ln\left[(\sqrt{2\pi})^k |K|^{1/2}\right] \tag{11.285}$$

$$= \int_{\mathcal{S}_{\mathbf{Y}}} \left[-\frac{1}{2}\mathbf{y}^\top K^{-1}\mathbf{y} - \ln\left[(\sqrt{2\pi})^k |K|^{1/2}\right] \right] f_{\mathbf{Y}}(\mathbf{y})d\mathbf{y} \tag{11.286}$$

$$= \int_{\mathcal{S}_{\mathbf{Y}}} [\ln f_{\mathbf{Y}^*}(\mathbf{y})] f_{\mathbf{Y}}(\mathbf{y})d\mathbf{y}. \tag{11.287}$$

In the above, (11.285) is justified because $\mathbf{Y}$ and $\mathbf{Y}^*$ have the same correlation matrix, and (11.286) is obtained by backtracking the manipulations from (11.281) to (11.284) with $f_{\mathbf{Y}}(\mathbf{y})$ in place of $f_{\mathbf{Y}^*}(\mathbf{y})$. The lemma is proved upon changing the base of the logarithm. □

Proof of Theorem 11.32. Let $\mathbf{Z}^* \sim \mathcal{N}(0, K)$ such that $\mathbf{Z}^*$ is independent of $\mathbf{X}^*$, and let

$$\mathbf{Y}^* = \mathbf{X}^* + \mathbf{Z}^*. \tag{11.288}$$

Obviously, the support of $\mathbf{Y}^*$ is $\Re^k$ because $\mathbf{Y}^*$ has a multivariate Gaussian distribution. Note that the support of $\mathbf{Y}$ is also $\Re^k$ regardless of the distribution of Z because the support of $\mathbf{X}^*$ is $\Re^k$. We need to prove that for any random. vector $\mathbf{Z}$ with correlation matrix K, where $\mathbf{Z}$ is independent of $\mathbf{X}^*$ and the joint pdf of $\mathbf{Z}$ exists,

$$I(\mathbf{X}^*; \mathbf{Y}^*) \leq I(\mathbf{X}^*; \mathbf{Y}). \tag{11.289}$$

Since $E\mathbf{Z}^* = 0$, $\tilde{K}_{\mathbf{Z}^*} = \dot{K}_{\mathbf{Z}^*} = K$. Therefore, $\mathbf{Z}^*$ and $\mathbf{Z}$ have the same correlation matrix. By noting that $\mathbf{X}^*$ has zero mean, we apply Lemma 11.33 to see that $\mathbf{Y}^*$ and $\mathbf{Y}$ have the same correlation matrix.

The inequality in (11.289) can be proved by considering

$$I(\mathbf{X}^*; \mathbf{Y}^*) - I(\mathbf{X}^*; \mathbf{Y})$$

$$\overset{a)}{=} h(\mathbf{Y}^*) - h(\mathbf{Z}^*) - h(\mathbf{Y}) + h(\mathbf{Z}) \tag{11.290}$$

$$= -\int f_{\mathbf{Y}^*}(\mathbf{y}) \log f_{\mathbf{Y}^*}(\mathbf{y}) d\mathbf{y} + \int f_{\mathbf{Z}^*}(\mathbf{z}) \log f_{\mathbf{Z}^*}(\mathbf{z}) d\mathbf{z}$$

$$+ \int f_{\mathbf{Y}}(\mathbf{y}) \log f_{\mathbf{Y}}(\mathbf{y}) d\mathbf{y} - \int_{S_{\mathbf{Z}}} f_{\mathbf{Z}}(\mathbf{z}) \log f_{\mathbf{Z}}(\mathbf{z}) d\mathbf{z} \tag{11.291}$$

$$\overset{b)}{=} -\int f_{\mathbf{Y}}(\mathbf{y}) \log f_{\mathbf{Y}^*}(\mathbf{y}) d\mathbf{y} + \int_{S_{\mathbf{Z}}} f_{\mathbf{Z}}(\mathbf{z}) \log f_{\mathbf{Z}^*}(\mathbf{z}) d\mathbf{z}$$

$$+ \int f_{\mathbf{Y}}(\mathbf{y}) \log f_{\mathbf{Y}}(\mathbf{y}) d\mathbf{y} - \int_{S_{\mathbf{Z}}} f_{\mathbf{Z}}(\mathbf{z}) \log f_{\mathbf{Z}}(\mathbf{z}) d\mathbf{z} \tag{11.292}$$

$$= \int \log \left(\frac{f_{\mathbf{Y}}(\mathbf{y})}{f_{\mathbf{Y}^*}(\mathbf{y})} \right) f_{\mathbf{Y}}(\mathbf{y}) d\mathbf{y} + \int_{S_{\mathbf{Z}}} \log \left(\frac{f_{\mathbf{Z}^*}(\mathbf{z})}{f_{\mathbf{Z}}(\mathbf{z})} \right) f_{\mathbf{Z}}(\mathbf{z}) d\mathbf{z} \tag{11.293}$$

$$\overset{c)}{=} \int_{S_{\mathbf{Z}}} \int \log \left(\frac{f_{\mathbf{Y}}(\mathbf{y}) f_{\mathbf{Z}^*}(\mathbf{z})}{f_{\mathbf{Y}^*}(\mathbf{y}) f_{\mathbf{Z}}(\mathbf{z})} \right) f_{\mathbf{YZ}}(\mathbf{y}, \mathbf{z}) d\mathbf{y} d\mathbf{z} \tag{11.294}$$

$$\overset{d)}{\leq} \log \left(\int_{S_{\mathbf{Z}}} \int \frac{f_{\mathbf{Y}}(\mathbf{y}) f_{\mathbf{Z}^*}(\mathbf{z})}{f_{\mathbf{Y}^*}(\mathbf{y}) f_{\mathbf{Z}}(\mathbf{z})} f_{\mathbf{YZ}}(\mathbf{y}, \mathbf{z}) d\mathbf{y} d\mathbf{z} \right) \tag{11.295}$$

$$\overset{e)}{=} \log \left(\int \left[\frac{1}{f_{\mathbf{Y}^*}(\mathbf{y})} \int_{S_{\mathbf{Z}}} f_{\mathbf{X}^*}(\mathbf{y} - \mathbf{z}) f_{\mathbf{Z}^*}(\mathbf{z}) d\mathbf{z} \right] f_{\mathbf{Y}}(\mathbf{y}) d\mathbf{y} \right) \tag{11.296}$$

$$\overset{f)}{\leq} \log \left(\int \frac{f_{\mathbf{Y}^*}(\mathbf{y})}{f_{\mathbf{Y}^*}(\mathbf{y})} f_{\mathbf{Y}}(\mathbf{y}) d\mathbf{y} \right) \tag{11.297}$$

$$= 0. \tag{11.298}$$

The above steps are explained as follows:

- We assume that the pdf of $\mathbf{Z}$ exists so that $h(\mathbf{Z})$ is defined. Moreover,

$$f_{\mathbf{Y}|\mathbf{X}^*}(\mathbf{y}|\mathbf{x}) = f_{\mathbf{Z}}(\mathbf{y} - \mathbf{x}). \tag{11.299}$$

Then by Proposition 10.24, $f_{\mathbf{Y}}(\mathbf{y})$ exists and hence $h(\mathbf{Y})$ is defined.
- In (b), we have replaced $f_{\mathbf{Y}^*}(\mathbf{y})$ by $f_{\mathbf{Y}}(\mathbf{y})$ in the first integral and replaced $f_{\mathbf{Z}^*}(\mathbf{z})$ by $f_{\mathbf{Z}}(\mathbf{z})$ in the second integral. The former is justified by an application of Lemma 11.34 to $\mathbf{Y}^*$ and $\mathbf{Y}$ by noting that $\mathbf{Y}^*$ is a zero-mean Gaussian random vector and $\mathbf{Y}^*$ and $\mathbf{Y}$ have the same correlation matrix. The latter is justified similarly.

- To justify (c), we need $\mathcal{S}_{\mathbf{YZ}} = \Re^k \times \mathcal{S}_{\mathbf{Z}}$, which can be seen by noting that

$$f_{\mathbf{YZ}}(\mathbf{y}, \mathbf{z}) = f_{\mathbf{Y}|\mathbf{Z}}(\mathbf{y}|\mathbf{z})f_{\mathbf{Z}}(\mathbf{z}) = f_{\mathbf{X}^*}(\mathbf{y} - \mathbf{z})f_{\mathbf{Z}}(\mathbf{z}) > 0 \qquad (11.300)$$

for all $\mathbf{y} \in \Re^k$ and all $\mathbf{z} \in \mathcal{S}_{\mathbf{Z}}$.
- (d) follows from Jensen's inequality and the concavity of the logarithmic function.
- (e) follows from (11.300).
- (f) follows because

$$\int_{\mathcal{S}_{\mathbf{z}}} f_{\mathbf{X}^*}(\mathbf{y} - \mathbf{z})f_{\mathbf{Z}^*}(\mathbf{z})d\mathbf{z} = \int_{\mathcal{S}_{\mathbf{z}}} f_{\mathbf{Y}^*|\mathbf{Z}^*}(\mathbf{y}|\mathbf{z})f_{\mathbf{Z}^*}(\mathbf{z})d\mathbf{z} \qquad (11.301)$$

$$\leq \int f_{\mathbf{Y}^*|\mathbf{Z}^*}(\mathbf{y}|\mathbf{z})f_{\mathbf{Z}^*}(\mathbf{z})d\mathbf{z} \qquad (11.302)$$

$$= f_{\mathbf{Y}^*}(\mathbf{y}). \qquad (11.303)$$

It remains to show that $\mathbf{Z} \sim \mathcal{N}(0, K)$ is a necessary and sufficient condition for $I(\mathbf{X}^*; \mathbf{Y})$ to be minimized. Toward this end, we need to show that equality holds in both (11.295) and (11.297) if and only if $\mathbf{Z} \sim \mathcal{N}(0, K)$. Assume that $\mathbf{Z} \sim \mathcal{N}(0, K)$. Then $f_{\mathbf{Z}}(\mathbf{z}) = f_{\mathbf{Z}^*}(\mathbf{z})$ for all $\mathbf{z} \in \mathcal{S}_{\mathbf{Z}} = \Re^k$. This implies $f_{\mathbf{Y}}(\mathbf{y}) = f_{\mathbf{Y}^*}(\mathbf{y})$ for all $\mathbf{y} \in \Re^k$ in view of (11.270) and (11.288), so that

$$f_{\mathbf{Y}}(\mathbf{y})f_{\mathbf{Z}^*}(\mathbf{z}) = f_{\mathbf{Y}^*}(\mathbf{y})f_{\mathbf{Z}}(\mathbf{z}) \quad \text{for all } \mathbf{y} \in \Re^k, \mathbf{z} \in \mathcal{S}_{\mathbf{Z}}. \qquad (11.304)$$

Then the iterated integral in (11.294) is evaluated to zero. Therefore, it follows from (11.294) to (11.298) that equality holds in both (11.295) and (11.297).

Conversely, assume that equality holds in both (11.295) and (11.297). In view of (11.302), equality holds in (11.297) if and only if $\Re^k \backslash \mathcal{S}_{\mathbf{Z}}$ has zero Lebesgue measure. On the other hand, equality holds in (11.295) if and only if

$$f_{\mathbf{Y}}(\mathbf{y})f_{\mathbf{Z}^*}(\mathbf{z}) = c f_{\mathbf{Y}^*}(\mathbf{y})f_{\mathbf{Z}}(\mathbf{z}) \qquad (11.305)$$

a.e. on $\Re^k \times \mathcal{S}_{\mathbf{Z}}$ for some constant c. Integrating with respect to $d\mathbf{y}$ over $\Re^k$, we obtain

$$f_{\mathbf{Z}^*}(\mathbf{z}) \int f_{\mathbf{Y}}(\mathbf{y})d\mathbf{y} = c f_{\mathbf{Z}}(\mathbf{z}) \int f_{\mathbf{Y}^*}(\mathbf{y})d\mathbf{y}, \qquad (11.306)$$

which implies

$$f_{\mathbf{Z}^*}(\mathbf{z}) = c f_{\mathbf{Z}}(\mathbf{z}) \qquad (11.307)$$

a.e. on $\mathcal{S}_{\mathbf{Z}}$. Further integrating with respect to $d\mathbf{z}$ over $\mathcal{S}_{\mathbf{Z}}$, we obtain

$$\int_{\mathcal{S}_{\mathbf{z}}} f_{\mathbf{Z}^*}(\mathbf{z})d\mathbf{z} = c \int_{\mathcal{S}_{\mathbf{z}}} f_{\mathbf{Z}}(\mathbf{z})d\mathbf{z} = c. \qquad (11.308)$$

Since $\Re^k \backslash \mathcal{S}_{\mathbf{Z}}$ has zero Lebesgue measure,

$$\int_{\mathcal{S}_{\mathbf{z}}} f_{\mathbf{Z}^*}(\mathbf{z})d\mathbf{z} = \int f_{\mathbf{Z}^*}(\mathbf{z})d\mathbf{z} = 1, \qquad (11.309)$$

which together with (11.308) implies $c = 1$. It then follows from (11.307) that $f_{\mathbf{Z}^*}(\mathbf{z}) = f_{\mathbf{Z}}(\mathbf{z})$ a.e. on $\mathcal{S}_{\mathbf{Z}}$ and hence on $\Re^k$, i.e., $\mathbf{Z} \sim \mathcal{N}(0, K)$.

Therefore, we conclude that $\mathbf{Z} \sim \mathcal{N}(0, K)$ is a necessary and sufficient condition for $I(\mathbf{X}^*; \mathbf{Y})$ to be minimized. The theorem is proved.   □

Consider the system of correlated Gaussian channels discussed in the last section. Denote the noise vector by $\mathbf{Z}^*$ and its correlation matrix by K. Note that K is also the covariance matrix of $\mathbf{Z}^*$ because $\mathbf{Z}^*$ has zero mean. In other words, $\mathbf{Z}^* \sim \mathcal{N}(0, K)$. Refer to this system as the zero-mean Gaussian system and let C^* be its capacity. Then consider another system with exactly the same specification except that the noise vector, denoted by $\mathbf{Z}$, may be neither zero-mean nor Gaussian. We, however, require that the joint pdf of $\mathbf{Z}$ exists. Refer to this system as the alternative system and let C be its capacity.

We now apply Theorem 11.32 to show that $C \geq C^*$. Let $\mathbf{X}^*$ be the input random vector that achieves the capacity of the zero-mean Gaussian system. We have mentioned at the end of Section 11.6 that $\mathbf{X}^*$ is a zero-mean Gaussian random vector. Let $\mathbf{Y}^*$ and $\mathbf{Y}$ be defined in (11.288) and (11.270), which correspond to the outputs of the zero-mean Gaussian system and the alternative system, respectively, when $\mathbf{X}^*$ is the input of both systems. Then

$$C \geq I(\mathbf{X}^*; \mathbf{Y}) \geq I(\mathbf{X}^*; \mathbf{Y}^*) = C^*, \qquad (11.310)$$

where the second inequality follows from (11.289) in the proof of Theorem 11.32. Hence, we conclude that the zero-mean Gaussian noise is indeed the worst additive noise subject to a constraint on the correlation matrix.

Chapter Summary

Capacity of Continuous Memoryless Channel: For a continuous memoryless channel $f(y|x)$ with average input constraint (κ, P), namely the requirement that

$$\frac{1}{n} \sum_{i=1}^{n} \kappa(x_i) \leq P$$

for any codeword $(x_1, x_2, \cdots, x_n)$ transmitted over the channel,

$$C(P) = \sup_{F(x): E\kappa(X) \leq P} I(X; Y),$$

where $F(x)$ is the input distribution of the channel. $C(P)$ is non-decreasing, concave, and left-continuous.

Mutually Typical Set: For a joint distribution $F(x, y)$,

$$\Psi^n_{[XY]\delta} = \left\{ (\mathbf{x}, \mathbf{y}) \in \mathcal{X}^n \times \mathcal{Y}^n : \left| \frac{1}{n} \log \frac{f(\mathbf{y}|\mathbf{x})}{f(\mathbf{y})} - I(X; Y) \right| \leq \delta \right\}.$$

Lemma: For any $\delta > 0$ and sufficiently large n,

$$\Pr\{(\mathbf{X}, \mathbf{Y}) \in \Psi^n_{[XY]\delta}\} \geq 1 - \delta.$$

Lemma: Let X and Y be a pair of random variables, and $(\mathbf{X}', \mathbf{Y}')$ be n i.i.d. copies of a pair of generic random variables (X', Y') where X' and Y' are independent and have the same marginal distributions as X and Y, respectively. Then

$$\Pr\{(\mathbf{X}', \mathbf{Y}') \in \Psi^n_{[XY]\delta}\} \leq 2^{-n(I(X;Y)-\delta)}.$$

Channel Coding Theorem: A message drawn uniformly from the set $\{1, 2, \cdots, 2^{n(R-\epsilon)}\}$ can be transmitted through a continuous memoryless channel with negligible probability of error as $n \to \infty$ if and only if $R \leq C$.

Capacity of Memoryless Gaussian Channel:

$$C = \frac{1}{2} \log \left(1 + \frac{P}{N}\right).$$

Capacity of Parallel Gaussian Channels: For a system of parallel Gaussian channels with noise variable $Z_i \sim \mathcal{N}(0, N_i)$ for the ith channel and total input power constraint P,

$$C = \frac{1}{2} \sum_{i=1}^{k} \log \left(1 + \frac{(\nu - N_i)^+}{N_i}\right),$$

where ν satisfies $\sum_{i=1}^{k} (\nu - N_i)^+ = P$.

Capacity of Correlated Gaussian Channels: For a system of correlated Gaussian channels with noise vector $\mathbf{Z} \sim \mathcal{N}(0, K_{\mathbf{Z}})$ and total input power constraint P,

$$C = \frac{1}{2} \sum_{i=1}^{k} \log \left(1 + \frac{(\nu - \lambda_i)^+}{\lambda_i}\right),$$

where $K_{\mathbf{Z}}$ is diagonalizable as $Q \Lambda Q^\top$ with λ_i being the ith diagonal element of Λ, and ν satisfies $\sum_{i=1}^{k} (\nu - \lambda_i)^+ = P$.

Capacity of the Bandlimited White Gaussian Channel:

$$C = W \log \left(1 + \frac{P}{N_0 W}\right) \quad \text{bits per unit time.}$$

Capacity of the Bandlimited Colored Gaussian Channel:

$$\frac{1}{2} \int_{-W}^{W} \log \left(1 + \frac{(\nu - S_Z(f))^+}{S_Z(f)}\right) df \quad \text{bits per unit time,}$$

where ν satisfies $\int_{-W}^{W} (\nu - S_Z(f))^+ df = P$.

Zero-Mean Gaussian Is the Worst Additive Noise: For a system of channels with additive noise, if the correlation matrix of the noise vector is given, the capacity is minimized when the noise vector is zero-mean and Gaussian.

Problems

In the following, $\mathbf{X}$, $\mathbf{Y}$, $\mathbf{Z}$, etc., denote vectors of random variables.

1. Verify the three properties in Theorem 11.8 for the capacity of the memoryless Gaussian channel.
2. Let X and Y be two jointly distributed random variables with Y being continuous. The random variable Y is related to the random variable X through a conditional pdf $f(y|x)$ defined for all x (cf. Definition 10.22). Prove that $I(X;Y)$ is concave in $F(x)$.
3. Refer to Lemma 11.18 and prove that

$$\Pr\{(\mathbf{X}', \mathbf{Y}') \in \Psi^n_{[XY]\delta}\} \geq (1 - \delta)2^{-n(I(X;Y)-\delta)}$$

   for n sufficiently large.
4. Show that the capacity of a continuous memoryless channel is not changed if (11.23) is replaced by

$$E\left[\frac{1}{n}\sum_{i=1}^{n}\kappa(x_i(W))\right] \leq P,$$

   i.e., the average input constraint is satisfied on the average by a randomly selected codeword instead of by every codeword in the codebook.
5. Show that $R_{ii'}(0)$ in (11.234) vanishes if and only if $i \neq i'$.
6. Consider a system of Gaussian channels with noise vector $\mathbf{Z} \sim (0, K_{\mathbf{Z}})$ and input power constraint equal to 3. Determine the capacity of the system for the following two cases:

   a)

$$K_{\mathbf{Z}} = \begin{bmatrix} 4 & 0 & 0 \\ 0 & 5 & 0 \\ 0 & 0 & 2 \end{bmatrix};$$

   b)

$$K_{\mathbf{Z}} = \begin{bmatrix} 7/4 & \sqrt{2}/4 & -3/4 \\ \sqrt{2}/4 & 5/2 & -\sqrt{2}/4 \\ -3/4 & -\sqrt{2}/4 & 7/4 \end{bmatrix}.$$

   For (b), you may use the results in Problem 5 in Chapter 10.
7. In the system of correlated Gaussian channels, let $K_{\mathbf{Z}}$ be diagonalizable as $Q\Lambda Q^\top$. Let A^* be the $k \times k$ diagonal matrix with a_i^* being the ith diagonal element, where a_i^* is prescribed in the discussion following (11.193). Show that $\mathcal{N}(0, QA^*Q^\top)$ is the optimal input distribution.
8. Show that for a wide-sense stationary process $X(t)$, $S_X(f) = S_X(-f)$ for all f.

9. Consider a zero-mean white Gaussian noise process $Z(t)$. Let $h_1(t)$ and $h_2(t)$ be two impulse responses such that the supports of $H_1(f)$ and $H_2(f)$ do not overlap.

    a) Show that for any t and t', the two random variables $Z(t) * h_1(t)$ and $Z(t) * h_2(t')$ are independent.

    b) Show that the two processes $Z(t) * h_1(t)$ and $Z(t) * h_2(t)$ are independent.

    c) Repeat (a) and (b) if $Z(t)$ is a zero-mean colored Gaussian noise process. Hint: Regard $Z(t)$ as obtained by passing a zero-mean white Gaussian noise process through a coloring filter.

10. Interpret the bandpass white Gaussian channel as a special case of the bandlimited colored Gaussian channel in terms of the channel capacity.

11. *Independent Gaussian noise is the worst.* Let C be the capacity of a system of k Gaussian channels with $Z_i \sim \mathcal{N}(0, N_i)$. By ignoring the possible correlation among the noise variables, we can use the channels in the system independently as parallel Gaussian channels. Thus C is lower bounded by the expression in (11.180). In this sense, a Gaussian noise vector is the worst if its components are uncorrelated. Justify this claim analytically. Hint: Show that $I(\mathbf{X}; \mathbf{Y}) \geq \sum_i I(X_i; Y_i)$ if X_i are independent.

Historical Notes

Channels with additive Gaussian noise were first analyzed by Shannon in [322], where the formula for the capacity of the bandlimited white Gaussian channel was given. The form of the channel coding theorem for the continuous memoryless channel presented in this chapter was first proved in the book by Gallager [129]. A rigorous proof of the capacity formula for the bandlimited white Gaussian channel was obtained by Wyner [387]. The water-filling solution to the capacity of the bandlimited colored Gaussian channel was developed by Shannon in [324] and was proved rigorously by Pinsker [292]. The discussion in this chapter on the continuous memoryless channel with an average input constraint is adapted from the discussions in the book by Gallager [129] and the book by Ihara [180], where in the former a comprehensive treatment of waveform channels can also be found. The Gaussian noise being the worst additive noise was proved by Ihara [179]. The proof presented here is due to Diggavi and Cover [92].

12

Markov Structures

We have proved in Section 3.5 that if $X_1 \to X_2 \to X_3 \to X_4$ forms a Markov chain, the I-Measure μ^* always vanishes on the five atoms

$$\tilde{X}_1 \cap \tilde{X}_2^c \cap \tilde{X}_3 \cap \tilde{X}_4^c,$$
$$\tilde{X}_1 \cap \tilde{X}_2^c \cap \tilde{X}_3 \cap \tilde{X}_4,$$
$$\tilde{X}_1 \cap \tilde{X}_2^c \cap \tilde{X}_3^c \cap \tilde{X}_4, \quad (12.1)$$
$$\tilde{X}_1 \cap \tilde{X}_2 \cap \tilde{X}_3^c \cap \tilde{X}_4,$$
$$\tilde{X}_1^c \cap \tilde{X}_2 \cap \tilde{X}_3^c \cap \tilde{X}_4.$$

Consequently, the I-Measure μ^* is completely specified by the values of μ^* on the other ten nonempty atoms of $\mathcal{F}_4$, and the information diagram for four random variables forming a Markov chain can be displayed in two dimensions as in Figure 3.11.

Figure 12.1 is a graph which represents the Markov chain $X_1 \to X_2 \to X_3 \to X_4$. The observant reader would notice that μ^* always vanishes on a nonempty atom A of $\mathcal{F}_4$ if and only if the graph in Figure 12.1 becomes disconnected upon removing all the vertices corresponding to the complemented set variables in A. For example, μ^* always vanishes on the atom $\tilde{X}_1 \cap \tilde{X}_2^c \cap \tilde{X}_3 \cap \tilde{X}_4^c$, and the graph in Figure 12.1 becomes disconnected upon removing vertices 2 and 4. On the other hand, μ^* does not necessarily vanish on the atom $\tilde{X}_1^c \cap \tilde{X}_2 \cap \tilde{X}_3 \cap \tilde{X}_4^c$, and the graph in Figure 12.1 remains connected upon removing vertices 1 and 4. This observation will be explained in a more general setting in the subsequent sections.

Fig. 12.1. The graph representing the Markov chain $X_1 \to X_2 \to X_3 \to X_4$.

The theory of I-Measure establishes a one-to-one correspondence between Shannon's information measures and set theory. Based on this theory, we develop in this chapter a set-theoretic characterization of a Markov structure called *full conditional mutual independence*. A Markov chain, and more generally a Markov random field, is a collection of full conditional mutual independencies. We will show that if a collection of random variables forms a Markov random field, then the structure of μ^* can be simplified. In particular, when the random variables form a Markov chain, μ^* exhibits a very simple structure so that the information diagram can be displayed in two dimensions regardless of the length of the Markov chain, and μ^* is always nonnegative. (See also Sections 3.5 and 3.6.)

The topics to be covered in this chapter are fundamental. Unfortunately, the proofs of the results are very heavy. At the first reading, the reader should understand the theorems through the examples instead of getting into the details of the proofs.

12.1 Conditional Mutual Independence

In this section, we explore the effect of conditional mutual independence on the structure of the I-Measure μ^*. We begin with a simple example.

Example 12.1. Let X, Y, and Z be mutually independent random variables. Then

$$I(X;Y) = I(X;Y;Z) + I(X;Y|Z) = 0. \tag{12.2}$$

Since $I(X;Y|Z) \geq 0$, we let

$$I(X;Y|Z) = a \geq 0, \tag{12.3}$$

so that

$$I(X;Y;Z) = -a. \tag{12.4}$$

Similarly,

$$I(Y;Z) = I(X;Y;Z) + I(Y;Z|X) = 0 \tag{12.5}$$

and

$$I(X;Z) = I(X;Y;Z) + I(X;Z|Y) = 0. \tag{12.6}$$

Then from (12.4), we obtain

$$I(Y;Z|X) = I(X;Z|Y) = a. \tag{12.7}$$

The relations (12.3), (12.4), and (12.7) are shown in the information diagram in Figure 12.2, which indicates that X, Y, and Z are pairwise independent.

We have proved in Theorem 2.39 that X, Y, and Z are mutually independent if and only if

$$H(X,Y,Z) = H(X) + H(Y) + H(Z). \tag{12.8}$$

By counting atoms in the information diagram, we see that

$$0 = H(X) + H(Y) + H(Z) - H(X, Y, Z) \tag{12.9}$$
$$= I(X; Y|Z) + I(Y; Z|X) + I(X; Z|Y) + 2I(X; Y; Z) \tag{12.10}$$
$$= a. \tag{12.11}$$

Thus $a = 0$, which implies

$$I(X; Y|Z), I(Y; Z|X), I(X; Z|Y), I(X; Y; Z) \tag{12.12}$$

are all equal to 0. Equivalently, μ^* vanishes on

$$\tilde{X} \cap \tilde{Y} - \tilde{Z}, \tilde{Y} \cap \tilde{Z} - \tilde{X}, \tilde{X} \cap \tilde{Z} - \tilde{Y}, \tilde{X} \cap \tilde{Y} \cap \tilde{Z}, \tag{12.13}$$

which are precisely the atoms in the intersection of any two of the set variables $\tilde{X}, \tilde{Y}$, and $\tilde{Z}$.

Conversely, if μ^* vanishes on the sets in (12.13), then we see from (12.10) that (12.8) holds, i.e., X, Y, and Z are mutually independent. Therefore, X, Y, and Z are mutually independent if and only if μ^* vanishes on the sets in (12.13). This is shown in the information diagram in Figure 12.3.

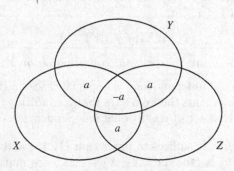

Fig. 12.2. X, Y, and Z are pairwise independent.

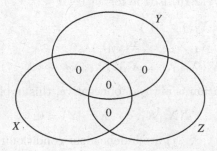

Fig. 12.3. X, Y, and Z are mutually independent.

The theme of this example will be extended to conditional mutual independence among collections of random variables in Theorem 12.9, which is the main result in this section. In the rest of the section, we will develop the necessary tools for proving this theorem. At first reading, the reader should try to understand the results by studying the examples without getting into the details of the proofs.

In Theorem 2.39, we have proved that $X_1, X_2, \cdots, X_n$ are mutually independent if and only if

$$H(X_1, X_2, \cdots, X_n) = \sum_{i=1}^{n} H(X_i). \tag{12.14}$$

By conditioning on a random variable Y, one can readily prove the following.

Theorem 12.2. $X_1, X_2, \cdots, X_n$ *are mutually independent conditioning on* Y *if and only if*

$$H(X_1, X_2, \cdots, X_n|Y) = \sum_{i=1}^{n} H(X_i|Y). \tag{12.15}$$

We now prove two alternative characterizations of conditional mutual independence.

Theorem 12.3. $X_1, X_2, \cdots, X_n$ *are mutually independent conditioning on* Y *if and only if for all* $1 \leq i \leq n$,

$$I(X_i; X_j, j \neq i|Y) = 0, \tag{12.16}$$

i.e., X_i *and* $(X_j, j \neq i)$ *are independent conditioning on* Y.

Remark A conditional independency is a special case of a conditional mutual independency. However, this theorem says that a conditional mutual independency is equivalent to a set of conditional independencies.

Proof of Theorem 12.3. It suffices to prove that (12.15) and (12.16) are equivalent. Assume (12.15) is true, i.e., $X_1, X_2, \cdots, X_n$ are mutually independent conditioning on Y. Then for all i, X_i is independent of $(X_j, j \neq i)$ conditioning on Y. This proves (12.16).

Now assume that (12.16) is true for all $1 \leq i \leq n$. Consider

$$0 = I(X_i; X_j, j \neq i|Y) \tag{12.17}$$
$$= I(X_i; X_1, X_2, \cdots, X_{i-1}|Y)$$
$$+ I(X_i; X_{i+1}, \cdots, X_n|Y, X_1, X_2, \cdots, X_{i-1}). \tag{12.18}$$

Since mutual information is always nonnegative, this implies

$$I(X_i; X_1, \cdots, X_{i-1}|Y) = 0 \tag{12.19}$$

or X_i and $(X_1, X_2, \cdots, X_{i-1})$ are independent conditioning on Y. Therefore, $X_1, X_2, \cdots, X_n$ are mutually independent conditioning on Y (see the proof of Theorem 2.39), proving (12.15). Hence, the theorem is proved.   □

Theorem 12.4. $X_1, X_2, \cdots, X_n$ *are mutually independent conditioning on* Y *if and only if*

$$H(X_1, X_2, \cdots, X_n|Y) = \sum_{i=1}^{n} H(X_i|Y, X_j, j \neq i). \qquad (12.20)$$

Proof. It suffices to prove that (12.15) and (12.20) are equivalent. Assume (12.15) is true, i.e., $X_1, X_2, \cdots, X_n$ are mutually independent conditioning on Y. Since for all i, X_i is independent of $X_j, j \neq i$ conditioning on Y,

$$H(X_i|Y) = H(X_i|Y, X_j, j \neq i). \qquad (12.21)$$

Therefore, (12.15) implies (12.20).

Now assume that (12.20) is true. Consider

$$H(X_1, X_2, \cdots, X_n|Y)$$

$$= \sum_{i=1}^{n} H(X_i|Y, X_1, \cdots, X_{i-1}) \qquad (12.22)$$

$$= \sum_{i=1}^{n} [H(X_i|Y, X_j, j \neq i) + I(X_i; X_{i+1}, \cdots, X_n|Y, X_1, \cdots, X_{i-1})]$$

$$\qquad (12.23)$$

$$= \sum_{i=1}^{n} H(X_i|Y, X_j, j \neq i) + \sum_{i=1}^{n} I(X_i; X_{i+1}, \cdots, X_n|Y, X_1, \cdots, X_{i-1}).$$

$$\qquad (12.24)$$

Then (12.20) implies

$$\sum_{i=1}^{n} I(X_i; X_{i+1}, \cdots, X_n|Y, X_1, \cdots, X_{i-1}) = 0. \qquad (12.25)$$

Since all the terms in the above summation are nonnegative, they must all be equal to 0. In particular, for $i = 1$, we have

$$I(X_1; X_2, \cdots, X_n|Y) = 0. \qquad (12.26)$$

By symmetry, it can be shown that

$$I(X_i; X_j, j \neq i|Y) = 0 \qquad (12.27)$$

for all $1 \leq i \leq n$. Then this implies (12.15) by the last theorem, completing the proof. $\square$

Theorem 12.5. *Let C and Q_i be disjoint index sets and W_i be a subset of Q_i for $1 \leq i \leq k$, where $k \geq 2$. Assume that there exist at least two i such that $W_i \neq \emptyset$. Let $X_{Q_i} = (X_l, l \in Q_i), 1 \leq i \leq k$, and $X_C = (X_l, l \in C)$ be collections of random variables. If $X_{Q_i}, 1 \leq i \leq k$, are mutually independent conditioning on X_C, then X_{W_i} such that $W_i \neq \emptyset$ are mutually independent conditioning on $(X_C, X_{Q_i - W_i}, 1 \leq i \leq k)$.*

We first give an example before we prove the theorem.

Example 12.6. Suppose $X_1, (X_2, X_3, X_4)$, and (X_5, X_6) are mutually independent conditioning on X_7. By Theorem 12.5, X_1, X_2, and (X_5, X_6) are mutually independent conditioning on (X_3, X_4, X_7).

Proof of Theorem 12.5. Assume $X_{Q_i}, 1 \leq i \leq k$, are mutually independent conditioning on X_C, i.e.,

$$H(X_{Q_i}, 1 \leq i \leq k | X_C) = \sum_{i=1}^{k} H(X_{Q_i} | X_C). \tag{12.28}$$

Consider

$$
\begin{aligned}
&H(X_{W_i}, 1 \leq i \leq k | X_C, X_{Q_i - W_i}, 1 \leq i \leq k) \\
&= H(X_{Q_i}, 1 \leq i \leq k | X_C) - H(X_{Q_i - W_i}, 1 \leq i \leq k | X_C) \quad\quad (12.29) \\
&= \sum_{i=1}^{k} H(X_{Q_i} | X_C) \\
&\quad - \sum_{i=1}^{k} H(X_{Q_i - W_i} | X_C, X_{Q_j - W_j}, 1 \leq j \leq i - 1) \quad\quad (12.30) \\
&\geq \sum_{i=1}^{k} H(X_{Q_i} | X_C, X_{Q_j - W_j}, 1 \leq j \leq i - 1) \\
&\quad - \sum_{i=1}^{k} H(X_{Q_i - W_i} | X_C, X_{Q_j - W_j}, 1 \leq j \leq i - 1) \quad\quad (12.31) \\
&= \sum_{i=1}^{k} H(X_{W_i} | X_C, X_{Q_j - W_j}, 1 \leq j \leq i) \quad\quad (12.32) \\
&\geq \sum_{i=1}^{k} H(X_{W_i} | X_C, X_{Q_j - W_j}, 1 \leq j \leq k). \quad\quad (12.33)
\end{aligned}
$$

In the second step we have used (12.28), and the two inequalities follow because conditioning does not increase entropy. On the other hand, by the chain rule for entropy, we have

$$H(X_{W_i}, 1 \leq i \leq k | X_C, X_{Q_i - W_i}, 1 \leq i \leq k)$$

$$= \sum_{i=1}^{k} H(X_{W_i} | X_C, (X_{Q_j - W_j}, 1 \leq j \leq k), (X_{W_l}, 1 \leq l \leq i - 1)).$$

$$(12.34)$$

Therefore, it follows from (12.33) that

$$\sum_{i=1}^{k} H(X_{W_i} | X_C, X_{Q_j - W_j}, 1 \leq j \leq k) \tag{12.35}$$

$$\leq H(X_{W_i}, 1 \leq i \leq k | X_C, X_{Q_i - W_i}, 1 \leq i \leq k) \tag{12.36}$$

$$= \sum_{i=1}^{k} H(X_{W_i} | X_C, (X_{Q_j - W_j}, 1 \leq j \leq k), (X_{W_l}, 1 \leq l \leq i - 1)).$$

$$(12.37)$$

However, since conditioning does not increase entropy, the ith term in the summation in (12.35) is lower bounded by the ith term in the summation in (12.37). Thus we conclude that the inequality in (12.36) is an equality. Hence, the conditional entropy in (12.36) is equal to the summation in (12.35), i.e.,

$$H(X_{W_i}, 1 \leq i \leq k | X_C, X_{Q_i - W_i}, 1 \leq i \leq k) \tag{12.38}$$

$$= \sum_{i=1}^{k} H(X_{W_i} | X_C, X_{Q_j - W_j}, 1 \leq j \leq k). \tag{12.39}$$

The theorem is proved. □

Theorem 12.5 specifies a set of conditional mutual independencies (CMIs) which is implied by a CMI. This theorem is crucial for understanding the effect of a CMI on the structure of the I-Measure μ^*, which we discuss next.

Lemma 12.7. *Let* $(Z_{i1}, \cdots, Z_{it_i}), 1 \leq i \leq r$ *be* r *collections of random variables, where* $r \geq 2$, *and let* Y *be a random variable such that* $(Z_{i1}, \cdots, Z_{it_i})$, $1 \leq i \leq r$ *are mutually independent conditioning on* Y. *Then*

$$\mu^* \left(\bigcap_{i=1}^{r} \bigcap_{j=1}^{t_i} \tilde{Z}_{ij} - \tilde{Y} \right) = 0. \tag{12.40}$$

We first prove the following set identity which will be used for proving the above lemma.

Lemma 12.8. *Let* S *and* T *be disjoint index sets, and* A_i *and* B *be sets. Let* μ *be a set-additive function. Then*

$$\mu\left(\left(\bigcap_{i\in S}A_i\right)\cap\left(\bigcap_{j\in T}A_j\right)-B\right)$$

$$=\sum_{S'\subset S}\sum_{T'\subset T}(-1)^{|S'|+|T'|}\left(\mu(A_{S'}-B)+\mu(A_{T'}-B)-\mu(A_{S'\cup T'}-B)\right),$$

$$(12.41)$$

where $A_{S'}$ denotes $\cup_{i\in S'}A_i$.

Proof. The right-hand side of (12.41) is equal to

$$\sum_{S'\subset S}\sum_{T'\subset T}(-1)^{|S'|+|T'|}\mu(A_{S'}-B)+\sum_{S'\subset S}\sum_{T'\subset T}(-1)^{|S'|+|T'|}\mu(A_{T'}-B)$$

$$-\sum_{S'\subset S}\sum_{T'\subset T}(-1)^{|S'|+|T'|}\mu(A_{S'\cup T'}-B).\qquad(12.42)$$

Now

$$\sum_{S'\subset S}\sum_{T'\subset T}(-1)^{|S'|+|T'|}\mu(A_{S'}-B)=\sum_{S'\subset S}(-1)^{|S'|}\mu(A_{S'}-B)\sum_{T'\subset T}(-1)^{|T'|}.$$

$$(12.43)$$

Since

$$\sum_{T'\subset T}(-1)^{|T'|}=\sum_{k=0}^{|T|}\binom{|T|}{k}(-1)^k=0\qquad(12.44)$$

by the binomial formula,[1] we conclude that

$$\sum_{S'\subset S}\sum_{T'\subset T}(-1)^{|S'|+|T'|}\mu(A_{S'}-B)=0.\qquad(12.45)$$

Similarly,

$$\sum_{S'\subset S}\sum_{T'\subset T}(-1)^{|S'|+|T'|}\mu(A_{T'}-B)=0.\qquad(12.46)$$

Therefore, (12.41) is equivalent to

$$\mu\left(\left(\bigcap_{i\in S}A_i\right)\cap\left(\bigcap_{j\in T}A_j\right)-B\right)=\sum_{S'\subset S}\sum_{T'\subset T}(-1)^{|S'|+|T'|+1}\mu(A_{S'\cup T'}-B)$$

$$(12.47)$$

---

[1] This can be obtained by letting $a=1$ and $b=-1$ in the binomial formula

$$(a+b)^{|T|}=\sum_{k=0}^{|T|}\binom{|T|}{k}a^k b^{|T|-k}.$$

which can readily be obtained from Theorem 3.19. Hence, the lemma is proved.
□

Proof of Lemma 12.7. We first prove the lemma for $r = 2$. By Lemma 12.8,

$$\mu^* \left(\bigcap_{i=1}^{2} \bigcap_{j=1}^{t_i} \tilde{Z}_{ij} - \tilde{Y} \right) =$$

$$\sum_{S' \subset \{1, \cdots, t_1\}} \sum_{T' \subset \{1, \cdots, t_2\}} (-1)^{|S'|+|T'|} \left[\mu^* \left(\bigcup_{j \in S'} \tilde{Z}_{1j} - \tilde{Y} \right) \right.$$

$$\left. + \mu^* \left(\bigcup_{k \in T'} \tilde{Z}_{2k} - \tilde{Y} \right) - \mu^* \left(\left(\bigcup_{j \in S'} \tilde{Z}_{1j} \right) \cup \left(\bigcup_{k \in T'} \tilde{Z}_{2k} \right) - \tilde{Y} \right) \right]. \quad (12.48)$$

The expression in the square bracket is equal to

$$H(Z_{1j}, j \in S'|Y) + H(Z_{2k}, k \in T'|Y)$$

$$-H((Z_{1j}, j \in S'), (Z_{2k}, k \in T')|Y), \quad (12.49)$$

which vanishes because $(Z_{1j}, j \in S')$ and $(Z_{2k}, k \in T')$ are independent conditioning on Y. Therefore the lemma is proved for $r = 2$.

For $r > 2$, we write

$$\mu^* \left(\bigcap_{i=1}^{r} \bigcap_{j=1}^{t_i} \tilde{Z}_{ij} - \tilde{Y} \right) = \mu^* \left(\left(\bigcap_{i=1}^{r-1} \bigcap_{j=1}^{t_i} \tilde{Z}_{ij} \right) \cap \left(\bigcap_{j=1}^{t_r} \tilde{Z}_{rj} \right) - \tilde{Y} \right). \quad (12.50)$$

Since $((Z_{i1}, \cdots, Z_{it_i}), 1 \leq i \leq r-1)$ and $(Z_{r1}, \cdots, Z_{rt_r})$ are independent conditioning on Y, upon applying the lemma for $r = 2$, we see that

$$\mu^* \left(\bigcap_{i=1}^{r} \bigcap_{j=1}^{t_i} \tilde{Z}_{ij} - \tilde{Y} \right) = 0. \quad (12.51)$$

The lemma is proved. □

Theorem 12.9. *Let T and $Q_i, 1 \leq i \leq k$, be disjoint index sets, where $k \geq 2$, and let $X_{Q_i} = (X_l, l \in Q_i), 1 \leq i \leq k$, and $X_T = (X_l, l \in T)$ be collections of random variables. Then $X_{Q_i}, 1 \leq i \leq k$, are mutually independent conditioning on X_T if and only if for any $W_1, W_2, \cdots, W_k$, where $W_i \subset Q_i, 1 \leq i \leq k$, if there exist at least two i such that $W_i \neq \emptyset$, then*

$$\mu^* \left(\left(\bigcap_{i=1}^{k} \bigcap_{j \in W_i} \tilde{X}_j \right) - \tilde{X}_{T \cup (\cup_{i=1}^{k} (Q_i - W_i))} \right) = 0. \quad (12.52)$$

We first give an example before proving this fundamental result. The reader should compare this example with Example 12.6.

Example 12.10. Suppose $X_1, (X_2, X_3, X_4)$, and (X_5, X_6) are mutually independent conditioning on X_7. By Theorem 12.9,

$$\mu^*(\tilde{X}_1 \cap \tilde{X}_2 \cap \tilde{X}_5 \cap \tilde{X}_6 - (\tilde{X}_3 \cup \tilde{X}_4 \cup \tilde{X}_7)) = 0. \tag{12.53}$$

However, the theorem does not say, for instance, that

$$\mu^*(\tilde{X}_2 \cap \tilde{X}_4 - (\tilde{X}_1 \cup \tilde{X}_3 \cup \tilde{X}_5 \cup \tilde{X}_6 \cup \tilde{X}_7)) \tag{12.54}$$

is equal to 0.

Proof of Theorem 12.9. We first prove the "if" part. Assume that for any $W_1, W_2, \cdots, W_k$, where $W_i \subset Q_i$, $1 \le i \le k$, if there exist at least two i such that $W_i \neq \emptyset$, then (12.52) holds. Consider

$$H(X_{Q_i}, 1 \le i \le k | X_T) = \mu^* \left(\tilde{X}_{\cup_{i=1}^k Q_i} - \tilde{X}_T \right) \tag{12.55}$$

$$= \sum_{B \in S} \mu^*(B), \tag{12.56}$$

where S consists of sets of the form

$$\left(\bigcap_{i=1}^k \bigcap_{j \in W_i} \tilde{X}_j \right) - \tilde{X}_{T \cup (\cup_{i=1}^k (Q_i - W_i))} \tag{12.57}$$

with $W_i \subset Q_i$ for $1 \le i \le k$ and there exists at least one i such that $W_i \neq \emptyset$. By our assumption, if $B \in S$ such that there exist at least two i for which $W_i \neq \emptyset$, then $\mu^*(B) = 0$. Therefore, if $\mu^*(B)$ is possibly nonzero, then B must be such that there exists a unique i for which $W_i \neq \emptyset$. Now for $1 \le i \le k$, let S_l be the set consisting of sets of the form in (12.57) with $W_i \subset Q_i$, $W_i \neq \emptyset$, and $W_l = \emptyset$ for $l \neq i$. In other words, S_i consists of atoms of the form

$$\left(\bigcap_{j \in W_i} \tilde{X}_j \right) - \tilde{X}_{T \cup (\cup_{l \neq i} Q_l) \cup (Q_i - W_i)} \tag{12.58}$$

with $W_i \subset Q_i$ and $W_i \neq \emptyset$. Then

$$\sum_{B \in S} \mu^*(B) = \sum_{i=1}^k \sum_{B \in S_i} \mu^*(B). \tag{12.59}$$

Now

$$\tilde{X}_{Q_i} - \tilde{X}_{T \cup (\cup_{l \neq i} Q_l)}$$

$$= \bigcup_{\substack{W_i \subset Q_i \\ W_i \neq \emptyset}} \left[\left(\bigcap_{j \in W_i} \tilde{X}_j \right) - \tilde{X}_{T \cup (\cup_{l \neq i} Q_l) \cup (Q_i - W_i)} \right] \qquad (12.60)$$

$$= \bigcup_{B \in S_i} B. \qquad (12.61)$$

Since μ^* is set-additive, we have

$$\mu^* \left(\tilde{X}_{Q_i} - \tilde{X}_{T \cup (\cup_{l \neq i} Q_l)} \right) = \sum_{B \in S_i} \mu^*(B). \qquad (12.62)$$

Hence, from (12.56) and (12.59), we have

$$H(X_{Q_i}, 1 \leq i \leq k | X_T)$$

$$= \sum_{i=1}^k \sum_{B \in S_i} \mu^*(B) \qquad (12.63)$$

$$= \sum_{i=1}^k \mu^* \left(\tilde{X}_{Q_i} - \tilde{X}_{T \cup (\cup_{l \neq i} Q_l)} \right) \qquad (12.64)$$

$$= \sum_{i=1}^k H(X_{Q_i} | X_T, X_{Q_l}, l \neq i), \qquad (12.65)$$

where (12.64) follows from (12.62). By Theorem 12.4, $X_{Q_i}, 1 \leq i \leq k$, are mutually independent conditioning on X_T.

We now prove the "only if" part. Assume $X_{Q_i}, 1 \leq i \leq k$, are mutually independent conditioning on X_T. For any collection of sets $W_1, W_2, \cdots, W_k$, where $W_i \subset Q_i$, $1 \leq i \leq k$, if there exist at least two i such that $W_i \neq \emptyset$, by Theorem 12.5, $X_{W_i}, 1 \leq i \leq k$, are mutually independent conditioning on $(X_T, X_{Q_i - W_i}, 1 \leq i \leq k)$. By Lemma 12.7, we obtain (12.52). The theorem is proved. □

12.2 Full Conditional Mutual Independence

Definition 12.11. *A conditional mutual independency on $X_1, X_2, \cdots, X_n$ is full if all $X_1, X_2, \cdots, X_n$ are involved. Such a conditional mutual independency is called a full conditional mutual independency (FCMI).*

Example 12.12. For $n = 5$,

X_1, X_2, X_4, and X_5 are mutually independent conditioning on X_3

is an FCMI. However,

X_1, X_2, and X_5 are mutually independent conditioning on X_3

is not an FCMI because X_4 is not involved.

As in the previous chapters, we let

$$\mathcal{N}_n = \{1, 2, \cdots, n\}. \tag{12.66}$$

In Theorem 12.9, if

$$T \cup \left(\bigcup_{i=1}^{k} Q_i \right) = \mathcal{N}_n, \tag{12.67}$$

then the tuple $(T, Q_i, 1 \leq i \leq k)$ defines the following FCMI on $X_1, X_2, \cdots, X_n$:

$K : X_{Q_1}, X_{Q_2}, \cdots, X_{Q_k}$ are mutually independent conditioning on X_T.

We will denote K by $(T, Q_i, 1 \leq i \leq k)$.

Definition 12.13. Let $K = (T, Q_i, 1 \leq i \leq k)$ be an FCMI on $X_1, X_2, \cdots, X_n$. The image of K, denoted by $Im(K)$, is the set of all atoms of $\mathcal{F}_n$ which has the form of the set in (12.57), where $W_i \subset Q_i$, $1 \leq i \leq k$, and there exist at least two i such that $W_i \neq \emptyset$.

Recall from Chapter 3 that $\mathcal{A}$ is the set of all nonempty atoms of $\mathcal{F}_n$.

Proposition 12.14. Let $K = (T, Q_1, Q_2)$ be an FCI (full conditional independency) on $X_1, X_2, \cdots, X_n$. Then

$$Im(K) = \{A \in \mathcal{A} : A \subset (\tilde{X}_{Q_1} \cap \tilde{X}_{Q_2} - \tilde{X}_T)\}. \tag{12.68}$$

Proposition 12.15. Let $K = (T, Q_i, 1 \leq i \leq k)$ be an FCMI on $X_1, X_2, \cdots, X_n$. Then

$$Im(K) = \left\{ A \in \mathcal{A} : A \subset \bigcup_{1 \leq i < j \leq k} (\tilde{X}_{Q_i} \cap \tilde{X}_{Q_j} - \tilde{X}_T) \right\}. \tag{12.69}$$

These two propositions greatly simplify the description of $Im(K)$. Their proofs are elementary and they are left as an exercise. We first illustrate these two propositions in the following example.

Example 12.16. Consider $n = 4$ and FCMIs $K_1 = (\{3\}, \{1\}, \{2, 4\})$ and $K_2 = (\emptyset, \{1\}, \{2, 3\}, \{4\})$. Then

$$Im(K_1) = \{A \in \mathcal{A} : A \subset (\tilde{X}_1 \cap \tilde{X}_{\{2,4\}} - \tilde{X}_3)\} \tag{12.70}$$

and

$$Im(K_2) = \{A \in \mathcal{A} : A \subset (\tilde{X}_1 \cap \tilde{X}_{\{2,3\}}) \cup (\tilde{X}_{\{2,3\}} \cap \tilde{X}_4) \cup (\tilde{X}_1 \cap \tilde{X}_4)\}. \tag{12.71}$$

Theorem 12.17. *Let K be an FCMI on $X_1, X_2, \cdots, X_n$. Then K holds if and only if $\mu^*(A) = 0$ for all $A \in Im(K)$.*

Proof. First, (12.67) is true if K is an FCMI. Then the set in (12.57) can be written as

$$\left(\bigcap_{j \in \cup_{i=1}^{k} W_i} \tilde{X}_j \right) - \tilde{X}_{\mathcal{N}_n - \cup_{i=1}^{k} W_i}, \tag{12.72}$$

which is seen to be an atom of $\mathcal{F}_n$. The theorem can then be proved by a direct application of Theorem 12.9 to the FCMI K. $\quad\square$

Let $A = \cap_{i=1}^{n} \tilde{Y}_i$ be a nonempty atom of $\mathcal{F}_n$. Define the set

$$U_A = \{i \in \mathcal{N}_n : \tilde{Y}_i = \tilde{X}_i^c\}. \tag{12.73}$$

Note that A is uniquely specified by U_A because

$$A = \left(\bigcap_{i \in \mathcal{N}_n - U_A} \tilde{X}_i \right) \cap \left(\bigcap_{i \in U_A} \tilde{X}_i^c \right) = \left(\bigcap_{i \in \mathcal{N}_n - U_A} \tilde{X}_i \right) - \tilde{X}_{U_A}. \tag{12.74}$$

Define $w(A) = n - |U_A|$ as the *weight* of the atom A, the number of $\tilde{X}_i$ in A which are not complemented. We now show that an FCMI $K = (T, Q_i, 1 \le i \le k)$ is uniquely specified by $Im(K)$. First, by letting $W_i = Q_i$ for $1 \le i \le k$ in Definition 12.13, we see that the atom

$$\left(\bigcap_{j \in \cup_{i=1}^{k} Q_i} \tilde{X}_j \right) - \tilde{X}_T \tag{12.75}$$

is in $Im(K)$, and it is the unique atom in $Im(K)$ with the largest weight. From this atom, T can be determined. To determine $Q_i, 1 \le i \le k$, we define a relation q on $T^c = \mathcal{N}_n \backslash T$ as follows. For $l, l' \in T^c$, (l, l') is in q if and only if

   i) $l = l'$; or

   ii) there exists an atom of the form

$$\tilde{X}_l \cap \tilde{X}_{l'} \cap \bigcap_{\substack{1 \le j \le n \\ j \ne l, l'}} \tilde{Y}_j \tag{12.76}$$

in $\mathcal{A} - Im(K)$, where $\tilde{Y}_j = \tilde{X}_j$ or $\tilde{X}_j^c$.

Recall that $\mathcal{A}$ is the set of nonempty atoms of $\mathcal{F}_n$. The idea of (ii) is that (l, l') is in q if and only if $l, l' \in Q_i$ for some $1 \le i \le k$. Then q is reflexive and symmetric by construction and is transitive by virtue of the structure of $Im(K)$. In other words, q is an *equivalence relation* which partitions T^c into $\{Q_i, 1 \le i \le k\}$. Therefore, K and $Im(K)$ uniquely specify each other.

The image of an FCMI K completely characterizes the effect of K on the I-Measure for $X_1, X_2, \cdots, X_n$. The joint effect of more than one FCMI can easily be described in terms of the images of the individual FCMIs. Let

$$\Pi = \{K_l, 1 \le l \le m\} \tag{12.77}$$

be a set of FCMIs. By Theorem 12.9, K_l holds if and only if μ^* vanishes on the atoms in $Im(K_l)$. Then $K_l, 1 \le l \le m$ hold simultaneously if and only if μ^* vanishes on the atoms in $\cup_{l=1}^k Im(K_l)$. This is summarized as follows.

Definition 12.18. *The image of a set of FCMI's $\Pi = \{K_l, 1 \le l \le m\}$ is defined as*

$$Im(\Pi) = \bigcup_{l=1}^k Im(K_l). \tag{12.78}$$

Theorem 12.19. *Let Π be a set of FCMIs on $X_1, X_2, \cdots, X_n$. Then Π holds if and only if $\mu^*(A) = 0$ for all $A \in Im(\Pi)$.*

In probability problems, we are often given a set of conditional independencies and we need to see whether another given conditional independency is logically implied. This is called the *implication problem* which will be discussed in detail in Section 13.5. The next theorem renders a solution to this problem if only FCMIs are involved.

Theorem 12.20. *Let Π_1 and Π_2 be two sets of FCMIs. Then Π_1 implies Π_2 if and only if $Im(\Pi_2) \subset Im(\Pi_1)$.*

Proof. We first prove that if $Im(\Pi_2) \subset Im(\Pi_1)$, then Π_1 implies Π_2. Assume $Im(\Pi_2) \subset Im(\Pi_1)$ and Π_1 holds. Then by Theorem 12.19, $\mu^*(A) = 0$ for all $A \in Im(\Pi_1)$. Since $Im(\Pi_2) \subset Im(\Pi_1)$, this implies that $\mu^*(A) = 0$ for all $A \in Im(\Pi_2)$. Again by Theorem 12.19, this implies Π_2 also holds. Therefore, if $Im(\Pi_2) \subset Im(\Pi_1)$, then Π_1 implies Π_2.

We now prove that if Π_1 implies Π_2, then $Im(\Pi_2) \subset Im(\Pi_1)$. To prove this, we assume that Π_1 implies Π_2 but $Im(\Pi_2) \not\subset Im(\Pi_1)$, and we will show that this leads to a contradiction. Fix a nonempty atom $A \in Im(\Pi_2) - Im(\Pi_1)$. By Theorem 3.11, we can construct random variables $X_1, X_2, \cdots, X_n$ such that μ^* vanishes on all the atoms of $\mathcal{F}_n$ except for A. Then μ^* vanishes on all the atoms in $Im(\Pi_1)$ but not on all the atoms in $Im(\Pi_2)$. By Theorem 12.19, this implies that for $X_1, X_2, \cdots, X_n$ so constructed, Π_1 holds but Π_2 does not hold. Therefore, Π_1 does not imply Π_2, which is a contradiction. The theorem is proved. $\square$

Remark In the course of proving this theorem and all its preliminaries, we have used nothing more than the basic inequalities. Therefore, we have shown that the basic inequalities are a sufficient set of tools to solve the implication problem if only FCMIs are involved.

Corollary 12.21. *Two sets of FCMIs are equivalent if and only if their images are identical.*

Proof. Two set of FCMIs Π_1 and Π_2 are equivalent if and only if

$$\Pi_1 \Rightarrow \Pi_2 \quad \text{and} \quad \Pi_2 \Rightarrow \Pi_1. \tag{12.79}$$

Then by the last theorem, this is equivalent to $Im(\Pi_2) \subset Im(\Pi_1)$ and $Im(\Pi_1) \subset Im(\Pi_2)$, i.e., $Im(\Pi_2) = Im(\Pi_1)$. The corollary is proved. $\square$

Thus a set of FCMIs is completely characterized by its image. A set of FCMIs is a set of probabilistic constraints, but the characterization by its image is purely set-theoretic! This characterization offers an intuitive set-theoretic interpretation of the joint effect of FCMIs on the I-Measure for $X_1, X_2, \cdots, X_n$. For example, $Im(K_1) \cap Im(K_2)$ is interpreted as the effect commonly due to K_1 and K_2, $Im(K_1) - Im(K_2)$ is interpreted as the effect due to K_1 but not K_2, etc. We end this section with an example.

Example 12.22. Consider $n = 4$. Let

$$K_1 = (\emptyset, \{1,2,3\}, \{4\}), \quad K_2 = (\emptyset, \{1,2,4\}, \{3\}), \tag{12.80}$$

$$K_3 = (\emptyset, \{1,2\}, \{3,4\}), \quad K_4 = (\emptyset, \{1,3\}, \{2,4\}), \tag{12.81}$$

and let $\Pi_1 = \{K_1, K_2\}$ and $\Pi_2 = \{K_3, K_4\}$. Then

$$Im(\Pi_1) = Im(K_1) \cup Im(K_2) \tag{12.82}$$

and

$$Im(\Pi_2) = Im(K_3) \cup Im(K_4), \tag{12.83}$$

where

$$Im(K_1) = \{A \in \mathcal{A} : A \subset (\tilde{X}_{\{1,2,3\}} \cap \tilde{X}_4)\}, \tag{12.84}$$

$$Im(K_2) = \{A \in \mathcal{A} : A \subset (\tilde{X}_{\{1,2,4\}} \cap \tilde{X}_3)\}, \tag{12.85}$$

$$Im(K_3) = \{A \in \mathcal{A} : A \subset (\tilde{X}_{\{1,2\}} \cap \tilde{X}_{\{3,4\}})\}, \tag{12.86}$$

$$Im(K_4) = \{A \in \mathcal{A} : A \subset (\tilde{X}_{\{1,3\}} \cap \tilde{X}_{\{2,4\}})\}. \tag{12.87}$$

It can readily be seen by using an information diagram that $Im(\Pi_1) \subset Im(\Pi_2)$. Therefore, Π_2 implies Π_1. Note that no probabilistic argument is involved in this proof.

12.3 Markov Random Field

A *Markov random field* is a generalization of a discrete-time Markov chain in the sense that the time index for the latter, regarded as a chain, is replaced by a general graph for the former. Historically, the study of Markov random field stems from statistical physics. The classical Ising model, which is defined on a rectangular lattice, was used to explain certain empirically observed facts about ferromagnetic materials. In this section, we explore the structure of the I-Measure for a Markov random field.

We refer the reader to textbooks on graph theory (e.g., [45]) for formal definitions of the graph-theoretic terminologies to be used in the rest of the chapter. Let $G = (V, E)$ be an undirected graph, where V is the set of vertices and E is the set of edges. We assume that there is no *loop* in G, i.e., there is no edge in G which connects a vertex to itself. For any (possibly empty) subset U of V, denote by $G \backslash U$ the graph obtained from G by eliminating all the vertices in U and all the edges joining a vertex in U.

The connectivity of a graph partitions the graph into subgraphs called *components*, i.e., two vertices are in the same component if and only if they are connected. Let $s(U)$ be the number of distinct components in $G \backslash U$. Denote the sets of vertices of these components by $V_1(U), V_2(U), \cdots, V_{s(U)}(U)$. If $s(U) > 1$, we say that U is a cutset in G.

Definition 12.23 (Markov Random Field). *Let $G = (V, E)$ be an undirected graph with $V = \mathcal{N}_n = \{1, 2, \cdots, n\}$, and let X_i be a random variable corresponding to vertex i. The random variables $X_1, X_2, \cdots, X_n$ form a Markov random field represented by G if for all cutsets U in G, the sets of random variables $X_{V_1(U)}, X_{V_2(U)}, \cdots, X_{V_{s(U)}(U)}$ are mutually independent conditioning on X_U.*

This definition of a Markov random field is referred to as the *global Markov property* in the literature. If $X_1, X_2, \cdots, X_n$ form a Markov random field represented by a graph G, we also say that $X_1, X_2, \cdots, X_n$ form a *Markov graph* G. When G is a chain, we say that $X_1, X_2, \cdots, X_n$ form a Markov chain.

In the definition of a Markov random field, each cutset U in G specifies an FCMI on $X_1, X_2, \cdots, X_n$, denoted by $[U]$. Formally,

$$[U] : X_{V_1(U)}, \cdots, X_{V_{s(U)}(U)} \text{ are mutually independent conditioning on } X_U.$$

For a collection of cutsets $U_1, U_2, \cdots, U_k$ in G, we introduce the notation

$$[U_1, U_2, \cdots, U_k] = [U_1] \wedge [U_2] \wedge \cdots \wedge [U_k], \tag{12.88}$$

where "$\wedge$" denotes "logical AND." Using this notation, $X_1, X_2, \cdots, X_n$ form a Markov graph G if and only if

$$[U \subset V : U \neq V \text{ and } s(U) > 1] \tag{12.89}$$

holds. Therefore, a Markov random field is simply a collection of FCMIs induced by a graph.

We now define two types of nonempty atoms of $\mathcal{F}_n$ with respect to a graph G. Recall the definition of the set U_A for a nonempty atom A of $\mathcal{F}_n$ in (12.73).

Definition 12.24. *For a nonempty atom A of $\mathcal{F}_n$, if $s(U_A) = 1$, i.e., $G \backslash U_A$ is connected, then A is a Type I atom, otherwise A is a Type II atom. The sets of all Type I and Type II atoms of $\mathcal{F}_n$ are denoted by T_1 and T_2, respectively.*

Theorem 12.25. *$X_1, X_2, \cdots, X_n$ form a Markov graph G if and only if μ^* vanishes on all the Type II atoms.*

Before we prove this theorem, we first state the following proposition which is the graph-theoretic analog of Theorem 12.5. The proof is trivial and is omitted. This proposition and Theorem 12.5 together establish an analogy between the structure of conditional mutual independence and the connectivity of a graph. This analogy will play a key role in proving Theorem 12.25.

Proposition 12.26. *Let C and Q_i be disjoint subsets of the vertex set V of a graph G and W_i be a subset of Q_i for $1 \le i \le k$, where $k \ge 2$. Assume that there exist at least two i such that $W_i \ne \emptyset$. If $Q_i, 1 \le i \le k$, are disconnected in $G \backslash C$, then those W_i which are nonempty are disconnected in $G \backslash (C \cup \bigcup_{i=1}^{k} (Q_i - W_i))$.*

Example 12.27. In the graph G in Figure 12.4, $\{1\}$, $\{2, 3, 4\}$, and $\{5, 6\}$ are disjoint in $G \backslash \{7\}$. Then Proposition 12.26 says that $\{1\}$, $\{2\}$, and $\{5, 6\}$ are disjoint in $G \backslash \{3, 4, 7\}$.

Proof of Theorem 12.25. We note that $\{U_A, A \in \mathcal{A}\}$ contains precisely all the proper subsets of $\mathcal{N}_n$. Thus the set of FCMIs specified by the graph G can be written as

$$[U_A : A \in \mathcal{A} \text{ and } s(U_A) > 1] \tag{12.90}$$

(cf. (12.89)). By Theorem 12.19, it suffices to prove that

$$Im([U_A : A \in \mathcal{A} \text{ and } s(U_A) > 1]) = T_2, \tag{12.91}$$

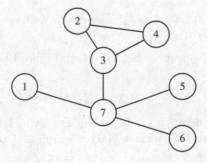

Fig. 12.4. The graph G in Example 12.27.

where T_2 is defined in Definition 12.24.

We first prove that

$$T_2 \subset Im([U_A : A \in \mathcal{A} \text{ and } s(U_A) > 1]). \tag{12.92}$$

Consider an atom $A \in T_2$. Then $s(U_A) > 1$. In Definition 12.13, let $T = U_A$, $k = s(U_A)$, and $Q_i = V_i(U_A)$ for $1 \le i \le s(U_A)$. By considering $W_i = V_i(U_A)$ for $1 \le i \le s(U_A)$, we see that $A \in Im([U_A])$. Therefore,

$$T_2 = \{A \in \mathcal{A} : s(U_A) > 1\} \tag{12.93}$$

$$\subset \bigcup_{A \in \mathcal{A}: s(U_A) > 1} Im([U_A]) \tag{12.94}$$

$$= Im([U_A : A \in \mathcal{A} \text{ and } s(U_A) > 1]). \tag{12.95}$$

We now prove that

$$Im([U_A : A \in \mathcal{A} \text{ and } s(U_A) > 1]) \subset T_2. \tag{12.96}$$

Consider $A \in Im([U_A : A \in \mathcal{A} \text{ and } s(U_A) > 1])$. Then there exists $A^* \in \mathcal{A}$ with $s(U_{A^*}) > 1$ such that $A \in Im([U_{A^*}])$. From Definition 12.13,

$$A = \left(\bigcap_{j \in \cup_{i=1}^{s(U_{A^*})} W_i} \tilde{X}_j \right) - \tilde{X}_{U_{A^*} \cup \left(\cup_{i=1}^{s(U_{A^*})} (V_i(U_{A^*}) - W_i) \right)}, \tag{12.97}$$

where $W_i \subset V_i(U_{A^*})$, $1 \le i \le s(U_{A^*})$, and there exist at least two i such that $W_i \ne \emptyset$. It follows from (12.97) and the definition of U_A that

$$U_A = U_{A^*} \cup \bigcup_{i=1}^{s(U_{A^*})} (V_i(U_{A^*}) - W_i). \tag{12.98}$$

With U_{A^*} playing the role of C and $V_i(U_{A^*})$ playing the role of Q_i in Proposition 12.26, we see by applying the proposition that those (at least two) W_i which are nonempty are disjoint in

$$G \setminus \left(U_{A^*} \cup \left(\bigcup_{i=1}^{s(U_{A^*})} (V_i(U_{A^*}) - W_i) \right) \right) = G \setminus U_A. \tag{12.99}$$

This implies $s(U_A) > 1$, i.e., $A \in T_2$. Therefore, we have proved (12.96), and hence the theorem is proved. □

Example 12.28. With respect to the graph G in Figure 12.5, the Type II atoms are

$$\tilde{X}_1 \cap \tilde{X}_2 \cap \tilde{X}_3^c \cap \tilde{X}_4, \ \tilde{X}_1^c \cap \tilde{X}_2 \cap \tilde{X}_3^c \cap \tilde{X}_4, \ \tilde{X}_1 \cap \tilde{X}_2^c \cap \tilde{X}_3^c \cap \tilde{X}_4, \tag{12.100}$$

while the other 12 nonempty atoms of $\mathcal{F}_4$ are Type I atoms. The random variables X_1, X_2, X_3, and X_4 form a Markov graph G if and only if $\mu^*(A) = 0$ for all Type II atoms A.

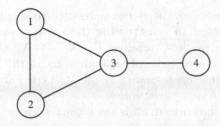

Fig. 12.5. The graph G in Example 12.28.

12.4 Markov Chain

When the graph G representing a Markov random field is a chain, the Markov random field becomes a Markov chain. In this section, we will show that the information diagram for a Markov chain can be displayed in two dimensions. We will also show that the I-Measure μ^* for a Markov chain is always nonnegative. This characteristic of μ^* facilitates the use of the information diagram because if B is seen to be a subset of B' in the information diagram, then

$$\mu^*(B') = \mu^*(B) + \mu^*(B' - B) \geq \mu^*(B). \qquad (12.101)$$

These two properties are not possessed by a general Markov random field.

Without loss of generality, we assume that the Markov chain is represented by the graph G in Figure 12.6. This corresponds to the Markov chain $X_1 \to X_2 \to \cdots \to X_n$. We first prove the following characterization of a Type I atom for a Markov chain.

Lemma 12.29. *For the Markov chain represented by the graph G in Figure 12.6, a nonempty atom A of $\mathcal{F}_n$ is a Type I atom if and only if*

$$\mathcal{N}_n \backslash U_A = \{l, l+1, \cdots, u\}, \qquad (12.102)$$

where $1 \leq l \leq u \leq n$, i.e., the indices of the set variables in A which are not complemented are consecutive.

Proof. It is easy to see that for a nonempty atom A, if (12.102) is satisfied, then $G \backslash U_A$ is connected, i.e., $s(U_A) = 1$. Therefore, A is a Type I atom of $\mathcal{F}_n$. On the other hand, if (12.102) is not satisfied, then $G \backslash U_A$ is not connected, i.e., $s(U_A) > 1$, or A is a Type II atom of $\mathcal{F}_n$. The lemma is proved. $\square$

We now show how the information diagram for a Markov chain with any length $n \geq 3$ can be constructed in two dimensions. Since μ^* vanishes on

Fig. 12.6. The graph G representing the Markov chain $X_1 \to X_2 \to \cdots \to X_n$.

all the Type II atoms of $\mathcal{F}_n$, it is not necessary to display these atoms in the information diagram. In constructing the information diagram, the regions representing the random variables $X_1, X_2, \cdots, X_n$ should overlap with each other such that the regions corresponding to all the Type II atoms are empty, while the regions corresponding to all the Type I atoms are nonempty. Figure 12.7 shows such a construction.

We have already shown that μ^* is nonnegative for a Markov chain with length 3 or 4. Toward proving that this is true for any length $n \geq 3$, it suffices to show that $\mu^*(A) \geq 0$ for all Type I atoms A of $\mathcal{F}_n$ because $\mu^*(A) = 0$ for all Type II atoms A of $\mathcal{F}_n$. We have seen in Lemma 12.29 that for a Type I atom A of $\mathcal{F}_n$, U_A has the form prescribed in (12.102). Consider any such atom A. Then an inspection of the information diagram in Figure 12.7 reveals that

$$\mu^*(A) = \mu^*(\tilde{X}_l \cap \tilde{X}_{l+1} \cap \cdots \cap \tilde{X}_u - \tilde{X}_{U_A}) \tag{12.103}$$

$$= I(X_l; X_u | X_{U_A}) \tag{12.104}$$

$$\geq 0. \tag{12.105}$$

This shows that μ^* is always nonnegative. However, since Figure 12.7 involves an indefinite number of random variables, we give a formal proof of this result in the following theorem.

Theorem 12.30. *For a Markov chain $X_1 \to X_2 \to \cdots \to X_n$, μ^* is nonnegative.*

Proof. Since $\mu^*(A) = 0$ for all Type II atoms A of $\mathcal{F}_n$, it suffices to show that $\mu^*(A) \geq 0$ for all Type I atoms A of $\mathcal{F}_n$. We have seen in Lemma 12.29 that for a Type I atom A of $\mathcal{F}_n$, U_A has the form prescribed in (12.102). Consider any such atom A and define the set

$$W = \{l+1, \cdots, u-1\}. \tag{12.106}$$

Then

$$I(X_l; X_u | X_{U_A})$$
$$= \mu^*(\tilde{X}_l \cap \tilde{X}_u - \tilde{X}_{U_A}) \tag{12.107}$$

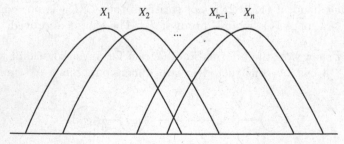

Fig. 12.7. The information diagram for the Markov chain $X_1 \to X_2 \to \cdots \to X_n$.

$$= \mu^* \left(\bigcup_{S \subset W} \left(\tilde{X}_l \cap \left(\bigcap_{t \in S} \tilde{X}_t \right) \cap \tilde{X}_u - \tilde{X}_{U_{A \cup (W \setminus S)}} \right) \right) \qquad (12.108)$$

$$= \sum_{S \subset W} \mu^* \left(\tilde{X}_l \cap \left(\bigcap_{t \in S} \tilde{X}_t \right) \cap \tilde{X}_u - \tilde{X}_{U_{A \cup (W \setminus S)}} \right). \qquad (12.109)$$

In the above summation, except for the atom corresponding to $S = W$, namely $(\tilde{X}_l \cap \tilde{X}_{l+1} \cap \cdots \cap \tilde{X}_u - \tilde{X}_{U_A})$, all the atoms are Type II atoms. Therefore,

$$I(X_l; X_u | X_{U_A}) = \mu^*(\tilde{X}_l \cap \tilde{X}_{l+1} \cap \cdots \cap \tilde{X}_u - \tilde{X}_{U_A}). \qquad (12.110)$$

Hence,

$$\mu^*(A) = \mu^*(\tilde{X}_l \cap \tilde{X}_{l+1} \cap \cdots \cap \tilde{X}_u - \tilde{X}_{U_A}) \qquad (12.111)$$
$$= I(X_l; X_u | X_{U_A}) \qquad (12.112)$$
$$\geq 0. \qquad (12.113)$$

The theorem is proved. $\square$

Chapter Summary

In the following, $\mathcal{N}_n = \{1, 2, \cdots, n\}$ and $\mathcal{A}$ is the set of all nonempty atoms of $\mathcal{F}_n$.

Full Conditional Mutual Independency (FCMI): For a partition $\{T, Q_i, 1 \leq i \leq k\}$ of $\mathcal{N}_n$, the tuple $(T, Q_i, 1 \leq i \leq k)$ specifies the following FCMI on $X_1, X_2, \cdots, X_n$:

$X_{Q_1}, X_{Q_2}, \cdots, X_{Q_k}$ are mutually independent conditioning on X_T.

Image of an FCMI: For an FCMI $K = (T, Q_i, 1 \leq i \leq k)$ on $X_1, X_2, \cdots, X_n$,

$$Im(K) = \left\{ A \in \mathcal{A} : A \subset \bigcup_{1 \leq i < j \leq k} (\tilde{X}_{Q_i} \cap \tilde{X}_{Q_j} - \tilde{X}_T) \right\}.$$

Characterization of an FCMI: An FCMI K on $X_1, X_2, \cdots, X_n$ holds if and only if $\mu^*(A) = 0$ for all $A \in Im(K)$.

Image of a Set of FCMIs: For a set of FCMIs $\Pi = \{K_l, 1 \leq l \leq m\}$, $Im(\Pi) = \bigcup_{l=1}^{k} Im(K_l)$.

Characterization of a Set of FCMIs: A set of FCMIs Π on $X_1, X_2, \cdots, X_n$ holds if and only if $\mu^*(A) = 0$ for all $A \in Im(\Pi)$.

Set-Theoretic Characterization of FCMI: Π_1 implies Π_2 if and only if $Im(\Pi_2) \subset Im(\Pi_1)$.

Markov Random Field (Markov Graph): Let $G = (V, E)$ be an undirected graph with $V = \mathcal{N}_n$, and X_i be a random variable corresponding to vertex i. $X_1, X_2, \cdots, X_n$ form a Markov graph G if for all cutsets U in G, $X_{V_1(U)}, X_{V_2(U)}, \cdots, X_{V_{s(U)}(U)}$ are mutually independent conditioning on X_U, where $V_1(U), V_2(U), \cdots, V_{s(U)}(U)$ are the components in $G \backslash U$.

Type I and Type II Atoms: For an atom $A = \cap_{i=1}^n \tilde{Y}_i$ in $\mathcal{A}$, $U_A = \{i \in \mathcal{N}_n : \tilde{Y}_i = \tilde{X}_i^c\}$. For an undirected graph $G = (V, E)$ with $V = \mathcal{N}_n$, an atom $A \in \mathcal{A}$ is Type I if $G \backslash U_A$ is connected, otherwise it is Type II.

I-Measure Characterization of Markov Random Field: $X_1, X_2, \cdots, X_n$ form a Markov graph G if and only if μ^* vanishes on all the Type II atoms.

I-Measure for Markov Chain:

1. μ^* is always nonnegative.
2. The information diagram can be displayed in two dimensions.

Problems

1. Prove Proposition 12.14 and Proposition 12.15.
2. In Example 12.22, it was shown that Π_2 implies Π_1. Show that Π_1 does not imply Π_2. Hint: Use an information diagram to determine $Im(\Pi_2) \backslash Im(\Pi_1)$.
3. *Alternative definition of the global Markov property.* For any partition $\{U, V_1, V_2\}$ of V such that the sets of vertices V_1 and V_2 are disconnected in $G \backslash U$, the sets of random variables X_{V_1} and X_{V_2} are independent conditioning on X_U.
   Show that this definition is equivalent to the global Markov property in Definition 12.23.
4. *The local Markov property.* For $1 \leq i \leq n$, X_i and X_{V-N_i-i} are independent conditioning on X_{N_i}, where N_i is the set of neighbors[2] of vertex i in G.
   a) Show that the global Markov property implies the local Markov property.
   b) Show that the local Markov property does not imply the global Markov property by giving a counterexample. Hint: Consider a joint distribution which is not strictly positive.
5. Construct a Markov random field whose I-Measure μ^* can take negative values. Hint: Consider a Markov "star."

---

[2] Vertices i and j in an undirected graph are neighbors if i and j are connected by an edge.

6. a) Show that X_1, X_2, X_3, and X_4 are mutually independent if and only if

$$X_1 \perp (X_2, X_3, X_4), \ \ X_2 \perp (X_3, X_4)|X_1, \ \ X_3 \perp X_4|(X_1, X_2).$$

   Hint: Use an information diagram.
   b) Generalize the result in (a) to n random variables.
7. Determine the Markov random field with four random variables X_1, X_2, X_3, and X_4 which is characterized by the following conditional independencies:

$$(X_1, X_2, X_5) \perp X_4|X_3,$$
$$X_2 \perp (X_4, X_5)|(X_1, X_3),$$
$$X_1 \perp (X_3, X_4)|(X_2, X_5).$$

What are the other conditional independencies pertaining to this Markov random field?

Historical Notes

A Markov random field can be regarded as a generalization of a discrete-time Markov chain. Historically, the study of Markov random field stems from statistical physics. The classical Ising model, which is defined on a rectangular lattice, was used to explain certain empirically observed facts about ferromagnetic materials. The foundation of the theory of Markov random fields can be found in Preston [296] or Spitzer [343].

The structure of the I-Measure for a Markov chain was first investigated in the unpublished work of Kawabata [196]. Essentially the same result was independently obtained by R. W. Yeung 11 years later in the context of the I-Measure, and the result was eventually published in Kawabata and Yeung [197]. Full conditional independencies were shown to be axiomatizable by Malvestuto [243]. The results in this chapter are due to Yeung et al. [407], where they obtained a set-theoretic characterization of full conditional independencies and investigated the structure of the I-Measure for a Markov random field. In this paper, they also obtained a hypergraph characterization of a Markov random field based on the I-Measure characterization in Theorem 12.25. Ge and Ye [131] have applied these results to characterize a class of graphical models for conditional independence of random variables.

13

Information Inequalities

An information expression f refers to a *linear combination*[1] of Shannon's information measures involving a finite number of random variables. For example,

$$H(X,Y) + 2I(X;Z) \tag{13.1}$$

and

$$I(X;Y) - I(X;Y|Z) \tag{13.2}$$

are information expressions. An information inequality has the form

$$f \geq c, \tag{13.3}$$

where the constant c is usually equal to zero. We consider non-strict inequalities only because these are usually the form of inequalities in information theory. Likewise, an information identity has the form

$$f = c. \tag{13.4}$$

We point out that an information identity $f = c$ is equivalent to the pair of information inequalities $f \geq c$ and $f \leq c$.

An information inequality or identity is said to *always hold* if it holds for any joint distribution for the random variables involved. For example, we say that the information inequality

$$I(X;Y) \geq 0 \tag{13.5}$$

always holds because it holds for any joint distribution $p(x,y)$. On the other hand, we say that an information inequality does not always hold if there exists a joint distribution for which the inequality does not hold. Consider the information inequality

---

[1] More generally, an information expression can be nonlinear, but they do not appear to be useful in information theory.

$$I(X;Y) \leq 0. \tag{13.6}$$

Since

$$I(X;Y) \geq 0 \tag{13.7}$$

always holds, (13.6) is equivalent to

$$I(X;Y) = 0, \tag{13.8}$$

which holds if and only if X and Y are independent. In other words, (13.6) does not hold if X and Y are not independent. Therefore, we say that (13.6) does not always hold.

As we have seen in the previous chapters, information inequalities are the major tools for proving converse coding theorems. These inequalities govern the impossibilities in information theory. More precisely, information inequalities imply that certain things cannot happen. For this reason, they are sometimes referred to as the *laws of information theory*.

The basic inequalities form the most important set of information inequalities. In fact, almost all the information inequalities known to date are implied by the basic inequalities. These are called *Shannon-type inequalities*. On the other hand, if an information inequality always holds but is not implied by the basic inequalities, then it is called a *non-Shannon-type inequality*. We have not yet explained what it means by the fact that an inequality is or is not implied by the basic inequalities, but this will become clear later in the chapter.

Let us now rederive the inequality obtained in Example 3.15 (Imperfect Secrecy Theorem) without using an information diagram. In this example, three random variables X, Y, and Z are involved, and the setup of the problem is equivalent to the constraint

$$H(X|Y,Z) = 0. \tag{13.9}$$

Then

$$
\begin{aligned}
I(X;Y) \\
&= H(X) + H(Y) - H(X,Y) \tag{13.10} \\
&= H(X) + H(Y) - [H(X,Y,Z) - H(Z|X,Y)] \tag{13.11} \\
&\geq H(X) + H(Y) - H(X,Y,Z) \tag{13.12} \\
&= H(X) + H(Y) - [H(Z) + H(Y|Z) + H(X|Y,Z)] \tag{13.13} \\
&= H(X) - H(Z) + I(Y;Z) - H(X|Y,Z) \tag{13.14} \\
&\geq H(X) - H(Z), \tag{13.15}
\end{aligned}
$$

where we have used

$$H(Z|X,Y) \geq 0 \tag{13.16}$$

in obtaining (13.12), and

$$I(Y;Z) \geq 0 \tag{13.17}$$

and (13.9) in obtaining (13.15). This derivation is less transparent than the one we presented in Example 3.15, but the point here is that the final inequality we obtain in (13.15) can be proved by invoking the basic inequalities (13.16) and (13.17). In other words, (13.15) is implied by the basic inequalities. Therefore, it is a (constrained) Shannon-type inequality.

We are motivated to ask the following two questions:

1. How can Shannon-type inequalities be characterized? That is, given an information inequality, how can we tell whether it is implied by the basic inequalities?
2. Are there any non-Shannon-type information inequalities?

These two are very fundamental questions in information theory. We point out that the first question naturally comes before the second question because if we cannot characterize all Shannon-type inequalities, even if we are given a non-Shannon-type inequality, we cannot tell that it actually is one.

In this chapter, we develop a geometric framework for information inequalities which enables them to be studied systematically. This framework naturally leads to an answer to the first question, which makes machine-proving of all Shannon-type inequalities possible. This will be discussed in the next chapter. The second question will be answered positively in Chapter 15. In other words, there *do* exist laws in information theory beyond those laid down by Shannon.

13.1 The Region Γ_n^*

Let

$$\mathcal{N}_n = \{1, 2, \cdots, n\}, \tag{13.18}$$

where $n \geq 2$, and let

$$\Theta = \{X_i, i \in \mathcal{N}_n\} \tag{13.19}$$

be any collection of n random variables. Associated with Θ are

$$k = 2^n - 1 \tag{13.20}$$

joint entropies. For example, for $n = 3$, the 7 joint entropies associated with random variables X_1, X_2, and X_3 are

$$H(X_1), H(X_2), H(X_3), H(X_1, X_2),$$
$$H(X_2, X_3), H(X_1, X_3), H(X_1, X_2, X_3). \tag{13.21}$$

Let $\Re$ denote the set of real numbers. For any nonempty subset α of $\mathcal{N}_n$, let

$$X_\alpha = (X_i, i \in \alpha) \tag{13.22}$$

and

$$H_\Theta(\alpha) = H(X_\alpha). \tag{13.23}$$

For a fixed Θ, we can then view H_Θ as a set function from $2^{\mathcal{N}_n}$ to $\Re$ with $H_\Theta(\emptyset) = 0$, i.e., we adopt the convention that the entropy of an empty set of random variable is equal to zero. For this reason, we call H_Θ the *entropy function* of Θ.

Let $\mathcal{H}_n$ be the k-dimensional Euclidean space with the coordinates labeled by $h_\alpha, \alpha \in 2^{\mathcal{N}_n} \setminus \{\emptyset\}$, where h_α corresponds to the value of $H_\Theta(\alpha)$ for any collection Θ of n random variables. We will refer to $\mathcal{H}_n$ as the *entropy space* for n random variables. Then an entropy function H_Θ can be represented by a column vector in $\mathcal{H}_n$. On the other hand, a column vector $\mathbf{h} \in \mathcal{H}_n$ is called *entropic* if $\mathbf{h}$ is equal to the entropy function H_Θ of some collection Θ of n random variables. We are motivated to define the following region in $\mathcal{H}_n$:

$$\Gamma_n^* = \{\mathbf{h} \in \mathcal{H}_n : \mathbf{h} \text{ is entropic}\}. \tag{13.24}$$

For convenience, the vectors in Γ_n^* will also be referred to as entropy functions. As an example, for $n = 3$, the coordinates of $\mathcal{H}_3$ are labeled by

$$h_1, h_2, h_3, h_{12}, h_{13}, h_{23}, h_{123}, \tag{13.25}$$

where h_{123} denotes $h_{\{1,2,3\}}$, etc, and Γ_3^* is the region in $\mathcal{H}_3$ of all entropy functions for 3 random variables.

While further characterizations of Γ_n^* will be given later, we first point out a few basic properties of Γ_n^*:

1. Γ_n^* contains the origin.
2. $\overline{\Gamma}_n^*$, the closure of Γ_n^*, is convex.
3. Γ_n^* is in the nonnegative orthant of the entropy space $\mathcal{H}_n$.[2]

The origin of the entropy space corresponds to the entropy function of n degenerate random variables taking constant values. Hence, Property 1 follows. Property 2 will be proved in Chapter 15. Properties 1 and 2 imply that $\overline{\Gamma}_n^*$ is a convex cone. Property 3 is true because the coordinates in the entropy space $\mathcal{H}_n$ correspond to joint entropies, which are always nonnegative.

13.2 Information Expressions in Canonical Form

Any Shannon's information measure other than a joint entropy can be expressed as a linear combination of joint entropies by application of one of the following information identities:

$$H(X|Y) = H(X,Y) - H(Y) \tag{13.26}$$

$$I(X;Y) = H(X) + H(Y) - H(X,Y) \tag{13.27}$$

$$I(X;Y|Z) = H(X,Z) + H(Y,Z) - H(X,Y,Z) - H(Z). \tag{13.28}$$

---

[2] The nonnegative orthant of $\mathcal{H}^n$ is the region $\{\mathbf{h} \in \mathcal{H}_n : h_\alpha \geq 0 \text{ for all } \alpha \in 2^{\mathcal{N}_n} \setminus \{\emptyset\}\}$.

The first and the second identity are special cases of the third identity, which has already been proved in Lemma 3.8. Thus any information expression which involves n random variables can be expressed as a linear combination of the k associated joint entropies. We call this the *canonical form* of an information expression. When we write an information expression f as $f(\mathbf{h})$, it means that f is in canonical form. Since an information expression in canonical form is a linear combination of the joint entropies, it has the form

$$\mathbf{b}^\top \mathbf{h} \tag{13.29}$$

where $\mathbf{b}^\top$ denotes the transpose of a constant column vector $\mathbf{b}$ in $\Re^k$.

The identities in (13.26–13.28) provide a way to express every information expression in canonical form. However, it is not clear whether such a canonical form is unique. To illustrate the point, we consider obtaining the canonical form of $H(X|Y)$ in two ways. First,

$$H(X|Y) = H(X,Y) - H(Y). \tag{13.30}$$

Second,

$$\begin{align}
H(X|Y) &= H(X) - I(X;Y) \tag{13.31} \\
&= H(X) - (H(Y) - H(Y|X)) \tag{13.32} \\
&= H(X) - (H(Y) - H(X,Y) + H(X)) \tag{13.33} \\
&= H(X,Y) - H(Y). \tag{13.34}
\end{align}$$

Thus it turns out that we can obtain the same canonical form for $H(X|Y)$ via two different expansions. This is not accidental, as it is implied by the uniqueness of the canonical form which we will prove shortly.

Recall from the proof of Theorem 3.6 that the vector $\mathbf{h}$ represents the values of the I-Measure μ^* on the unions in $\mathcal{F}_n$. Moreover, $\mathbf{h}$ is related to the values of μ^* on the atoms of $\mathcal{F}_n$, represented as $\mathbf{u}$, by

$$\mathbf{h} = C_n \mathbf{u} \tag{13.35}$$

where C_n is a unique $k \times k$ matrix (cf. (3.27)). We now state the following lemma which is a rephrase of Theorem 3.11. This lemma is essential for proving the next theorem which implies the uniqueness of the canonical form.

Lemma 13.1. *Let*

$$\Psi_n^* = \{\mathbf{u} \in \Re^k : C_n\mathbf{u} \in \Gamma_n^*\}. \tag{13.36}$$

Then the nonnegative orthant of $\Re^k$ is a subset of Ψ_n^*.

Theorem 13.2. *Let f be an information expression. Then the unconstrained information identity $f = 0$ always holds if and only if f is the zero function.*

Proof. Without loss of generality, assume f is in canonical form and let

$$f(\mathbf{h}) = \mathbf{b}^\top \mathbf{h}. \tag{13.37}$$

Assume $f = 0$ always holds and f is not the zero function, i.e., $\mathbf{b} \neq 0$. We will show that this leads to a contradiction.

First, $f = 0$, or more precisely the set

$$\{\mathbf{h} \in \mathcal{H}_n : \mathbf{b}^\top \mathbf{h} = 0\}, \tag{13.38}$$

is a hyperplane[3] in the entropy space which has zero Lebesgue measure.[4] We claim that Γ_n^* is contained in the hyperplane $f = 0$. If this is not true, then there exists $\mathbf{h}_0 \in \Gamma_n^*$ which is not on $f = 0$, i.e., $f(\mathbf{h}_0) \neq 0$. Since $\mathbf{h}_0 \in \Gamma_n^*$, it corresponds to the entropy function of some joint distribution. This means that there exists a joint distribution such that $f(\mathbf{h}) = 0$ does not hold, which is a contradiction to our assumption that $f = 0$ always holds. This proves our claim.

If Γ_n^* has positive Lebesgue measure, it cannot be contained in the hyperplane $f = 0$ which has zero Lebesgue measure. Therefore, it suffices to show that Γ_n^* has positive Lebesgue measure. To this end, we see from Lemma 13.1 that the nonnegative orthant of $\mathcal{H}_n$, which has positive Lebesgue measure, is a subset of Ψ_n^*. Thus Ψ_n^* has positive Lebesgue measure. Since Γ_n^* is an invertible linear transformation of Ψ_n^*, its Lebesgue measure is also positive.

Therefore, Γ_n^* is not contained in the hyperplane $f = 0$, which implies that there exists a joint distribution for which $f = 0$ does not hold. This leads to a contradiction because we have assumed that $f = 0$ always holds. Hence, we have proved that if $f = 0$ always holds, then f must be the zero function.

Conversely, if f is the zero function, then it is trivial that $f = 0$ always holds. The theorem is proved. $\square$

Corollary 13.3. *The canonical form of an information expression is unique.*

Proof. Let f_1 and f_2 be canonical forms of an information expression g. Since

$$g = f_1 \tag{13.39}$$

and

$$g = f_2 \tag{13.40}$$

always hold,

$$f_1 - f_2 = 0 \tag{13.41}$$

always holds. By the above theorem, $f_1 - f_2$ is the zero function, which implies that f_1 and f_2 are identical. The corollary is proved. $\square$

---

[3] If $\mathbf{b} = 0$, then $\{\mathbf{h} \in \mathcal{H}_n : \mathbf{b}^\top \mathbf{h} = 0\}$ is equal to $\mathcal{H}_n$.

[4] The Lebesgue measure can be thought of as "volume" in the Euclidean space if the reader is not familiar with measure theory.

Due to the uniqueness of the canonical form of an information expression, it is an easy matter to check whether for two information expressions f_1 and f_2 the unconstrained information identity

$$f_1 = f_2 \tag{13.42}$$

always holds. All we need to do is to express $f_1 - f_2$ in canonical form. Then (13.42) always holds if and only if all the coefficients are zero.

13.3 A Geometrical Framework

In the last section, we have seen the role of the region Γ_n^* in proving unconstrained information identities. In this section, we explain the geometrical meanings of unconstrained information inequalities, constrained information inequalities, and constrained information identities in terms of Γ_n^*. Without loss of generality, we assume that all information expressions are in canonical form.

13.3.1 Unconstrained Inequalities

Consider an unconstrained information inequality $f \geq 0$, where $f(\mathbf{h}) = \mathbf{b}^\top \mathbf{h}$. Then $f \geq 0$ corresponds to the set

$$\{\mathbf{h} \in \mathcal{H}_n : \mathbf{b}^\top \mathbf{h} \geq 0\} \tag{13.43}$$

which is a half-space in the entropy space $\mathcal{H}_n$ containing the origin. Specifically, for any $\mathbf{h} \in \mathcal{H}_n$, $f(\mathbf{h}) \geq 0$ if and only if $\mathbf{h}$ belongs to this set. For simplicity, we will refer to this set as the half-space $f \geq 0$. As an example, for $n = 2$, the information inequality

$$I(X_1; X_2) = H(X_1) + H(X_2) - H(X_1, X_2) \geq 0, \tag{13.44}$$

written as

$$h_1 + h_2 - h_{12} \geq 0, \tag{13.45}$$

corresponds to the half-space

$$\{\mathbf{h} \in \mathcal{H}_n : h_1 + h_2 - h_{12} \geq 0\} \tag{13.46}$$

in the entropy space $\mathcal{H}_2$.

Since an information inequality always holds if and only if it is satisfied by the entropy function of any joint distribution for the random variables involved, we have the following geometrical interpretation of an information inequality:

$f \geq 0$ always holds if and only if $\Gamma_n^* \subset \{\mathbf{h} \in \mathcal{H}_n : f(\mathbf{h}) \geq 0\}$.

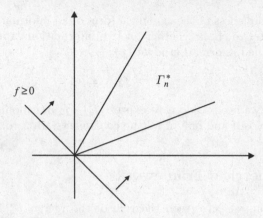

Fig. 13.1. An illustration for $f \geq 0$ always holds.

This gives a complete characterization of all unconstrained inequalities in terms of Γ_n^*. If Γ_n^* is known, we in principle can determine whether any information inequality involving n random variables always holds.

The two possible cases for $f \geq 0$ are illustrated in Figure 13.1 and Figure 13.2. In Figure 13.1, Γ_n^* is completely included in the half-space $f \geq 0$, so $f \geq 0$ always holds. In Figure 13.2, there exists a vector $\mathbf{h}_0 \in \Gamma_n^*$ such that $f(\mathbf{h}_0) < 0$. Thus the inequality $f \geq 0$ does not always hold.

13.3.2 Constrained Inequalities

In information theory, we very often deal with information inequalities (identities) with certain constraints on the joint distribution for the random variables involved. These are called constrained information inequalities (identities), and the constraints on the joint distribution can usually be expressed as linear constraints on the entropies. The following are such examples:

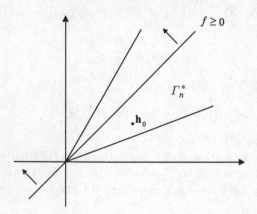

Fig. 13.2. An illustration for $f \geq 0$ not always holds.

1. X_1, X_2, and X_3 are mutually independent if and only if $H(X_1, X_2, X_3) = H(X_1) + H(X_2) + H(X_3)$.
2. X_1, X_2, and X_3 are pairwise independent if and only if $I(X_1; X_2) = I(X_2; X_3) = I(X_1; X_3) = 0$.
3. X_1 is a function of X_2 if and only if $H(X_1|X_2) = 0$.
4. $X_1 \to X_2 \to X_3 \to X_4$ forms a Markov chain if and only if $I(X_1; X_3|X_2) = 0$ and $I(X_1, X_2; X_4|X_3) = 0$.

Suppose there are q linear constraints on the entropies given by

$$Q\mathbf{h} = 0, \tag{13.47}$$

where Q is a $q \times k$ matrix. Here we do not assume that the q constraints are linearly independent, so Q is not necessarily full rank. Let

$$\Phi = \{\mathbf{h} \in \mathcal{H}_n : Q\mathbf{h} = 0\}. \tag{13.48}$$

In other words, the q constraints confine $\mathbf{h}$ to a linear subspace Φ in the entropy space. Parallel to our discussion on unconstrained inequalities, we have the following geometrical interpretation of a constrained inequality:

Under the constraint Φ, $f \geq 0$ always holds if and only if $(\Gamma_n^* \cap \Phi) \subset \{\mathbf{h} \in \mathcal{H}_n : f(\mathbf{h}) \geq 0\}$.

This gives a complete characterization of all constrained inequalities in terms of Γ_n^*. Note that $\Phi = \mathcal{H}_n$ when there is no constraint on the entropies. In this sense, an unconstrained inequality is a special case of a constrained inequality.

The two cases of $f \geq 0$ under the constraint Φ are illustrated in Figure 13.3 and Figure 13.4. Figure 13.3 shows the case when $f \geq 0$ always holds under

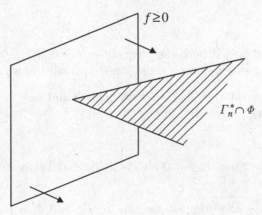

Fig. 13.3. An illustration for $f \geq 0$ always holds under the constraint Φ.

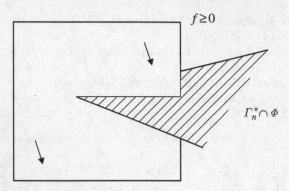

Fig. 13.4. An illustration for $f \geq 0$ not always holds under the constraint Φ.

the constraint Φ. Note that $f \geq 0$ may or may not always hold when there is no constraint. Figure 13.4 shows the case when $f \geq 0$ does not always hold under the constraint Φ. In this case, $f \geq 0$ also does not always hold when there is no constraint, because

$$(\Gamma_n^* \cap \Phi) \not\subset \{\mathbf{h} \in \mathcal{H}_n : f(\mathbf{h}) \geq 0\} \tag{13.49}$$

implies

$$\Gamma_n^* \not\subset \{\mathbf{h} \in \mathcal{H}_n : f(\mathbf{h}) \geq 0\}. \tag{13.50}$$

13.3.3 Constrained Identities

As we have pointed out at the beginning of the chapter, an identity

$$f = 0 \tag{13.51}$$

always holds if and only if both the inequalities $f \geq 0$ and $f \leq 0$ always hold. Then following our discussion on constrained inequalities, we have

Under the constraint Φ, $f = 0$ always holds if and only if $(\Gamma_n^* \cap \Phi) \subset \{\mathbf{h} \in \mathcal{H}_n : f(\mathbf{h}) \geq 0\} \cap \{\mathbf{h} \in \mathcal{H}_n : f(\mathbf{h}) \leq 0\}$

or

Under the constraint Φ, $f = 0$ always holds if and only if $(\Gamma_n^* \cap \Phi) \subset \{\mathbf{h} \in \mathcal{H}_n : f(\mathbf{h}) = 0\}$.

This condition says that the intersection of Γ_n^* and Φ is contained in the hyperplane $f = 0$.

13.4 Equivalence of Constrained Inequalities

When there is no constraint on the entropies, two information inequalities

$$\mathbf{b}^\top \mathbf{h} \geq 0 \tag{13.52}$$

and

$$\mathbf{c}^\top \mathbf{h} \geq 0 \tag{13.53}$$

are equivalent if and only if $\mathbf{c} = a\mathbf{b}$, where a is a positive constant. However, this is not the case under a nontrivial constraint $\Phi \neq \mathcal{H}_n$. This situation is illustrated in Figure 13.5. In this figure, although the inequalities in (13.52) and (13.53) correspond to different half-spaces in the entropy space, they actually impose the same constraint on $\mathbf{h}$ when $\mathbf{h}$ is confined to Φ.

In this section, we present a characterization of (13.52) and (13.53) being equivalent under a set of linear constraints Φ. The reader may skip this section at the first reading.

Let r be the rank of Q in (13.47). Since $\mathbf{h}$ is in the null space of Q, we can write

$$\mathbf{h} = \tilde{Q}\mathbf{h}', \tag{13.54}$$

where $\tilde{Q}$ is a $k \times (k-r)$ matrix such that the rows of $\tilde{Q}^\top$ form a basis of the orthogonal complement of the row space of Q, and $\mathbf{h}'$ is a column $(k-r)$-vector. Then using (13.54), (13.52) and (13.53) can be written as

$$\mathbf{b}^\top \tilde{Q}\mathbf{h}' \geq 0 \tag{13.55}$$

and

$$\mathbf{c}^\top \tilde{Q}\mathbf{h}' \geq 0, \tag{13.56}$$

respectively, in terms of the set of basis given by the columns of $\tilde{Q}$. Then (13.55) and (13.56) are equivalent if and only if

$$\mathbf{c}^\top \tilde{Q} = a\mathbf{b}^\top \tilde{Q}, \tag{13.57}$$

where a is a positive constant, or

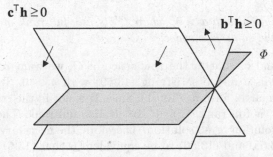

Fig. 13.5. Equivalence of $\mathbf{b}^\top \mathbf{h} \geq 0$ and $\mathbf{c}^\top \mathbf{h} \geq 0$ under the constraint Φ.

$$(\mathbf{c} - a\mathbf{b})^\top \tilde{Q} = 0. \tag{13.58}$$

In other words, $(\mathbf{c} - a\mathbf{b})^\top$ is in the orthogonal complement of the row space of $\tilde{Q}^\top$, i.e., $(\mathbf{c} - a\mathbf{b})^\top$ is in the row space of Q. Let Q' be any $r \times k$ matrix such that Q' and Q have the same row space. (Q can be taken as Q' if Q is full rank.) Since the rank of Q is r and Q' has r rows, the rows of Q' form a basis for the row space of Q, and Q' is full rank. Then from (13.58), (13.55) and (13.56) are equivalent under the constraint Φ if and only if

$$\mathbf{c} = a\mathbf{b} + (Q')^\top \mathbf{e} \tag{13.59}$$

for some positive constant a and some column r-vector $\mathbf{e}$.

Suppose for given $\mathbf{b}$ and $\mathbf{c}$, we want to see whether (13.55) and (13.56) are equivalent under the constraint Φ. We first consider the case when either $\mathbf{b}^\top$ or $\mathbf{c}^\top$ is in the row space of Q. This is actually not an interesting case because if $\mathbf{b}^\top$, for example, is in the row space of Q, then

$$\mathbf{b}^\top \tilde{Q} = 0 \tag{13.60}$$

in (13.55), which means that (13.55) imposes no additional constraint under the constraint Φ.

Theorem 13.4. *If either $\mathbf{b}^\top$ or $\mathbf{c}^\top$ is in the row space of Q, then $\mathbf{b}^\top \mathbf{h} \geq 0$ and $\mathbf{c}^\top \mathbf{h} \geq 0$ are equivalent under the constraint Φ if and only if both $\mathbf{b}^\top$ and $\mathbf{c}^\top$ are in the row space of Q.*

The proof of this theorem is left as an exercise. We now turn to the more interesting case when neither $\mathbf{b}^\top$ nor $\mathbf{c}^\top$ is in the row space of Q. The following theorem gives an explicit condition for (13.55) and (13.56) to be equivalent under the constraint Φ.

Theorem 13.5. *If neither $\mathbf{b}^\top$ nor $\mathbf{c}^\top$ is in the row space of Q, then $\mathbf{b}^\top \mathbf{h} \geq 0$ and $\mathbf{c}^\top \mathbf{h} \geq 0$ are equivalent under the constraint Φ if and only if*

$$\left[\, (Q')^\top \ \ \mathbf{b} \, \right] \begin{bmatrix} \mathbf{e} \\ a \end{bmatrix} = \mathbf{c} \tag{13.61}$$

has a unique solution with $a > 0$, where Q' is any full-rank matrix such that Q' and Q have the same row space.

Proof. For $\mathbf{b}^\top$ and $\mathbf{c}^\top$ not in the row space of Q, we want to see when we can find unknowns a and $\mathbf{e}$ satisfying (13.59) with $a > 0$. To this end, we write (13.59) in matrix form as (13.61). Since $\mathbf{b}$ is not in the column space of $(Q')^\top$ and $(Q')^\top$ is full rank, $\left[\, (Q')^\top \ \ \mathbf{b} \, \right]$ is also full rank. Then (13.61) has either a unique solution or no solution. Therefore, the necessary and sufficient condition for (13.55) and (13.56) to be equivalent is that (13.61) has a unique solution and $a > 0$. The theorem is proved. □

Example 13.6. Consider three random variables X_1, X_2, and X_3 with the Markov constraint

$$I(X_1; X_3|X_2) = 0, \tag{13.62}$$

which is equivalent to

$$H(X_1, X_2) + H(X_2, X_3) - H(X_1, X_2, X_3) - H(X_2) = 0. \tag{13.63}$$

In terms of the coordinates in the entropy space $\mathcal{H}_3$, this constraint is written as

$$Q\mathbf{h} = 0, \tag{13.64}$$

where

$$Q = [\,0 \;{-1}\; 0 \; 1 \; 1 \; 0 \;{-1}\,] \tag{13.65}$$

and

$$\mathbf{h} = [\,h_1 \; h_2 \; h_3 \; h_{12} \; h_{23} \; h_{13} \; h_{123}\,]^{\mathsf{T}}. \tag{13.66}$$

We now show that under the constraint in (13.64), the inequalities

$$H(X_1|X_3) - H(X_1|X_2) \geq 0 \tag{13.67}$$

and

$$I(X_1; X_2|X_3) \geq 0 \tag{13.68}$$

are in fact equivalent. Toward this end, we write (13.67) and (13.68) as $\mathbf{b}^{\mathsf{T}}\mathbf{h} \geq 0$ and $\mathbf{c}^{\mathsf{T}}\mathbf{h} \geq 0$, respectively, where

$$\mathbf{b} = [\,0 \; 1 \;{-1}\;{-1}\; 0 \; 1 \; 0\,]^{\mathsf{T}} \tag{13.69}$$

and

$$\mathbf{c} = [\,0 \; 0 \;{-1}\; 0 \; 1 \; 1 \;{-1}\,]^{\mathsf{T}}. \tag{13.70}$$

Since Q is full rank, we may take $Q' = Q$. Upon solving

$$[\, Q^{\mathsf{T}} \; \mathbf{b} \,] \begin{bmatrix} \mathbf{e} \\ a \end{bmatrix} = \mathbf{c}, \tag{13.71}$$

we obtain the unique solution $a = 1 > 0$ and $\mathbf{e} = 1$ ($\mathbf{e}$ is a 1×1 matrix). Therefore, (13.67) and (13.68) are equivalent under the constraint in (13.64).

Under the constraint Φ, if neither $\mathbf{b}^{\mathsf{T}}$ nor $\mathbf{c}^{\mathsf{T}}$ is in the row space of Q, it can be shown that the identities

$$\mathbf{b}^{\mathsf{T}}\mathbf{h} = 0 \tag{13.72}$$

and

$$\mathbf{c}^{\mathsf{T}}\mathbf{h} = 0 \tag{13.73}$$

are equivalent if and only if (13.61) has a unique solution. We leave the proof as an exercise.

13.5 The Implication Problem of Conditional Independence

We use $X_\alpha \perp X_\beta | X_\gamma$ to denote the conditional independency (CI)

$$X_\alpha \text{ and } X_\beta \text{ are conditionally independent given } X_\gamma.$$

We have proved in Theorem 2.34 that $X_\alpha \perp X_\beta | X_\gamma$ is equivalent to

$$I(X_\alpha; X_\beta | X_\gamma) = 0. \tag{13.74}$$

When $\gamma = \emptyset$, $X_\alpha \perp X_\beta | X_\gamma$ becomes an unconditional independency which we regard as a special case of a conditional independency. When $\alpha = \beta$, (13.74) becomes

$$H(X_\alpha | X_\gamma) = 0, \tag{13.75}$$

which we see from Proposition 2.36 that X_α is a function of X_γ. For this reason, we also regard functional dependency as a special case of conditional independency.

In probability problems, we are often given a set of CIs and we need to determine whether another given CI is logically implied. This is called the *implication problem*, which is one of the most basic problems in probability theory. We have seen in Section 12.2 that the implication problem has a solution if only full conditional mutual independencies are involved. However, the general problem is extremely difficult, and it has been solved only up to four random variables [256].

We end this section by explaining the relation between the implication problem and the region Γ_n^*. A CI involving random variables $X_1, X_2, \cdots, X_n$ has the form

$$X_\alpha \perp X_\beta | X_\gamma, \tag{13.76}$$

where $\alpha, \beta, \gamma \subset \mathcal{N}_n$. Since $I(X_\alpha; X_\beta | X_\gamma) = 0$ is equivalent to

$$H(X_{\alpha \cup \gamma}) + H(X_{\beta \cup \gamma}) - H(X_{\alpha \cup \beta \cup \gamma}) - H(X_\gamma) = 0, \tag{13.77}$$

$X_\alpha \perp X_\beta | X_\gamma$ corresponds to the hyperplane

$$\{\mathbf{h} \in \mathcal{H}_n : h_{\alpha \cup \gamma} + h_{\beta \cup \gamma} - h_{\alpha \cup \beta \cup \gamma} - h_\gamma = 0\}. \tag{13.78}$$

For a CI K, we denote the hyperplane in $\mathcal{H}_n$ corresponding to K by $\mathcal{E}(K)$.

Let $\Pi = \{K_l\}$ be a collection of CIs, and we want to determine whether Π implies a given CI K. This would be the case if and only if the following is true:

$$\text{For all } \mathbf{h} \in \Gamma_n^*, \text{ if } \mathbf{h} \in \bigcap_l \mathcal{E}(K_l), \text{ then } \mathbf{h} \in \mathcal{E}(K).$$

Equivalently,

Π implies K if and only if $\left(\bigcap_l \mathcal{E}(K_l)\right) \cap \Gamma_n^* \subset \mathcal{E}(K)$.

Therefore, the implication problem can be solved if Γ_n^* can be characterized. Hence, the region Γ_n^* is not only of fundamental importance in information theory, but is also of fundamental importance in probability theory.

Chapter Summary

Entropy Space: The entropy space $\mathcal{H}_n$ is the $(2^n - 1)$-dimensional Euclidean space with the coordinates labeled by h_α, $\alpha \in 2^{\mathcal{N}_n} \setminus \{\emptyset\}$, where $\mathcal{N}_n = \{1, 2, \cdots, n\}$.

The Region Γ_n^* is the subset of $\mathcal{H}_n$ of all entropy functions for n discrete random variables.

Basic Properties of Γ_n^*:

1. Γ_n^* contains the origin.
2. $\overline{\Gamma}_n^*$, the closure of Γ_n^*, is convex.
3. Γ_n^* is in the nonnegative orthant of the entropy space $\mathcal{H}_n$.

Canonical Form of an Information Expression: Any information expression can be expressed as a linear combination of joint entropies, called the canonical form. The canonical form of an information expression is unique.

Unconstrained Information Identities: $\mathbf{b}^\top \mathbf{h} = 0$ always holds if and only if $\mathbf{b} = 0$.

Unconstrained Information Inequalities: $\mathbf{b}^\top \mathbf{h} \geq 0$ always holds if and only if $\Gamma_n^* \subset \{\mathbf{h} \in \mathcal{H}_n : \mathbf{b}^\top \mathbf{h} \geq 0\}$.

Constrained Information Inequalities: Under the constraint $\Phi = \{\mathbf{h} \in \mathcal{H}_n : Q\mathbf{h} = 0\}$, $\mathbf{b}^\top \mathbf{h} \geq 0$ always holds if and only if $(\Gamma_n^* \cap \Phi) \subset \{\mathbf{h} \in \mathcal{H}_n : \mathbf{b}^\top \mathbf{h} \geq 0\}$.

Equivalence of Constrained Inequalities (Identities): Under the constraint $\Phi = \{\mathbf{h} \in \mathcal{H}_n : Q\mathbf{h} = 0\}$, $\mathbf{b}^\top \mathbf{h} \geq 0$ and $\mathbf{c}^\top \mathbf{h} \geq 0$ ($\mathbf{b}^\top \mathbf{h} = 0$ and $\mathbf{c}^\top \mathbf{h} = 0$) are equivalent if and only if one of the following holds:

1. Both $\mathbf{b}^\top$ and $\mathbf{c}^\top$ are in the row space of Q.
2. Neither $\mathbf{b}^\top$ nor $\mathbf{c}^\top$ is in the row space of Q, and

$$\left[\ (Q')^\top\ \mathbf{b}\ \right] \begin{bmatrix} \mathbf{e} \\ a \end{bmatrix} = \mathbf{c}$$

has a unique solution with $a > 0$ (has a unique solution), where Q' is any full-rank matrix such that Q' and Q have the same row space.

Problems

1. *Symmetrical information expressions.* An information expression is said to be symmetrical if it is identical under every permutation of the random variables involved. However, sometimes a symmetrical information expression cannot be readily recognized symbolically. For example, $I(X_1; X_2) - I(X_1; X_2|X_3)$ is symmetrical in X_1, X_2, and X_3 but it is not symmetrical symbolically. Devise a general method for recognizing symmetrical information expressions.

2. The canonical form of an information expression is unique when there is no constraint on the random variables involved. Show by an example that this does not hold when certain constraints are imposed on the random variables involved.

3. *Alternative canonical form.* Denote $\cap_{i \in G} \tilde{X}_i$ by $\check{X}_G$ and let

$$\mathcal{C} = \left\{ \check{X}_G : \ G \text{ is a nonempty subset of } \mathcal{N}_n \right\}.$$

   a) Prove that a signed measure μ on $\mathcal{F}_n$ is completely specified by $\{\mu(C), \ C \in \mathcal{C}\}$, which can be any set of real numbers.

   b) Prove that an information expression involving $X_1, X_2, \cdots, X_n$ can be expressed uniquely as a linear combination of $\mu^*(\check{X}_G)$, where G are nonempty subsets of $\mathcal{N}_n$.

4. *Uniqueness of the canonical form for nonlinear information expressions.* Consider a function $f : \Re^k \to \Re$, where $k = 2^n - 1$ such that $\{\mathbf{h} \in \Re^k : f(\mathbf{h}) = 0\}$ has zero Lebesgue measure.

   a) Prove that f cannot be identically zero on Γ_n^*.

   b) Use the result in (a) to show the uniqueness of the canonical form for the class of information expressions of the form $g(\mathbf{h})$ where g is a polynomial.

   (Yeung [401]).

5. Prove that under the constraint $Q\mathbf{h} = 0$, if neither $\mathbf{b}^\top$ nor $\mathbf{c}^\top$ is in the row space of Q, the identities $\mathbf{b}^\top\mathbf{h} = 0$ and $\mathbf{c}^\top\mathbf{h} = 0$ are equivalent if and only if (13.61) has a unique solution.

Historical Notes

The uniqueness of the canonical form for linear information expressions was first proved by Han [144]. The same result was independently obtained in the book by Csiszár and Körner [83]. The geometrical framework for information inequalities is due to Yeung [401]. The characterization of equivalent constrained inequalities in Section 13.4 first appeared in the book by Yeung [402].

14

Shannon-Type Inequalities

The basic inequalities form the most important set of information inequalities. In fact, almost all the information inequalities known to date are implied by the basic inequalities. These are called *Shannon-type inequalities*. In this chapter, we show that verification of Shannon-type inequalities can be formulated as a linear programming problem, thus enabling machine-proving of all such inequalities.

14.1 The Elemental Inequalities

Consider the conditional mutual information

$$I(X, Y; X, Z, U | Z, T), \tag{14.1}$$

in which the random variables X and Z appear more than once. It is readily seen that $I(X, Y; X, Z, U | Z, T)$ can be written as

$$H(X|Z, T) + I(Y; U | X, Z, T), \tag{14.2}$$

where in both $H(X|Z, T)$ and $I(Y; U | X, Z, T)$, each random variable appears only once.

A Shannon's information measure is said to be *reducible* if there exists a random variable which appears more than once in the information measure; otherwise the information measure is said to be *irreducible*. Without loss of generality, we will consider irreducible Shannon's information measures only, because a reducible Shannon's information measure can always be written as the sum of irreducible Shannon's information measures.

The nonnegativity of all Shannon's information measures forms a set of inequalities called the basic inequalities. The set of basic inequalities, however, is not minimal in the sense that some basic inequalities are implied by the others. For example,

$$H(X|Y) \geq 0 \tag{14.3}$$

and

$$I(X;Y) \geq 0, \tag{14.4}$$

which are both basic inequalities involving random variables X and Y, imply

$$H(X) = H(X|Y) + I(X;Y) \geq 0, \tag{14.5}$$

again a basic inequality involving X and Y.

Let $\mathcal{N}_n = \{1, 2, \cdots, n\}$, where $n \geq 2$. Unless otherwise specified, all information expressions in this chapter involve some or all of the random variables $X_1, X_2, \cdots, X_n$. The value of n will be specified when necessary. Through application of the identities

$$H(X) = H(X|Y) + I(X;Y), \tag{14.6}$$
$$H(X,Y) = H(X) + H(Y|X), \tag{14.7}$$
$$I(X;Y,Z) = I(X;Y) + I(X;Z|Y), \tag{14.8}$$
$$H(X|Z) = H(X|Y,Z) + I(X;Y|Z), \tag{14.9}$$
$$H(X,Y|Z) = H(X|Z) + H(Y|X,Z), \tag{14.10}$$
$$I(X;Y,Z|T) = I(X;Y|T) + I(X;Z|Y,T), \tag{14.11}$$

any Shannon's information measure can be expressed as the sum of Shannon's information measures of the following two *elemental forms*:

i) $H(X_i|X_{\mathcal{N}_n-\{i\}}), i \in \mathcal{N}_n,$
ii) $I(X_i; X_j|X_K)$, where $i \neq j$ and $K \subset \mathcal{N}_n - \{i,j\}$.

This will be illustrated in the next example. It is not difficult to check that the total number of the two elemental forms of Shannon's information measures for n random variables is equal to

$$m = n + \binom{n}{2} 2^{n-2}. \tag{14.12}$$

The proof of (14.12) is left as an exercise.

Example 14.1. We can expand $H(X_1, X_2)$ into a sum of elemental forms of Shannon's information measures for $n = 3$ by applying the identities in (14.6) to (14.11) as follows:

$$
\begin{aligned}
&H(X_1, X_2) \\
&= H(X_1) + H(X_2|X_1) \tag{14.13} \\
&= H(X_1|X_2, X_3) + I(X_1; X_2, X_3) + H(X_2|X_1, X_3) \\
&\quad + I(X_2; X_3|X_1) \tag{14.14} \\
&= H(X_1|X_2, X_3) + I(X_1; X_2) + I(X_1; X_3|X_2) \\
&\quad + H(X_2|X_1, X_3) + I(X_2; X_3|X_1). \tag{14.15}
\end{aligned}
$$

The nonnegativity of the two elemental forms of Shannon's information measures forms a proper subset of the set of basic inequalities. We call the m inequalities in this smaller set the *elemental inequalities*. They are equivalent to the basic inequalities because each basic inequality which is not an elemental inequality can be obtained as the sum of a set of elemental inequalities in view of (14.6–14.11). This will be illustrated in the next example. The proof for the minimality of the set of elemental inequalities is deferred to Section 14.6.

Example 14.2. In the last example, we expressed $H(X_1, X_2)$ as

$$H(X_1|X_2, X_3) + I(X_1; X_2) + I(X_1; X_3|X_2)$$
$$+H(X_2|X_1, X_3) + I(X_2; X_3|X_1). \tag{14.16}$$

All the five Shannon's information measures in the above expression are in elemental form for $n = 3$. Then the basic inequality

$$H(X_1, X_2) \geq 0 \tag{14.17}$$

can be obtained as the sum of the following elemental inequalities:

$$H(X_1|X_2, X_3) \geq 0, \tag{14.18}$$
$$I(X_1; X_2) \geq 0, \tag{14.19}$$
$$I(X_1; X_3|X_2) \geq 0, \tag{14.20}$$
$$H(X_2|X_1, X_3) \geq 0, \tag{14.21}$$
$$I(X_2; X_3|X_1) \geq 0. \tag{14.22}$$

14.2 A Linear Programming Approach

Recall from Section 13.2 that any information expression can be expressed uniquely in canonical form, i.e., a linear combination of the $k = 2^n - 1$ joint entropies involving some or all of the random variables $X_1, X_2, \cdots, X_n$. If the elemental inequalities are expressed in canonical form, they become linear inequalities in the entropy space $\mathcal{H}_n$. Denote this set of inequalities by $G\mathbf{h} \geq 0$, where G is an $m \times k$ matrix, and define

$$\Gamma_n = \{\mathbf{h} \in \mathcal{H}_n : G\mathbf{h} \geq 0\}. \tag{14.23}$$

We first show that Γ_n is a pyramid in the nonnegative orthant of the entropy space $\mathcal{H}_n$. Evidently, Γ_n contains the origin. Let $\mathbf{e}_j, 1 \leq j \leq k$, be the column k-vector whose jth component is equal to 1 and all the other components are equal to 0. Then the inequality

$$\mathbf{e}_j^\top \mathbf{h} \geq 0 \tag{14.24}$$

corresponds to the nonnegativity of a joint entropy, which is a basic inequality. Since the set of elemental inequalities is equivalent to the set of basic inequalities, if $\mathbf{h} \in \Gamma_n$, i.e., $\mathbf{h}$ satisfies all the elemental inequalities, then $\mathbf{h}$ also satisfies the basic inequality in (14.24). In other words,

$$\Gamma_n \subset \{\mathbf{h} \in \mathcal{H}_n : \mathbf{e}_j^\top \mathbf{h} \geq 0\} \tag{14.25}$$

for all $1 \leq j \leq k$. This implies that Γ_n is in the nonnegative orthant of the entropy space. Since Γ_n contains the origin and the constraints $G\mathbf{h} \geq 0$ are linear, we conclude that Γ_n is a pyramid in the nonnegative orthant of $\mathcal{H}_n$.

Since the elemental inequalities are satisfied by the entropy function of any n random variables $X_1, X_2, \cdots, X_n$, for any $\mathbf{h}$ in Γ_n^*, $\mathbf{h}$ is also in Γ_n, i.e.,

$$\Gamma_n^* \subset \Gamma_n. \tag{14.26}$$

Therefore, for any unconstrained inequality $f \geq 0$, if

$$\Gamma_n \subset \{\mathbf{h} \in \mathcal{H}_n : f(\mathbf{h}) \geq 0\}, \tag{14.27}$$

then

$$\Gamma_n^* \subset \{\mathbf{h} \in \mathcal{H}_n : f(\mathbf{h}) \geq 0\}, \tag{14.28}$$

i.e., $f \geq 0$ always holds. In other words, (14.27) is a sufficient condition for $f \geq 0$ to always hold. Moreover, an inequality $f \geq 0$ such that (14.27) is satisfied is implied by the basic inequalities, because if $\mathbf{h}$ satisfies the basic inequalities, i.e., $\mathbf{h} \in \Gamma_n$, then $\mathbf{h}$ satisfies $f(\mathbf{h}) \geq 0$.

For constrained inequalities, following our discussion in Section 13.3, we impose the constraint

$$Q\mathbf{h} = 0 \tag{14.29}$$

and let

$$\Phi = \{\mathbf{h} \in \mathcal{H}_n : Q\mathbf{h} = 0\}. \tag{14.30}$$

For an inequality $f \geq 0$, if

$$(\Gamma_n \cap \Phi) \subset \{\mathbf{h} \in \mathcal{H}_n : f(\mathbf{h}) \geq 0\}, \tag{14.31}$$

then by (14.26),

$$(\Gamma_n^* \cap \Phi) \subset \{\mathbf{h} \in \mathcal{H}_n : f(\mathbf{h}) \geq 0\}, \tag{14.32}$$

i.e., $f \geq 0$ always holds under the constraint Φ. In other words, (14.31) is a sufficient condition for $f \geq 0$ to always hold under the constraint Φ. Moreover, an inequality $f \geq 0$ under the constraint Φ such that (14.31) is satisfied is implied by the basic inequalities and the constraint Φ, because if $\mathbf{h} \in \Phi$ and $\mathbf{h}$ satisfies the basic inequalities, i.e., $\mathbf{h} \in \Gamma_n \cap \Phi$, then $\mathbf{h}$ satisfies $f(\mathbf{h}) \geq 0$.

14.2.1 Unconstrained Inequalities

To check whether an unconstrained inequality $\mathbf{b}^\top \mathbf{h} \geq 0$ is a Shannon-type inequality, we need to check whether Γ_n is a subset of $\{\mathbf{h} \in \mathcal{H}_n : \mathbf{b}^\top \mathbf{h} \geq 0\}$. The following theorem induces a computational procedure for this purpose.

Theorem 14.3. $\mathbf{b}^\top \mathbf{h} \geq 0$ *is a Shannon-type inequality if and only if the minimum of the problem*

$$Minimize\ \mathbf{b}^\top \mathbf{h},\ subject\ to\ G\mathbf{h} \geq 0 \tag{14.33}$$

is zero. In this case, the minimum occurs at the origin.

Remark The idea of this theorem is illustrated in Figures 14.1 and 14.2. In Figure 14.1, Γ_n is contained in $\{\mathbf{h} \in \mathcal{H}_n : \mathbf{b}^\top \mathbf{h} \geq 0\}$. The minimum of

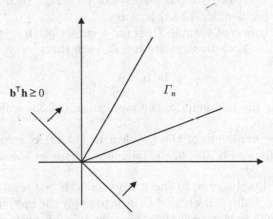

Fig. 14.1. Γ_n is contained in $\{\mathbf{h} \in \mathcal{H}_n : \mathbf{b}^\top \mathbf{h} \geq 0\}$.

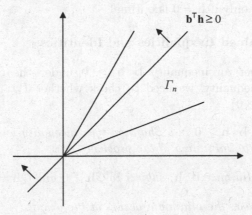

Fig. 14.2. Γ_n is not contained in $\{\mathbf{h} \in \mathcal{H}_n : \mathbf{b}^\top \mathbf{h} \geq 0\}$.

$\mathbf{b}^\top \mathbf{h}$ subject to Γ_n occurs at the origin with the minimum equal to 0. In Figure 14.2, Γ_n is not contained in $\{\mathbf{h} \in \mathcal{H}_n : \mathbf{b}^\top \mathbf{h} \geq 0\}$. The minimum of $\mathbf{b}^\top \mathbf{h}$ subject to Γ_n is $-\infty$. A formal proof of the theorem is given next.

Proof of Theorem 14.3. We have to prove that Γ_n is a subset of $\{\mathbf{h} \in \mathcal{H}_n : \mathbf{b}^\top \mathbf{h} \geq 0\}$ if and only if the minimum of the problem in (14.33) is zero. First of all, since $0 \in \Gamma_n$ and $\mathbf{b}^\top 0 = 0$ for any $\mathbf{b}$, the minimum of the problem in (14.33) is at most 0. Assume Γ_n is a subset of $\{\mathbf{h} \in \mathcal{H}_n : \mathbf{b}^\top \mathbf{h} \geq 0\}$ and the minimum of the problem in (14.33) is negative. Then there exists $\mathbf{h} \in \Gamma_n$ such that

$$\mathbf{b}^\top \mathbf{h} < 0, \tag{14.34}$$

which implies

$$\Gamma_n \not\subset \{\mathbf{h} \in \mathcal{H}_n : \mathbf{b}^\top \mathbf{h} \geq 0\}, \tag{14.35}$$

a contradiction. Therefore, if Γ_n is a subset of $\{\mathbf{h} \in \mathcal{H}_n : \mathbf{b}^\top \mathbf{h} \geq 0\}$, then the minimum of the problem in (14.33) is zero.

To prove the converse, assume Γ_n is not a subset of $\{\mathbf{h} \in \mathcal{H}_n : \mathbf{b}^\top \mathbf{h} \geq 0\}$, i.e., (14.35) is true. Then there exists $\mathbf{h} \in \Gamma_n$ such that

$$\mathbf{b}^\top \mathbf{h} < 0. \tag{14.36}$$

This implies that the minimum of the problem in (14.33) is negative, i.e., it is not equal to zero.

Finally, if the minimum of the problem in (14.33) is zero, since the Γ_n contains the origin and $\mathbf{b}^\top 0 = 0$, the minimum occurs at the origin.   □

By virtue of this theorem, to check whether $\mathbf{b}^\top \mathbf{h} \geq 0$ is an unconstrained Shannon-type inequality, all we need to do is to apply the *optimality test* of the *simplex method* [90] to check whether the point $\mathbf{h} = 0$ is optimal for the minimization problem in (14.33). Then $\mathbf{b}^\top \mathbf{h} \geq 0$ is an unconstrained Shannon-type inequality if and only if $\mathbf{h} = 0$ is optimal.

14.2.2 Constrained Inequalities and Identities

To check whether an inequality $\mathbf{b}^\top \mathbf{h} \geq 0$ under the constraint Φ is a Shannon-type inequality, we need to check whether $\Gamma_n \cap \Phi$ is a subset of $\{\mathbf{h} \in \mathcal{H}_n : \mathbf{b}^\top \mathbf{h} \geq 0\}$.

Theorem 14.4. $\mathbf{b}^\top \mathbf{h} \geq 0$ *is a Shannon-type inequality under the constraint* Φ *if and only if the minimum of the problem*

$$\text{Minimize } \mathbf{b}^\top \mathbf{h}, \text{ subject to } G\mathbf{h} \geq 0 \text{ and } Q\mathbf{h} = 0 \tag{14.37}$$

is zero. In this case, the minimum occurs at the origin.

The proof of this theorem is similar to that for Theorem 14.3, so it is omitted. By taking advantage of the linear structure of the constraint Φ, we can reformulate the minimization problem in (14.37) as follows. Let r be the rank of Q. Since $\mathbf{h}$ is in the null space of Q, we can write

$$\mathbf{h} = \tilde{Q}\mathbf{h}', \tag{14.38}$$

where $\tilde{Q}$ is a $k \times (k-r)$ matrix such that the rows of $\tilde{Q}^{\top}$ form a basis of the orthogonal complement of the row space of Q, and $\mathbf{h}'$ is a column $(k-r)$-vector. Then the elemental inequalities can be expressed as

$$G\tilde{Q}\mathbf{h}' \geq 0, \tag{14.39}$$

and in terms of $\mathbf{h}'$, Γ_n becomes

$$\Gamma_n' = \{\mathbf{h}' \in \Re^{k-r} : G\tilde{Q}\mathbf{h}' \geq 0\}, \tag{14.40}$$

which is a pyramid in $\Re^{k-r}$ (but not necessarily in the nonnegative orthant). Likewise, $\mathbf{b}^{\top}\mathbf{h}$ can be expressed as $\mathbf{b}^{\top}\tilde{Q}\mathbf{h}'$.

With all the information expressions in terms of $\mathbf{h}'$, the problem in (14.37) becomes

$$\text{Minimize } \mathbf{b}^{\top}\tilde{Q}\mathbf{h}', \text{ subject to } G\tilde{Q}\mathbf{h}' \geq 0. \tag{14.41}$$

Therefore, to check whether $\mathbf{b}^{\top}\mathbf{h} \geq 0$ is a Shannon-type inequality under the constraint Φ, all we need to do is to apply the optimality test of the simplex method to check whether the point $\mathbf{h}' = 0$ is optimal for the problem in (14.41). Then $\mathbf{b}^{\top}\mathbf{h} \geq 0$ is a Shannon-type inequality under the constraint Φ if and only if $\mathbf{h}' = 0$ is optimal.

By imposing the constraint Φ, the number of elemental inequalities remains the same, while the dimension of the problem decreases from k to $k - r$.

Finally, to verify that $\mathbf{b}^{\top}\mathbf{h} = 0$ is a Shannon-type identity under the constraint Φ, i.e., $\mathbf{b}^{\top}\mathbf{h} = 0$ is implied by the basic inequalities, all we need to do is to verify that both $\mathbf{b}^{\top}\mathbf{h} \geq 0$ and $\mathbf{b}^{\top}\mathbf{h} \leq 0$ are Shannon-type inequalities under the constraint Φ.

14.3 A Duality

A *nonnegative linear combination* is a linear combination whose coefficients are all nonnegative. It is clear that a nonnegative linear combination of basic inequalities is a Shannon-type inequality. However, it is not clear that all Shannon-type inequalities are of this form. By applying the *duality theorem* in linear programming [336], we will see that this is in fact the case.

The *dual* of the *primal* linear programming problem in (14.33) is

$$\text{Maximize } \mathbf{y}^{\top} \cdot 0 \text{ subject to } \mathbf{y} \geq 0 \text{ and } \mathbf{y}^{\top}G \leq \mathbf{b}^{\top}, \tag{14.42}$$

where

$$\mathbf{y} = [\, y_1 \, \cdots \, y_m \,]^\top. \tag{14.43}$$

By the duality theorem, if the minimum of the primal problem is zero, which happens when $\mathbf{b}^\top \mathbf{h} \geq 0$ is a Shannon-type inequality, the maximum of the dual problem is also zero. Since the cost function in the dual problem is zero, the maximum of the dual problem is zero if and only if the feasible region

$$\Psi = \{\mathbf{y} \in \Re^m : \mathbf{y} \geq 0 \text{ and } \mathbf{y}^\top G \leq \mathbf{b}^\top\} \tag{14.44}$$

is nonempty.

Theorem 14.5. $\mathbf{b}^\top \mathbf{h} \geq 0$ *is a Shannon-type inequality if and only if* $\mathbf{b}^\top = \mathbf{x}^\top G$ *for some* $\mathbf{x} \geq 0$, *where* $\mathbf{x}$ *is a column m-vector, i.e.,* $\mathbf{b}^\top$ *is a nonnegative linear combination of the rows of G.*

Proof. We have to prove that Ψ is nonempty if and only if $\mathbf{b}^\top = \mathbf{x}^\top G$ for some $\mathbf{x} \geq 0$. The feasible region Ψ is nonempty if and only if

$$\mathbf{b}^\top \geq \mathbf{z}^\top G \tag{14.45}$$

for some $\mathbf{z} \geq 0$, where $\mathbf{z}$ is a column m-vector. Consider any $\mathbf{z}$ which satisfies (14.45), and let

$$\mathbf{s}^\top = \mathbf{b}^\top - \mathbf{z}^\top G \geq 0. \tag{14.46}$$

Denote by $\mathbf{e}_j$ the column k-vector whose jth component is equal to 1 and all the other components are equal to 0, $1 \leq j \leq k$. Then $\mathbf{e}_j^\top \mathbf{h}$ is a joint entropy. Since every joint entropy can be expressed as the sum of elemental forms of Shannon's information measures, $\mathbf{e}_j^\top$ can be expressed as a nonnegative linear combination of the rows of G. Write

$$\mathbf{s} = [\, s_1 \, s_2 \, \cdots \, s_k \,]^\top, \tag{14.47}$$

where $s_j \geq 0$ for all $1 \leq j \leq k$. Then

$$\mathbf{s}^\top = \sum_{j=1}^{k} s_j \mathbf{e}_j^\top \tag{14.48}$$

can also be expressed as a nonnegative linear combinations of the rows of G, i.e.,

$$\mathbf{s}^\top = \mathbf{w}^\top G \tag{14.49}$$

for some $\mathbf{w} \geq 0$. From (14.46), we see that

$$\mathbf{b}^\top = (\mathbf{w}^\top + \mathbf{z}^\top)G = \mathbf{x}^\top G, \tag{14.50}$$

where $\mathbf{x} \geq 0$. The proof is accomplished. $\square$

From this theorem, we see that all Shannon-type inequalities are actually trivially implied by the basic inequalities! However, the verification of a Shannon-type inequality requires a computational procedure as described in the last section.

14.4 Machine Proving – ITIP

Theorems 14.3 and 14.4 transform the problem of verifying a Shannon-type inequality into a linear programming problem. This enables machine-proving of all Shannon-type inequalities. A software package called ITIP[1] has been developed for this purpose. The most updated versions of ITIP can be downloaded from the World Wide Web [409].

Using ITIP is very simple and intuitive. The following examples illustrate the use of ITIP:

```
1. >> ITIP('H(XYZ) <= H(X) + H(Y) + H(Z)')
   True
2. >> ITIP('I(X;Z) = 0','I(X;Z|Y) = 0','I(X;Y) = 0')
   True
3. >> ITIP('I(Z;U) - I(Z;U|X) - I(Z;U|Y) <=
     0.5 I(X;Y) + 0.25 I(X;ZU) + 0.25 I(Y;ZU)')
   Not provable by ITIP
```

In the first example, we prove an unconstrained inequality. In the second example, we prove that X and Z are independent if $X \to Y \to Z$ forms a Markov chain and X and Y are independent. The first identity is what we want to prove, while the second and third expressions specify the Markov chain $X \to Y \to Z$ and the independency of X and Y, respectively. In the third example, ITIP returns the clause "Not provable by ITIP," which means that the inequality is not a Shannon-type inequality. This, however, does not mean that the inequality to be proved cannot always hold. In fact, this inequality is one of the known non-Shannon-type inequalities which will be discussed in Chapter 15.

We note that most of the results we have previously obtained by using information diagrams can also be proved by ITIP. However, the advantage of using information diagrams is that one can visualize the structure of the problem. Therefore, the use of information diagrams and ITIP very often complements each other. In the rest of the section, we give a few examples which demonstrate the use of ITIP.

Example 14.6. By Proposition 2.10, the long Markov chain $X \to Y \to Z \to T$ implies the two short Markov chains $X \to Y \to Z$ and $Y \to Z \to T$. We want to see whether the two short Markov chains also imply the long Markov chain. If so, they are equivalent to each other.

Using ITIP, we have

```
>> ITIP('X/Y/Z/T', 'X/Y/Z', 'Y/Z/T')
Not provable by ITIP
```

In the above, we have used a macro in ITIP to specify the three Markov chains. The above result from ITIP says that the long Markov chain cannot

---

[1] ITIP stands for *Information-Theoretic Inequality Prover*.

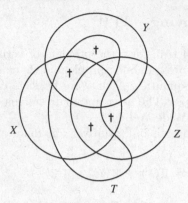

Fig. 14.3. The information diagram for X, Y, Z, and T in Example 14.6.

be proved from the two short Markov chains by means of the basic inequalities. This strongly suggests that the two short Markov chains are weaker than the long Markov chain. However, in order to prove that this is in fact the case, we need an explicit construction of a joint distribution for X, Y, Z, and T which satisfies the two short Markov chains but not the long Markov chain. Toward this end, we resort to the information diagram in Figure 14.3. The Markov chain $X \to Y \to Z$ is equivalent to $I(X; Z|Y) = 0$, i.e.,

$$\mu^*(\tilde{X} \cap \tilde{Y}^c \cap \tilde{Z} \cap \tilde{T}) + \mu^*(\tilde{X} \cap \tilde{Y}^c \cap \tilde{Z} \cap \tilde{T}^c) = 0. \tag{14.51}$$

Similarly, the Markov chain $Y \to Z \to T$ is equivalent to

$$\mu^*(\tilde{X} \cap \tilde{Y} \cap \tilde{Z}^c \cap \tilde{T}) + \mu^*(\tilde{X}^c \cap \tilde{Y} \cap \tilde{Z}^c \cap \tilde{T}) = 0. \tag{14.52}$$

The four atoms involved in the constraints (14.51) and (14.52) are marked by a dagger in Figure 14.3. In Section 3.5, we have seen that the Markov chain $X \to Y \to Z \to T$ holds if and only if μ^* takes zero value on the set of atoms in Figure 14.4 which are marked with an asterisk.[2] Comparing Figure 14.3 and Figure 14.4, we see that the only atom marked in Figure 14.4 but not in Figure 14.3 is $\tilde{X} \cap \tilde{Y}^c \cap \tilde{Z}^c \cap \tilde{T}$. Thus if we can construct a μ^* such that it takes zero value on all the atoms except for $\tilde{X} \cap \tilde{Y}^c \cap \tilde{Z}^c \cap \tilde{T}$, then the corresponding joint distribution satisfies the two short Markov chains but not the long Markov chain. This would show that the two short Markov chains are in fact weaker than the long Markov chain. Following Theorem 3.11, such a μ^* can be constructed.

In fact, the required joint distribution can be obtained by simply letting $X = T = U$, where U is any random variable such that $H(U) > 0$, and letting Y and Z be degenerate random variables taking constant values. Then it is easy to see that $X \to Y \to Z$ and $Y \to Z \to T$ hold, while $X \to Y \to Z \to T$ does not hold.

---

[2] This information diagram is essentially a reproduction of Figure 3.8.

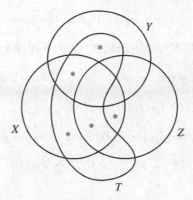

Fig. 14.4. The atoms of $\mathcal{F}_4$ on which μ^* vanishes when $X \to Y \to Z \to T$ forms a Markov chain.

Example 14.7. The data processing theorem says that if $X \to Y \to Z \to T$ forms a Markov chain, then

$$I(Y;Z) \geq I(X;T). \tag{14.53}$$

We want to see whether this inequality holds under the weaker condition that $X \to Y \to Z$ and $Y \to Z \to T$ form two short Markov chains. By using ITIP, we can show that (14.53) is not a Shannon-type inequality under the Markov conditions

$$I(X;Z|Y) = 0 \tag{14.54}$$

and

$$I(Y;T|Z) = 0. \tag{14.55}$$

This strongly suggests that (14.53) does not always hold under the constraint of the two short Markov chains. However, this has to be proved by an explicit construction of a joint distribution for $X, Y, Z,$ and T which satisfies (14.54) and (14.55) but not (14.53). The construction at the end of the last example serves this purpose.

Example 14.8 (Secret Sharing [42][321]). Let S be a secret to be encoded into three pieces, $X, Y,$ and Z. We need to design a scheme that satisfies the following two requirements:

1. No information about S can be obtained from any one of the three encoded pieces.
2. S can be recovered from any two of the three encoded pieces.

This is called a (1,2)-threshold secret sharing scheme. The first requirement of the scheme is equivalent to the constraints

$$I(S;X) = I(S;Y) = I(S;Z) = 0, \tag{14.56}$$

while the second requirement is equivalent to the constraints

$$H(S|X,Y) = H(S|Y,Z) = H(S|X,Z) = 0. \tag{14.57}$$

Since the secret S can be recovered if all X, Y, and Z are known,

$$H(X) + H(Y) + H(Z) \geq H(S). \tag{14.58}$$

We are naturally interested in the maximum constant c that satisfies

$$H(X) + H(Y) + H(Z) \geq cH(S). \tag{14.59}$$

We can explore the possible values of c by ITIP. After a few trials, we find that ITIP returns a "True" for all $c \leq 3$ and returns the clause "Not provable by ITIP" for any c slightly larger than 3, say 3.0001. This means that the maximum value of c is lower bounded by 3. This lower bound is in fact tight, as we can see from the following construction. Let S and N be mutually independent ternary random variables uniformly distributed on $\{0,1,2\}$, and define

$$X = N, \tag{14.60}$$
$$Y = S + N \bmod 3, \tag{14.61}$$

and

$$Z = S + 2N \bmod 3. \tag{14.62}$$

Then it is easy to verify that

$$S = Y - X \bmod 3 \tag{14.63}$$
$$= 2Y - Z \bmod 3 \tag{14.64}$$
$$= Z - 2X \bmod 3. \tag{14.65}$$

Thus the requirements in (14.57) are satisfied. It is also readily verified that the requirements in (14.56) are satisfied. Finally, all S, X, Y, and Z are distributed uniformly on $\{0,1,2\}$. Therefore,

$$H(X) + H(Y) + H(Z) = 3H(S). \tag{14.66}$$

This proves that the maximum constant c which satisfies (14.59) is 3.

Using the approach in this example, almost all information-theoretic bounds reported in the literature for this class of problems can be obtained when a definite number of random variables are involved.

14.5 Tackling the Implication Problem

We have already mentioned in Section 13.5 that the implication problem of conditional independence is extremely difficult except for the special case that only full conditional mutual independencies are involved. In this section, we employ the tools we have developed in this chapter to tackle this problem.

In Bayesian networks (see [287]), the following four axioms are often used for proving implications of conditional independencies:

- *Symmetry:*

$$X \perp Y|Z \; \Leftrightarrow \; Y \perp X|Z. \tag{14.67}$$

- *Decomposition:*

$$X \perp (Y,T)|Z \; \Rightarrow \; X \perp Y|Z \; \wedge \; X \perp T|Z. \tag{14.68}$$

- *Weak Union:*

$$X \perp (Y,T)|Z \; \Rightarrow \; X \perp Y|(Z,T). \tag{14.69}$$

- *Contraction:*

$$X \perp Y|Z \; \wedge \; X \perp T|(Y,Z) \; \Rightarrow \; X \perp (Y,T)|Z. \tag{14.70}$$

These axioms form a system called *semi-graphoid* and were first proposed in [92] as heuristic properties of conditional independence.

The axiom of symmetry is trivial in the context of probability.[3] The other three axioms can be summarized by

$$X \perp (Y,T)|Z \; \Leftrightarrow \; X \perp Y|Z \; \wedge \; X \perp T|(Y,Z). \tag{14.71}$$

This can easily be proved as follows. Consider the identity

$$I(X;Y,T|Z) = I(X;Y|Z) + I(X;T|Y,Z). \tag{14.72}$$

Since conditional mutual informations are always nonnegative by the basic inequalities, if $I(X;Y,T|Z)$ vanishes, then $I(X;Y|Z)$ and $I(X;T|Y,Z)$ also vanish and vice versa. This proves (14.71). In other words, (14.71) is the result of a specific application of the basic inequalities. Therefore, any implication which can be proved by invoking these four axioms are provable by ITIP.

In fact, ITIP is considerably more powerful than the above four axioms. This will be shown in the next example in which we give an implication which can be proved by ITIP but not by these four axioms.[4] We will see some implications which cannot be proved by ITIP when we discuss non-Shannon-type inequalities in the next chapter.

---

[3] These four axioms may be applied beyond the context of probability.

[4] This example is due to Zhen Zhang, private communication.

For a number of years, researchers in Bayesian networks generally believed that the semi-graphoidal axioms form a complete set of axioms for conditional independence until it was refuted by Studený [346]. See Problem 10 for a discussion.

Example 14.9. We will show that

$$\left.\begin{array}{l} I(X;Y|Z) = 0 \\ I(X;T|Z) = 0 \\ I(X;T|Y) = 0 \\ I(X;Z|Y) = 0 \\ I(X;Z|T) = 0 \end{array}\right\} \Rightarrow I(X;Y|T) = 0 \tag{14.73}$$

can be proved by invoking the basic inequalities. First, we write

$$I(X;Y|Z) = I(X;Y|Z,T) + I(X;Y;T|Z). \tag{14.74}$$

Since $I(X;Y|Z) = 0$ and $I(X;Y|Z,T) \geq 0$, we let

$$I(X;Y|Z,T) = a \tag{14.75}$$

for some nonnegative real number a, so that

$$I(X;Y;T|Z) = -a \tag{14.76}$$

from (14.74). In the information diagram in Figure 14.5, we mark the atom $I(X;Y|Z,T)$ by a "+" and the atom $I(X;Y;T|Z)$ by a "−." Then we write

$$I(X;T|Z) = I(X;Y;T|Z) + I(X;T|Y,Z). \tag{14.77}$$

Since $I(X;T|Z) = 0$ and $I(X;Y;T|Z) = -a$, we obtain

$$I(X;T|Y,Z) = a. \tag{14.78}$$

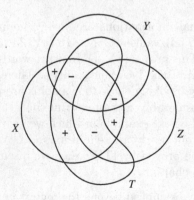

Fig. 14.5. The information diagram for X, Y, Z, and T.

In the information diagram, we mark the atom $I(X;T|Y,Z)$ with a "+." Continuing in this fashion, the five CIs on the left-hand side of (14.73) imply that all the atoms marked with a "+" in the information diagram take the value a, while all the atoms marked with a "−" take the value $-a$. From the information diagram, we see that

$$I(X;Y|T) = I(X;Y;Z|T) + I(X;Y|Z,T) = (-a) + a = 0, \qquad (14.79)$$

which proves our claim. Since we base our proof on the basic inequalities, this implication can also be proved by ITIP.

Due to the form of the five given CIs in (14.73), none of the axioms in (14.68)–(14.70) can be applied. Thus we conclude that the implication in (14.73) cannot be proved by invoking the four axioms in (14.67)–(14.70).

14.6 Minimality of the Elemental Inequalities

We have already seen in Section 14.1 that the set of basic inequalities is not minimal in the sense that in the set, some inequalities are implied by the others. We then showed that the set of basic inequalities is equivalent to the smaller set of elemental inequalities. Again, we can ask whether the set of elemental inequalities is minimal.

In this section, we prove that the set of elemental inequalities is minimal. This result is important for efficient implementation of ITIP because it says that we cannot consider a smaller set of inequalities. The proof, however, is rather technical. The reader may skip this proof without missing the essence of this chapter.

The elemental inequalities in set-theoretic notations have one of the following two forms:

1. $\mu(\tilde{X}_i - \tilde{X}_{\mathcal{N}_n-\{i\}}) \geq 0$,
2. $\mu(\tilde{X}_i \cap \tilde{X}_j - \tilde{X}_K) \geq 0$, $i \neq j$ and $K \subset \mathcal{N}_n - \{i,j\}$,

where μ denotes a set-additive function defined on $\mathcal{F}_n$. They will be referred to as α-inequalities and β-inequalities, respectively.

We are to show that all the elemental inequalities are nonredundant, i.e., none of them is implied by the others. For an α-inequality

$$\mu(\tilde{X}_i - \tilde{X}_{\mathcal{N}_n-\{i\}}) \geq 0, \qquad (14.80)$$

since it is the only elemental inequality which involves the atom $\tilde{X}_i - \tilde{X}_{\mathcal{N}_n-\{i\}}$, it is clearly not implied by the other elemental inequalities. Therefore we only need to show that all β-inequalities are nonredundant. To show that a β-inequality is nonredundant, it suffices to show that there exists a measure $\hat{\mu}$ on $\mathcal{F}_n$ which satisfies all other elemental inequalities except for that β-inequality.

We will show that the β-inequality

$$\mu(\tilde{X}_i \cap \tilde{X}_j - \tilde{X}_K) \geq 0 \tag{14.81}$$

is nonredundant. To facilitate our discussion, we denote $\mathcal{N}_n - K - \{i, j\}$ by $L(i, j, K)$, and we let $C_{ij|K}(S), S \subset L(i, j, K)$ be the atoms in $\tilde{X}_i \cap \tilde{X}_j - \tilde{X}_K$, where

$$C_{ij|K}(S) = \tilde{X}_i \cap \tilde{X}_j \cap \tilde{X}_S \cap \tilde{X}_K^c \cap \tilde{X}_{L(i,j,K)-S}^c. \tag{14.82}$$

We first consider the case when $L(i, j, K) = \emptyset$, i.e., $K = \mathcal{N}_n - \{i, j\}$. We construct a measure $\hat{\mu}$ by

$$\hat{\mu}(A) = \begin{cases} -1 \text{ if } A = \tilde{X}_i \cap \tilde{X}_j - \tilde{X}_K \\ 1 \quad \text{otherwise} \end{cases}, \tag{14.83}$$

where $A \in \mathcal{A}$. In other words, $\tilde{X}_i \cap \tilde{X}_j - \tilde{X}_K$ is the only atom with measure -1; all other atoms have measure 1. Then $\hat{\mu}(\tilde{X}_i \cap \tilde{X}_j - \tilde{X}_K) < 0$ is trivially true. It is also trivial to check that for any $i' \in \mathcal{N}_n$,

$$\hat{\mu}(\tilde{X}_{i'} - \tilde{X}_{\mathcal{N}_n - \{i'\}}) = 1 \geq 0, \tag{14.84}$$

and for any $(i', j', K') \neq (i, j, K)$ such that $i' \neq j'$ and $K' \subset \mathcal{N}_n - \{i', j'\}$,

$$\hat{\mu}(\tilde{X}_{i'} \cap \tilde{X}_{j'} - \tilde{X}_{K'}) = 1 \geq 0 \tag{14.85}$$

if $K' = \mathcal{N}_n - \{i', j'\}$. On the other hand, if K' is a proper subset of $\mathcal{N}_n - \{i', j'\}$, then $\tilde{X}_{i'} \cap \tilde{X}_{j'} - \tilde{X}_{K'}$ contains at least two atoms, and therefore

$$\hat{\mu}(\tilde{X}_{i'} \cap \tilde{X}_{j'} - \tilde{X}_{K'}) \geq 0. \tag{14.86}$$

This completes the proof for the β-inequality in (14.81) to be nonredundant when $L(i, j, K) = \phi$.

We now consider the case when $L(i, j, K) \neq \phi$ or $|L(i, j, K)| \geq 1$. We construct a measure $\hat{\mu}$ as follows. For the atoms in $\tilde{X}_i \cap \tilde{X}_j - \tilde{X}_K$, let

$$\hat{\mu}(C_{ij|K}(S)) = \begin{cases} (-1)^{|S|} - 1 & S = L(i, j, K) \\ (-1)^{|S|} & S \neq L(i, j, K) \end{cases}. \tag{14.87}$$

For $C_{ij|K}(S)$, if $|S|$ is odd, it is referred to as an *odd atom* of $\tilde{X}_i \cap \tilde{X}_j - \tilde{X}_K$, and if $|S|$ is even, it is referred to as an *even atom* of $\tilde{X}_i \cap \tilde{X}_j - \tilde{X}_K$. For any atom $A \notin \tilde{X}_i \cap \tilde{X}_j - \tilde{X}_K$, we let

$$\hat{\mu}(A) = 1. \tag{14.88}$$

This completes the construction of $\hat{\mu}$.

We first prove that

$$\hat{\mu}(\tilde{X}_i \cap \tilde{X}_j - \tilde{X}_K) < 0. \tag{14.89}$$

Consider

$$\hat{\mu}(\tilde{X}_i \cap \tilde{X}_j - \tilde{X}_K) = \sum_{S \subset L(i,j,K)} \hat{\mu}(C_{ij|K}(S))$$

$$= \left(\sum_{r=0}^{|L(i,j,K)|} \binom{|L(i,j,K)|}{r} (-1)^r \right) - 1$$

$$= -1,$$

where the last equality follows from the binomial formula

$$\sum_{r=0}^{n} \binom{n}{r} (-1)^r = 0 \tag{14.90}$$

for $n \geq 1$. This proves (14.89).

Next we prove that $\hat{\mu}$ satisfies all α-inequalities. We note that for any $i' \in \mathcal{N}_n$, the atom $\tilde{X}_{i'} - \tilde{X}_{\mathcal{N}_n - \{i'\}}$ is not in $\tilde{X}_i \cap \tilde{X}_j - \tilde{X}_K$. Thus

$$\hat{\mu}(\tilde{X}_{i'} - \tilde{X}_{\mathcal{N}_n - \{i\}}) = 1 \geq 0. \tag{14.91}$$

It remains to prove that $\hat{\mu}$ satisfies all β-inequalities except for (14.81), i.e., for any $(i', j', K') \neq (i, j, K)$ such that $i' \neq j'$ and $K' \subset \mathcal{N}_n - \{i', j'\}$,

$$\hat{\mu}(\tilde{X}_{i'} \cap \tilde{X}_{j'} - \tilde{X}_{K'}) \geq 0. \tag{14.92}$$

Consider

$$\hat{\mu}(\tilde{X}_{i'} \cap \tilde{X}_{j'} - \tilde{X}_{K'})$$
$$= \hat{\mu}((\tilde{X}_{i'} \cap \tilde{X}_{j'} - \tilde{X}_{K'}) \cap (\tilde{X}_i \cap \tilde{X}_j - \tilde{X}_K))$$
$$+ \hat{\mu}((\tilde{X}_{i'} \cap \tilde{X}_{j'} - \tilde{X}_{K'}) - (\tilde{X}_i \cap \tilde{X}_j - \tilde{X}_K)). \tag{14.93}$$

The nonnegativity of the second term above follows from (14.88). For the first term,

$$(\tilde{X}_{i'} \cap \tilde{X}_{j'} - \tilde{X}_{K'}) \cap (\tilde{X}_i \cap \tilde{X}_j - \tilde{X}_K) \tag{14.94}$$

is nonempty if and only if

$$\{i', j'\} \cap K = \phi \quad \text{and} \quad \{i, j\} \cap K' = \phi. \tag{14.95}$$

If this condition is not satisfied, then the first term in (14.93) becomes $\hat{\mu}(\phi) = 0$, and (14.92) follows immediately.

Let us assume that the condition in (14.95) is satisfied. Then by simple counting, we see that the number atoms in

$$(\tilde{X}_{i'} \cap \tilde{X}_{j'} - \tilde{X}_{K'}) \cap (\tilde{X}_i \cap \tilde{X}_j - \tilde{X}_K) \tag{14.96}$$

is equal to 2^φ, where

$$\varphi = n - |\{i, j\} \cup \{i', j'\} \cup K \cup K'|. \tag{14.97}$$

For example, for $n = 6$, there are $4 = 2^2$ atoms in

$$(\tilde{X}_1 \cap \tilde{X}_2) \cap (\tilde{X}_1 \cap \tilde{X}_3 - \tilde{X}_4), \tag{14.98}$$

namely $\tilde{X}_1 \cap \tilde{X}_2 \cap \tilde{X}_3 \cap \tilde{X}_4^c \cap Y_5 \cap Y_6$, where $Y_i = \tilde{X}_i$ or $\tilde{X}_i^c$ for $i = 5, 6$. We check that

$$\varphi = 6 - |\{1, 2\} \cup \{1, 3\} \cup \phi \cup \{4\}| = 2. \tag{14.99}$$

We first consider the case when $\varphi = 0$, i.e.,

$$\mathcal{N}_n = \{i, j\} \cup \{i', j'\} \cup K \cup K'. \tag{14.100}$$

Then

$$(\tilde{X}_{i'} \cap \tilde{X}_{j'} - \tilde{X}_{K'}) \cap (\tilde{X}_i \cap \tilde{X}_j - \tilde{X}_K) \tag{14.101}$$

contains exactly one atom. If this atom is an even atom of $\tilde{X}_i \cap \tilde{X}_j - \tilde{X}_K$, then the first term in (14.93) is either 0 or 1 (cf., (14.87)), and (14.92) follows immediately. If this atom is an odd atom of $\tilde{X}_i \cap \tilde{X}_j - \tilde{X}_K$, then the first term in (14.93) is equal to -1. This happens if and only if $\{i, j\}$ and $\{i', j'\}$ have one common element, which implies that $(\tilde{X}_{i'} \cap \tilde{X}_{j'} - \tilde{X}_{K'}) - (\tilde{X}_i \cap \tilde{X}_j - \tilde{X}_K)$ is nonempty. Therefore the second term in (14.93) is at least 1, and hence (14.92) follows.

Finally, we consider the case when $\varphi \geq 1$. Using the binomial formula in (14.90), we see that the number of odd atoms and even atoms of $\tilde{X}_i \cap \tilde{X}_j - \tilde{X}_K$ in

$$(\tilde{X}_{i'} \cap \tilde{X}_{j'} - \tilde{X}_{K'}) \cap (\tilde{X}_i \cap \tilde{X}_j - \tilde{X}_K) \tag{14.102}$$

are the same. Therefore the first term in (14.93) is equal to -1 if

$$C_{ij|K}(L(i, j, K)) \in \tilde{X}_{i'} \cap \tilde{X}_{j'} - \tilde{X}_{K'} \tag{14.103}$$

and is equal to 0 otherwise. The former is true if and only if $K' \subset K$, which implies that $(\tilde{X}_{i'} \cap \tilde{X}_{j'} - \tilde{X}_{K'}) - (\tilde{X}_i \cap \tilde{X}_j - \tilde{X}_K)$ is nonempty or that the second term is at least 1. Thus in either case (14.92) is true. This completes the proof that (14.81) is nonredundant.

Appendix 14.A:
The Basic Inequalities and the Polymatroidal Axioms

In this appendix, we show that the basic inequalities for a collection of n random variables $\Theta = \{X_i, i \in \mathcal{N}_n\}$ are equivalent to the following polymatroidal axioms: For all $\alpha, \beta \subset \mathcal{N}_n$,

    P1. $H_\Theta(\emptyset) = 0$.
    P2. $H_\Theta(\alpha) \leq H_\Theta(\beta)$ if $\alpha \subset \beta$.
    P3. $H_\Theta(\alpha) + H_\Theta(\beta) \geq H_\Theta(\alpha \cap \beta) + H_\Theta(\alpha \cup \beta)$.

We first show that the polymatroidal axioms imply the basic inequalities. From P1 and P2, since $\emptyset \subset \alpha$ for any $\alpha \subset \mathcal{N}_n$, we have

$$H_\Theta(\alpha) \geq H_\Theta(\emptyset) = 0 \tag{14.104}$$

or

$$H(X_\alpha) \geq 0. \tag{14.105}$$

This shows that entropy is nonnegative.

In P2, letting $\gamma = \beta \backslash \alpha$, we have

$$H_\Theta(\alpha) \leq H_\Theta(\alpha \cup \gamma) \tag{14.106}$$

or

$$H(X_\gamma | X_\alpha) \geq 0. \tag{14.107}$$

Here, γ and α are disjoint subsets of $\mathcal{N}_n$.

In P3, letting $\gamma = \beta \backslash \alpha$, $\delta = \alpha \cap \beta$, and $\sigma = \alpha \backslash \beta$, we have

$$H_\Theta(\sigma \cup \delta) + H_\Theta(\gamma \cup \delta) \geq H_\Theta(\delta) + H_\Theta(\sigma \cup \delta \cup \gamma) \tag{14.108}$$

or

$$I(X_\sigma; X_\gamma | X_\delta) \geq 0. \tag{14.109}$$

Again, σ, δ, and γ are disjoint subsets of $\mathcal{N}_n$. When $\delta = \emptyset$, from P3, we have

$$I(X_\sigma; X_\gamma) \geq 0. \tag{14.110}$$

Thus P1–P3 imply that entropy is nonnegative, and that conditional entropy, mutual information, and conditional mutual information are nonnegative provided that they are irreducible. However, it has been shown in Section 14.1 that a reducible Shannon's information measure can always be written as the sum of irreducible Shannon's information measures. Therefore, we have shown that the polymatroidal axioms P1–P3 imply the basic inequalities. The converse is trivial and the proof is omitted.

Chapter Summary

Shannon-Type Inequalities are information inequalities implied by the basic inequalities.

Elemental Form of Shannon's Information Measures: Any Shannon's information measure involving random variables $X_1, X_2, \cdots, X_n$ can be expressed as the sum of the following two element forms:

i) $H(X_i | X_{\mathcal{N}_n - \{i\}}), i \in \mathcal{N}_n$.

ii) $I(X_i; X_j | X_K)$, where $i \neq j$ and $K \subset \mathcal{N}_n - \{i, j\}$.

Elemental Inequalities: For a set of random variables, the nonnegativity of the two elemental forms of Shannon's information measures are called the elemental inequalities. The elemental inequalities are equivalent to the basic inequalities for the same set of random variables, and they form the minimal such subset of the basic inequalities.

The region $\Gamma_n = \{\mathbf{h} \in \mathcal{H}_n : G\mathbf{h} \geq 0\}$ is the subset of $\mathcal{H}_n$ defined by the basic inequalities for n random variables, and $\Gamma_n^* \subset \Gamma_n$.

Unconstrained Shannon-Type Inequalities: $\mathbf{b}^\top \mathbf{h} \geq 0$ is a Shannon-type inequality if and only if one of the following is true:

1. $\Gamma_n \subset \{\mathbf{h} \in \mathcal{H}_n : \mathbf{b}^\top \mathbf{h} \geq 0\}$.
2. The minimum of the problem "Minimize $\mathbf{b}^\top \mathbf{h}$, subject to $G\mathbf{h} \geq 0$" is zero.

Constrained Shannon-Type Inequalities: Under the constraint $\Phi = \{\mathbf{h} \in \mathcal{H}_n : Q\mathbf{h} = 0\}$, $\mathbf{b}^\top \mathbf{h} \geq 0$ is a Shannon-type inequality if and only if

1. $(\Gamma_n \cap \Phi) \subset \{\mathbf{h} \in \mathcal{H}_n : \mathbf{b}^\top \mathbf{h} \geq 0\}$.
2. The minimum of the problem "Minimize $\mathbf{b}^\top \mathbf{h}$, subject to $G\mathbf{h} \geq 0$ and $Q\mathbf{h} = 0$" is zero.

Duality: An unconstrained Shannon-type inequality is a nonnegative linear combination of the elemental inequalities for the same set of random variables.

ITIP is a software package running on MATLAB for proving Shannon-type inequalities.

Problems

1. Prove (14.12) for the total number of elemental forms of Shannon's information measures for n random variables.
2. Shannon-type inequalities for n random variables $X_1, X_2, \cdots, X_n$ refer to all information inequalities implied by the basic inequalities for these n random variables. Show that no new information inequality can be generated by considering the basic inequalities for more than n random variables.
3. Show by an example that the decomposition of an information expression into a sum of elemental forms of Shannon's information measures is not unique.
4. *Elemental forms of conditional independencies.* Consider random variables $X_1, X_2, \cdots, X_n$. A conditional independency is said to be *elemental* if it corresponds to setting an elemental form of Shannon's information measure to zero. Show that any conditional independency involving $X_1, X_2, \cdots, X_n$ is equivalent to a collection of elemental conditional independencies.

5. *Symmetrical information inequalities.*

   a) Show that every symmetrical information expression (cf. Problem 1 in Chapter 13) involving random variable $X_1, X_2, \cdots, X_n$ can be written in the form

   $$E = \sum_{k=0}^{n-1} a_k c_k^{(n)},$$

   where

   $$c_0^{(n)} = \sum_{i=1}^{n} H(X_i | X_{N-i})$$

   and for $1 \le k \le n-1$,

   $$c_k^{(n)} = \sum_{\substack{1 \le i < j \le n \\ K \subset N - \{i,j\}, |K| = k-1}} I(X_i; X_j | X_K).$$

   Note that $c_0^{(n)}$ is the sum of all Shannon's information measures of the first elemental form, and for $1 \le k \le n-1$, $c_k^{(n)}$ is the sum of all Shannon's information measures of the second elemental form conditioning on $k-1$ random variables.

   b) Show that $E \ge 0$ always holds if $a_k \ge 0$ for all k.

   c) Show that if $E \ge 0$ always holds, then $a_k \ge 0$ for all k. Hint: Construct random variables $X_1, X_2, \cdots, X_n$ for each $0 \le k \le n-1$ such that $c_k^{(n)} > 0$ and $c_{k'}^{(n)} = 0$ for all $0 \le k' \le n-1$ and $k' \ne k$.

   (Han [147].)

6. *Strictly positive probability distributions.* It was shown in Proposition 2.12 that

   $$\left. \begin{array}{l} X_1 \perp X_4 | (X_2, X_3) \\ X_1 \perp X_3 | (X_2, X_4) \end{array} \right\} \Rightarrow X_1 \perp (X_3, X_4) | X_2$$

   if $p(x_1, x_2, x_3, x_4) > 0$ for all x_1, x_2, x_3, and x_4. Show by using ITIP that this implication is not implied by the basic inequalities. This strongly suggests that this implication does not hold in general, which was shown to be the case by the construction following Proposition 2.12.

7. a) Verify by ITIP that

   $$I(X_1, X_2; Y_1, Y_2) \le I(X_1; Y_1) + I(X_2; Y_2)$$

   under the constraint $H(Y_1, Y_2 | X_1, X_2) = H(Y_1 | X_1) + H(Y_2 | X_2)$. This constrained inequality was used in Problem 10 in Chapter 7 to obtain the capacity of two parallel channels.

   b) Verify by ITIP that

   $$I(X_1, X_2; Y_1, Y_2) \ge I(X_1; Y_1) + I(X_2; Y_2)$$

   under the constraint $I(X_1; X_2) = 0$. This constrained inequality was used in Problem 4 in Chapter 8 to obtain the rate-distortion function for a product source.

8. Verify by ITIP the information identity in Example 3.18.
9. Repeat Problem 13 in Chapter 3 with the help of ITIP.
10. Prove the implications in Problem 15 in Chapter 3 by ITIP and show that they cannot be deduced from the semi-graphoidal axioms. (Studený [346].)

Historical Notes

For almost half a century, all information inequalities known in the literature are consequences of the basic inequalities due to Shannon [322]. Fujishige [126] showed that the entropy function is a polymatroid (see Appendix 14.A). Yeung [401] showed that verification of all such inequalities, referred to Shannon-type inequalities, can be formulated as a linear programming problem if the number of random variables involved is fixed. ITIP, a software package for this purpose, was developed by Yeung and Yan [409]. Non-Shannon-type inequalities, which were first discovered in the late 1990s, will be discussed in the next chapter.

15

Beyond Shannon-Type Inequalities

In Chapter 13, we introduced the regions Γ_n^* and Γ_n in the entropy space $\mathcal{H}_n$ for n random variables. From Γ_n^*, one in principle can determine whether any information inequality always holds. The region Γ_n, defined by the set of all basic inequalities (equivalently all elemental inequalities) involving n random variables, is an outer bound on Γ_n^*. From Γ_n, one can determine whether any information inequality is implied by the basic inequalities. If so, it is called a Shannon-type inequality. Since the basic inequalities always hold, so do all Shannon-type inequalities. In the last chapter, we have shown how machine-proving of all Shannon-type inequalities can be made possible by taking advantage of the linear structure of Γ_n.

If the two regions Γ_n^* and Γ_n are identical, then all information inequalities which always hold are Shannon-type inequalities, and hence all information inequalities can be completely characterized. However, if Γ_n^* is a proper subset of Γ_n, then there exist constraints on an entropy function which are not implied by the basic inequalities. Such a constraint, if in the form of an inequality, is referred to as a non-Shannon-type inequality.

There is a point here which needs further explanation. The fact that $\Gamma_n^* \neq \Gamma_n$ does not necessarily imply the existence of a non-Shannon-type inequality. As an example, suppose Γ_n contains all but an isolated point in Γ_n^*. Then this does not lead to the existence of a non-Shannon-type inequality for n random variables.

In this chapter, we present characterizations of Γ_n^* which are more refined than Γ_n. These characterizations lead to the existence of non-Shannon-type inequalities for $n \geq 4$.

15.1 Characterizations of Γ_2^*, Γ_3^*, and $\overline{\Gamma}_n^*$

Recall from the proof of Theorem 3.6 that the vector $\mathbf{h}$ represents the values of the I-Measure μ^* on the unions in $\mathcal{F}_n$. Moreover, $\mathbf{h}$ is related to the values of μ^* on the atoms of $\mathcal{F}_n$, represented as $\mathbf{u}$, by

$$\mathbf{h} = C_n \mathbf{u}, \tag{15.1}$$

where C_n is a unique $k \times k$ matrix with $k = 2^n - 1$ (cf. (3.27)).

Let $\mathcal{I}_n$ be the k-dimensional Euclidean space with the coordinates labeled by the components of $\mathbf{u}$. Note that each coordinate in $\mathcal{I}_n$ corresponds to the value of μ^* on a nonempty atom of $\mathcal{F}_n$. Recall from Lemma 13.1 the definition of the region

$$\Psi_n^* = \{\mathbf{u} \in \mathcal{I}_n : C_n \mathbf{u} \in \Gamma_n^*\}, \tag{15.2}$$

which is obtained from the region Γ_n^* via the linear transformation induced by C_n^{-1}. Analogously, we define the region

$$\Psi_n = \{\mathbf{u} \in \mathcal{I}_n : C_n \mathbf{u} \in \Gamma_n\}. \tag{15.3}$$

The region Γ_n^*, as we will see, is extremely difficult to characterize for a general n. Therefore, we start our discussion with the simplest case, namely $n = 2$.

Theorem 15.1. $\Gamma_2^* = \Gamma_2$.

Proof. For $n = 2$, the elemental inequalities are

$$H(X_1|X_2) = \mu^*(\tilde{X}_1 - \tilde{X}_2) \geq 0, \tag{15.4}$$
$$H(X_2|X_1) = \mu^*(\tilde{X}_2 - \tilde{X}_1) \geq 0, \tag{15.5}$$
$$I(X_1; X_2) = \mu^*(\tilde{X}_1 \cap \tilde{X}_2) \geq 0. \tag{15.6}$$

Note that the quantities on the left-hand sides above are precisely the values of μ^* on the atoms of $\mathcal{F}_2$. Therefore,

$$\Psi_2 = \{\mathbf{u} \in \mathcal{I}_2 : \mathbf{u} \geq 0\}, \tag{15.7}$$

i.e., Ψ_2 is the nonnegative orthant of $\mathcal{I}_2$. Since $\Gamma_2^* \subset \Gamma_2$, $\Psi_2^* \subset \Psi_2$. On the other hand, $\Psi_2 \subset \Psi_2^*$ by Lemma 13.1. Thus $\Psi_2^* = \Psi_2$, which implies $\Gamma_2^* = \Gamma_2$. The proof is accomplished. $\square$

Next, we prove that Theorem 15.1 cannot even be generalized to $n = 3$.

Theorem 15.2. $\Gamma_3^* \neq \Gamma_3$.

Proof. For $n = 3$, the elemental inequalities are

$$H(X_i|X_j, X_k) = \mu^*(\tilde{X}_i - \tilde{X}_j - \tilde{X}_k) \geq 0, \tag{15.8}$$
$$I(X_i; X_j|X_k) = \mu^*(\tilde{X}_i \cap \tilde{X}_j - \tilde{X}_k) \geq 0, \tag{15.9}$$

and

$$I(X_i; X_j) = \mu^*(\tilde{X}_i \cap \tilde{X}_j) \tag{15.10}$$
$$= \mu^*(\tilde{X}_i \cap \tilde{X}_j \cap \tilde{X}_k) + \mu^*(\tilde{X}_i \cap \tilde{X}_j - \tilde{X}_k) \tag{15.11}$$
$$\geq 0 \tag{15.12}$$

for $1 \leq i < j < k \leq 3$. For $\mathbf{u} \in \mathcal{I}_3$, let

$$\mathbf{u} = (u_1, u_2, u_3, u_4, u_5, u_6, u_7), \tag{15.13}$$

where $u_i, 1 \leq i \leq 7$ correspond to the values

$$\mu^*(\tilde{X}_1 - \tilde{X}_2 - \tilde{X}_3),\ \mu^*(\tilde{X}_2 - \tilde{X}_1 - \tilde{X}_3),\ \mu^*(\tilde{X}_3 - \tilde{X}_1 - \tilde{X}_2),$$
$$\mu^*(\tilde{X}_1 \cap \tilde{X}_2 - \tilde{X}_3),\ \mu^*(\tilde{X}_1 \cap \tilde{X}_3 - \tilde{X}_2),\ \mu^*(\tilde{X}_2 \cap \tilde{X}_3 - \tilde{X}_1), \tag{15.14}$$
$$\mu^*(\tilde{X}_1 \cap \tilde{X}_2 \cap \tilde{X}_3),$$

respectively. These are the values of μ^* on the nonempty atoms of $\mathcal{F}_3$. Then from (15.8), (15.9), and (15.12), we see that

$$\Psi_3 = \{\mathbf{u} \in \mathcal{I}_3 : u_i \geq 0,\ 1 \leq i \leq 6;\ u_j + u_7 \geq 0,\ 4 \leq j \leq 6\}. \tag{15.15}$$

It is easy to check that the point $(0, 0, 0, a, a, a, -a)$ for any $a \geq 0$ is in Ψ_3. This is illustrated in Figure 15.1, and it is readily seen that the relations

$$H(X_i | X_j, X_k) = 0 \tag{15.16}$$

and

$$I(X_i; X_j) = 0 \tag{15.17}$$

for $1 < i < j < k \leq 3$ are satisfied, i.e., each random variable is a function of the other two, and the three random variables are pairwise independent.

Let $\mathcal{S}_{X_i}$ be the support of X_i, $i = 1, 2, 3$. For any $x_1 \in \mathcal{S}_{X_1}$ and $x_2 \in \mathcal{S}_{X_2}$, since X_1 and X_2 are independent, we have

$$p(x_1, x_2) = p(x_1)p(x_2) > 0. \tag{15.18}$$

Since X_3 is a function of X_1 and X_2, there is a unique $x_3 \in \mathcal{S}_{X_3}$ such that

$$p(x_1, x_2, x_3) = p(x_1, x_2) = p(x_1)p(x_2) > 0. \tag{15.19}$$

Since X_2 is a function of X_1 and X_3, and X_1 and X_3 are independent, we can write

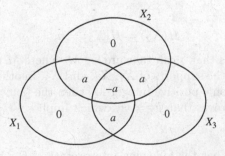

Fig. 15.1. The set-theoretic structure of the point $(0, 0, 0, a, a, a, -a)$ in Ψ_3.

$$p(x_1, x_2, x_3) = p(x_1, x_3) = p(x_1)p(x_3). \qquad (15.20)$$

Equating (15.19) and (15.20), we have

$$p(x_2) = p(x_3). \qquad (15.21)$$

Now consider any $x_2' \in S_{X_2}$ such that $x_2' \neq x_2$. Since X_2 and X_3 are independent, we have

$$p(x_2', x_3) = p(x_2')p(x_3) > 0. \qquad (15.22)$$

Since X_1 is a function of X_2 and X_3, there is a unique $x_1' \in S_{X_1}$ such that

$$p(x_1', x_2', x_3) = p(x_2', x_3) = p(x_2')p(x_3) > 0. \qquad (15.23)$$

Since X_2 is a function of X_1 and X_3, and X_1 and X_3 are independent, we can write

$$p(x_1', x_2', x_3) = p(x_1', x_3) = p(x_1')p(x_3). \qquad (15.24)$$

Similarly, since X_3 is a function of X_1 and X_2, and X_1 and X_2 are independent, we can write

$$p(x_1', x_2', x_3) = p(x_1', x_2') = p(x_1')p(x_2'). \qquad (15.25)$$

Equating (15.24) and (15.25), we have

$$p(x_2') = p(x_3), \qquad (15.26)$$

and from (15.21), we have

$$p(x_2') = p(x_2). \qquad (15.27)$$

Therefore, X_2 must have a uniform distribution on its support. The same can be proved for X_1 and X_3. Now from Figure 15.1,

$$
\begin{aligned}
H(X_1) &= H(X_1|X_2, X_3) + I(X_1; X_2|X_3) + I(X_1; X_3|X_2) \\
&\quad + I(X_1; X_2; X_3) &(15.28)\\
&= 0 + a + a + (-a) &(15.29)\\
&= a, &(15.30)
\end{aligned}
$$

and similarly

$$H(X_2) = H(X_3) = a. \qquad (15.31)$$

Then the only values that a can take are $\log M$, where M (a positive integer) is the cardinality of the supports of X_1, X_2, and X_3. In other words, if a is not equal to $\log M$ for some positive integer M, then the point $(0, 0, 0, a, a, a, -a)$ is not in Ψ_3^*. This proves that $\Psi_3^* \neq \Psi_3$, which implies $\Gamma_3^* \neq \Gamma_3$. The theorem is proved. $\square$

The proof above has the following interpretation. For $\mathbf{h} \in \mathcal{H}_3$, let

$$\mathbf{h} = (h_1, h_2, h_3, h_{12}, h_{13}, h_{23}, h_{123}). \tag{15.32}$$

From Figure 15.1, we see that the point $(0, 0, 0, a, a, a, -a)$ in Ψ_3 corresponds to the point $(a, a, a, 2a, 2a, 2a, 2a)$ in Γ_3. Evidently, the point $(a, a, a, 2a, 2a, 2a, 2a)$ in Γ_3 satisfies the six elemental inequalities given in (15.8) and (15.12) for $1 \le i < j < k \le 3$ with equality. Since Γ_3 is defined by all the elemental inequalities, the set

$$\{(a, a, a, 2a, 2a, 2a, 2a) \in \Gamma_3 : a \ge 0\} \tag{15.33}$$

is in the intersection of six hyperplanes in $\mathcal{H}_3$ (i.e., $\Re^7$) defining the boundary of Γ_3, and hence it defines an extreme direction of Γ_3. Then the proof says that along this extreme direction of Γ_3, only certain discrete points, namely those points with a equals $\log M$ for some positive integer M, are entropic. This is illustrated in Figure 15.2. As a consequence, the region Γ_3^* is not convex.

Having proved that $\Gamma_3^* \ne \Gamma_3$, it is natural to conjecture that the gap between Γ_3^* and Γ_3 has zero Lebesgue measure. In other words, $\overline{\Gamma}_3^* = \Gamma_3$, where $\overline{\Gamma}_3^*$ is the closure of Γ_3. This is indeed the case and will be proved at the end of the section.

More generally, we are interested in characterizing $\overline{\Gamma}_n^*$, the closure of Γ_n^*. Although the region $\overline{\Gamma}_n^*$ is not sufficient for characterizing all information inequalities, it is actually sufficient for characterizing all unconstrained information inequalities. This can be seen as follows. Following the discussion in Section 13.3.1, an unconstrained information inequality $f \ge 0$ involving n random variables always hold if and only if

$$\Gamma_n^* \subset \{\mathbf{h} : f(\mathbf{h}) \ge 0\}. \tag{15.34}$$

Since $\{\mathbf{h} : f(\mathbf{h}) \ge 0\}$ is closed, upon taking closure on both sides, we have

$$\overline{\Gamma}_n^* \subset \{\mathbf{h} : f(\mathbf{h}) \ge 0\}. \tag{15.35}$$

On the other hand, if $f \ge 0$ satisfies (15.35), then

$$\Gamma_n^* \subset \overline{\Gamma}_n^* \subset \{\mathbf{h} : f(\mathbf{h}) \ge 0\}. \tag{15.36}$$

Therefore, (15.34) and (15.35) are equivalent, and hence $\overline{\Gamma}_n^*$ is sufficient for characterizing all unconstrained information inequalities.

We will prove in the next theorem an important property of the region $\overline{\Gamma}_n^*$ for all $n \ge 2$. This result will be used in the proof for $\overline{\Gamma}_3^* = \Gamma_3$. Further, this result will be used in Chapter 16 when we establish a fundamental relation between information theory and group theory.

We first prove a simple lemma. In the following, we use $\mathcal{N}_n$ to denote the set $\{1, 2, \cdots, n\}$.

Fig. 15.2. The values of a for which $(a, a, a, 2a, 2a, 2a, 2a)$ is in Γ_3.

Lemma 15.3. *If* $\mathbf{h}$ *and* $\mathbf{h}'$ *are in* Γ_n^*, *then* $\mathbf{h} + \mathbf{h}'$ *is in* Γ_n^*.

Proof. Consider $\mathbf{h}$ and $\mathbf{h}'$ in Γ_n^*. Let $\mathbf{h}$ represents the entropy function for random variables $X_1, X_2, \cdots, X_n$, and let $\mathbf{h}'$ represents the entropy function for random variables $X_1', X_2', \cdots, X_n'$. Let $(X_1, X_2, \cdots, X_n)$ and $(X_1', X_2', \cdots, X_n')$ be independent, and define random variables $Y_1, Y_2, \cdots, Y_n$ by

$$Y_i = (X_i, X_i') \tag{15.37}$$

for all $i \in \mathcal{N}_n$. Then for any subset α of $\mathcal{N}_n$,

$$H(Y_\alpha) = H(X_\alpha) + H(X_\alpha') = h_\alpha + h_\alpha'. \tag{15.38}$$

Therefore, $\mathbf{h} + \mathbf{h}'$, which represents the entropy function for $Y_1, Y_2, \cdots, Y_n$, is in Γ_n^*. The lemma is proved. $\square$

Corollary 15.4. *If* $\mathbf{h} \in \Gamma_n^*$, *then* $k\mathbf{h} \in \Gamma_n^*$ *for any positive integer* k.

Proof. It suffices to write

$$k\mathbf{h} = \underbrace{\mathbf{h} + \mathbf{h} + \cdots + \mathbf{h}}_{k} \tag{15.39}$$

and apply Lemma 15.3. $\square$

Theorem 15.5. $\overline{\Gamma}_n^*$ *is a convex cone.*

Proof. Consider the entropy function for random variables $X_1, X_2, \cdots, X_n$ all taking constant values with probability 1. Then for all subset α of $\mathcal{N}_n$,

$$H(X_\alpha) = 0. \tag{15.40}$$

Therefore, Γ_n^* contains the origin in $\mathcal{H}_n$.

Let $\mathbf{h}$ and $\mathbf{h}'$ in Γ_n^* be the entropy functions for any two sets of random variables $Y_1, Y_2, \cdots, Y_n$ and $Z_1, Z_2, \cdots, Z_n$, respectively. In view of Corollary 15.4, in order to prove that $\overline{\Gamma}_n^*$ is a convex cone, we only need to show that if $\mathbf{h}$ and $\mathbf{h}'$ are in Γ_n^*, then $b\mathbf{h} + \overline{b}\mathbf{h}'$ is in $\overline{\Gamma}_n^*$ for all $0 < b < 1$, where $\overline{b} = 1 - b$.

Let $(\mathbf{Y}_1, \mathbf{Y}_2, \cdots, \mathbf{Y}_n)$ be k independent copies of $(Y_1, Y_2, \cdots, Y_n)$ and $(\mathbf{Z}_1, \mathbf{Z}_2, \cdots, \mathbf{Z}_n)$ be k independent copies of $(Z_1, Z_2, \cdots, Z_n)$. Let U be a ternary random variable independent of all other random variables such that

$$\Pr\{U = 0\} = 1 - \delta - \mu, \ \Pr\{U = 1\} = \delta, \ \Pr\{U = 2\} = \mu.$$

Now construct random variables $X_1, X_2, \cdots, X_n$ by letting

$$X_i = \begin{cases} 0 & \text{if } U = 0 \\ \mathbf{Y}_i & \text{if } U = 1 \\ \mathbf{Z}_i & \text{if } U = 2 \end{cases}.$$

Note that $H(U) \to 0$ as $\delta, \mu \to 0$. Then for any nonempty subset α of $\mathcal{N}_n$,

$$H(X_\alpha) \leq H(X_\alpha, U) \tag{15.41}$$
$$= H(U) + H(X_\alpha | U) \tag{15.42}$$
$$= H(U) + \delta k H(Y_\alpha) + \mu k H(Z_\alpha). \tag{15.43}$$

On the other hand,

$$H(X_\alpha) \geq H(X_\alpha | U) = \delta k H(Y_\alpha) + \mu k H(Z_\alpha). \tag{15.44}$$

Combining the above, we have

$$0 \leq H(X_\alpha) - (\delta k H(Y_\alpha) + \mu k H(Z_\alpha)) \leq H(U). \tag{15.45}$$

Now take

$$\delta = \frac{b}{k} \tag{15.46}$$

and

$$\mu = \frac{\bar{b}}{k} \tag{15.47}$$

to obtain

$$0 \leq H(X_\alpha) - (b H(Y_\alpha) + \bar{b} H(Z_\alpha)) \leq H(U). \tag{15.48}$$

By letting k be sufficiently large, the upper bound can be made arbitrarily small. This shows that $b\mathbf{h} + \bar{b}\mathbf{h}' \in \overline{\Gamma}_n^*$. The theorem is proved. $\quad\square$

In the next theorem, we prove that Γ_3^* and Γ_3 are almost identical. Analogous to $\overline{\Gamma}_n^*$, we will use $\overline{\Psi}_n^*$ to denote the closure of Ψ_n^*.

Theorem 15.6. $\overline{\Gamma}_3^* = \Gamma_3$.

Proof. We first note that $\overline{\Gamma}_3^* = \Gamma_3$ if and only if

$$\overline{\Psi}_3^* = \Psi_3. \tag{15.49}$$

Since

$$\Gamma_3^* \subset \Gamma_3 \tag{15.50}$$

and Γ_3 is closed, by taking closure on both sides in the above, we obtain $\overline{\Gamma}_3^* \subset \Gamma_3$. This implies that $\overline{\Psi}_3^* \subset \Psi_3$. Therefore, in order to prove the theorem, it suffices to show that $\Psi_3 \subset \overline{\Psi}_3^*$.

We first show that the point $(0, 0, 0, a, a, a, -a)$ is in $\overline{\Psi}_3^*$ for all $a > 0$. Let random variables X_1, X_2, and X_3 be defined as in Example 3.10, i.e., X_1 and X_2 are two independent binary random variables taking values in $\{0, 1\}$ according to the uniform distribution, and

$$X_3 = X_1 + X_2 \bmod 2. \tag{15.51}$$

Let $\mathbf{h} \in \Gamma_3^*$ represents the entropy function for X_1, X_2, and X_3, and let

$$\mathbf{u} = C_3^{-1}\mathbf{h}. \tag{15.52}$$

As in the proof of Theorem 15.2, we let u_i, $1 \le i \le 7$, be the coordinates of $\mathcal{I}_3$ which correspond to the values of the quantities in (15.14), respectively. From Example 3.10, we have

$$u_i = \begin{cases} 0 & \text{for } i = 1, 2, 3 \\ 1 & \text{for } i = 4, 5, 6 \\ -1 & \text{for } i = 7 \end{cases}. \tag{15.53}$$

Thus the point $(0, 0, 0, 1, 1, 1, -1)$ is in Ψ_3^*, and the I-Measure μ^* for X_1, X_2, and X_3 is shown in Figure 15.3. Then by Corollary 15.4, $(0, 0, 0, k, k, k, -k)$ is in Ψ_3^* and hence in $\overline{\Psi}_3^*$ for all positive integers k. Since $\overline{\Gamma}_3^*$ contains the origin, $\overline{\Psi}_3^*$ also contains the origin. By Theorem 15.5, $\overline{\Gamma}_3^*$ is convex. This implies $\overline{\Psi}_3^*$ is also convex. Therefore, $(0, 0, 0, a, a, a, -a)$ is in $\overline{\Psi}_3^*$ for all $a > 0$.

Consider any $\mathbf{u} \in \Psi_3$. Referring to (15.15), we have

$$u_i \ge 0 \tag{15.54}$$

for $1 \le i \le 6$. Thus u_7 is the only component of $\mathbf{u}$ which can possibly be negative. We first consider the case when $u_7 \ge 0$. Then $\mathbf{u}$ is in the nonnegative orthant of $\mathcal{I}_3$, and by Lemma 13.1, $\mathbf{u}$ is in Ψ_3^*. Next, consider the case when $u_7 < 0$. Let

$$\mathbf{t} = (0, 0, 0, -u_7, -u_7, -u_7, u_7). \tag{15.55}$$

Then

$$\mathbf{u} = \mathbf{w} + \mathbf{t}, \tag{15.56}$$

where

$$\mathbf{w} = (u_1, u_2, u_3, u_4 + u_7, u_5 + u_7, u_6 + u_7, 0). \tag{15.57}$$

Since $-u_7 > 0$, we see from the foregoing that $\mathbf{t} \in \overline{\Psi}_3^*$. From (15.15), we have

$$u_i + u_7 \ge 0 \tag{15.58}$$

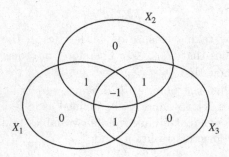

Fig. 15.3. The I-Measure μ^* for X_1, X_2, and X_3 in the proof of Theorem 15.6.

for $i = 4, 5, 6$. Thus $\mathbf{w}$ is in the nonnegative orthant in $\mathcal{I}_3$ and hence in Ψ_3^* by Lemma 13.1. Now for any $\epsilon > 0$, let $\mathbf{t}' \in \Psi_3^*$ such that

$$\|\mathbf{t} - \mathbf{t}'\| < \epsilon, \tag{15.59}$$

where $\|\mathbf{t} - \mathbf{t}'\|$ denotes the Euclidean distance between $\mathbf{t}$ and $\mathbf{t}'$, and let

$$\mathbf{u}' = \mathbf{w} + \mathbf{t}'. \tag{15.60}$$

Since both $\mathbf{w}$ and $\mathbf{t}'$ are in Ψ_3^*, by Lemma 15.3, $\mathbf{u}'$ is also in Ψ_3^*, and

$$\|\mathbf{u} - \mathbf{u}'\| = \|\mathbf{t} - \mathbf{t}'\| < \epsilon. \tag{15.61}$$

Therefore, $\mathbf{u} \in \overline{\Psi}_3^*$. Hence, $\Psi_3 \subset \overline{\Psi}_3^*$, and the theorem is proved. $\square$

Remark 1 Han [148] has found that Γ_3 is the smallest cone that contains Γ_3^*. This result together with Theorem 15.5 implies Theorem 15.6. Theorem 15.6 was also obtained by Golić [137], and it is a consequence of the theorem in Matúš [254].

Remark 2 We have shown that the region $\overline{\Gamma}_n^*$ completely characterizes all unconstrained information inequalities involving n random variables. Since $\overline{\Gamma}_3^* = \Gamma_3$, it follows that there exists no unconstrained information inequalities involving three random variables other than the Shannon-type inequalities. Matúš [258] has obtained piecewise-linear constrained non-Shannon-type inequalities for three random variables that generalize the construction in the proof of Theorem 15.2.

15.2 A Non-Shannon-Type Unconstrained Inequality

We have proved in Theorem 15.6 at the end of the last section that $\overline{\Gamma}_3^* = \Gamma_3$. It is natural to conjecture that this theorem can be generalized to $n \geq 4$. If this conjecture is true, then it follows that all unconstrained information inequalities involving a finite number of random variables are Shannon-type inequalities, and they can all be proved by ITIP running on a sufficiently powerful computer. However, it turns out that this is not the case even for $n = 4$.

We will prove in the next theorem an unconstrained information inequality involving four random variables. Then we will show that this inequality is a non-Shannon-type inequality, and that $\overline{\Gamma}_4^* \neq \Gamma_4$.

Theorem 15.7. *For any four random variables X_1, X_2, X_3, and X_4,*

$$2I(X_3; X_4) \leq I(X_1; X_2) + I(X_1; X_3, X_4)$$
$$+ 3I(X_3; X_4|X_1) + I(X_3; X_4|X_2). \tag{15.62}$$

Toward proving this theorem, we introduce two auxiliary random variables $\tilde{X}_1$ and $\tilde{X}_2$ jointly distributed with X_1, X_2, X_3, and X_4 such that $\tilde{\mathcal{X}}_1 = \mathcal{X}_1$ and $\tilde{\mathcal{X}}_2 = \mathcal{X}_2$. To simplify notation, we will use $p_{1234\tilde{1}\tilde{2}}(x_1, x_2, x_3, x_4, \tilde{x}_1, \tilde{x}_2)$ to denote $p_{X_1 X_2 X_3 X_4 \tilde{X}_1 \tilde{X}_2}(x_1, x_2, x_3, x_4, \tilde{x}_1, \tilde{x}_2)$, etc. The joint distribution for the six random variables $X_1, X_2, X_3, X_4, \tilde{X}_1$, and $\tilde{X}_2$ is defined by

$$
p_{1234\tilde{1}\tilde{2}}(x_1, x_2, x_3, x_4, \tilde{x}_1, \tilde{x}_2)
$$
$$
= \begin{cases} \frac{p_{1234}(x_1, x_2, x_3, x_4) p_{1234}(\tilde{x}_1, \tilde{x}_2, x_3, x_4)}{p_{34}(x_3, x_4)} & \text{if } p_{34}(x_3, x_4) > 0 \\ 0 & \text{if } p_{34}(x_3, x_4) = 0 \end{cases} . \tag{15.63}
$$

Lemma 15.8.
$$
(X_1, X_2) \to (X_3, X_4) \to (\tilde{X}_1, \tilde{X}_2) \tag{15.64}
$$
forms a Markov chain. Moreover, (X_1, X_2, X_3, X_4) *and* $(\tilde{X}_1, \tilde{X}_2, X_3, X_4)$ *have the same marginal distribution.*

Proof. The Markov chain in (15.64) is readily seen by invoking Proposition 2.5. The second part of the lemma is readily seen to be true by noting that in (15.63), $p_{1234\tilde{1}\tilde{2}}$ is symmetrical in X_1 and $\tilde{X}_1$ and in X_2 and $\tilde{X}_2$. $\square$

From the above lemma, we see that the pair of auxiliary random variables $(\tilde{X}_1, \tilde{X}_2)$ corresponds to the pair of random variables (X_1, X_2) in the sense that $(\tilde{X}_1, \tilde{X}_2, X_3, X_4)$ have the same marginal distribution as (X_1, X_2, X_3, X_4). We need to prove two inequalities regarding these six random variables before we prove Theorem 15.7.

Lemma 15.9. *For any four random variables* X_1, X_2, X_3, *and* X_4 *and auxiliary random variables* $\tilde{X}_1$ *and* $\tilde{X}_2$ *as defined in (15.63),*

$$
I(X_3; X_4) - I(X_3; X_4 | X_1) - I(X_3; X_4 | X_2) \leq I(X_1; \tilde{X}_2). \tag{15.65}
$$

Proof. Consider

$$
I(X_3; X_4) - I(X_3; X_4 | X_1) - I(X_3; X_4 | X_2)
$$
$$
\overset{a)}{=} [I(X_3; X_4) - I(X_3; X_4 | X_1)] - I(X_3; X_4 | \tilde{X}_2) \tag{15.66}
$$
$$
= I(X_1; X_3; X_4) - I(X_3; X_4 | \tilde{X}_2) \tag{15.67}
$$
$$
= [I(X_1; X_3; X_4; \tilde{X}_2) + I(X_1; X_3; X_4 | \tilde{X}_2)] - I(X_3; X_4 | \tilde{X}_2) \tag{15.68}
$$
$$
= I(X_1; X_3; X_4; \tilde{X}_2) - [I(X_3; X_4 | \tilde{X}_2) - I(X_1; X_3; X_4 | \tilde{X}_2)] \tag{15.69}
$$
$$
= I(X_1; X_3; X_4; \tilde{X}_2) - I(X_3; X_4 | X_1, \tilde{X}_2) \tag{15.70}
$$
$$
= [I(X_1; X_4; \tilde{X}_2) - I(X_1; X_4; \tilde{X}_2 | X_3)] - I(X_3; X_4 | X_1, \tilde{X}_2) \tag{15.71}
$$
$$
= [I(X_1; \tilde{X}_2) - I(X_1; \tilde{X}_2 | X_4)] - [I(X_1; \tilde{X}_2 | X_3)
$$
$$
- I(X_1; \tilde{X}_2 | X_3, X_4)] - I(X_3; X_4 | X_1, \tilde{X}_2) \tag{15.72}
$$

$$\overset{b)}{=} I(X_1; \tilde{X}_2) - I(X_1; \tilde{X}_2|X_4) - I(X_1; \tilde{X}_2|X_3)$$

$$-I(X_3; X_4|X_1, \tilde{X}_2) \tag{15.73}$$

$$\leq I(X_1; \tilde{X}_2), \tag{15.74}$$

where (a) follows because we see from Lemma 15.8 that (X_2, X_3, X_4) and $(\tilde{X}_2, X_3, X_4)$ have the same marginal distribution, and (b) follows because

$$I(X_1; \tilde{X}_2|X_3, X_4) = 0 \tag{15.75}$$

from the Markov chain in (15.64). The lemma is proved. □

Lemma 15.10. *For any four random variables X_1, X_2, X_3, and X_4 and auxiliary random variables $\tilde{X}_1$ and $\tilde{X}_2$ as defined in (15.63),*

$$I(X_3; X_4) - 2I(X_3; X_4|X_1) \leq I(X_1; \tilde{X}_1). \tag{15.76}$$

Proof. Notice that (15.76) can be obtained from (15.65) by replacing X_2 by X_1 and $\tilde{X}_2$ by $\tilde{X}_1$ in (15.65). The inequality (15.76) can be proved by replacing X_2 by X_1 and $\tilde{X}_2$ by $\tilde{X}_1$ in (15.66) through (15.74) in the proof of the last lemma. The details are omitted. □

Proof of Theorem 15.7. By adding (15.65) and (15.76), we have

$$2I(X_3; X_4) - 3I(X_3; X_4|X_1) - I(X_3; X_4|X_2)$$

$$\leq I(X_1; \tilde{X}_2) + I(X_1; \tilde{X}_1) \tag{15.77}$$

$$= I(X_1; \tilde{X}_2) + [I(X_1; \tilde{X}_1|\tilde{X}_2) + I(X_1; \tilde{X}_1; \tilde{X}_2)] \tag{15.78}$$

$$= [I(X_1; \tilde{X}_2) + I(X_1; \tilde{X}_1|\tilde{X}_2)] + I(X_1; \tilde{X}_1; \tilde{X}_2) \tag{15.79}$$

$$= I(X_1; \tilde{X}_1, \tilde{X}_2) + I(X_1; \tilde{X}_1; \tilde{X}_2) \tag{15.80}$$

$$= I(X_1; \tilde{X}_1, \tilde{X}_2) + [I(\tilde{X}_1; \tilde{X}_2) - I(\tilde{X}_1; \tilde{X}_2|X_1)] \tag{15.81}$$

$$\leq I(X_1; \tilde{X}_1, \tilde{X}_2) + I(\tilde{X}_1; \tilde{X}_2) \tag{15.82}$$

$$\overset{a)}{\leq} I(X_1; X_3, X_4) + I(\tilde{X}_1; \tilde{X}_2) \tag{15.83}$$

$$\overset{b)}{=} I(X_1; X_3, X_4) + I(X_1; X_2), \tag{15.84}$$

where (a) follows from the Markov chain in (15.64), and (b) follows because we see from Lemma 15.8 that $(\tilde{X}_1, \tilde{X}_2)$ and (X_1, X_2) have the same marginal distribution. Note that the auxiliary random variables $\tilde{X}_1$ and $\tilde{X}_2$ disappear in (15.84) after the sequence of manipulations. The theorem is proved. □

Theorem 15.11. *The inequality (15.62) is a non-Shannon-type inequality, and $\overline{\Gamma}_4^* \neq \Gamma_4$.*

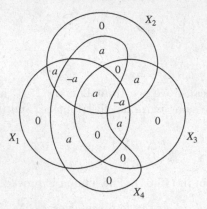

Fig. 15.4. The set-theoretic structure of $\tilde{\mathbf{h}}(a)$.

Proof. Consider for any $a > 0$ the point $\tilde{\mathbf{h}}(a) \in \mathcal{H}_4$, where

$$\tilde{h}_1(a) = \tilde{h}_2(a) = \tilde{h}_3(a) = \tilde{h}_4(a) = 2a,$$
$$\tilde{h}_{12}(a) = 4a, \ \tilde{h}_{13}(a) = \tilde{h}_{14}(a) = 3a,$$
$$\tilde{h}_{23}(a) = \tilde{h}_{24}(a) = \tilde{h}_{34}(a) = 3a,$$
$$\tilde{h}_{123}(a) = \tilde{h}_{124}(a) = \tilde{h}_{134}(a) = \tilde{h}_{234}(a) = \tilde{h}_{1234}(a) = 4a. \tag{15.85}$$

The set-theoretic structure of $\tilde{\mathbf{h}}(a)$ is illustrated by the information diagram in Figure 15.4. The reader should check that this information diagram correctly represents $\tilde{\mathbf{h}}(a)$ as defined. It is also easy to check from this diagram that $\tilde{\mathbf{h}}(a)$ satisfies all the elemental inequalities for four random variables, and therefore $\tilde{\mathbf{h}}(a) \in \Gamma_4$. However, upon substituting the corresponding values in (15.62) for $\mathbf{h}(a)$ with the help of Figure 15.4, we have

$$2a \le 0 + a + 0 + 0 = a, \tag{15.86}$$

which is a contradiction because $a > 0$. In other words, $\tilde{\mathbf{h}}(a)$ does not satisfy (15.62). Equivalently,

$$\tilde{\mathbf{h}}(a) \notin \{\mathbf{h} \in \mathcal{H}_4 : \mathbf{h} \text{ satisfies } (15.62)\}. \tag{15.87}$$

Since $\tilde{\mathbf{h}}(a) \in \Gamma_4$, we conclude that

$$\Gamma_4 \not\subset \{\mathbf{h} \in \mathcal{H}_4 : \mathbf{h} \text{ satisfies } (15.62)\}, \tag{15.88}$$

i.e., (15.62) is not implied by the basic inequalities for four random variables. Hence, (15.62) is a non-Shannon-type inequality.

Since (15.62) is satisfied by all entropy functions for four random variables, we have

$$\Gamma_4^* \subset \{\mathbf{h} \in \mathcal{H}_4 : \mathbf{h} \text{ satisfies } (15.62)\}, \tag{15.89}$$

and upon taking closure on both sides, we have

$$\overline{\Gamma}_4^* \subset \{\mathbf{h} \in \mathcal{H}_4 : \mathbf{h} \text{ satisfies } (15.62)\}. \tag{15.90}$$

Then (15.87) implies $\tilde{\mathbf{h}}(a) \notin \overline{\Gamma}_4^*$. Since $\tilde{\mathbf{h}}(a) \in \Gamma_4$ and $\tilde{\mathbf{h}}(a) \notin \overline{\Gamma}_4^*$, we conclude that $\overline{\Gamma}_4^* \neq \Gamma_4$. The theorem is proved. $\square$

Remark We have shown in the proof of Theorem 15.11 that the inequality (15.62) cannot be proved by invoking the basic inequalities for four random variables. However, (15.62) can be proved by invoking the basic inequalities for the six random variables $X_1, X_2, X_3, X_4, \tilde{X}_1$, and $\tilde{X}_2$ with the joint probability distribution $p_{1234\tilde{1}\tilde{2}}$ as constructed in (15.63).

The inequality (15.62) remains valid when the indices 1, 2, 3, and 4 are permuted. Since (15.62) is symmetrical in X_3 and X_4, $4!/2! = 12$ distinct versions of (15.62) can be obtained by permuting the indices, and all these 12 inequalities are simultaneously satisfied by the entropy function of any set of random variables X_1, X_2, X_3, and X_4. We will denote these 12 inequalities collectively by $\langle 15.62 \rangle$. Now define the region

$$\tilde{\Gamma}_4 = \{\mathbf{h} \in \Gamma_4 : \mathbf{h} \text{ satisfies } \langle 15.62 \rangle\}. \tag{15.91}$$

Evidently,

$$\Gamma_4^* \subset \tilde{\Gamma}_4 \subset \Gamma_4. \tag{15.92}$$

Since both $\tilde{\Gamma}_4$ and Γ_4 are closed, upon taking closure, we also have

$$\overline{\Gamma}_4^* \subset \tilde{\Gamma}_4 \subset \Gamma_4. \tag{15.93}$$

Since $\langle 15.62 \rangle$ are non-Shannon-type inequalities as we have proved in the last theorem, $\tilde{\Gamma}_4$ is a proper subset of Γ_4 and hence a tighter outer bound on Γ_4^* and $\overline{\Gamma}_4^*$ than Γ_4.

In the course of proving that (15.62) is of non-Shannon-type, it was shown in the proof of Theorem 15.11 that there exists $\tilde{\mathbf{h}}(a) \in \Gamma_4$ as defined in (15.85) which does not satisfy (15.62). By investigating the geometrical relation between $\tilde{\mathbf{h}}(a)$ and Γ_4, we prove in the next theorem that (15.62) in fact induces a class of $2^{14} - 1$ non-Shannon-type constrained inequalities. Applications of some of these inequalities will be discussed in Section 15.4.

Theorem 15.12. *The inequality (15.62) is a non-Shannon-type inequality conditioning on setting any nonempty subset of the following 14 Shannon's information measures to zero:*

$$I(X_1; X_2), I(X_1; X_2|X_3), I(X_1; X_2|X_4), I(X_1; X_3|X_4),$$
$$I(X_1; X_4|X_3), I(X_2; X_3|X_4), I(X_2; X_4|X_3), I(X_3; X_4|X_1),$$
$$I(X_3; X_4|X_2), I(X_3; X_4|X_1, X_2), H(X_1|X_2, X_3, X_4), \tag{15.94}$$
$$H(X_2|X_1, X_3, X_4), H(X_3|X_1, X_2, X_4), H(X_4|X_1, X_2, X_3).$$

Proof. It is easy to verify from Figure 15.4 that $\tilde{\mathbf{h}}(a)$ lies in exactly 14 hyperplanes in $\mathcal{H}_4$ (i.e., $\Re^{15}$) defining the boundary of Γ_4 which corresponds to setting the 14 Shannon's measures in (15.94) to zero. Therefore, $\tilde{\mathbf{h}}(a)$ for $a \geq 0$ defines an extreme direction of Γ_4.

Now for any linear subspace Φ of $\mathcal{H}_4$ containing $\tilde{\mathbf{h}}(a)$, where $a > 0$, we have

$$\tilde{\mathbf{h}}(a) \in \Gamma_4 \cap \Phi \tag{15.95}$$

and $\tilde{\mathbf{h}}(a)$ does not satisfy (15.62). Therefore,

$$(\Gamma_4 \cap \Phi) \not\subset \{\mathbf{h} \in \mathcal{H}_4 : \mathbf{h} \text{ satisfies (15.62)}\}. \tag{15.96}$$

This means that (15.62) is a non-Shannon-type inequality under the constraint Φ. From the above, we see that Φ can be taken to be the intersection of any nonempty subset of the 14 hyperplanes containing $\tilde{\mathbf{h}}(a)$. Thus (15.62) is a non-Shannon-type inequality conditioning on any nonempty subset of the 14 Shannon's measures in (15.94) being equal to zero. Hence, (15.62) induces a class of $2^{14} - 1$ non-Shannon-type constrained inequalities. The theorem is proved. □

Remark It is not true that the inequality (15.62) is of non-Shannon-type under any constraint. Suppose we impose the constraint

$$I(X_3; X_4) = 0. \tag{15.97}$$

Then the left-hand side of (15.62) becomes zero, and the inequality is trivially implied by the basic inequalities because only mutual informations with positive coefficients appear on the right-hand side. Then (15.62) becomes a Shannon-type inequality under the constraint in (15.97).

15.3 A Non-Shannon-Type Constrained Inequality

In the last section, we proved a non-Shannon-type unconstrained inequality for four random variables which implies $\overline{\Gamma}_4^* \neq \Gamma_4$. This inequality induces a region $\tilde{\Gamma}_4$ which is a tighter outer bound on Γ_4^* and $\overline{\Gamma}_4^*$ then Γ_4. We further showed that this inequality induces a class of $2^{14} - 1$ non-Shannon-type constrained inequalities for four random variables.

In this section, we prove a non-Shannon-type constrained inequality for four random variables. Unlike the non-Shannon-type unconstrained inequality we proved in the last section, this constrained inequality is not strong enough to imply that $\Gamma_4^* \neq \Gamma_4$. However, the latter is not implied by the former.

Lemma 15.13. *Let $p(x_1, x_2, x_3, x_4)$ be any probability distribution. Then*

$$\tilde{p}(x_1, x_2, x_3, x_4) = \begin{cases} \frac{p(x_1, x_3, x_4) p(x_2, x_3, x_4)}{p(x_3, x_4)} & \text{if } p(x_3, x_4) > 0 \\ 0 & \text{if } p(x_3, x_4) = 0 \end{cases} \tag{15.98}$$

is also a probability distribution. Moreover,

$$\tilde{p}(x_1, x_3, x_4) = p(x_1, x_3, x_4) \tag{15.99}$$

and

$$\tilde{p}(x_2, x_3, x_4) = p(x_2, x_3, x_4) \tag{15.100}$$

for all x_1, x_2, x_3, and x_4.

Proof. The proof for the first part of the lemma is straightforward (see Problem 4 in Chapter 2). The details are omitted here.

To prove the second part of the lemma, it suffices to prove (15.99) for all x_1, x_3, and x_4 because $\tilde{p}(x_1, x_2, x_3, x_4)$ is symmetrical in x_1 and x_2. We first consider x_1, x_3, and x_4 such that $p(x_3, x_4) > 0$. From (15.98), we have

$$\tilde{p}(x_1, x_3, x_4) = \sum_{x_2} \tilde{p}(x_1, x_2, x_3, x_4) \tag{15.101}$$

$$= \sum_{x_2} \frac{p(x_1, x_3, x_4) p(x_2, x_3, x_4)}{p(x_3, x_4)} \tag{15.102}$$

$$= \frac{p(x_1, x_3, x_4)}{p(x_3, x_4)} \sum_{x_2} p(x_2, x_3, x_4) \tag{15.103}$$

$$= \left[\frac{p(x_1, x_3, x_4)}{p(x_3, x_4)} \right] p(x_3, x_4) \tag{15.104}$$

$$= p(x_1, x_3, x_4). \tag{15.105}$$

For x_1, x_3, and x_4 such that $p(x_3, x_4) = 0$, we have

$$0 \leq p(x_1, x_3, x_4) \leq p(x_3, x_4) = 0, \tag{15.106}$$

which implies

$$p(x_1, x_3, x_4) = 0. \tag{15.107}$$

Therefore, from (15.98), we have

$$\tilde{p}(x_1, x_3, x_4) = \sum_{x_2} \tilde{p}(x_1, x_2, x_3, x_4) \tag{15.108}$$

$$= \sum_{x_2} 0 \tag{15.109}$$

$$= 0 \tag{15.110}$$

$$= p(x_1, x_3, x_4). \tag{15.111}$$

Thus we have proved (15.99) for all x_1, x_3, and x_4, and the lemma is proved.
□

Theorem 15.14. *For any four random variables* $X_1, X_2, X_3,$ *and* $X_4,$ *if*

$$I(X_1; X_2) = I(X_1; X_2|X_3) = 0, \tag{15.112}$$

then

$$I(X_3; X_4) \leq I(X_3; X_4|X_1) + I(X_3; X_4|X_2). \tag{15.113}$$

Proof. Consider

$$I(X_3; X_4) - I(X_3; X_4|X_1) - I(X_3; X_4|X_2)$$

$$= \sum_{\substack{x_1, x_2, x_3, x_4: \\ p(x_1, x_2, x_3, x_4) > 0}} p(x_1, x_2, x_3, x_4) \log \frac{p(x_3, x_4)p(x_1, x_3)p(x_1, x_4)p(x_2, x_3)p(x_2, x_4)}{p(x_3)p(x_4)p(x_1)p(x_2)p(x_1, x_3, x_4)p(x_2, x_3, x_4)}$$

$$= E_p \log \frac{p(X_3, X_4)p(X_1, X_3)p(X_1, X_4)p(X_2, X_3)p(X_2, X_4)}{p(X_3)p(X_4)p(X_1)p(X_2)p(X_1, X_3, X_4)p(X_2, X_3, X_4)}, \tag{15.114}$$

where we have used E_p to denote expectation with respect to $p(x_1, x_2, x_3, x_4)$. We claim that the above expectation is equal to

$$E_{\tilde{p}} \log \frac{p(X_3, X_4)p(X_1, X_3)p(X_1, X_4)p(X_2, X_3)p(X_2, X_4)}{p(X_3)p(X_4)p(X_1)p(X_2)p(X_1, X_3, X_4)p(X_2, X_3, X_4)}, \tag{15.115}$$

where $\tilde{p}(x_1, x_2, x_3, x_4)$ is defined in (15.98).

Toward proving that the claim is correct, we note that (15.115) is the sum of a number of expectations with respect to $\tilde{p}$. Let us consider one of these expectations, say

$$E_{\tilde{p}} \log p(X_1, X_3) = \sum_{\substack{x_1, x_2, x_3, x_4: \\ \tilde{p}(x_1, x_2, x_3, x_4) > 0}} \tilde{p}(x_1, x_2, x_3, x_4) \log p(x_1, x_3). \tag{15.116}$$

Note that in the above summation, if $\tilde{p}(x_1, x_2, x_3, x_4) > 0$, then from (15.98), we see that

$$p(x_1, x_3, x_4) > 0, \tag{15.117}$$

and hence

$$p(x_1, x_3) > 0. \tag{15.118}$$

Therefore, the summation in (15.116) is always well defined. Further, it can be written as

$$\sum_{x_1, x_3, x_4} \log p(x_1, x_3) \sum_{x_2: \tilde{p}(x_1, x_2, x_3, x_4) > 0} \tilde{p}(x_1, x_2, x_3, x_4)$$

$$= \sum_{x_1, x_3, x_4} \tilde{p}(x_1, x_3, x_4) \log p(x_1, x_3). \tag{15.119}$$

Thus $E_{\tilde{p}} \log p(X_1, X_3)$ depends on $\tilde{p}(x_1, x_2, x_3, x_4)$ only through $\tilde{p}(x_1, x_3, x_4)$, which by Lemma 15.13 is equal to $p(x_1, x_3, x_4)$. It then follows that

$$E_{\tilde{p}} \log p(X_1, X_3)$$

$$= \sum_{x_1, x_3, x_4} \tilde{p}(x_1, x_3, x_4) \log p(x_1, x_3) \qquad (15.120)$$

$$= \sum_{x_1, x_3, x_4} p(x_1, x_3, x_4) \log p(x_1, x_3) \qquad (15.121)$$

$$= E_p \log p(X_1, X_3). \qquad (15.122)$$

In other words, the expectation on $\log p(X_1, X_3)$ can be taken with respect to either $\tilde{p}(x_1, x_2, x_3, x_4)$ or $p(x_1, x_2, x_3, x_4)$ without affecting its value. By observing that all the marginals of p in the logarithm in (15.115) involve only subsets of either $\{X_1, X_3, X_4\}$ or $\{X_2, X_3, X_4\}$, we see that similar conclusions can be drawn for all the other expectations in (15.115), and hence the claim is proved.

Thus the claim implies that

$$I(X_3; X_4) - I(X_3; X_4|X_1) - I(X_3; X_4|X_2)$$

$$= E_{\tilde{p}} \log \frac{p(X_3, X_4)p(X_1, X_3)p(X_1, X_4)p(X_2, X_3)p(X_2, X_4)}{p(X_3)p(X_4)p(X_1)p(X_2)p(X_1, X_3, X_4)p(X_2, X_3, X_4)}$$

$$= \sum_{\substack{x_1, x_2, x_3, x_4: \\ \tilde{p}(x_1, x_2, x_3, x_4) > 0}} \tilde{p}(x_1, x_2, x_3, x_4) \log \frac{p(x_3, x_4)p(x_1, x_3)p(x_1, x_4)p(x_2, x_3)p(x_2, x_4)}{p(x_3)p(x_4)p(x_1)p(x_2)p(x_1, x_3, x_4)p(x_2, x_3, x_4)}$$

$$= -\sum_{\substack{x_1, x_2, x_3, x_4: \\ \tilde{p}(x_1, x_2, x_3, x_4) > 0}} \tilde{p}(x_1, x_2, x_3, x_4) \log \frac{\tilde{p}(x_1, x_2, x_3, x_4)}{\hat{p}(x_1, x_2, x_3, x_4)}, \qquad (15.123)$$

where

$$\hat{p}(x_1, x_2, x_3, x_4) =$$

$$\begin{cases} \frac{p(x_1, x_3)p(x_1, x_4)p(x_2, x_3)p(x_2, x_4)}{p(x_1)p(x_2)p(x_3)p(x_4)} & \text{if } p(x_1), p(x_2), p(x_3), p(x_4) > 0 \\ 0 & \text{otherwise} \end{cases}.$$

$$(15.124)$$

The equality in (15.123) is justified by observing that if x_1, x_2, x_3, and x_4 are such that $\tilde{p}(x_1, x_2, x_3, x_4) > 0$, then

$$p(x_1, x_3), p(x_1, x_4), p(x_2, x_3), p(x_2, x_4), p(x_1), p(x_2), p(x_3), p(x_4) \qquad (15.125)$$

are all strictly positive, and we see from (15.124) that $\hat{p}(x_1, x_2, x_3, x_4) > 0$.

To complete the proof, we only need to show that $\hat{p}(x_1, x_2, x_3, x_4)$ is a probability distribution. Once this is proven, the conclusion of the theorem follows immediately because the summation in (15.123), which is identified as the divergence between $\tilde{p}(x_1, x_2, x_3, x_4)$ and $\hat{p}(x_1, x_2, x_3, x_4)$, is always non-negative by the divergence inequality (Theorem 2.31). Toward this end, we notice that for x_1, x_2, and x_3 such that $p(x_3) > 0$,

$$p(x_1, x_2, x_3) = \frac{p(x_1, x_3)p(x_2, x_3)}{p(x_3)} \tag{15.126}$$

by the assumption

$$I(X_1; X_2|X_3) = 0, \tag{15.127}$$

and for all x_1 and x_2,

$$p(x_1, x_2) = p(x_1)p(x_2) \tag{15.128}$$

by the assumption

$$I(X_1; X_2) = 0. \tag{15.129}$$

Then

$$\sum_{x_1, x_2, x_3, x_4} \hat{p}(x_1, x_2, x_3, x_4)$$

$$= \sum_{\substack{x_1, x_2, x_3, x_4: \\ \hat{p}(x_1, x_2, x_3, x_4) > 0}} \hat{p}(x_1, x_2, x_3, x_4) \tag{15.130}$$

$$= \sum_{\substack{x_1, x_2, x_3, x_4: \\ p(x_1), p(x_2), p(x_3), p(x_4) > 0}} \frac{p(x_1, x_3)p(x_1, x_4)p(x_2, x_3)p(x_2, x_4)}{p(x_1)p(x_2)p(x_3)p(x_4)} \tag{15.131}$$

$$\overset{a)}{=} \sum_{\substack{x_1, x_2, x_3, x_4: \\ p(x_1), p(x_2), p(x_3), p(x_4) > 0}} \frac{p(x_1, x_2, x_3)p(x_1, x_4)p(x_2, x_4)}{p(x_1)p(x_2)p(x_4)} \tag{15.132}$$

$$\overset{b)}{=} \sum_{\substack{x_1, x_2, x_3, x_4: \\ p(x_1), p(x_2), p(x_3), p(x_4) > 0}} \frac{p(x_1, x_2, x_3)p(x_1, x_4)p(x_2, x_4)}{p(x_1, x_2)p(x_4)} \tag{15.133}$$

$$= \sum_{\substack{x_1, x_2, x_4: \\ p(x_1), p(x_2), p(x_4) > 0}} \frac{p(x_1, x_4)p(x_2, x_4)}{p(x_4)} \sum_{x_3: p(x_3) > 0} p(x_3|x_1, x_2) \tag{15.134}$$

$$= \sum_{\substack{x_1, x_2, x_4: \\ p(x_1), p(x_2), p(x_4) > 0}} \frac{p(x_1, x_4)p(x_2, x_4)}{p(x_4)} \tag{15.135}$$

$$= \sum_{\substack{x_2, x_4: \\ p(x_2), p(x_4) > 0}} p(x_2, x_4) \sum_{x_1: p(x_1) > 0} p(x_1|x_4) \tag{15.136}$$

$$\overset{c)}{=} \sum_{\substack{x_2, x_4: \\ p(x_2), p(x_4) > 0}} p(x_2, x_4) \tag{15.137}$$

$$\overset{d)}{=} 1, \tag{15.138}$$

where (a) and (b) follow from (15.126) and (15.128), respectively. The equality in (c) is justified as follows. For x_1 such that $p(x_1) = 0$,

$$p(x_1|x_4) = \frac{p(x_1)p(x_4|x_1)}{p(x_4)} = 0. \qquad (15.139)$$

Therefore

$$\sum_{x_1:p(x_1)>0} p(x_1|x_4) = \sum_{x_1} p(x_1|x_4) = 1. \qquad (15.140)$$

Finally, the equality in (d) is justified as follows. For x_2 and x_4 such that $p(x_2)$ or $p(x_4)$ vanishes, $p(x_2, x_4)$ must vanish because

$$0 \le p(x_2, x_4) \le p(x_2) \qquad (15.141)$$

and

$$0 \le p(x_2, x_4) \le p(x_4). \qquad (15.142)$$

Therefore,

$$\sum_{\substack{x_2,x_4:\\p(x_2),p(x_4)>0}} p(x_2, x_4) = \sum_{x_2,x_4} p(x_2, x_4) = 1. \qquad (15.143)$$

The theorem is proved. $\square$

Theorem 15.15. *The constrained inequality in Theorem 15.14 is a non-Shannon-type inequality.*

Proof. The theorem can be proved by considering the point $\tilde{\mathbf{h}}(a) \in \mathcal{H}_4$ for $a > 0$ as in the proof of Theorem 15.11. The details are left as an exercise. $\square$

The constrained inequality in Theorem 15.14 has the following geometrical interpretation. The constraints in (15.112) correspond to the intersection of two hyperplanes in $\mathcal{H}_4$ which define the boundary of Γ_4. Then the inequality (15.62) says that a certain region on the boundary of Γ_4 is not in Γ_4^*. It can further be proved by computation[1] that the constrained inequality in Theorem 15.14 is not implied by the 12 distinct versions of the unconstrained inequality in Theorem 15.7 (i.e., $\langle 15.62 \rangle$) together with the basic inequalities.

We have proved in the last section that the non-Shannon-type inequality (15.62) implies a class of $2^{14} - 1$ constrained non-Shannon-type inequalities. We end this section by proving a similar result for the non-Shannon-type constrained inequality in Theorem 15.14.

Theorem 15.16. *The inequality*

$$I(X_3; X_4) \le I(X_3; X_4|X_1) + I(X_3; X_4|X_2) \qquad (15.144)$$

is a non-Shannon-type inequality conditioning on setting both $I(X_1; X_2)$ and $I(X_1; X_2|X_3)$ and any subset of the following 12 Shannon's information measures to zero:

---

[1] Ying-On Yan, private communication.

$$I(X_1; X_2|X_4), I(X_1; X_3|X_4), I(X_1; X_4|X_3),$$
$$I(X_2; X_3|X_4), I(X_2; X_4|X_3), I(X_3; X_4|X_1),$$
$$I(X_3; X_4|X_2), I(X_3; X_4|X_1, X_2), H(X_1|X_2, X_3, X_4), \qquad (15.145)$$
$$H(X_2|X_1, X_3, X_4), H(X_3|X_1, X_2, X_4), H(X_4|X_1, X_2, X_3).$$

Proof. The proof of this theorem is very similar to the proof of Theorem 15.12. We first note that $I(X_1; X_2)$ and $I(X_1; X_2|X_3)$ together with the 12 Shannon's information measures in (15.145) are exactly the 14 Shannon's information measures in (15.94). We have already shown in the proof of Theorem 15.12 that $\tilde{h}(a)$ (cf. Figure 15.4) lies in exactly 14 hyperplanes defining the boundary of Γ_4 which corresponds to setting these 14 Shannon's information measures to zero. We have also shown that $\tilde{h}(a)$ for $a \geq 0$ defines an extreme direction of Γ_4.

Denote by Φ_0 the intersection of the two hyperplanes in $\mathcal{H}_4$ which correspond to setting $I(X_1; X_2)$ and $I(X_1; X_2|X_3)$ to zero. Since $\tilde{h}(a)$ for any $a > 0$ satisfies

$$I(X_1; X_2) = I(X_1; X_2|X_3) = 0, \qquad (15.146)$$

$\tilde{h}(a)$ is in Φ_0. Now for any linear subspace Φ of $\mathcal{H}_4$ containing $\tilde{h}(a)$ such that $\Phi \subset \Phi_0$, we have

$$\tilde{h}(a) \in \Gamma_4 \cap \Phi. \qquad (15.147)$$

Upon substituting the corresponding values in (15.113) for $\tilde{h}(a)$ with the help of Figure 15.4, we have

$$a \leq 0 + 0 = 0, \qquad (15.148)$$

which is a contradiction because $a > 0$. Therefore, $\tilde{h}(a)$ does not satisfy (15.113). Therefore,

$$(\Gamma_4 \cap \Phi) \not\subset \{\mathbf{h} \in \mathcal{H}_4 : \mathbf{h} \text{ satisfies (15.113)}\}. \qquad (15.149)$$

This means that (15.113) is a non-Shannon-type inequality under the constraint Φ. From the above, we see that Φ can be taken to be the intersection of Φ_0 and any subset of the 12 hyperplanes which correspond to setting the 12 Shannon's information measures in (15.145) to zero. Hence, (15.113) is a non-Shannon-type inequality conditioning on $I(X_1; X_2)$, $I(X_1; X_2|X_3)$, and any subset of the 12 Shannon's information measures in (15.145) being equal to zero. In other words, the constrained inequality in Theorem 15.14 in fact induces a class of 2^{12} constrained non-Shannon-type inequalities. The theorem is proved. □

15.4 Applications

As we have mentioned in Chapter 13, information inequalities govern the impossibilities in information theory. In this section, we give several applications of the non-Shannon-type inequalities we have proved in this chapter in probability theory and information theory. An application in group theory of the unconstrained inequality proved in Section 15.2 will be discussed in Chapter 16. Non-Shannon-type inequalities also find applications in network coding theory to be discussed in Part II of this book.

Example 15.17. For the constrained inequality in Theorem 15.14, if we further impose the constraints

$$I(X_3; X_4|X_1) = I(X_3; X_4|X_2) = 0, \qquad (15.150)$$

then the right-hand side of (15.113) becomes zero. This implies

$$I(X_3; X_4) = 0 \qquad (15.151)$$

because $I(X_3; X_4)$ is nonnegative. This means that

$$\left. \begin{array}{l} X_1 \perp X_2 \\ X_1 \perp X_2|X_3 \\ X_3 \perp X_4|X_1 \\ X_3 \perp X_4|X_2 \end{array} \right\} \Rightarrow X_3 \perp X_4. \qquad (15.152)$$

We leave it as an exercise for the reader to show that this implication cannot be deduced from the basic inequalities.

Example 15.18. If we impose the constraints

$$I(X_1; X_2) = I(X_1; X_3, X_4) = I(X_3; X_4|X_1) = I(X_3; X_4|X_2) = 0, \quad (15.153)$$

then the right-hand side of (15.62) becomes zero, which implies

$$I(X_3; X_4) = 0. \qquad (15.154)$$

This means that

$$\left. \begin{array}{l} X_1 \perp X_2 \\ X_1 \perp (X_3, X_4) \\ X_3 \perp X_4|X_1 \\ X_3 \perp X_4|X_2 \end{array} \right\} \Rightarrow X_3 \perp X_4. \qquad (15.155)$$

Note that (15.152) and (15.155) differ only in the second constraint. Again, we leave it as an exercise for the reader to show that this implication cannot be deduced from the basic inequalities.

Example 15.19. Consider a fault-tolerant data storage system consisting of random variables X_1, X_2, X_3, X_4 such that any three random variables can recover the remaining one, i.e.,

$$H(X_i|X_j, j \neq i) = 0, \quad 1 \leq i, j \leq 4. \qquad (15.156)$$

We are interested in the set of all entropy functions subject to these constraints, denoted by Υ, which characterizes the amount of joint information which can possibly be stored in such a data storage system. Let

$$\Phi = \{\mathbf{h} \in \mathcal{H}_4 : \mathbf{h} \text{ satisfies } (15.156)\}. \tag{15.157}$$

Then the set Υ is equal to the intersection between Γ_4^* and Φ, i.e., $\Gamma_4^* \cap \Phi$.

Since each constraint in (15.156) is one of the 14 constraints specified in Theorem 15.12, we see that (15.62) is a non-Shannon-type inequality under the constraints in (15.156). Then $\tilde{\Gamma}_4 \cap \Phi$ (cf. (15.91)) is a tighter outer bound on Υ than $\Gamma_4 \cap \Phi$.

Example 15.20. Consider four random variables X_1, X_2, X_3, and X_4 such that $X_3 \to (X_1, X_2) \to X_4$ forms a Markov chain. This Markov condition is equivalent to

$$I(X_3; X_4 | X_1, X_2) = 0. \tag{15.158}$$

It can be proved by invoking the basic inequalities (using ITIP) that

$$I(X_3; X_4) \le I(X_3; X_4 | X_1) + I(X_3; X_4 | X_2) + 0.5 I(X_1; X_2)$$
$$+ c I(X_1; X_3, X_4) + (1 - c) I(X_2; X_3, X_4), \tag{15.159}$$

where $0.25 \le c \le 0.75$, and this is the best possible.

Now observe that the Markov condition (15.158) is one of the 14 constraints specified in Theorem 15.12. Therefore, (15.62) is a non-Shannon-type inequality under this Markov condition. By replacing X_1 and X_2 by each other in (15.62), we obtain

$$2 I(X_3; X_4) \le I(X_1; X_2) + I(X_2; X_3, X_4)$$
$$+ 3 I(X_3; X_4 | X_2) + I(X_3; X_4 | X_1). \tag{15.160}$$

Upon adding (15.62) and (15.160) and dividing by 4, we obtain

$$I(X_3; X_4) \le I(X_3; X_4 | X_1) + I(X_3; X_4 | X_2) + 0.5 I(X_1; X_2)$$
$$+ 0.25 I(X_1; X_3, X_4) + 0.25 I(X_2; X_3, X_4). \tag{15.161}$$

Comparing the last two terms in (15.159) and the last two terms in (15.161), we see that (15.161) is a sharper upper bound than (15.159).

The Markov chain $X_3 \to (X_1, X_2) \to X_4$ arises in many communication situations. As an example, consider a person listening to an audio source. Then the situation can be modeled by this Markov chain with X_3 being the sound wave generated at the source, X_1 and X_2 being the sound waves received at the two ear drums, and X_4 being the nerve impulses which eventually arrive at the brain. The inequality (15.161) gives an upper bound on $I(X_3; X_4)$ which is tighter than what can be implied by the basic inequalities.

There is some resemblance between the constrained inequality (15.161) and the data processing theorem, but they do not appear to be directly related.

Chapter Summary

Characterizations of Γ_n^* and $\overline{\Gamma}_n^*$:

1. $\Gamma_2^* = \Gamma_2$.
2. $\Gamma_3^* \neq \Gamma_3$, but $\overline{\Gamma}_3^* = \Gamma_3$.
3. $\overline{\Gamma}_4^* \neq \Gamma_4$.
4. $\overline{\Gamma}_n^*$ is a convex cone.

An Unconstrained Non-Shannon-Type Inequality:

$$2I(X_3; X_4) \leq I(X_1; X_2) + I(X_1; X_3, X_4) + 3I(X_3; X_4|X_1) + I(X_3; X_4|X_2).$$

A Constrained Non-Shannon-Type Inequality: If $I(X_1; X_2) = I(X_1; X_2|X_3) = 0$, then

$$I(X_3; X_4) \leq I(X_3; X_4|X_1) + I(X_3; X_4|X_2).$$

Problems

1. Verify by ITIP that the unconstrained information inequality in Theorem 15.7 is of non-Shannon-type.
2. Verify by ITIP and prove analytically that the constrained information inequality in Theorem 15.14 is of non-Shannon-type.
3. Use ITIP to verify the unconstrained information inequality in Theorem 15.7. Hint: Create two auxiliary random variables as in the proof of Theorem 15.7 and impose appropriate constraints on the random variables.
4. Verify by ITIP that the implications in Examples 15.17 and 15.18 cannot be deduced from the basic inequalities.
5. Can you show that the sets of constraints in Examples 15.17 and 15.18 are in fact different?
6. Consider an information inequality involving random variables $X_1, X_2, \cdots, X_n$, which can be written as

$$\sum_{\alpha \in 2^{\mathcal{N}_n} \setminus \{\emptyset\}} c_\alpha H(X_\alpha) \geq 0,$$

where $\mathcal{N}_n = \{1, 2, \cdots, n\}$. For $i \in \mathcal{N}_n$, let

$$r_i = \sum_{\alpha \in 2^{\mathcal{N}_n} \setminus \{\emptyset\}} c_\alpha n_\alpha(i),$$

where $n_\alpha(i)$ is equal to 1 if $i \in \alpha$ and is equal to 0 otherwise.

a) Show that r_i is the coefficient of $H(X_i|X_{\mathcal{N}_n-\{i\}})$ when the information inequality is expressed in terms of the elemental forms of Shannon's information measures for n random variables.

b) Show that if the information inequality always holds, then $r_i \geq 0$ for all $i \in \mathcal{N}_n$.

(Chan [60].)

7. Let $X_i, i = 1, 2, \cdots, n$, Z, and T be discrete random variables.

a) Prove that

$$nI(Z;T) - \sum_{j=1}^{n} I(Z;T|X_j) - nI(Z;T|X_i)$$

$$\leq I(X_i; Z, T) + \sum_{j=1}^{n} H(X_j) - H(X_1, X_2, \cdots, X_n).$$

Hint: When $n = 2$, this inequality reduces to the unconstrained non-Shannon-type inequality in Theorem 15.7.

b) Prove that

$$nI(Z;T) - 2\sum_{j=1}^{n} I(Z;T|X_j)$$

$$\leq \frac{1}{n} \sum_{i=1}^{n} I(X_i; Z, T) + \sum_{j=1}^{n} H(X_j) - H(X_1, X_2, \cdots, X_n).$$

(Zhang and Yeung [416].)

8. Let $p(x_1, x_2, x_3, x_4)$ be the joint distribution for random variables X_1, X_2, X_3, and X_4 such that $I(X_1; X_2|X_3) = I(X_2; X_4|X_3) = 0$, and let $\check{p}$ be defined in (15.98).

a) Show that

$$\check{p}(x_1, x_2, x_3, x_4)$$
$$= \begin{cases} c \cdot \frac{p(x_1,x_2,x_3)p(x_1,x_4)p(x_2,x_4)}{p(x_1,x_2)p(x_4)} & \text{if } p(x_1, x_2), p(x_4) > 0 \\ 0 & \text{otherwise} \end{cases}$$

defines a probability distribution for an appropriate $c \geq 1$.

b) Prove that $\check{p}(x_1, x_2, x_3) = p(x_1, x_2, x_3)$ for all x_1, x_2, and x_3.

c) By considering $D(\check{p}\|\check{p}) \geq 0$, prove that

$$H(X_{13}) + H(X_{14}) + H(X_{23}) + H(X_{24}) + H(X_{34})$$
$$\geq H(X_3) + H(X_4) + H(X_{12}) + H(X_{134}) + H(X_{234}),$$

where $H(X_{134})$ denotes $H(X_1, X_3, X_4)$, etc.

d) Prove that under the constraints in (15.112), the inequality in (15.113) is equivalent to the inequality in (c).

The inequality in (c) is referred to as the *Ingleton inequality* for entropy in the literature. For the origin of the Ingleton inequality, see Problem 9 in Chapter 16. (Matúš [256].)

Historical Notes

In 1986, Pippenger [294] asked whether there exist constraints on the entropy function other than the polymatroidal axioms, which are equivalent to the basic inequalities. He called the constraints on the entropy function the *laws of information theory*. The problem had been open until Zhang and Yeung discovered for four random variables first a constrained non-Shannon-type inequality [415] and then an unconstrained non-Shannon-type inequality [416] in the late 1990s. The inequality reported in [416] has been further generalized by Makarychev et al. [242] and Zhang [413]. The existence of these inequalities implies that there are laws in information theory beyond those laid down by Shannon [322].

The non-Shannon-type inequalities that have been discovered induce outer bounds on the region Γ_4^* which are tighter than Γ_4. Matúš and Studený [261] showed that an entropy function in Γ_4 is entropic if it satisfies the Ingleton inequality (see Problem 9 in Chapter 16). This gives an inner bound on $\overline{\Gamma}_4^*$. A more explicit proof of this inner bound can be found in [416], where the bound was shown not to be tight. Matúš [259] has obtained asymptotically tight inner bounds on $\overline{\Gamma}_n^*$ by constructing entropy functions from matroids.

Dougherty et al. [97] discovered a host of unconstrained non-Shannon-type inequalities by means of a computer search based on ITIP and the Markov chain construction in [416] (see Problem 3). Recently, Matúš [260] proved an infinite class of unconstrained non-Shannon-type inequalities, implying that $\overline{\Gamma}_n^*$ is not a pyramid.

Chan [60] proved a characterization for an inequality for differential entropy in terms of its discrete version. Lněnička [237] proved that the tightness of the continuous version of the unconstrained non-Shannon-type inequality reported in [416] can be achieved by a multivariate Gaussian distribution.

In the 1990s, Matúš and Studený [254][261][255] studied the structure of conditional independence (which subsumes the implication problem) of random variables. Matúš [256] finally settled the problem for four random variables by means of a constrained non-Shannon-type inequality which is a variation of the inequality reported in [415].

The von Neumann entropy is an extension of classical entropy (as discussed in this book) to the field of quantum mechanics. The strong subadditivity of the von Neumann entropy proved by Lieb and Ruskai [233] plays the same role as the basic inequalities for classical entropy. Pippenger [295] proved that for a three-party system, there exists no inequality for the von Neumann entropy beyond strong subadditivity. Subsequently, Linden and Winter [235] discovered for a four-party system a constrained inequality for the von Neumann entropy which is independent of strong subadditivity. We refer the reader to the book by Nielsen and Chuang [274] for an introduction to quantum information theory.

Along a related direction, Hammer et al. [144] have shown that all linear inequalities that always hold for Kolmogorov complexity also always hold for entropy and vice versa. This establishes a one-to-one correspondence between entropy and Kolmogorov complexity.

16

Entropy and Groups

The *group* is the first major mathematical structure in abstract algebra, while entropy is the most basic measure of information. Group theory and information theory are two seemingly unrelated subjects which turn out to be intimately related to each other. This chapter explains this intriguing relation between these two fundamental subjects. Those readers who have no knowledge in group theory may skip this introduction and go directly to the next section.

Let X_1 and X_2 be any two random variables. Then

$$H(X_1) + H(X_2) \geq H(X_1, X_2), \tag{16.1}$$

which is equivalent to the basic inequality

$$I(X_1; X_2) \geq 0. \tag{16.2}$$

Let G be any finite group and G_1 and G_2 be subgroups of G. We will show in Section 16.4 that

$$|G||G_1 \cap G_2| \geq |G_1||G_2|, \tag{16.3}$$

where $|G|$ denotes the *order* of G and $G_1 \cap G_2$ denotes the *intersection* of G_1 and G_2 ($G_1 \cap G_2$ is also a subgroup of G, see Proposition 16.13). By rearranging the terms, the above inequality can be written as

$$\log \frac{|G|}{|G_1|} + \log \frac{|G|}{|G_2|} \geq \log \frac{|G|}{|G_1 \cap G_2|}. \tag{16.4}$$

By comparing (16.1) and (16.4), one can easily identify the one-to-one correspondence between these two inequalities, namely that X_i corresponds to G_i, $i = 1, 2$, and (X_1, X_2) corresponds to $G_1 \cap G_2$. While (16.1) is true for any pair of random variables X_1 and X_2, (16.4) is true for any finite group G and subgroups G_1 and G_2.

Recall from Chapter 13 that the region Γ_n^* characterizes all information inequalities (involving n random variables). In particular, we have shown in Section 15.1 that the region $\overline{\Gamma}_n^*$ is sufficient for characterizing all unconstrained information inequalities, i.e., by knowing $\overline{\Gamma}_n^*$, one can determine whether any unconstrained information inequality always holds. The main purpose of this chapter is to obtain a characterization of $\overline{\Gamma}_n^*$ in terms of finite groups. An important consequence of this result is a one-to-one correspondence between unconstrained information inequalities and group inequalities. Specifically, for every unconstrained information inequality, there is a corresponding group inequality and vice versa. A special case of this correspondence has been given in (16.1) and (16.4).

By means of this result, unconstrained information inequalities can be proved by techniques in group theory, and a certain form of inequalities in group theory can be proved by techniques in information theory. In particular, the unconstrained non-Shannon-type inequality in Theorem 15.7 corresponds to the group inequality

$$|G_1 \cap G_3|^3 |G_1 \cap G_4|^3 |G_3 \cap G_4|^3 |G_2 \cap G_3||G_2 \cap G_4|$$

$$\leq |G_1||G_1 \cap G_2||G_3|^2|G_4|^2|G_1 \cap G_3 \cap G_4|^4|G_2 \cap G_3 \cap G_4|, \qquad (16.5)$$

where G_i are subgroups of a finite group G, $i = 1, 2, 3, 4$. The meaning of this inequality and its implications in group theory are yet to be understood.

16.1 Group Preliminaries

In this section, we present the definition and some basic properties of a group which are essential for subsequent discussions.

Definition 16.1. *A group is a set of objects G together with a binary operation on the elements of G, denoted by "$\circ$" unless otherwise specified, which satisfy the following four axioms:*

1. Closure *For every a, b in G, $a \circ b$ is also in G.*
2. Associativity *For every a, b, c in G, $a \circ (b \circ c) = (a \circ b) \circ c$.*
3. Existence of Identity *There exists an element e in G such that $a \circ e = e \circ a$ $= a$ for every a in G.*
4. Existence of Inverse *For every a in G, there exists an element b in G such that $a \circ b = b \circ a = e$.*

Proposition 16.2. *For any group G, the identity element is unique.*

Proof. Let both e and e' be identity elements in a group G. Since e is an identity element,

$$e' \circ e = e, \qquad (16.6)$$

and since e' is also an identity element,

$$e' \circ e = e'. \qquad (16.7)$$

It follows by equating the right-hand sides of (16.6) and (16.7) that $e = e'$, which implies the uniqueness of the identity element of a group. $\square$

Proposition 16.3. *For every element a in a group G, its inverse is unique.*

Proof. Let b and b' be inverses of an element a, so that

$$a \circ b = b \circ a = e \tag{16.8}$$

and

$$a \circ b' = b' \circ a = e. \tag{16.9}$$

Then

$$b = b \circ e \tag{16.10}$$
$$= b \circ (a \circ b') \tag{16.11}$$
$$= (b \circ a) \circ b' \tag{16.12}$$
$$= e \circ b' \tag{16.13}$$
$$= b', \tag{16.14}$$

where (16.11) and (16.13) follow from (16.9) and (16.8), respectively, and (16.12) is by associativity. Therefore, the inverse of a is unique. $\square$

Thus the inverse of a group element a is a function of a, and it will be denoted by a^{-1}.

Definition 16.4. *The number of elements of a group G is called the order of G, denoted by $|G|$. If $|G| < \infty$, G is called a finite group, otherwise it is called an infinite group.*

There is an unlimited supply of examples of groups. Some familiar examples are the integers under addition, the rationals excluding zero under multiplication, and the set of real-valued 2×2 matrices under addition, where addition and multiplication refer to the usual addition and multiplication for real numbers and matrices. In each of these examples, the operation (addition or multiplication) plays the role of the binary operation "$\circ$" in Definition 16.1.

All the above are examples of infinite groups. In this chapter, however, we are concerned with finite groups. In the following, we discuss two examples of finite groups in details.

Example 16.5 (Modulo 2 Addition). The trivial group consists of only the identity element. The simplest nontrivial group is the group of modulo 2 addition. The order of this group is 2, and the elements are $\{0, 1\}$. The binary operation, denoted by "$+$", is defined by following table:

+	0	1
0	0	1
1	1	0

The four axioms of a group simply say that certain constraints must hold in the above table. We now check that all these axioms are satisfied. First, the closure axiom requires that all the entries in the table are elements in the group, which is easily seen to be the case. Second, it is required that associativity holds. To this end, it can be checked in the above table that for all a, b, and c,

$$a + (b + c) = (a + b) + c. \tag{16.15}$$

For example,

$$0 + (1 + 1) = 0 + 0 = 0, \tag{16.16}$$

while

$$(0 + 1) + 1 = 1 + 1 = 0, \tag{16.17}$$

which is the same as $0 + (1 + 1)$. Third, the element 0 is readily identified as the unique identity. Fourth, it is readily seen that an inverse exists for each element in the group. For example, the inverse of 1 is 1, because

$$1 + 1 = 0. \tag{16.18}$$

Thus the above table defines a group of order 2. It happens in this example that the inverse of each element is the element itself, which is not true for a group in general.

We remark that in the context of a group, the elements in the group should be regarded strictly as *symbols* only. In particular, one should not associate group elements with *magnitudes* as we do for real numbers. For instance, in the above example, one should not think of 0 as being less than 1. The element 0, however, is a special symbol which plays the role of the identity of the group.

We also notice that for the group in the above example, $a + b$ is equal to $b + a$ for all group elements a and b. A group with this property is called a *commutative* group or an *Abelian* group.[1]

Example 16.6 (Symmetric Group). Consider a permutation of the components of a vector

$$\mathbf{x} = (x_1, x_2, \cdots, x_r) \tag{16.19}$$

given by

$$\sigma[\mathbf{x}] = (x_{\sigma(1)}, x_{\sigma(2)}, \cdots, x_{\sigma(r)}), \tag{16.20}$$

where

$$\sigma : \{1, 2, \cdots, r\} \rightarrow \{1, 2, \cdots, r\} \tag{16.21}$$

is a one-to-one mapping. The one-to-one mapping σ is called a permutation on $\{1, 2, \cdots, r\}$, which is represented by

$$\sigma = (\sigma(1), \sigma(2), \cdots, \sigma(r)). \tag{16.22}$$

---

[1] The Abelian group is named after the Norwegian mathematician Niels Henrik Abel (1802–1829).

For two permutations σ_1 and σ_2, define $\sigma_1 \circ \sigma_2$ as the composite function of σ_1 and σ_2. For example, for $r = 4$, suppose

$$\sigma_1 = (2, 1, 4, 3) \tag{16.23}$$

and

$$\sigma_2 = (1, 4, 2, 3). \tag{16.24}$$

Then $\sigma_1 \circ \sigma_2$ is given by

$$\begin{aligned}
\sigma_1 \circ \sigma_2(1) &= \sigma_1(\sigma_2(1)) = \sigma_1(1) = 2, \\
\sigma_1 \circ \sigma_2(2) &= \sigma_1(\sigma_2(2)) = \sigma_1(4) = 3, \\
\sigma_1 \circ \sigma_2(3) &= \sigma_1(\sigma_2(3)) = \sigma_1(2) = 1, \\
\sigma_1 \circ \sigma_2(4) &= \sigma_1(\sigma_2(4)) = \sigma_1(3) = 4,
\end{aligned} \tag{16.25}$$

or

$$\sigma_1 \circ \sigma_2 = (2, 3, 1, 4). \tag{16.26}$$

The reader can easily check that

$$\sigma_2 \circ \sigma_1 = (4, 1, 2, 3), \tag{16.27}$$

which is different from $\sigma_1 \circ \sigma_2$. Therefore, the operation "$\circ$" is not commutative.

We now show that the set of all permutations on $\{1, 2, \cdots, r\}$ and the operation "$\circ$" form a group, called the symmetric group on $\{1, 2, \cdots, r\}$. First, for two permutations σ_1 and σ_2, since both σ_1 and σ_2 are one-to-one mappings, so is $\sigma_1 \circ \sigma_2$. Therefore, the closure axiom is satisfied. Second, for permutations σ_1, σ_2, and σ_3,

$$\sigma_1 \circ (\sigma_2 \circ \sigma_3)(i) = \sigma_1(\sigma_2 \circ \sigma_3(i)) \tag{16.28}$$
$$= \sigma_1(\sigma_2(\sigma_3(i))) \tag{16.29}$$
$$= \sigma_1 \circ \sigma_2(\sigma_3(i)) \tag{16.30}$$
$$= (\sigma_1 \circ \sigma_2) \circ \sigma_3(i) \tag{16.31}$$

for $1 \leq i \leq r$. Therefore, associativity is satisfied. Third, it is clear that the identity map is the identity element. Fourth, for a permutation σ, it is clear that its inverse is σ^{-1}, the inverse mapping of σ which is defined because σ is one-to-one. Therefore, the set of all permutations on $\{1, 2, \cdots, r\}$ and the operation "$\circ$" form a group. The order of this group is evidently equal to $(r!)$.

Definition 16.7. *Let G be a group with operation "$\circ$", and S be a subset of G. If S is a group with respect to the operation "$\circ$", then S is called a subgroup of G.*

Definition 16.8. *Let S be a subgroup of a group G and a be an element of G. The left coset of S with respect to a is the set $a \circ S = \{a \circ s : s \in S\}$. Similarly, the right coset of S with respect to a is the set $S \circ a = \{s \circ a : s \in S\}$.*

In the sequel, only the left coset will be used. However, any result which applies to the left coset also applies to the right coset and vice versa. For simplicity, $a \circ S$ will be denoted by aS.

Proposition 16.9. *For a_1 and a_2 in G, $a_1 S$ and $a_2 S$ are either identical or disjoint. Further, $a_1 S$ and $a_2 S$ are identical if and only if a_1 and a_2 belong to the same left coset of S.*

Proof. Suppose $a_1 S$ and $a_2 S$ are not disjoint. Then there exists an element b in $a_1 S \cap a_2 S$ such that

$$b = a_1 \circ s_1 = a_2 \circ s_2, \tag{16.32}$$

for some s_i in S, $i = 1, 2$. Then

$$a_1 = (a_2 \circ s_2) \circ s_1^{-1} = a_2 \circ (s_2 \circ s_1^{-1}) = a_2 \circ t, \tag{16.33}$$

where $t = s_2 \circ s_1^{-1}$ is in S. We now show that $a_1 S \subset a_2 S$. For an element $a_1 \circ s$ in $a_1 S$, where $s \in S$,

$$a_1 \circ s = (a_2 \circ t) \circ s = a_2 \circ (t \circ s) = a_2 \circ u, \tag{16.34}$$

where $u = t \circ s$ is in S. This implies that $a_1 \circ s$ is in $a_2 S$. Thus, $a_1 S \subset a_2 S$. By symmetry, $a_2 S \subset a_1 S$. Therefore, $a_1 S = a_2 S$. Hence, if $a_1 S$ and $a_2 S$ are not disjoint, then they are identical. Equivalently, $a_1 S$ and $a_2 S$ are either identical or disjoint. This proves the first part of the proposition.

We now prove the second part of the proposition. Since S is a group, it contains e, the identity element. Then for any group element a, $a = a \circ e$ is in aS because e is in S. If $a_1 S$ and $a_2 S$ are identical, then $a_1 \in a_1 S$ and $a_2 \in a_2 S = a_1 S$. Therefore, a_1 and a_2 belong to the same left coset of S.

To prove the converse, assume a_1 and a_2 belong to the same left coset of S. From the first part of the proposition, we see that a group element belongs to one and only one left coset of S. Since a_1 is in $a_1 S$ and a_2 is in $a_2 S$, and a_1 and a_2 belong to the same left coset of S, we see that $a_1 S$ and $a_2 S$ are identical. The proposition is proved. $\square$

Proposition 16.10. *Let S be a subgroup of a group G and a be an element of G. Then $|aS| = |S|$, i.e., the numbers of elements in all the left cosets of S are the same, and they are equal to the order of S.*

Proof. Consider two elements $a \circ s_1$ and $a \circ s_2$ in $a \circ S$, where s_1 and s_2 are in S such that

$$a \circ s_1 = a \circ s_2. \tag{16.35}$$

Then

$$a^{-1} \circ (a \circ s_1) = a^{-1} \circ (a \circ s_2), \tag{16.36}$$

$$(a^{-1} \circ a) \circ s_1 = (a^{-1} \circ a) \circ s_2, \tag{16.37}$$

$$e \circ s_1 = e \circ s_2, \tag{16.38}$$

$$s_1 = s_2. \tag{16.39}$$

Thus each element in S corresponds to a unique element in aS. Therefore, $|aS| = |S|$ for all $a \in G$. $\square$

We are just one step away from obtaining the celebrated *Lagrange's theorem* stated below.

Theorem 16.11 (Lagrange's Theorem). *If S is a subgroup of a finite group G, then $|S|$ divides $|G|$.*

Proof. Since $a \in aS$ for every $a \in G$, every element of G belongs to a left coset of S. Then from Proposition 16.9, we see that the distinct left cosets of S partition G. Therefore $|G|$, the total number of elements in G, is equal to the number of distinct cosets of S multiplied by the number of elements in each left coset, which is equal to $|S|$ by Proposition 16.10. This implies that $|S|$ divides $|G|$, proving the theorem. $\square$

The following corollary is immediate from the proof of Lagrange's Theorem.

Corollary 16.12. *Let S be a subgroup of a group G. The number of distinct left cosets of S is equal to $\frac{|G|}{|S|}$.*

16.2 Group-Characterizable Entropy Functions

Recall from Chapter 13 that the region Γ_n^* consists of all the entropy functions in the entropy space $\mathcal{H}_n$ for n random variables. As a first step toward establishing the relation between entropy and groups, we discuss in this section entropy functions in Γ_n^* which can be described by a finite group G and subgroups $G_1, G_2, \cdots, G_n$. Such entropy functions are said to be *group-characterizable*. The significance of this class of entropy functions will become clear in the next section.

In the sequel, we will make use of the intersections of subgroups extensively. We first prove that the intersection of two subgroups is also a subgroup.

Proposition 16.13. *Let G_1 and G_2 be subgroups of a group G. Then $G_1 \cap G_2$ is also a subgroup of G.*

Proof. It suffices to show that $G_1 \cap G_2$ together with the operation "$\circ$" satisfy all the axioms of a group. First, consider two elements a and b of G in $G_1 \cap G_2$. Since both a and b are in G_1, $(a \circ b)$ is in G_1. Likewise, $(a \circ b)$ is in G_2. Therefore, $a \circ b$ is in $G_1 \cap G_2$. Thus the closure axiom holds for $G_1 \cap G_2$. Second, associativity for $G_1 \cap G_2$ inherits from G. Third, G_1 and G_2 both contain the identity element because they are groups. Therefore, the identity element is in $G_1 \cap G_2$. Fourth, for an element $a \in G_i$, since G_i is a group, a^{-1} is in G_i, $i = 1, 2$. Thus for an element $a \in G_1 \cap G_2$, a^{-1} is also in $G_1 \cap G_2$. Therefore, $G_1 \cap G_2$ is a group and hence a subgroup of G. $\square$

Corollary 16.14. *Let $G_1, G_2, \cdots, G_n$ be subgroups of a group G. Then $\cap_{i=1}^{n} G_i$ is also a subgroup of G.*

In the rest of the chapter, we let $\mathcal{N}_n = \{1, 2, \cdots, n\}$ and denote $\cap_{i \in \alpha} G_i$ by G_α, where α is a nonempty subset of $\mathcal{N}_n$.

Lemma 16.15. *Let G_i be subgroups of a group G and a_i be elements of G, $i \in \alpha$. Then*

$$|\cap_{i \in \alpha} a_i G_i| = \begin{cases} |G_\alpha| & \text{if } \bigcap_{i \in \alpha} a_i G_i \neq \emptyset \\ 0 & \text{otherwise} \end{cases}. \tag{16.40}$$

Proof. For the special case that α is a singleton, i.e., $\alpha = \{i\}$ for some $i \in \mathcal{N}_n$, (16.40) reduces to

$$|a_i G_i| = |G_i|, \tag{16.41}$$

which has already been proved in Proposition 16.10.

Let α be any nonempty subset of $\mathcal{N}_n$. If $\bigcap_{i \in \alpha} a_i G_i = \emptyset$, then (16.40) is obviously true. If $\bigcap_{i \in \alpha} a_i G_i \neq \emptyset$, then there exists $x \in \bigcap_{i \in \alpha} a_i G_i$ such that for all $i \in \alpha$,

$$x = a_i \circ s_i, \tag{16.42}$$

where $s_i \in G_i$. For any $i \in \alpha$ and for any $y \in G_\alpha$, consider

$$x \circ y = (a_i \circ s_i) \circ y = a_i \circ (s_i \circ y). \tag{16.43}$$

Since both s_i and y are in G_i, $s_i \circ y$ is in G_i. Thus $x \circ y$ is in $a_i G_i$ for all $i \in \alpha$ or $x \circ y$ is in $\bigcap_{i \in \alpha} a_i G_i$. Moreover, for $y, y' \in G_\alpha$, if $x \circ y = x \circ y'$, then $y = y'$. Therefore, each element in G_α corresponds to a unique element in $\bigcap_{i \in \alpha} a_i G_i$. Hence,

$$|\cap_{i \in \alpha} a_i G_i| = |G_\alpha|, \tag{16.44}$$

proving the lemma. $\square$

The relation between a finite group G and subgroups G_1 and G_2 is illustrated by the *membership table* in Figure 16.1. In this table, an element of G is represented by a dot. The first column represents the subgroup G_1, with the dots in the first column being the elements in G_1. The other columns represent the left cosets of G_1. By Proposition 16.10, all the columns have the same number of dots. Similarly, the first row represents the subgroup G_2 and the other rows represent the left cosets of G_2. Again, all the rows have the same number of dots.

The upper left entry in the table represents the subgroup $G_1 \cap G_2$. There are $|G_1 \cap G_2|$ dots in this entry, with one of them representing the identity element. Any other entry represents the intersection between a left coset of G_1 and a left coset of G_2, and by Lemma 16.15, the number of dots in each of these entries is either equal to $|G_1 \cap G_2|$ or zero.

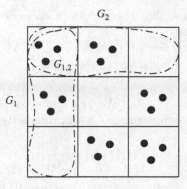

Fig. 16.1. The membership table for a finite group G and subgroups G_1 and G_2.

Since all the columns have the same numbers of dots and all the rows have the same number of dots, we say that the table in Figure 16.1 exhibits a *quasi-uniform* structure. We have already seen a similar structure in Figure 6.2 for the two-dimensional strong joint typicality array, which we reproduce in Figure 16.2. In this array, when n is large, all the columns have approximately the same number of dots and all the rows have approximately the same number of dots. For this reason, we say that the two-dimensional strong typicality array exhibits an *asymptotic* quasi-uniform structure. In a strong typicality array, however, each entry can contain only one dot, while in a membership table, each entry can contain multiple dots.

One can make a similar comparison between a strong joint typicality array for any $n \geq 2$ random variables and the membership table for a finite group with n subgroups. The details are omitted here.

Theorem 16.16. *Let $G_i, i \in \mathcal{N}_n$ be subgroups of a group G. Then $\mathbf{h} \in \mathcal{H}_n$ defined by*

$$h_\alpha = \log \frac{|G|}{|G_\alpha|} \tag{16.45}$$

for all nonempty subsets α of $\mathcal{N}_n$ is entropic, i.e., $\mathbf{h} \in \Gamma_n^*$.

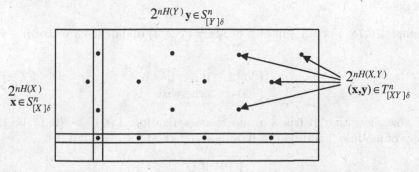

Fig. 16.2. A two-dimensional strong typicality array.

Proof. It suffices to show that there exists a collection of random variables $X_1, X_2, \cdots, X_n$ such that

$$H(X_\alpha) = \log \frac{|G|}{|G_\alpha|} \tag{16.46}$$

for all nonempty subsets α of $\mathcal{N}_n$. We first introduce a uniform random variable Λ defined on the sample space G with probability mass function

$$\Pr\{\Lambda = a\} = \frac{1}{|G|} \tag{16.47}$$

for all $a \in G$. For any $i \in \mathcal{N}_n$, let random variable X_i be a function of Λ such that $X_i = aG_i$ if $\Lambda = a$.

Let α be a nonempty subset of $\mathcal{N}_n$. Since $X_i = a_iG_i$ for all $i \in \alpha$ if and only if Λ is equal to some $b \in \cap_{i \in \alpha} a_i G_i$,

$$\Pr\{X_i = a_iG_i : i \in \alpha\} = \frac{|\bigcap_{i \in \alpha} a_iG_i|}{|G|} \tag{16.48}$$

$$= \begin{cases} \frac{|G_\alpha|}{|G|} & \text{if } \bigcap_{i \in \alpha} a_iG_i \neq \emptyset \\ 0 & \text{otherwise} \end{cases} \tag{16.49}$$

by Lemma 16.15. In other words, $(X_i, i \in \alpha)$ is distributed uniformly on its support whose cardinality is $\frac{|G|}{|G_\alpha|}$. Then (16.46) follows and the theorem is proved. □

Definition 16.17. *Let G be a finite group and $G_1, G_2, \cdots, G_n$ be subgroups of G. Let $\mathbf{h}$ be a vector in $\mathcal{H}_n$. If $h_\alpha = \log \frac{|G|}{|G_\alpha|}$ for all nonempty subsets α of $\mathcal{N}_n$, then $(G, G_1, \cdots, G_n)$ is a group characterization of $\mathbf{h}$.*

Theorem 16.16 asserts that certain entropy functions in Γ_n^* have a group characterization. These are called group-characterizable entropy functions, which will be used in the next section to obtain a group characterization of the region $\overline{\Gamma}_n^*$. We end this section by giving a few examples of such entropy functions.

Example 16.18. Fix any subset β of $\mathcal{N}_3 = \{1, 2, 3\}$ and define a vector $\mathbf{h} \in \mathcal{H}_3$ by

$$h_\alpha = \begin{cases} \log 2 & \text{if } \alpha \cap \beta \neq \emptyset \\ 0 & \text{otherwise} \end{cases} . \tag{16.50}$$

We now show that $\mathbf{h}$ has a group characterization. Let $G = \{0, 1\}$ be the group of modulo 2 addition in Example 16.5, and for $i = 1, 2, 3$, let

$$G_i = \begin{cases} \{0\} & \text{if } i \in \beta \\ G & \text{otherwise} \end{cases} . \tag{16.51}$$

Then for a nonempty subset α of $\mathcal{N}_3$, if $\alpha \cap \beta \neq \emptyset$, there exists an i in α such that i is also in β, and hence by definition $G_i = \{0\}$. Thus,

$$G_\alpha = \bigcap_{i \in \alpha} G_i = \{0\}. \tag{16.52}$$

Therefore,

$$\log \frac{|G|}{|G_\alpha|} = \log \frac{|G|}{|\{0\}|} = \log \frac{2}{1} = \log 2. \tag{16.53}$$

If $\alpha \cap \beta = \emptyset$, then $G_i = G$ for all $i \in \alpha$, and

$$G_\alpha = \bigcap_{i \in \alpha} G_i = G. \tag{16.54}$$

Therefore,

$$\log \frac{|G|}{|G_\alpha|} = \log \frac{|G|}{|G|} = \log 1 = 0. \tag{16.55}$$

Then we see from (16.50), (16.53), and (16.55) that

$$h_\alpha = \log \frac{|G|}{|G_\alpha|} \tag{16.56}$$

for all nonempty subsets α of $\mathcal{N}_3$. Hence, (G, G_1, G_2, G_3) is a group characterization of $\mathbf{h}$.

Example 16.19. This is a generalization of the last example. Fix any nonempty subset β of $\mathcal{N}_n$ and define a vector $\mathbf{h} \in \mathcal{H}_n$ by

$$h_\alpha = \begin{cases} \log 2 & \text{if } \alpha \cap \beta \neq \emptyset \\ 0 & \text{otherwise} \end{cases}. \tag{16.57}$$

Then $(G, G_1, G_2, \cdots, G_n)$ is a group characterization of $\mathbf{h}$, where G is the group of modulo 2 addition, and

$$G_i = \begin{cases} \{0\} & \text{if } i \in \beta \\ G & \text{otherwise} \end{cases}. \tag{16.58}$$

By letting $\beta = \emptyset$, we have $\mathbf{h} = 0$. Thus we see that $(G, G_1, G_2, \cdots, G_n)$ is a group characterization of the origin of $\mathcal{H}_n$, with $G = G_1 = G_2 = \cdots = G_n$.

Example 16.20. Define a vector $\mathbf{h} \in \mathcal{H}_3$ as follows:

$$h_\alpha = \min(|\alpha|, 2). \tag{16.59}$$

Let F be the group of modulo 2 addition, $G = F \times F$, and

$$G_1 = \{(0,0), (1,0)\}, \tag{16.60}$$
$$G_2 = \{(0,0), (0,1)\}, \tag{16.61}$$
$$G_3 = \{(0,0), (1,1)\}. \tag{16.62}$$

Then (G, G_1, G_2, G_3) is a group characterization of $\mathbf{h}$.

16.3 A Group Characterization of $\overline{\Gamma}_n^*$

We have introduced in the last section the class of entropy functions in Γ_n^* which have a group characterization. However, an entropy function $\mathbf{h} \in \Gamma_n^*$ may not have a group characterization due to the following observation. Suppose $\mathbf{h} \in \Gamma_n^*$. Then there exists a collection of random variables $X_1, X_2, \cdots, X_n$ such that

$$h_\alpha = H(X_\alpha) \tag{16.63}$$

for all nonempty subsets α of $\mathcal{N}_n$. If $(G, G_1, \cdots, G_n)$ is a group characterization of $\mathbf{h}$, then

$$H(X_\alpha) = \log \frac{|G|}{|G_\alpha|} \tag{16.64}$$

for all nonempty subsets of $\mathcal{N}_n$. Since both $|G|$ and $|G_\alpha|$ are integers, $H(X_\alpha)$ must be the logarithm of a rational number. However, the joint entropy of a set of random variables in general is not necessarily the logarithm of a rational number (see Corollary 2.44). Therefore, it is possible to construct an entropy function $\mathbf{h} \in \Gamma_n^*$ which has no group characterization.

Although $\mathbf{h} \in \Gamma_n^*$ does not imply $\mathbf{h}$ has a group characterization, it turns out that the set of all $\mathbf{h} \in \Gamma_n^*$ which have a group characterization is almost good enough to characterize the region Γ_n^*, as we will see next.

Definition 16.21. *Define the following region in* $\mathcal{H}_n$:

$$\Upsilon_n = \{\mathbf{h} \in \mathcal{H}_n : \mathbf{h} \text{ has a group characterization}\}. \tag{16.65}$$

By Theorem 16.16, if $\mathbf{h} \in \mathcal{H}_n$ has a group characterization, then $\mathbf{h} \in \Gamma_n^*$. Therefore, $\Upsilon_n \subset \Gamma_n^*$. We will prove as a corollary of the next theorem that $\overline{con}(\Upsilon_n)$, the convex closure of Υ_n, is in fact equal to $\overline{\Gamma}_n^*$, the closure of Γ_n^*.

Theorem 16.22. *For any* $\mathbf{h} \in \overline{\Gamma}_n^*$, *there exists a sequence* $\{\mathbf{f}^{(r)}\}$ *in* Υ_n *such that* $\lim_{r \to \infty} \frac{1}{r} \mathbf{f}^{(r)} = \mathbf{h}$.

We need the following lemma to prove this theorem. The proof of this lemma resembles the proof of the strong conditional AEP (Theorem 6.10). Nevertheless, we give a sketch of the proof for the sake of completeness.

Lemma 16.23. *Let X be a random variable such that $|\mathcal{X}| < \infty$ and the distribution $\{p(x)\}$ is rational, i.e., $p(x)$ is a rational number for all $x \in \mathcal{X}$. Without loss of generality, assume $p(x)$ is a rational number with denominator q for all $x \in \mathcal{X}$. Then for $r = q, 2q, 3q, \cdots$,*

$$\lim_{r \to \infty} \frac{1}{r} \log \frac{r!}{\prod_x (rp(x))!} = H(X). \tag{16.66}$$

Proof. Applying Lemma 6.11, we can obtain

$$\frac{1}{r}\ln\frac{r!}{\prod_x(rp(x))!}$$

$$\leq -\sum_x p(x)\ln p(x) + \frac{r+1}{r}\ln(r+1) - \ln r \tag{16.67}$$

$$= H_e(X) + \frac{1}{r}\ln r + \left(1+\frac{1}{r}\right)\ln\left(1+\frac{1}{r}\right). \tag{16.68}$$

This upper bound tends to $H_e(X)$ as $r \to \infty$. On the other hand, we can obtain

$$\frac{1}{r}\ln\frac{r!}{\prod_x(rp(x))!}$$

$$\geq -\sum_x \left(p(x)+\frac{1}{r}\right)\ln\left(p(x)+\frac{1}{r}\right) - \frac{\ln r}{r}. \tag{16.69}$$

This lower bound also tends to $H_e(X)$ as $r \to \infty$. Then the proof is completed by changing the base of the logarithm if necessary. $\square$

Proof of Theorem 16.22. For any $\mathbf{h} \in \Gamma_n^*$, there exists a collection of random variables $X_1, X_2, \cdots, X_n$ such that

$$h_\alpha = H(X_\alpha) \tag{16.70}$$

for all nonempty subsets α of $\mathcal{N}_n$. We first consider the special case that $|\mathcal{X}_i| < \infty$ for all $i \in \mathcal{N}_n$ and the joint distribution of $X_1, X_2, \cdots, X_n$ is rational. We want to show that there exists a sequence $\{\mathbf{f}^{(r)}\}$ in Υ_n such that $\lim_{r\to\infty}\frac{1}{r}\mathbf{f}^{(r)} = \mathbf{h}$.

Denote $\prod_{i\in\alpha}\mathcal{X}_i$ by $\mathcal{X}_\alpha$. For any nonempty subset α of $\mathcal{N}_n$, let Q_α be the marginal distribution of X_α. Assume without loss of generality that for any nonempty subset α of $\mathcal{N}_n$ and for all $a \in \mathcal{X}_\alpha$, $Q_\alpha(a)$ is a rational number with denominator q.

For each $r = q, 2q, 3q, \cdots$, fix a sequence

$$\mathbf{x}_{\mathcal{N}_n} = (x_{\mathcal{N}_n,1}, x_{\mathcal{N}_n,2}, \cdots x_{\mathcal{N}_n,r}),$$

where for all $j = 1, 2, \cdots, r$, $x_{\mathcal{N}_n,j} = (x_{i,j} : i \in \mathcal{N}_n) \in \mathcal{X}_{\mathcal{N}_n}$, such that $N(a; \mathbf{x}_{\mathcal{N}_n})$, the number of occurrences of a in sequence $\mathbf{x}_{\mathcal{N}_n}$, is equal to $rQ_{\mathcal{N}_n}(a)$ for all $a \in \mathcal{X}_{\mathcal{N}_n}$. The existence of such a sequence is guaranteed by that all the values of the joint distribution of $X_{\mathcal{N}_n}$ are rational numbers with denominator q. Also, we denote the sequence of r elements of $\mathcal{X}_\alpha$, $(x_{\alpha,1}, x_{\alpha,2}, \cdots x_{\alpha,r})$, where $x_{\alpha,j} = (x_{i,j} : i \in \alpha)$, by $\mathbf{x}_\alpha$. Let $a \in \mathcal{X}_\alpha$. It is easy to check that $N(a; \mathbf{x}_\alpha)$, the number of occurrences of a in the sequence $\mathbf{x}_\alpha$, is equal to $rQ_\alpha(a)$ for all $a \in \mathcal{X}_\alpha$.

Let G be the group of all permutations on $\{1, 2, \cdots, r\}$, i.e., the symmetric group on $\{1, 2, \cdots, r\}$ (cf. Example 16.6). The group G depends on r, but for simplicity, we do not state this dependency explicitly. For any $i \in \mathcal{N}_n$, define

$$G_i = \{\sigma \in G : \sigma[\mathbf{x}_i] = \mathbf{x}_i\},$$

where

$$\sigma[\mathbf{x}_i] = (x_{i,\sigma(1)}, x_{i,\sigma(2)}, \cdots, x_{i,\sigma(r)}). \tag{16.71}$$

It is easy to check that G_i is a subgroup of G.

Let α be a nonempty subset of $\mathcal{N}_n$. Then

$$G_\alpha = \bigcap_{i \in \alpha} G_i \tag{16.72}$$

$$= \bigcap_{i \in \alpha} \{\sigma \in G : \sigma[\mathbf{x}_i] = \mathbf{x}_i\} \tag{16.73}$$

$$= \{\sigma \in G : \sigma[\mathbf{x}_i] = \mathbf{x}_i \text{ for all } i \in \alpha\} \tag{16.74}$$

$$= \{\sigma \in G : \sigma[\mathbf{x}_\alpha] = \mathbf{x}_\alpha\}, \tag{16.75}$$

where

$$\sigma[\mathbf{x}_\alpha] = (x_{\alpha,\sigma(1)}, x_{\alpha,\sigma(2)}, \cdots, x_{\alpha,\sigma(r)}). \tag{16.76}$$

For any $a \in \mathcal{X}_\alpha$, define the set

$$L_{\mathbf{x}_\alpha}(a) = \{j \in \{1, 2, \cdots, r\} : x_{\alpha,j} = a\}. \tag{16.77}$$

$L_{\mathbf{x}_\alpha}(a)$ contains the "locations" of a in $\mathbf{x}_\alpha$. Then $\sigma[\mathbf{x}_\alpha] = \mathbf{x}_\alpha$ if and only if for all $a \in \mathcal{X}_\alpha$, $j \in L_{\mathbf{x}_\alpha}(a)$ implies $\sigma(j) \in L_{\mathbf{x}_\alpha}(a)$. Since

$$|L_{\mathbf{x}_\alpha}(a)| = N(a; \mathbf{x}_\alpha) = rQ_\alpha(a), \tag{16.78}$$

$$|G_\alpha| = \prod_{a \in \mathcal{X}_\alpha} (rQ_\alpha(a))! \tag{16.79}$$

and therefore

$$\frac{|G|}{|G_\alpha|} = \frac{r!}{\prod_{a \in \mathcal{X}_\alpha} (rQ_\alpha(a))!}. \tag{16.80}$$

By Lemma 16.23,

$$\lim_{r \to \infty} \frac{1}{r} \log \frac{|G|}{|G_\alpha|} = H(X_\alpha) = h_\alpha. \tag{16.81}$$

Recall that G and hence all its subgroups depend on r. Define $\mathbf{f}^{(r)}$ by

$$f_\alpha^{(r)} = \log \frac{|G|}{|G_\alpha|} \tag{16.82}$$

for all nonempty subsets α of $\mathcal{N}_n$. Then $\mathbf{f}^{(r)} \in \Upsilon_n$ and

$$\lim_{r \to \infty} \frac{1}{r} \mathbf{f}^{(r)} = \mathbf{h}. \tag{16.83}$$

We have already proved the theorem for the special case that $\mathbf{h}$ is the entropy function of a collection of random variables $X_1, X_2, \cdots, X_n$ with finite alphabets and a rational joint distribution. To complete the proof, we only have to note that for any $\mathbf{h} \in \Gamma_n^*$, it is always possible to construct a sequence $\{\mathbf{h}^{(k)}\}$ in Γ_n^* such that $\lim_{k \to \infty} \mathbf{h}^{(k)} = \mathbf{h}$, where $\mathbf{h}^{(k)}$ is the entropy function of a collection of random variables $X_1^{(k)}, X_2^{(k)}, \cdots, X_n^{(k)}$ with finite alphabets and a rational joint distribution. This can be proved by techniques similar to those used in Appendix 2.A together with the continuity of the entropy function for a fixed finite support (Section 2.3). The details are omitted here. $\quad\square$

Corollary 16.24. $\overline{con}(\Upsilon_n) = \overline{\Gamma}_n^*$.

Proof. First of all, $\Upsilon_n \subset \Gamma_n^*$. By taking convex closure, we have $\overline{con}(\Upsilon_n) \subset \overline{con}(\Gamma_n^*)$. By Theorem 15.5, $\overline{\Gamma}_n^*$ is convex. Therefore, $\overline{con}(\Gamma_n^*) = \overline{\Gamma}_n^*$, and we have $\overline{con}(\Upsilon_n) \subset \overline{\Gamma}_n^*$. On the other hand, we have shown in Example 16.19 that the origin of $\mathcal{H}_n$ has a group characterization and therefore is in Υ_n. It then follows from Theorem 16.22 that $\overline{\Gamma}_n^* \subset \overline{con}(\Upsilon_n)$. Hence, we conclude that $\overline{\Gamma}_n^* = \overline{con}(\Upsilon_n)$, completing the proof. $\quad\square$

16.4 Information Inequalities and Group Inequalities

We have proved in Section 15.1 that an unconstrained information inequality

$$\mathbf{b}^\top \mathbf{h} \geq 0 \tag{16.84}$$

always holds if and only if

$$\overline{\Gamma}_n^* \subset \{\mathbf{h} \in \mathcal{H}_n : \mathbf{b}^\top \mathbf{h} \geq 0\}. \tag{16.85}$$

In other words, all unconstrained information inequalities are fully characterized by $\overline{\Gamma}_n^*$. We have also proved at the end of the last section that $\overline{con}(\Upsilon_n) = \overline{\Gamma}_n^*$. Since $\Upsilon_n \subset \Gamma_n^* \subset \overline{\Gamma}_n^*$, if (16.85) holds, then

$$\Upsilon_n \subset \{\mathbf{h} \in \mathcal{H}_n : \mathbf{b}^\top \mathbf{h} \geq 0\}. \tag{16.86}$$

On the other hand, if (16.86) holds, since $\{\mathbf{h} \in \mathcal{H}_n : \mathbf{b}^\top \mathbf{h} \geq 0\}$ is closed and convex, by taking convex closure in (16.86), we obtain

$$\overline{\Gamma}_n^* = \overline{con}(\Upsilon_n) \subset \{\mathbf{h} \in \mathcal{H}_n : \mathbf{b}^\top \mathbf{h} \geq 0\}. \tag{16.87}$$

Therefore, (16.85) and (16.86) are equivalent.

Now (16.86) is equivalent to

$$\mathbf{b}^\top \mathbf{h} \geq 0 \text{ for all } \mathbf{h} \in \Upsilon_n. \tag{16.88}$$

Since $\mathbf{h} \in \Upsilon_n$ if and only if

$$h_\alpha = \log \frac{|G|}{|G_\alpha|} \tag{16.89}$$

for all nonempty subsets α of $\mathcal{N}_n$ for some finite group G and subgroups G_1, $G_2, \cdots, G_n$, we see that the inequality (16.84) holds for all random variables $X_1, X_2, \cdots, X_n$ if and only if the inequality obtained from (16.84) by replacing h_α by $\log \frac{|G|}{|G_\alpha|}$ for all nonempty subsets α of $\mathcal{N}_n$ holds for any finite group G and subgroups $G_1, G_2, \cdots, G_n$. In other words, for every unconstrained information inequality, there is a corresponding group inequality and vice versa. Therefore, inequalities in information theory can be proved by methods in group theory, and inequalities in group theory can be proved by methods in information theory.

In the rest of the section, we explore this one-to-one correspondence between information theory and group theory. We first give a group-theoretic proof of the basic inequalities in information theory. At the end of the section, we will give an information-theoretic proof for the group inequality in (16.5).

Definition 16.25. *Let G_1 and G_2 be subgroups of a finite group G. Define*

$$G_1 \circ G_2 = \{a \circ b : a \in G_1 \text{ and } b \in G_2\}. \tag{16.90}$$

$G_1 \circ G_2$ is in general not a subgroup of G. However, it can be shown that $G_1 \circ G_2$ is a subgroup of G if G is Abelian (see Problem 1).

Proposition 16.26. *Let G_1 and G_2 be subgroups of a finite group G. Then*

$$|G_1 \circ G_2| = \frac{|G_1||G_2|}{|G_1 \cap G_2|}. \tag{16.91}$$

Proof. Fix $(a_1, a_2) \in G_1 \times G_2$, Then $a_1 \circ a_2$ is in $G_1 \circ G_2$. Consider any $(b_1, b_2) \in G_1 \times G_2$ such that

$$b_1 \circ b_2 = a_1 \circ a_2. \tag{16.92}$$

We will determine the number of (b_1, b_2) in $G_1 \times G_2$ which satisfies this relation. From (16.92), we have

$$b_1^{-1} \circ (b_1 \circ b_2) = b_1^{-1} \circ (a_1 \circ a_2), \tag{16.93}$$
$$(b_1^{-1} \circ b_1) \circ b_2 = b_1^{-1} \circ a_1 \circ a_2, \tag{16.94}$$
$$b_2 = b_1^{-1} \circ a_1 \circ a_2. \tag{16.95}$$

Then

$$b_2 \circ a_2^{-1} = b_1^{-1} \circ a_1 \circ (a_2 \circ a_2^{-1}) = b_1^{-1} \circ a_1. \tag{16.96}$$

Let k be this common element in G, i.e.,

$$k = b_2 \circ a_2^{-1} = b_1^{-1} \circ a_1. \tag{16.97}$$

Since $b_1^{-1} \circ a_1 \in G_1$ and $b_2 \circ a_2^{-1} \in G_2$, k is in $G_1 \cap G_2$. In other words, for given $(a_1, a_2) \in G_1 \times G_2$, if $(b_1, b_2) \in G_1 \times G_2$ satisfies (16.92), then (b_1, b_2)

satisfies (16.97) for some $k \in G_1 \cap G_2$. On the other hand, if $(b_1, b_2) \in G_1 \times G_2$ satisfies (16.97) for some $k \in G_1 \cap G_2$, then (16.96) is satisfied, which implies (16.92). Therefore, for given $(a_1, a_2) \in G_1 \times G_2$, $(b_1, b_2) \in G_1 \times G_2$ satisfies (16.92) if and only if (b_1, b_2) satisfies (16.97) for some $k \in G_1 \cap G_2$.

Now from (16.97), we obtain

$$b_1(k) = (k \circ a_1^{-1})^{-1} \tag{16.98}$$

and

$$b_2(k) = k \circ a_2, \tag{16.99}$$

where we have written b_1 and b_2 as $b_1(k)$ and $b_2(k)$ to emphasize their dependence on k. Now consider $k, k' \in G_1 \cap G_2$ such that

$$(b_1(k), b_2(k)) = (b_1(k'), b_2(k')). \tag{16.100}$$

Since $b_1(k) = b_1(k')$, from (16.98), we have

$$(k \circ a_1^{-1})^{-1} = (k' \circ a_1^{-1})^{-1}, \tag{16.101}$$

which implies

$$k = k'. \tag{16.102}$$

Therefore, each $k \in G_1 \cap G_2$ corresponds to a unique pair $(b_1, b_2) \in G_1 \times G_2$ which satisfies (16.92). Therefore, we see that the number of distinct elements in $G_1 \circ G_2$ is given by

$$|G_1 \circ G_2| = \frac{|G_1 \times G_2|}{|G_1 \cap G_2|} = \frac{|G_1||G_2|}{|G_1 \cap G_2|}, \tag{16.103}$$

completing the proof. $\square$

Theorem 16.27. *Let G_1, G_2, and G_3 be subgroups of a finite group G. Then*

$$|G_3||G_{123}| \geq |G_{13}||G_{23}|. \tag{16.104}$$

Proof. First of all,

$$G_{13} \cap G_{23} = (G_1 \cap G_3) \cap (G_2 \cap G_3) = G_1 \cap G_2 \cap G_3 = G_{123}. \tag{16.105}$$

By Proposition 16.26, we have

$$|G_{13} \circ G_{23}| = \frac{|G_{13}||G_{23}|}{|G_{123}|}. \tag{16.106}$$

It is readily seen that $G_{13} \circ G_{23}$ is a subset of G_3, Therefore,

$$|G_{13} \circ G_{23}| = \frac{|G_{13}||G_{23}|}{|G_{123}|} \leq |G_3|. \tag{16.107}$$

The theorem is proved. $\square$

Corollary 16.28. *For random variables X_1, X_2, and X_3,*

$$I(X_1; X_2 | X_3) \geq 0. \tag{16.108}$$

Proof. Let G_1, G_2, and G_3 be subgroups of a finite group G. Then

$$|G_3||G_{123}| \geq |G_{13}||G_{23}| \tag{16.109}$$

by Theorem 16.27 or

$$\frac{|G|^2}{|G_{13}||G_{23}|} \geq \frac{|G|^2}{|G_3||G_{123}|}. \tag{16.110}$$

This is equivalent to

$$\log \frac{|G|}{|G_{13}|} + \log \frac{|G|}{|G_{23}|} \geq \log \frac{|G|}{|G_3|} + \log \frac{|G|}{|G_{123}|}. \tag{16.111}$$

This group inequality corresponds to the information inequality

$$H(X_1, X_3) + H(X_2, X_3) \geq H(X_3) + H(X_1, X_2, X_3), \tag{16.112}$$

which is equivalent to

$$I(X_1; X_2 | X_3) \geq 0. \tag{16.113}$$

□

The above corollary shows that all the basic inequalities in information theory has a group-theoretic proof. Of course, Theorem 16.27 is also implied by the basic inequalities. As a remark, the inequality in (16.3) is seen to be a special case of Theorem 16.27 by letting $G_3 = G$.

We are now ready to prove the group inequality in (16.5). The unconstrained non-Shannon-type inequality we have proved in Theorem 15.7 can be expressed in canonical form as

$$\begin{aligned}
&H(X_1) + H(X_1, X_2) + 2H(X_3) + 2H(X_4) \\
&+4H(X_1, X_3, X_4) + H(X_2, X_3, X_4) \\
&\leq 3H(X_1, X_3) + 3H(X_1, X_4) + 3H(X_3, X_4) \\
&+H(X_2, X_3) + H(X_2, X_4),
\end{aligned} \tag{16.114}$$

which corresponds to the group inequality

$$\begin{aligned}
&\log \frac{|G|}{|G_1|} + \log \frac{|G|}{|G_{12}|} + 2\log \frac{|G|}{|G_3|} + 2\log \frac{|G|}{|G_4|} \\
&+4\log \frac{|G|}{|G_{134}|} + \log \frac{|G|}{|G_{234}|} \\
&\leq 3\log \frac{|G|}{|G_{13}|} + 3\log \frac{|G|}{|G_{14}|} + 3\log \frac{|G|}{|G_{34}|} + \log \frac{|G|}{|G_{23}|} \\
&+\log \frac{|G|}{|G_{24}|}.
\end{aligned} \tag{16.115}$$

Upon rearranging the terms, we obtain

$$|G_1 \cap G_3|^3 |G_1 \cap G_4|^3 |G_3 \cap G_4|^3 |G_2 \cap G_3||G_2 \cap G_4|$$
$$\leq |G_1||G_1 \cap G_2||G_3|^2 |G_4|^2 |G_1 \cap G_3 \cap G_4|^4 |G_2 \cap G_3 \cap G_4|, \qquad (16.116)$$

which is the group inequality in (16.5). The meaning of this inequality and its implications in group theory are yet to be understood.

Chapter Summary

In the following, $\mathcal{N}_n = \{1, 2, \cdots, n\}$.

Properties of Subgroups of a Finite Group:

1. Lagrange's Theorem: If S is a subgroup of a finite group G, then $|S|$ divides $|G|$.
2. Let G_i be subgroups of a finite group G and a_i be elements of G, $i \in \alpha$. Then

$$|\cap_{i \in \alpha} a_i G_i| = \begin{cases} |G_\alpha| & \text{if } \bigcap_{i \in \alpha} a_i G_i \neq \emptyset \\ 0 & \text{otherwise} \end{cases},$$

where $a_i G_i$ is the left coset of G_i containing a_i and $G_\alpha = \cap_{i \in \alpha} G_i$.

Group Characterization of an Entropy Function: Let G be a finite group and $G_1, G_2, \cdots, G_n$ be subgroups of G. For a vector $\mathbf{h} \in \mathcal{H}_n$, if $h_\alpha = \log \frac{|G|}{|G_\alpha|}$ for all nonempty subsets α of $\mathcal{N}_n$, then $(G, G_1, \cdots, G_n)$ is a group characterization of $\mathbf{h}$. A vector $\mathbf{h}$ that has a group characterization is entropic.

Group Characterization of $\overline{\Gamma}_n^*$: $\overline{con}(\Upsilon_n) = \overline{\Gamma}_n^*$, where $\Upsilon_n = \{\mathbf{h} \in \mathcal{H}_n : \mathbf{h}$ has a group characterization$\}$.

Information Inequalities and Group Inequalities: An unconstrained inequality $\mathbf{b}^\top \mathbf{h} \geq 0$ involving random variables $X_1, X_2, \cdots, X_n$, where $\mathbf{h} \in \mathcal{H}_n$, always holds if and only if the inequality obtained by replacing h_α by $\log \frac{|G|}{|G_\alpha|}$ for all nonempty subsets α of $\mathcal{N}_n$ holds for any finite group G and subgroups $G_1, G_2, \cdots, G_n$.

A "Non-Shannon-Type" Group Inequality:

$$|G_1 \cap G_3|^3 |G_1 \cap G_4|^3 |G_3 \cap G_4|^3 |G_2 \cap G_3||G_2 \cap G_4|$$
$$\leq |G_1||G_1 \cap G_2||G_3|^2 |G_4|^2 |G_1 \cap G_3 \cap G_4|^4 |G_2 \cap G_3 \cap G_4|.$$

Problems

1. Let G_1 and G_2 be subgroups of a finite group G. Show that $G_1 \circ G_2$ is a subgroup if G is Abelian.
2. Let $\mathbf{g}_1$ and $\mathbf{g}_2$ be group-characterizable entropy functions.
   a) Prove that $m_1\mathbf{g}_1 + m_2\mathbf{g}_2$ is group characterizable, where m_1 and m_2 are any positive integers.
   b) For any positive real numbers a_1 and a_2, construct a sequence of group-characterizable entropy functions $\mathbf{f}^{(k)}$ for $k = 1, 2, \cdots$, such that

$$\lim_{k \to \infty} \frac{\mathbf{f}^{(k)}}{||\mathbf{f}^{(k)}||} = \frac{\mathbf{h}}{||\mathbf{h}||},$$

   where $\mathbf{h} = a_1\mathbf{g}_1 + a_2\mathbf{g}_2$.
3. Let $(G, G_1, G_2, \cdots, G_n)$ be a group characterization of $\mathbf{g} \in \Gamma_n^*$, where $\mathbf{g}$ is the entropy function for random variables $X_1, X_2, \cdots, X_n$. Fix any nonempty subset α of $\mathcal{N}_n$, and define $\mathbf{h}$ by

$$h_\beta = g_{\alpha \cup \beta} - g_\alpha$$

   for all nonempty subsets β of $\mathcal{N}_n$. It can easily be checked that $h_\beta = H(X_\beta | X_\alpha)$. Show that $(K, K_1, K_2, \cdots, K_n)$ is a group characterization of $\mathbf{h}$, where $K = G_\alpha$ and $K_i = G_i \cap G_\alpha$.
4. Let $(G, G_1, G_2, \cdots, G_n)$ be a group characterization of $\mathbf{g} \in \Gamma_n^*$, where $\mathbf{g}$ is the entropy function for random variables $X_1, X_2, \cdots, X_n$. Show that if X_i is a function of $(X_j : j \in \alpha)$, then G_α is a subgroup of G_i.
5. Let G_1, G_2, G_3 be subgroups of a finite group G. Prove that

$$|G||G_1 \cap G_2 \cap G_3|^2 \geq |G_1 \cap G_2||G_2 \cap G_3||G_1 \cap G_3|.$$

   Hint: Use the information-theoretic approach.

6. Let $\mathbf{h} \in \Gamma_2^*$ be the entropy function for random variables X_1 and X_2 such that $h_1 + h_2 = h_{12}$, i.e., X_1 and X_2 are independent. Let (G, G_1, G_2) be a group characterization of $\mathbf{h}$, and define a mapping $L : G_1 \times G_2 \to G$ by

$$L(a, b) = a \circ b.$$

   a) Prove that the mapping L is onto, i.e., for any element $c \in G$, there exists $(a, b) \in G_1 \times G_2$ such that $a \circ b = c$.
   b) Prove that $G_1 \circ G_2$ is a group.
7. Denote an entropy function $\mathbf{h} \in \Gamma_2^*$ by (h_1, h_2, h_{12}). Construct a group characterization for each of the following entropy functions:
   a) $\mathbf{h}_1 = (\log 2, 0, \log 2)$;
   b) $\mathbf{h}_2 = (0, \log 2, \log 2)$;
   c) $\mathbf{h}_3 = (\log 2, \log 2, \log 2)$.

Verify that Γ_2 is the minimal convex set containing the above three entropy functions.

8. Denote an entropy function $\mathbf{h} \in \Gamma_3^*$ by $(h_1, h_2, h_3, h_{12}, h_{23}, h_{13}, h_{123})$. Construct a group characterization for each of the following entropy functions:

   a) $\mathbf{h}_1 = (\log 2, 0, 0, \log 2, 0, \log 2, \log 2)$;
   b) $\mathbf{h}_2 = (\log 2, \log 2, 0, \log 2, \log 2, \log 2, \log 2)$;
   c) $\mathbf{h}_3 = (\log 2, \log 2, \log 2, \log 2, \log 2, \log 2, \log 2)$;
   d) $\mathbf{h}_4 = (\log 2, \log 2, \log 2, \log 4, \log 4, \log 4, \log 4)$.

9. *Ingleton inequality.* Let G be a finite Abelian group and G_1, G_2, G_3, and G_4 be subgroups of G. Let (G, G_1, G_2, G_3, G_4) be a group characterization of $\mathbf{g}$, where $\mathbf{g}$ is the entropy function for random variables X_1, X_2, X_3, and X_4. Prove the following statements:

   a)
   $$|(G_1 \cap G_3) \circ (G_1 \cap G_4)| \leq |G_1 \cap (G_3 \circ G_4)|$$

   Hint: Show that $(G_1 \cap G_3) \circ (G_1 \cap G_4) \subset G_1 \cap (G_3 \circ G_4)$.

   b)
   $$|G_1 \circ G_3 \circ G_4| \leq \frac{|G_1||G_3 \circ G_4||G_1 \cap G_3 \cap G_4|}{|G_1 \cap G_3||G_1 \cap G_4|}.$$

   c)
   $$|G_1 \circ G_2 \circ G_3 \circ G_4| \leq \frac{|G_1 \circ G_3 \circ G_4||G_2 \circ G_3 \circ G_4|}{|G_3 \circ G_4|}.$$

   d)
   $$|G_1 \circ G_2 \circ G_3 \circ G_4|$$
   $$\leq \frac{|G_1||G_2||G_3||G_4||G_1 \cap G_3 \cap G_4||G_2 \cap G_3 \cap G_4|}{|G_1 \cap G_3||G_1 \cap G_4||G_2 \cap G_3||G_2 \cap G_4||G_3 \cap G_4|}.$$

   e)
   $$|G_1 \cap G_3||G_1 \cap G_4||G_2 \cap G_3||G_2 \cap G_4||G_3 \cap G_4|$$
   $$\leq |G_3||G_4||G_1 \cap G_2||G_1 \cap G_3 \cap G_4||G_2 \cap G_3 \cap G_4|.$$

   f)
   $$H(X_{13}) + H(X_{14}) + H(X_{23}) + H(X_{24}) + H(X_{34})$$
   $$\geq H(X_3) + H(X_4) + H(X_{12}) + H(X_{134}) + H(X_{234}),$$

   where $H(X_{134})$ denotes $H(X_1, X_3, X_4)$, etc.

   g) Is the inequality in (f) implied by the basic inequalities? And does it always hold? Explain.

The Ingleton inequality [181] (see also [283]) was originally obtained as a constraint on the rank functions of vector spaces. The inequality in (e) was obtained in the same spirit by Chan [57] for subgroups of a finite group. The inequality in (f) is referred to as the Ingleton inequality for entropy in the literature. (See also Problem 8 in Chapter 15.)

Historical Notes

The results in this chapter are due to Chan and Yeung [63], whose work was inspired by a one-to-one correspondence between entropy and quasi-uniform arrays previously established by Chan [57] (also Chan [59]). Romashchenko et al. [311] have developed an interpretation of Kolmogorov complexity similar to the combinatorial interpretation of entropy in Chan [57].

The results in this chapter have been used by Chan [61] to construct codes for multi-source network coding to be discussed in Chapter 21.

Part II

Fundamentals of Network Coding

17

Introduction

For a point-to-point communication system, we see from Section 7.7 and Problem 6 in Chapter 8 that asymptotic optimality can be achieved by separating source coding and channel coding. Recall from Section 5.3 that the goal of source coding is to represent the information source in (almost) fair bits.[1] Then the role of channel coding is to enable the transmission of fair bits through the channel essentially free of error with no reference to the meaning of these fair bits. Thus a theme in classical information theory for point-to-point communication is that fair bits can be drawn equivalence to a *commodity*.

It is intuitively appealing that this theme in classical information theory would continue to hold in network communication where the network consists of *noiseless* point-to-point communication channels. If so, in order to multicast[2] information from a source node to possibly more than one sink node, we only need to compress the information at the source node into fair bits, organize them into data packets, and route the packets to the sink node through the intermediate nodes in the network. In the case when there are more than one sink node, the information needs to be replicated at certain intermediate nodes so that every sink node can receive a copy of the information. This method of transmitting information in a network is generally referred to as *store-and-forward* or *routing*. As a matter of fact, almost all computer networks built in the last few decades are based on this principle, where *routers* are deployed at the intermediate nodes to switch a data packet from an input channel to an output channel without processing the data content. The delivery of data packets in a computer network resembles mail delivery in a postal system. We refer the readers to textbooks on *data communication* [35][215] and *switching theory* [176][227].

---

[1] Fair bits refer to i.i.d. bits, each distributed uniformly on $\{0, 1\}$.

[2] Multicast means to transmit information from a source node to a specified set of sink nodes.

However, we will see very shortly that in network communication, it does not suffice to simply route and/or replicate information within the network. Specifically, coding generally needs to be employed at the intermediate nodes in order to achieve bandwidth optimality. This notion, called *network coding*, is the subject of discussion in Part II of this book.

17.1 The Butterfly Network

In this section, the advantage of network coding over routing is explained by means of a few simple examples. The application of network coding in wireless and satellite communication will be discussed in the next section.

We will use a finite directed graph to represent a point-to-point communication network. A node in the network corresponds to a vertex in the graph, while a communication channel in the network corresponds to an edge in the graph. We will not distinguish a node from a vertex nor a channel from an edge. In the graph, a node is represented by a circle, with the exception that the unique source node, denoted by s (if exists), is represented by a square. Each edge is labeled by a positive integer called the *capacity*[3] or the *rate constraint*, which gives the maximum number of information symbols taken from some finite alphabet that can be transmitted over the channel per unit time. In this section, we assume that the information symbol is binary. When there is only one edge from node a to node b, we denote the edge by (a, b).

Example 17.1 (Butterfly Network I). Consider the network in Figure 17.1(a). In this network, two bits b_1 and b_2 are generated at source node s, and they are to be multicast to two sink nodes t_1 and t_2. In Figure 17.1(b), we try to devise a routing scheme for this purpose. By symmetry, we send the two bits on different output channels at node s. Without loss of generality, b_1 is sent on channel $(s, 1)$ and b_2 is sent on channel $(s, 2)$. At nodes 1 and 2, the received bit is replicated and the copies are sent on the two output channels. At node 3, since both b_1 and b_2 are received but there is only one output channel, we have to choose one of the two bits to be sent on the output channel $(3, 4)$. Suppose we send b_1 as in Figure 17.1(b). Then the bit is replicated at node 4 and the two copies are sent to nodes t_1 and t_2, respectively. At node t_2, both b_1 and b_2 are received. However, at node t_1, two copies of b_1 are received and b_2 cannot be recovered. Thus this routing scheme does not work. Similarly, if b_2 instead of b_1 is sent on channel $(3, 4)$, then b_1 cannot be recovered at node t_2.

However, if network coding is allowed, it is actually possible to achieve our goal. Figure 17.1(c) shows a scheme which multicasts both b_1 and b_2 to nodes t_1 and t_2, where "+" denotes *modulo* 2 addition. At node t_1, b_1 is received, and b_2 can be recovered by adding b_1 and $b_1 + b_2$, because

---

[3] Here the term "capacity" is used in the sense of graph theory.

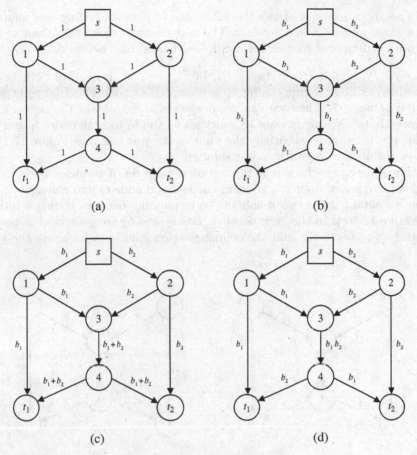

Fig. 17.1. Butterfly Network I.

$$b_1 + (b_1 + b_2) = (b_1 + b_1) + b_2 = 0 + b_2 = b_2. \qquad (17.1)$$

Similarly, b_2 is received at node t_2, and b_1 can be recovered by adding b_2 and $b_1 + b_2$.

In this scheme, b_1 and b_2 are encoded into the bit $b_1 + b_2$ which is then sent on channel $(3, 4)$. If network coding is not allowed, in order to multicast both b_1 and b_2 to nodes t_1 and t_2, at least one more bit has to be sent. Figure 17.1(d) shows such a scheme. In this scheme, however, the capacity of channel $(3, 4)$ is exceeded by 1 bit. If the capacity of channel $(3, 4)$ cannot be exceeded and network coding is not allowed, it can be shown that at most 1.5 bits can be multicast per unit time on the average (see Problem 3).

The above example shows the advantage of network coding over routing for a single multicast in a network. The next example shows the advantage of network coding over routing for multiple unicasts[4] in a network.

Example 17.2 (Butterfly Network II). In Figure 17.1, instead of both being generated at node s, suppose bit b_1 is generated at node 1 and bit b_2 is generated at node 2. Then we can remove node s and obtain the network in Figure 17.2(a). We again want to multicast b_1 and b_2 to both nodes t_1 and t_2. Since this network is essentially the same as the previous one, Figure 17.2(b) shows the obvious network coding solution.

There are two multicasts in this network. However, if we merge node 1 and node t_1 into a new node t_1' and merge node 2 and node t_2 into a new node t_2', then we obtain the network and the corresponding network coding solution in Figure 17.2(c). In this new network, bits b_1 and b_2 are generated at nodes t_1' and t_2', respectively, and the communication goal is to exchange the two

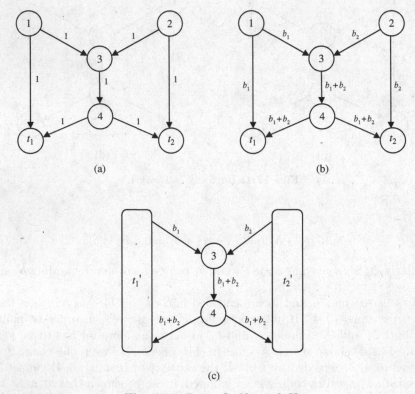

Fig. 17.2. Butterfly Network II.

---

[4] Unicast is the special case of multicast with one sink node.

bits through the network. In other words, the two multicasts in Figure 17.2(a) become two unicasts in Figure 17.2(c).

If network coding is not allowed, we need to route b_1 from node t_1' to node t_2' and to route b_2 from node t_2' to node t_1'. Since each of these routes has to go through node 3 and node 4, if b_1 and b_2 are routed simultaneously, the capacity of channel $(3,4)$ is exceeded. Therefore, we see the advantage of network coding over routing when there are multiple unicasts in the network.

For the network in Figure 17.2(b), the two sink nodes are required to recover both of the information sources, namely the bits b_1 and b_2. Even though they are generated at two different source nodes 1 and 2, they can be regarded as being generated at a *super source node s* connecting to nodes t_1 and t_2 as in Figure 17.1(c). Precisely, the network (network code) in Figure 17.2(b) can be obtained from the network (network code) in Figure 17.1(c) by removing node s and all its output channels. This observation will be further discussed in Example 19.26 in the context of single-source linear network coding.

17.2 Wireless and Satellite Communications

In wireless communication, when a node broadcasts, different noisy versions of the signal are received by the neighboring nodes. Under certain conditions, with suitable channel coding, we can assume the existence of an error-free channel between the broadcast node and the neighboring nodes such that each of the latter receives exactly the same information. Such an abstraction, though generally suboptimal, provides very useful tools for communication systems design.

Our model for network communication can be used for modeling the above broadcast scenario by imposing the following constraints on the broadcast node:

1. all the output channels have the same capacity;
2. the same symbol is sent on each of the output channels.

We will refer to these constraints as the *broadcast constraint*. Figure 17.3(a) is an illustration of a broadcast node b with two neighboring nodes n_1 and n_2, where the two output channels of node b have the same capacity.

In order to express the broadcast constraint in the usual graph-theoretic terminology, we need to establish the following simple fact about network coding.

Proposition 17.3. *Network coding is not necessary at a node if the node has only one input channel and the capacity of each output channel is the same as that of the input channel.*

Proof. Consider a node in the network as prescribed and denote the symbol(s) received on the input channel by x. (There is more than one symbol in x if

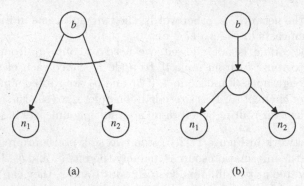

Fig. 17.3. A broadcast node b with two neighboring nodes n_1 and n_2.

the input channel has capacity larger than 1.) Let a coding scheme be given, and denote the symbol sent on the ith output channel by $g_i(x)$.

We now show that one may assume without loss of generality that x is sent on all the output channels. If x instead of $g_i(x)$ is sent on the ith output channel, then the receiving node can mimic the effect of receiving $g_i(x)$ by applying the function g_i on x upon receiving it. In other words, any coding scheme that does not send x on all the output channels can readily be converted into one which does. This proves the proposition. □

We now show that the broadcast constraint depicted in Figure 17.3(a) is logically equivalent to the usual graph representation in Figure 17.3(b). In this figure, the unlabeled node is a dummy node associated with the broadcast node which is inserted for the purpose of modeling the broadcast constraint, where the input channel and all the output channels of the dummy node have the same capacity as an output channel of the broadcast node b in Figure 17.3(a). Although no broadcast constraint is imposed on the dummy node in Figure 17.3(b), by Proposition 17.3, we may assume without loss of generality that the dummy node simply sends the symbol received on the input channel on each of the output channels. Then (a) and (b) of Figure 17.3 are logically equivalent to each other because a coding scheme for the former corresponds to a coding scheme for the latter and vice versa.

Example 17.4 (A Wireless/Satellite System). Consider a communication system with two wireless nodes t_1' and t_2' that generate two bits b_1 and b_2, respectively, and the two bits are to be exchanged through a relay node. Such a system can also be the model of a satellite communication system, where the relay node corresponds to a satellite and the two nodes t_1' and t_2' correspond to ground stations that communicate with each other through the satellite.

We make the usual assumption that a wireless node cannot simultaneously

1. transmit and receive;
2. receive the transmission from more than one neighboring node.

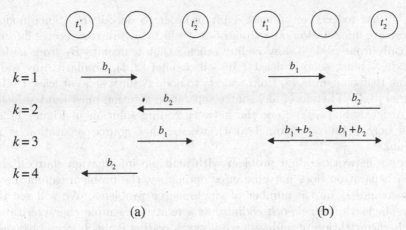

Fig. 17.4. A network coding application in wireless communication.

A straightforward routing scheme which takes a total of 4 time units to complete is shown in Figure 17.4(a), with k being the discrete-time index.

By taking into account the broadcast nature of the relay node, the system can be modeled by the network in Figure 17.2(c), where node 3 corresponds to the relay node and node 4 corresponds to the associated dummy node. Then the network coding solution is shown in Figure 17.4(b), which takes a total of 3 time units to complete. In other words, a very simple coding scheme at the relay node can save 50% of the downlink bandwidth.

17.3 Source Separation

In an error-free point-to-point communication system, suppose we want to transmit two information sources X and Y. If we compress the two sources separately, we need to transmit approximately $H(X) + H(Y)$ bits. If we compress the two sources jointly, we need to transmit approximately $H(X,Y)$ bits. If X and Y are independent, we have

$$H(X,Y) = H(X) + H(Y). \qquad (17.2)$$

In other words, if the information sources are independent, asymptotically there is no difference between coding them separately or jointly.

We will refer to coding independent information sources separately as *source separation*. Example 17.2 reveals the important fact that source separation is not necessarily optimal in network communication, which is explained as follows. Let B_1 and B_2 be random bits generated at nodes t'_1 and t'_2, respectively, where B_1 and B_2 are independent and each of them are distributed uniformly on $\{0,1\}$. With B_2 as side-information which is independent of B_1, node t_2' has to receive at least 1 bit in order to decode B_1. Since node t_2' can receive information only from node 4 which in turn can receive information only from node 3, any coding scheme that transmits B_1 from node t_1' to node t_2' must send at least 1 bit on channel $(3,4)$. Similarly, any coding scheme that transmits B_2 from node t_2' to node t_1' must send at least 1 bit on channel $(3,4)$. Therefore, any source separation solution must send at least 2 bits on channel $(3,4)$. Since the network coding solution in Figure 17.2(c) sends only 1 bit on channel $(3,4)$, we see that source separation is not optimal.

For a network coding problem with multiple information sources, since source separation does not guarantee optimality, the problem cannot always be decomposed into a number of single-source problems. We will see that while single-source network coding has a relatively simple characterization, the characterization of multi-source network coding is much more involved.

Chapter Summary

Advantage of Network Coding: For communication on a point-to-point network, store-and-forward may not be bandwidth optimal when

1. there is one information source to be multicast;
2. there are two or more independent information sources to be unicast (more generally multicast).

In general, network coding needs to be employed for bandwidth optimality.

Source Separation: For communication on a point-to-point network, when there are two or more independent information sources to be unicast (more generally multicast), source separation coding may not be bandwidth optimal.

Problems

In the following problems, the rate constraint for an edge is in bits per unit time.

1. Consider the following network.
   We want to multicast information to the sink nodes at the maximum rate without using network coding. Let $B = \{b_1, b_2, \ldots, b_\kappa\}$ be the set of bits to be multicast. Let B_i be the set of bits sent in edge (s, i), where $|B_i| = 2$, $i = 1, 2, 3$. At node i, the received bits are duplicated and sent in the two outgoing edges. Thus two bits are sent in each edge in the network.
   (a) Show that $B = B_i \cup B_j$ for any $1 \leq i < j \leq 3$.
   (b) Show that $B_3 \cup (B_1 \cap B_2) = B$.
   (c) Show that $|B_3 \cup (B_1 \cap B_2)| \leq |B_3| + |B_1| + |B_2| - |B_1 \cup B_2|$.

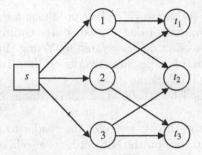

(d) Determine the maximum value of κ and devise a network code which achieves this maximum value.

(e) What is the percentage of improvement if network coding is used? (Ahlswede et al. [6].)

2. Consider the following butterfly network.

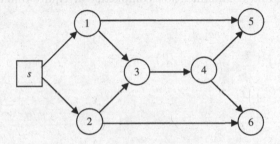

Devise a network coding scheme which multicasts two bits b_1 and b_2 from node s to all the other nodes such that nodes 3, 5, and 6 receive b_1 and b_2 after 1 unit of time and nodes 1, 2, and 4 receive b_1 and b_2 after 2 units of time. In other words, node i receives information at a rate equal to maxflow(s, i) for all $i \neq s$.

3. Determine the maximum rate at which information can be multicast to nodes 5 and 6 only in the network in Problem 2 if network coding is not used. Devise a network coding scheme which achieves this maximum rate.

Historical Notes

The concept of *network coding* was first introduced for satellite communication networks in Yeung and Zhang [411] and then fully developed in Ahlswede et al. [6], where in the latter the term "network coding" was coined. In this work, the advantage of network coding over store-and-forward was first demonstrated by the butterfly network, thus refuting the folklore that information transmission in a point-to-point network is equivalent to a commodity flow.

Prior to [411] and [6], network coding problems for special networks had been studied in the context of distributed source coding. The suboptimality of source separation was first demonstrated by Yeung [400]. Source separation was proved to be optimal for special networks by Hau [160], Roche et al. [308], and Yeung and Zhang [410]. Some other special cases of single-source network coding had been studied by Roche et al. [307], Rabin [297], Ayanoglu et al. [22], and Roche [306].

For a tutorial on the theory, we refer the reader to the unifying work by Yeung et al. [408]. Tutorials on the subject have also been written by Fragouli and Soljanin [123] and Chou and Wu [68] from the algorithm and application perspectives. We also refer the reader to the book by Ho and Lun [164]. For an update of the literature, the reader may visit the *Network Coding Homepage* [273].

By regarding communication as a special case of computation, it can be seen that network coding is in the spirit of communication complexity in computer science studied by Yao [394]. However, the problem formulations of network coding and communication complexity are quite different.

18

The Max-Flow Bound

In this chapter, we discuss an important bound for single-source network coding which has a strong connection with graph theory. This bound, called the max-flow min-cut bound, or simply the *max-flow bound*, gives a fundamental limit on the amount of information that can be multicast in a network.

The max-flow bound is established in a general setting where information can be transmitted within the network in some arbitrary manner. Toward this end, we first formally define a point-to-point network and a class of codes on such a network. In Chapters 19 and 20, we will prove the achievability of the max-flow bound by *linear* network coding.[1]

18.1 Point-to-Point Communication Networks

A point-to-point communication network is represented by a directed graph $G = (V, E)$, where V is the set of nodes in the network and E is the set of edges in G which represent the point-to-point channels. Parallel edges between a pair of nodes is allowed.[2] We assume that G is finite, i.e., $|E| < \infty$ (and hence $|V| < \infty$). The unique *source node* in the network, where information is generated, is denoted by s. All the other nodes are referred to as *non-source* nodes. The sets of input channels and output channels of a node i are denoted by $\text{In}(i)$ and $\text{Out}(i)$, respectively.

For a channel e, let R_e be the *rate constraint*, i.e., the maximum number of information symbols taken from a finite alphabet that can be sent on the channel per unit time. As before, we also refer to R_e as the capacity of channel e in the sense of graph theory. Let

$$\mathbf{R} = (R_e : e \in E) \tag{18.1}$$

---

[1] A more specific form of the max-flow bound will be proved in Theorem 19.10 for linear network coding.

[2] Such a graph is sometimes called a *multigraph*.

be the rate constraints for the graph G. To simplify our discussion, we assume that R_e are positive integers for all $e \in E$.

In the following, we introduce some notions in graph theory which will facilitate the characterization of a point-to-point network. Temporarily regard an edge in the graph G as a water pipe and G as a network of water pipes. Fix a node $t \neq s$ and call it the *sink node*. Suppose water is generated at a constant rate at node s. We assume that the rate of water flow in each pipe does not exceed its capacity. We also assume that there is no leakage in the network, so that water is conserved at every node other than s and t in the sense that the total rate of water flowing into the node is equal to the total rate of water flowing out of the node. The water generated at node s is eventually drained at node t.

A *flow*

$$\mathbf{F} = (F_e : e \in E) \tag{18.2}$$

in G from node s to node t with respect to rate constraints $\mathbf{R}$ is a valid assignment of a nonnegative integer F_e to every edge $e \in E$ such that F_e is equal to the rate of water flow in edge e under all the assumptions in the last paragraph. The integer F_e is referred to as the value of $\mathbf{F}$ on edge e. Specifically, $\mathbf{F}$ is a flow in G from node s to node t if for all $e \in E$,

$$0 \leq F_e \leq R_e, \tag{18.3}$$

and for all $i \in V$ except for s and t,

$$F_+(i) = F_-(i), \tag{18.4}$$

where

$$F_+(i) = \sum_{e \in \text{In}(i)} F_e \tag{18.5}$$

and

$$F_-(i) = \sum_{e \in \text{Out}(i)} F_e. \tag{18.6}$$

In the above, $F_+(i)$ is the total flow into node i and $F_-(i)$ is the total flow out of node i, and (18.4) is called the *conservation* conditions.

Since the conservation conditions require that the resultant flow out of any node other than s and t is zero, it is intuitively clear and not difficult to show that the resultant flow out of node s is equal to the resultant flow into node t. This common value is called the *value* of $\mathbf{F}$. $\mathbf{F}$ is a *max-flow* from node s to node t in G with respect to rate constraints $\mathbf{R}$ if $\mathbf{F}$ is a flow from node s to node t whose value is greater than or equal to the value of any other flow from node s to node t.

A *cut* between node s and node t is a subset U of V such that $s \in U$ and $t \notin U$. Let

$$E_U = \{e \in E : e \in \text{Out}(i) \cap \text{In}(j) \text{ for some } i \in U \text{ and } j \notin U\} \tag{18.7}$$

be the set of edges across the cut U. The *capacity* of the cut U with respect to rate constraints $\mathbf{R}$ is defined as the sum of the capacities of all the edges across the cut, i.e.,

$$\sum_{e\in E_U} R_e. \tag{18.8}$$

A cut U is a *min-cut* between node s and node t if it is a cut between node s and node t whose capacity is less than or equal to the capacity of any other cut between s and t.

A min-cut between node s and node t can be thought of as a *bottleneck* between node s and node t. Therefore, it is intuitively clear that the value of a max-flow from node s to node t cannot exceed the capacity of a min-cut between the two nodes. The following theorem, known as the *max-flow min-cut theorem*, states that the capacity of a min-cut is always achievable. This theorem will play a key role in the subsequent discussions.

Theorem 18.1 (Max-Flow Min-Cut Theorem [116]). *Let G be a graph with source node s, sink node t, and rate constraints $\mathbf{R}$. Then the value of a max-flow from node s to node t is equal to the capacity of a min-cut between the two nodes.*

The notions of max-flow and min-cut can be generalized to a collection of non-source nodes T. To define the max-flow and the min-cut from s to T, we expand the graph $G = (V, E)$ into $G' = (V', E')$ by installing a new node τ which is connected from every node in T by an edge. The capacity of an edge $(t, \tau), t \in T$, is set to infinity. Intuitively, node τ acts as a single sink node that collects all the flows into T. Then the max-flow and the min-cut from node s to T in graph G are defined as the max-flow and the min-cut from node s to node τ in graph G', respectively. This is illustrated in Figure 18.1(a).

The notions of max-flow and min-cut can be further generalized to a collection of edges ξ. For an edge $e \in \xi$, let the edge be from node v_e to node w_e. We modify the graph $G = (V, E)$ to obtain the graph $\tilde{G} = (\tilde{V}, \tilde{E})$ by installing a new node t_e for each edge $e \in \xi$ and replacing edge e by two new edges e' and e'', where e' is from node v_e to node t_e and e'' is from node t_e to node w_e. Let T_ξ be the set of nodes $t_e, e \in \xi$. Then the max-flow and the min-cut between

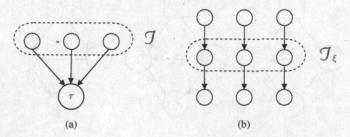

(a)   (b)

Fig. 18.1. Illustrations of the max-flow and the min-cut from the source node to (a) a collection of non-source nodes T and (b) a collection of edges ξ.

node s and the collection of edges ξ in graph G are defined as the max-flow and the min-cut between node s and the collection of nodes T_ξ in graph $\tilde{G}$, respectively. This is illustrated in Figure 18.1(b).

18.2 Examples Achieving the Max-Flow Bound

Let ω be the rate at which information is multicast from source node s to sink nodes $t_1, t_2, \cdots, t_L$ in a network G with rate constraints $\mathbf{R}$. We are naturally interested in the maximum possible value of ω. With a slight abuse of notation, we denote the value of a max-flow from source node s to a sink node t_l by $\mathrm{maxflow}(t_l)$. It is intuitive that

$$\omega \leq \mathrm{maxflow}(t_l) \tag{18.9}$$

for all $l = 1, 2, \cdots, L$, i.e.,

$$\omega \leq \min_l \mathrm{maxflow}(t_l). \tag{18.10}$$

This is called the *max-flow bound*, which will be formally established in the next two sections. In this section, we first show by a few examples that the max-flow bound can be achieved. In these examples, the unit of information is the bit.

First, we consider the network in Figure 18.2 which has one sink node. Figure 18.2(a) shows the capacity of each edge. By identifying the min-cut to be $\{s, 1, 2\}$ and applying the max-flow min-cut theorem, we see that

$$\mathrm{maxflow}(t_1) = 3. \tag{18.11}$$

Therefore the flow in Figure 18.2(b) is a max-flow. In Figure 18.2(c), we show how we can send three bits b_1, b_2, and b_3 from node s to node t_1 based on the max-flow in Figure 18.2(b). Evidently, the max-flow bound is achieved.

In fact, we can easily see that the max-flow bound can always be achieved when there is only one sink node in the network. In this case, we only need to

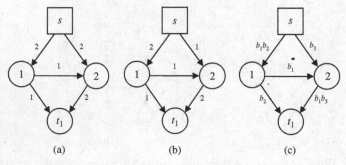

Fig. 18.2. A one-sink network.

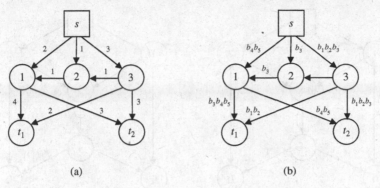

Fig. 18.3. A two-sink network without coding.

treat the information bits constituting the message as a commodity and route them through the network according to any fixed routing scheme. Eventually, all the bits will arrive at the sink node. Since the routing scheme is fixed, the sink node knows which bit is coming in from which edge, and the message can be recovered accordingly.

Next, we consider the network in Figure 18.3 which has two sink nodes. Figure 18.3(a) shows the capacity of each edge. It is easy to see that

$$\text{maxflow}(t_1) = 5 \tag{18.12}$$

and

$$\text{maxflow}(t_2) = 6. \tag{18.13}$$

So the max-flow bound asserts that we cannot send more than 5 bits to both t_1 and t_2. Figure 18.3(b) shows a scheme which sends 5 bits $b_1, b_2, b_3, b_4,$ and b_5 to t_1 and t_2 simultaneously. Therefore, the max-flow bound is achieved. In this scheme, b_1 and b_2 are replicated at node 3, b_3 is replicated at node s, while b_4 and b_5 are replicated at node 1. Note that each bit is replicated exactly once in the network because two copies of each bit are needed to be sent to the two sink nodes.

We now revisit the butterfly network reproduced in Figure 18.4(a), which again has two sink nodes. It is easy to see that

$$\text{maxflow}(t_l) = 2 \tag{18.14}$$

for $l = 1, 2$. So the max-flow bound asserts that we cannot send more than 2 bits to both sink nodes t_1 and t_2. We have already seen the network coding scheme in Figure 18.4(b) that achieves the max-flow bound. In this scheme, coding is required at node 3.

Finally, we consider the network in Figure 18.5 which has three sink nodes. Figure 18.5(a) shows the capacity of each edge. It is easy to see that

$$\text{maxflow}(t_l) = 2 \tag{18.15}$$

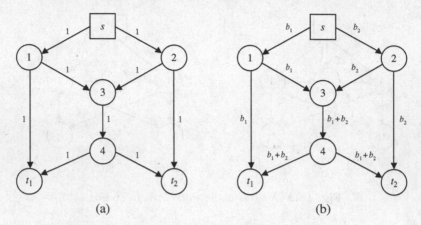

Fig. 18.4. Butterfly network I.

for all l. In Figure 18.5(b), we show how to multicast 2 bits b_1 and b_2 to all the sink nodes. Therefore, the max-flow bound is achieved. Again, it is necessary to code at the nodes in order to multicast the maximum number of bits to all the sink nodes.

The network in Figure 18.5 is of special interest in practice because it is a special case of the *diversity coding* scheme used in commercial *disk arrays*, which are a kind of fault-tolerant data storage system. For simplicity, assume the disk array has three disks which are represented by nodes 1, 2, and 3 in the network, and the information to be stored are the bits b_1 and b_2. The information is encoded into three pieces, namely b_1, b_2, and $b_1 + b_2$, which are stored on the disks represented by nodes 1, 2, and 3, respectively. In the system, there are three decoders, represented by sink nodes t_1, t_2, and t_3, such that each of them has access to a distinct set of two disks. The idea is that when any one disk is out of order, the information can still be recovered from the remaining two disks. For example, if the disk represented by node 1 is out of order, then

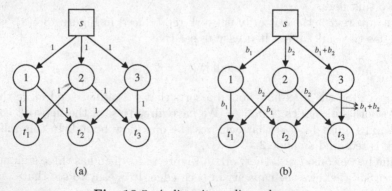

Fig. 18.5. A diversity coding scheme.

the information can be recovered by the decoder represented by sink node t_3 which has access to the disks represented by node 2 and node 3. When all the three disks are functioning, the information can be recovered by any decoder.

18.3 A Class of Network Codes

In this section, we introduce a general class of codes for the point-to-point network defined in Section 18.1. In the next section, the max-flow bound will be proved for this class of network codes.

Since the max-flow bound concerns only the values of max-flows from source node s to the sink nodes, we assume without loss of generality that there is no loop in the graph G, i.e., $\text{In}(i) \cap \text{Out}(i) = \emptyset$ for all $i \in V$, because such edges do not increase the value of a max-flow from node s to a sink node. For the same reason, we assume that there is no input edge at node s, i.e., $\text{In}(s) = \emptyset$.

We consider a block code of length n. Let X denote the information source and assume that x, the outcome of X, is obtained by selecting an index from a set $\mathcal{X}$ according to the uniform distribution. The elements in $\mathcal{X}$ are called messages. The information sent on an output channel of a node can depend only on the information previously received by that node. This constraint specifies the *causality* of any coding scheme on the network.

An $(n, (\eta_e : e \in E), \tau)$ network code on the graph G that multicasts information from source node s to sink nodes $t_1, t_2, \cdots, t_L$, where n is the block length, is defined by the components listed below; the construction of the code from these components will be described after their definitions are given.

1) A positive integer K.
2) Mappings

$$u : \{1, 2, \cdots, K\} \to V, \tag{18.16}$$
$$v : \{1, 2, \cdots, K\} \to V, \tag{18.17}$$

and

$$\hat{e} : \{1, 2, \cdots, K\} \to E, \tag{18.18}$$

such that $\hat{e}(k) \in \text{Out}(u(k))$ and $\hat{e}(k) \in \text{In}(v(k))$.
3) Index sets $A_k = \{1, 2, \cdots, |A_k|\}$, $1 \le k \le K$, such that

$$\prod_{k \in T_e} |A_k| = \eta_e, \tag{18.19}$$

where

$$T_e = \{1 \le k \le K : \hat{e}(k) = e\}. \tag{18.20}$$

4) (Encoding functions). If $u(k) = s$, then

$$f_k : \mathcal{X} \to A_k, \tag{18.21}$$

where

$$\mathcal{X} = \{1, 2, \cdots, \lceil 2^{n\tau} \rceil\}. \tag{18.22}$$

If $u(k) \neq s$, if

$$Q_k = \{1 \leq k' < k : v(k') = u(k)\} \tag{18.23}$$

is nonempty, then

$$f_k : \prod_{k' \in Q_k} A_{k'} \to A_k; \tag{18.24}$$

otherwise, let f_k be an arbitrary constant taken from A_k.

5) (Decoding functions). Mappings

$$g_l : \prod_{k' \in W_l} A_{k'} \to \mathcal{X} \tag{18.25}$$

for $l = 1, 2, \cdots, L$, where

$$W_l = \{1 \leq k \leq K : v(k) = t_l\} \tag{18.26}$$

such that for all $l = 1, 2, \cdots, L$,

$$\tilde{g}_l(x) = x \tag{18.27}$$

for all $x \in \mathcal{X}$, where $\tilde{g}_l : \mathcal{X} \to \mathcal{X}$ is the function induced inductively by $f_k, 1 \leq k \leq K$, and g_l, and $\tilde{g}_l(x)$ denotes the value of g_l as a function of x.

The quantity τ is the rate of the information source X, which is also the rate at which information is multicast from the source node to all the sink nodes. The $(n, (\eta_e : e \in E), \tau)$ code is constructed from the above components as follows. At the beginning of a coding session, the value of X is available to node s. During the coding session, there are K transactions which take place in chronological order, where each transaction refers to a node sending information to another node. In the kth transaction, node $u(k)$ encodes according to encoding function f_k and sends an index in A_k to node $v(k)$. The domain of f_k is the set of all possible information that can be received by node $u(k)$ just before the kth transaction, and we distinguish two cases. If $u(k) = s$, the domain of f_k is $\mathcal{X}$. If $u(k) \neq s$, Q_k gives the time indices of all the previous transactions for which information was sent to node $u(k)$, so the domain of f_k is $\prod_{k' \in Q_k} A_{k'}$. The set T_e gives the time indices of all the transactions for which information is sent on channel e, so η_e is the number of possible index tuples that can be sent on channel e during the coding session. Finally, W_l gives the indices of all the transactions for which information is sent to node t_l, and g_l is the decoding function at node t_l which recovers x with zero error.

18.4 Proof of the Max-Flow Bound

In this section, we state and prove the max-flow bound for the class of network codes defined in the last section.

Definition 18.2. *For a graph G with rate constraints $\mathbf{R}$, an information rate $\omega \geq 0$ is asymptotically achievable if for any $\epsilon > 0$, there exists for sufficiently large n an $(n, (\eta_e : e \in E), \tau)$ network code on G such that*

$$n^{-1} \log_2 \eta_e \leq R_e + \epsilon \tag{18.28}$$

for all $e \in E$, where $n^{-1} \log_2 \eta_e$ is the average bit rate of the code on channel e, and

$$\tau \geq \omega - \epsilon. \tag{18.29}$$

For brevity, an asymptotically achievable information rate will be referred to as an achievable information rate.

Remark It follows from the above definition that if $\omega \geq 0$ is achievable, then ω' is also achievable for all $0 \leq \omega' \leq \omega$. Also, if $\omega^{(k)}$ is achievable for all $k \geq 1$, then it can be shown that $\omega = \lim_{k \to \infty} \omega^{(k)}$, if exists, is also achievable. Therefore, the set of all achievable information rates is closed and fully characterized by the maximum value in the set.

Theorem 18.3 (Max-Flow Bound). *For a graph G with rate constraints $\mathbf{R}$, if ω is achievable, then*

$$\omega \leq \min_{l} \mathrm{maxflow}(t_l). \tag{18.30}$$

Proof. It suffices to prove that for a graph G with rate constraints $\mathbf{R}$, if for any $\epsilon > 0$ there exists for sufficiently large n an $(n, (\eta_e : e \in E), \tau)$ code on G such that

$$n^{-1} \log_2 \eta_e \leq R_e + \epsilon \tag{18.31}$$

for all $e \in E$ and

$$\tau \geq \omega - \epsilon, \tag{18.32}$$

then ω satisfies (18.30).

Consider such a code for a fixed ϵ and a sufficiently large n, and consider any $l = 1, 2, \cdots, L$ and any cut U between node s and node t_l. Let

$$w_j(x) = (\tilde{f}_k(x) : k \in \cup_{e \in \mathrm{In}(j)} T_e), \tag{18.33}$$

where $x \in \mathcal{X}$ and $\tilde{f}_k : \mathcal{X} \to A_k$ is the function induced inductively by $f_{k'}, 1 \leq k' \leq k$, and $\tilde{f}_k(x)$ denotes the value of f_k as a function of x. The tuple $w_j(x)$ is all the information known by node j during the whole coding session when the message is x. Since $\tilde{f}_k(x)$ is a function of the information previously received by node $u(k)$, it can be shown by induction (see Problem 3) that $w_{t_l}(x)$ is a function of $\tilde{f}_k(x), k \in \cup_{e \in E_U} T_e$, where E_U is the set of edges across the cut U as previously defined in (18.7). Since x can be determined at node t_l, we have

$$H(X) \leq H(X, w_{t_l}(X)) \tag{18.34}$$

$$= H(w_{t_l}(X)) \tag{18.35}$$

$$\overset{a)}{\leq} H\left(\tilde{f}_k(X), k \in \bigcup_{e \in E_U} T_e\right) \tag{18.36}$$

$$\overset{b)}{\leq} \sum_{e \in E_U} \sum_{k \in T_e} H(\tilde{f}_k(X)) \tag{18.37}$$

$$\overset{c)}{\leq} \sum_{e \in E_U} \sum_{k \in T_e} \log_2 |A_k| \tag{18.38}$$

$$= \sum_{e \in E_U} \log_2 \left(\prod_{k \in T_e} |A_k|\right) \tag{18.39}$$

$$\overset{d)}{\leq} \sum_{e \in E_U} \log_2 \eta_e, \tag{18.40}$$

where

- (a) follows because $w_{t_l}(x)$ is a function of $\tilde{f}_k(x), k \in \cup_{e \in E_U} T_e$;
- (b) follows from the independence bound for entropy (Theorem 2.39);
- (c) follows from (18.21) and Theorem 2.43;
- (d) follows from (18.19).

Thus

$$\omega - \epsilon \leq \tau \tag{18.41}$$

$$\leq n^{-1} \log_2 \lceil 2^{n\tau} \rceil \tag{18.42}$$

$$= n^{-1} \log_2 |\mathcal{X}| \tag{18.43}$$

$$= n^{-1} H(X) \tag{18.44}$$

$$\leq \sum_{e \in E_U} n^{-1} \log_2 \eta_e \tag{18.45}$$

$$\leq \sum_{e \in E_U} (R_e + \epsilon) \tag{18.46}$$

$$\leq \sum_{e \in E_U} R_e + |E|\epsilon, \tag{18.47}$$

where (18.45) follows from (18.40). Minimizing the right-hand side over all U, we have

$$\omega - \epsilon \leq \min_U \sum_{e \in E_U} R_e + |E|\epsilon. \tag{18.48}$$

The first term on the right-hand side is the capacity of a min-cut between node s and node t_l. By the max-flow min-cut theorem, it is equal to the value of a max-flow from node s to node t_l, i.e., maxflow(t_l). Letting $\epsilon \to 0$, we obtain

$$\omega \le \text{maxflow}(t_l). \tag{18.49}$$

Since this upper bound on ω holds for all $l = 1, 2, \cdots, L$,

$$\omega \le \min_l \text{maxflow}(t_l). \tag{18.50}$$

The theorem is proved.   □

Remark 1 In proving the max-flow bound, the time evolution and the causality of the network code have been taken into account.

Remark 2 Even if we allow an arbitrarily small probability of decoding error in the usual Shannon sense, by modifying our proof by means of a standard application of Fano's inequality, it can be shown that it is still necessary for ω to satisfy (18.50). The details are omitted here.

Chapter Summary

Max-Flow Min-Cut Bound: In a point-to-point communication network, if node t receives an information source from node s, then the value of a maximum flow from s to t, or equivalently the capacity of a minimum cut between s to t, is at least equal to the rate of the information source.

Problems

1. In a network, for a flow $\mathbf{F}$ from a source node s to a sink node t, show that $F_+(s) = F_-(t)$ provided that the conservation conditions in (18.4) hold.
2. For the class of codes defined in Section 18.3, show that if the rates $\omega^{(k)}$ are achievable for all $k \ge 1$, then $\omega = \lim_{k\to\infty} \omega^{(k)}$, if exists, is also achievable (see Definition 18.2).
3. Prove the claim in the proof of Theorem 18.3 that for any cut U between node s and node t_l, $w_{t_l}(x)$ is a function of $\tilde{f}_k(x), k \in \cup_{e \in E_U} T_e$. Hint: Define

$$w_{j,\kappa}(x) = (\tilde{f}_k(x) : k \in \cup_{e \in \text{In}(j)} T_e, \; k \le \kappa)$$

and prove by induction on κ that for all $1 \le \kappa \le K$, $(w_{j,\kappa}(x) : j \notin U)$ is a function of $(\tilde{f}_k(x) : k \in \cup_{e \in E_U} T_e, \; k \le \kappa)$.

4. *Probabilistic network code.* For a network code defined in Section 18.3, the kth transaction of the coding process is specified by a mapping f_k. Suppose instead of a mapping f_k, the kth transaction is specified by a transition probability matrix from the domain of f_k to the range of f_k. Also, instead of a mapping g_l, decoding at sink node t_l is specified by a transition probability matrix from the domain of g_l to the range of g_l, $1 \leq l \leq L$. Conditioning on the indices received by node $u(k)$ during $1 \leq k' < k$, the index sent from node $u(k)$ to node $v(k)$ in the kth transaction is independent of all the previously generated random variables. Similarly, conditioning on all the indices received by sink node t_l during the whole coding session, the decoding at t_l is independent of all the previously generated random variables.

   We refer to such a code as a probabilistic network code. Since a deterministic network code is a special case of a probabilistic network code, the latter can potentially multicast at a higher rate compared with the former. Prove that this is not possible.

5. Consider a probabilistic network code on the network below.

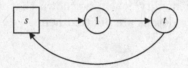

Let $X = (X_1, X_2)$ be uniformly distributed on $GF(2)^2$ and Z be independent of X and uniformly distributed on $GF(2)$. We use $\tilde{F}_k$ to denote the index transmitted in the kth transaction and W_{t_l} to denote $(\tilde{F}_k, k \in \cup_{e \in \mathrm{In}(t_l)} T_e)$. The probabilistic network code is specified by the following five transactions:

$$u(1) = s, \ v(1) = 1, \ \tilde{F}_1 = X_1,$$
$$u(2) = 1, \ v(2) = t, \ \tilde{F}_2 = X_1 + Z,$$
$$u(3) = t, \ v(3) = s, \ \tilde{F}_3 = X_1 + Z,$$
$$u(4) = s, \ v(4) = 1, \ \tilde{F}_4 = (X_1, X_2 + Z),$$
$$u(5) = 1, \ v(5) = t, \ \tilde{F}_5 = (X_1, X_2 + Z).$$

Note that the fourth transaction is possible because upon knowing X_1 and $X_1 + Z$, Z can be determined.

   a) Determine W_t.
   b) Verify that X can be recovered from W_t.
   c) Show that $X \to (\tilde{F}_1, \tilde{F}_4) \to W_t$ does not form a Markov chain.

   Here, $\tilde{F}_1$ and $\tilde{F}_4$ are all the random variables sent on edge $(s, 1)$ during the coding session. Although node t receives all the information through the edge $(s, 1)$, the Markov chain in (c) does not hold.

   (Ahlswede et al. [6].)

6. *Convolutional network code.* In the following network, $\text{maxflow}(s, t_l) = 3$ for $l = 1, 2, 3$. The max-flow bound asserts that 3 bits can be multicast to

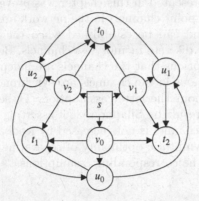

all the three sink nodes per unit time. We now describe a network coding scheme which achieves this. Let 3 bits $b_0(k), b_1(k), b_2(k)$ be generated at node s at time $k = 1, 2, \cdots$, where we assume without loss of generality that $b_l(k)$ is an element of the finite field $GF(2)$. We adopt the convention that $b_l(k) = 0$ for $k \leq 0$. At time $k \geq 1$, information transactions T1 to T11 occur in the following order:

  T1. s sends $b_l(k)$ to v_l, $l = 0, 1, 2$
  T2. v_l sends $b_l(k)$ to u_l, $t_{l \oplus 1}$, and $t_{l \oplus 2}$, $l = 0, 1, 2$
  T3. u_0 sends $b_0(k) + b_1(k-1) + b_2(k-1)$ to u_1
  T4. u_1 sends $b_0(k) + b_1(k-1) + b_2(k-1)$ to t_2
  T5. u_1 sends $b_0(k) + b_1(k) + b_2(k-1)$ to u_2
  T6. u_2 sends $b_0(k) + b_1(k) + b_2(k-1)$ to t_0
  T7. u_2 sends $b_0(k) + b_1(k) + b_2(k)$ to u_0
  T8. u_0 sends $b_0(k) + b_1(k) + b_2(k)$ to t_1
  T9. t_2 decodes $b_2(k-1)$
  T10. t_0 decodes $b_0(k)$
  T11. t_1 decodes $b_1(k)$

where "$\oplus$" denotes modulo 3 addition and "$+$" denotes modulo 2 addition.

a) Show that the information transactions T1 to T11 can be performed at time $k = 1$.

b) Show that T1 to T11 can be performed at any time $k \geq 1$ by induction on k.

c) Verify that at time k, nodes t_0 and t_1 can recover $b_0(k'), b_1(k')$, and $b_2(k')$ for all $k' \leq k$.

d) Verify that at time k, node t_2 can recover $b_0(k')$ and $b_1(k')$ for all $k' \leq k$, and $b_2(k')$ for all $k' \leq k - 1$. Note the unit time delay for t_2 to recover $b_2(k)$.

(Ahlswede et al. [6].)

Historical Notes

The max-flow bound presented in this chapter was proved by Ahlswede et al. [6], where the point-to-point channels in the network are noiseless.

The max-flow bound can be established when the point-to-point channels in the network are discrete memoryless channels. Borade [46] proved the bound with the assumptions that the channels are independent of each other and that the transmissions in the channels are synchronous. Song et al. [342] proved the bound without the latter assumption. These results are network generalizations of the result by Shannon [326] asserting that the capacity of a discrete memoryless channel is not increased by feedback (see Section 7.6), and they imply the asymptotic optimality of separating network coding and channel coding under the corresponding assumptions.

Single-Source Linear Network Coding:
Acyclic Networks

In the last chapter, we have established the max-flow bound as the fundamental bound for multicasting a single information source in a point-to-point communication network. In the next two chapters, we will construct *linear network codes* that achieve the max-flow bound at various levels of generality.

A finite field is a system of symbols on which one can perform operations corresponding to the four operations in arithmetic for real numbers, namely addition, subtraction, multiplication, and division. The set of real numbers and the associated operations together are referred to as the field of real numbers or simply the real field. Unlike the real field that has an infinite number of elements, a finite field has only a finite number of elements. For finite field theory, we refer the reader to [264]. For our discussions here, since we will not make use of the detailed structural properties of a finite field, the reader may by and large regard the algebra on a finite field and the algebra on the real field as the same.

In a linear network code, all the information symbols are regarded as elements of a finite field F called the *base field*. These include the symbols that comprise the information source as well as the symbols transmitted on the channels. For example, F is taken to be the binary field $GF(2)$ when the information unit is the bit. Furthermore, encoding and decoding are based on linear algebra defined on the based field, so that efficient algorithms for encoding and decoding as well as for code construction can be obtained.

In this chapter, we consider acyclic networks, i.e., networks with no directed cycle. We study the network coding problem in which a message consisting of a finite block of symbols is multicast. We make the ideal assumption that the propagation delay in the network, which includes the processing delay at the nodes and the transmission delay over the channels, is zero. In a general setting, a pipeline of messages may be multicast, and the propagation delay may be non-negligible. If the network is acyclic, then the operations in the network can be so synchronized that sequential messages are processed independent of each other. In this way, the network coding problem is independent of the propagation delay. Therefore, it suffices to study the network coding problem as described.

On the other hand, when a network contains directed cycles, the processing and transmission of sequential messages can convolve with together. Then the amount of delay incurred becomes part of the consideration in network coding. This will be discussed in the next chapter.

19.1 Acyclic Networks

Denote a directed network by $G = (V, E)$, where V and E are the sets of nodes and channels, respectively. A pair of channels $(d, e) \in E \times E$ is called an *adjacent pair* if there exists a node $t \in V$ such that $d \in \text{In}(t)$ and $e \in \text{Out}(t)$. A *directed path* in G is a sequence of channels

$$e_1, e_2, \cdots, e_m \tag{19.1}$$

such that (e_i, e_{i+1}) is an adjacent pair for all $1 \leq i < m$. Let $e_1 \in \text{Out}(t)$ and $e_m \in \text{In}(t')$. The sequence in (19.1) is called a directed path from e_1 to e_m or, equivalently, a directed path from node t to node t'. If $t = t'$, then the directed path is called a *directed cycle*. A directed network G is *cyclic* if it contains a directed cycle, otherwise G is *acyclic*.

Acyclic networks are easier to handle because the nodes in the network can be ordered in a way which allows encoding at the nodes to be carried out in a sequential and consistent manner. The following proposition and its proof describe such an order.

Proposition 19.1. *If G is a finite directed acyclic graph, then it is possible to order the nodes of G in a sequence such that if there is an edge from node i to node j, then node i appears before node j in the sequence.*

Proof. We partition the set V into subsets $V_1, V_2, \cdots$, such that node i is in V_k if and only if the length of a longest directed path ending at node i is equal to k. We first prove that if node i is in $V_{k'}$ and node j is in V_k such that there exists a directed path from node i to node j, then $k' < k$. Since the length of a longest directed path ending at node i is equal to k' and there exists a directed path from node i to node j (with length at least equal to 1), there exists a directed path ending at node j with length equal to $k' + 1$. As node j is in V_k, we have

$$k' + 1 \leq k, \tag{19.2}$$

so that

$$k' < k. \tag{19.3}$$

Hence, by listing the nodes of G in a sequence such that the nodes in $V_{k'}$ appear before the nodes in V_k if $k' < k$, where the order of the nodes within each V_k is arbitrary, we obtain an order of the nodes of G with the desired property. $\square$

Following the direction of the edges, we will refer to an order prescribed by Proposition 19.1 as an *upstream-to-downstream order*.[1] For a given acyclic network, such an order (not unique) is implicitly assumed. The nodes in the network encodes according to this order, referred to as the *encoding order*. Then whenever a node encodes, all the information needed would have already been received on the input channels of that node.

Example 19.2. Consider ordering the nodes in the butterfly network in Figure 17.1 by the sequence

$$s, 2, 1, 3, 4, t_2, t_1. \tag{19.4}$$

It is easy to check that in this sequence, if there is a directed path from node i to node j, then node i appears before node j.

19.2 Linear Network Codes

In this section, we formulate a linear network code on an acyclic network G. By allowing parallel channels between a pair of nodes, we assume without loss of generality that all the channels in the network have unit capacity, i.e., one symbol in the base field F can be transmitted on each channel. There exists a unique node s in G, called the source node, where a message consisting of ω symbols taken from the base field F is generated. To avoid trivially, we assume that every non-source node has at least one input channel.

As in Section 18.3, we assume that there is no loop in G, and there is no input channel at node s. To facilitate our discussion, however, we let $\text{In}(s)$ be a set of ω *imaginary channels* that terminate at node s but have no originating nodes. The reader may think of the ω symbols forming the message as being received by source node s on these ω imaginary channels. We emphasize that these imaginary channels are not part of the network, and the number of these channels is context dependent. Figure 19.1(a) illustrates the butterfly network with $\omega = 2$ imaginary channels appended at source node s.

Two directed paths P_1 and P_2 in G are *edge-disjoint* if the two paths do not share a common channel. It is not difficult to see from the conservation conditions in (18.4) that for a non-source node t, the maximum number of edge-disjoint paths from node s to node t is equal to $\text{maxflow}(t)$.

The message generated at source node s, consisting of ω symbols in the base field F, is represented by a row ω-vector $\mathbf{x} \in F^\omega$. Based on the value of $\mathbf{x}$, source node s transmits a symbol over each output channel. Encoding at the nodes in the network is carried out according to a certain upstream-to-downstream order. At a node in the network, the ensemble of received symbols

---

[1] Also called an ancestral order in graph theory.

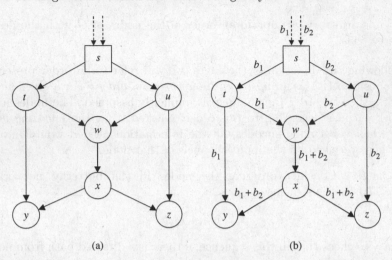

Fig. 19.1. (a) Two imaginary channels are appended to the source node of the butterfly network. (b) A two-dimensional network code for the butterfly network.

is mapped to a symbol in F specific to each output channel, and the symbol is sent on that channel. The following definition of a network code formally describes this mechanism. Since the code so defined is not necessarily linear, the base field F can be regarded in this context as any finite alphabet.

Definition 19.3 (Local Description of a Network Code). *An ω-dimensional network code on an acyclic network over a base field F consists of a local encoding mapping*

$$\tilde{k}_e : F^{|\mathrm{In}(t)|} \to F \tag{19.5}$$

for every channel e in the network, where $e \in \mathrm{Out}(t)$.

With the encoding mechanism as described, the local encoding mappings derive recursively the symbols transmitted over all channels e, denoted by $\tilde{f}_e(\mathbf{x})$. The above definition of a network code does not explicitly give the values of $\tilde{f}_e(\mathbf{x})$, whose mathematical properties are at the focus of the present discussion. Therefore, we also present an equivalent definition below, which describes a network code in terms of both the local encoding mechanisms as well as the recursively derived values $\tilde{f}_e(\mathbf{x})$.

Definition 19.4 (Global Description of a Network Code). *An ω-dimensional network code on an acyclic network over a base field F consists of a local encoding mapping*

$$\tilde{k}_e : F^{|\mathrm{In}(t)|} \to F \tag{19.6}$$

and a global encoding mapping

$$\tilde{f}_e : F^\omega \to F \tag{19.7}$$

for each channel e in the network, where $e \in \text{Out}(t)$, such that

(19.8) *For every node t and every channel $e \in \text{Out}(t)$, $\tilde{f}_e(\mathbf{x})$ is uniquely determined by $(\tilde{f}_d(\mathbf{x}) : d \in \text{In}(t))$ via the local encoding mapping $\tilde{k}_e$.*

(19.9) *The mappings $\tilde{f}_e$ for the ω imaginary channels $e \in \text{In}(s)$ project F^ω onto the distinct dimensions of F^ω.*

Example 19.5. Let $\mathbf{x} = [b_1 \, b_2]$ denote a generic row vector in $GF(2)^2$. Figure 19.1(b) shows a two-dimensional binary network code for the butterfly network with the following global encoding mappings:

$$\tilde{f}_e(x) = b_1 \quad \text{for } e = (o,s), (s,t), (t,w), (t,y), \tag{19.10}$$

$$\tilde{f}_e(x) = b_2 \quad \text{for } e = (o,s)', (s,u), (u,w), (u,z), \tag{19.11}$$

$$\tilde{f}_e(x) = b_1 + b_2 \quad \text{for } e = (w,x), (x,y), (x,z), \tag{19.12}$$

where (o,s) and $(o,s)'$ denote the two imaginary channels at node s. The corresponding local encoding mappings are

$$\tilde{k}_{(s,t)}(b_1, b_2) = b_1, \; \tilde{k}_{(s,u)}(b_1, b_2) = b_2, \tag{19.13}$$

$$\tilde{k}_{(t,w)}(b_1) = \tilde{k}_{(t,y)}(b_1) = b_1, \tag{19.14}$$

$$\tilde{k}_{(u,w)}(b_2) = \tilde{k}_{(u,z)}(b_2) = b_2, \; \tilde{k}_{(w,x)}(b_1, b_2) = b_1 + b_2, \tag{19.15}$$

etc.

When a global encoding mapping $\tilde{f}_e$ is linear, it corresponds to a column ω-vector $\mathbf{f}_e$ such that $\tilde{f}_e(x)$ is the product $\mathbf{x} \cdot \mathbf{f}_e$, where the row ω-vector $\mathbf{x}$ is the message generated at node s. Similarly, when a local encoding mapping $\tilde{k}_e$, where $e \in \text{Out}(t)$, is linear, it corresponds to a column $|\text{In}(t)|$-vector $\mathbf{k}_e$ such that $\tilde{k}_e(\mathbf{y}) = \mathbf{y} \cdot \mathbf{k}_e$, where $\mathbf{y} \in F^{|\text{In}(t)|}$ is the row vector representing the symbols received at node t. In an ω-dimensional network code on an acyclic network, if all the local encoding mappings are linear, then so are the global encoding mappings since they are functional compositions of the local encoding mappings. The converse is also true: If the global encoding mappings are all linear, then so are the local encoding mappings. We leave the proof as an exercise.

In the following, we formulate a linear network code as a network code whose local and global encoding mappings are all linear. Again, both the local and global descriptions are presented even though they are equivalent. The global description of a linear network code will be very useful when we construct such codes in Section 19.4.

Definition 19.6 (Local Description of a Linear Network Code). *An ω-dimensional linear network code on an acyclic network over a base field F consists of a scalar $k_{d,e}$, called the local encoding kernel, for every adjacent pair of channels (d, e) in the network. The $|\text{In}(t)| \times |\text{Out}(t)|$ matrix*

$$K_t = [k_{d,e}]_{d \in \text{In}(t), e \in \text{Out}(t)} \tag{19.16}$$

is called the local encoding kernel at node t.

Note that the matrix structure of K_t implicitly assumes an ordering among the channels.

Definition 19.7 (Global Description of a Linear Network Code). *An ω-dimensional linear network code on an acyclic network over a base field F consists of a scalar $k_{d,e}$ for every adjacent pair of channels (d, e) in the network as well as a column ω-vector $\mathbf{f}_e$ for every channel e such that*

(19.17) $\mathbf{f}_e = \sum_{d \in \text{In}(t)} k_{d,e} \mathbf{f}_d$ *for $e \in \text{Out}(t)$.*

(19.18) *The vectors $\mathbf{f}_e$ for the ω imaginary channels $e \in \text{In}(s)$ form the standard basis of the vector space F^ω.*

The vector $\mathbf{f}_e$ is called the global encoding kernel for channel e.

We now explain how the global description above specifies the linear network code. Initially, source node s generates a message $\mathbf{x}$ as a row ω-vector. In view of (19.18), the symbols in $\mathbf{x}$ are regarded as being received by source node s on the imaginary channels as $\mathbf{x} \cdot \mathbf{f}_d, d \in \text{In}(s)$. Starting at source node s, any node t in the network receives the symbols $\mathbf{x} \cdot \mathbf{f}_d, d \in \text{In}(t)$, from which it calculates the symbol $\mathbf{x} \cdot \mathbf{f}_e$ for sending on each channel $e \in \text{Out}(t)$ via the linear formula

$$\mathbf{x} \cdot \mathbf{f}_e = \mathbf{x} \sum_{d \in \text{In}(t)} k_{d,e} \mathbf{f}_d = \sum_{d \in \text{In}(t)} k_{d,e} (\mathbf{x} \cdot \mathbf{f}_d), \tag{19.19}$$

where the first equality follows from (19.17). In this way, the symbol $\mathbf{x} \cdot \mathbf{f}_e$ is transmitted on any channel e (which may be an imaginary channel) in the network.

Given the local encoding kernels for all the channels in an acyclic network, the global encoding kernels can be calculated recursively in any upstream-to-downstream order by (19.17), while (19.18) provides the boundary conditions.

Remark A partial analogy can be drawn between the global encoding kernels for the channels in a linear network code and the columns of a generator matrix of a linear block code in algebraic coding theory [234, 39, 378]. The former are indexed by the channels in the network, while the latter are indexed by "time." However, the global encoding kernels in a linear network code are constrained by the network topology via (19.17), while the columns in the generator matrix of a linear block code in general are not subject to any such constraint.

The following two examples illustrate the relation between the local encoding kernels and the global encoding kernels of a linear network code. The reader should understand these two examples thoroughly before proceeding to the next section.

Example 19.8. The network code in Figure 19.1(b) is in fact linear. Assume the alphabetical order among the channels $(o, s), (o, s)', (s, t), \cdots, (x, z)$. Then the local encoding kernels at the nodes are

$$K_s = \begin{bmatrix} 1 & 0 \\ 0 & 1 \end{bmatrix}, \; K_t = K_u = K_x = \begin{bmatrix} 1 & 1 \end{bmatrix}, \; K_w = \begin{bmatrix} 1 \\ 1 \end{bmatrix}. \tag{19.20}$$

The corresponding global encoding kernels are

$$\mathbf{f}_e = \begin{cases} \begin{bmatrix} 1 \\ 0 \end{bmatrix} & \text{for } e = (o, s), (s, t), (t, w), \text{ and } (t, y) \\[3ex] \begin{bmatrix} 0 \\ 1 \end{bmatrix} & \text{for } e = (o, s)', (s, u), (u, w), \text{ and } (u, z) \; . \\[3ex] \begin{bmatrix} 1 \\ 1 \end{bmatrix} & \text{for } e = (w, x), (x, y), \text{ and } (x, z) \end{cases} \tag{19.21}$$

The local/global encoding kernels are summarized in Figure 19.2. In fact, they describe a two-dimensional linear network code regardless of the choice of the base field.

Example 19.9. For a general two-dimensional linear network code on the network in Figure 19.2, the local encoding kernels at the nodes can be expressed as

$$K_s = \begin{bmatrix} a & c \\ b & d \end{bmatrix}, \; K_t = \begin{bmatrix} e & f \end{bmatrix}, \; K_u = \begin{bmatrix} g & h \end{bmatrix}, \tag{19.22}$$

$$K_w = \begin{bmatrix} i \\ j \end{bmatrix}, \; K_x = \begin{bmatrix} k & l \end{bmatrix}, \tag{19.23}$$

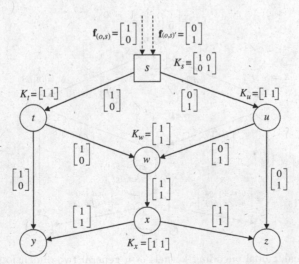

Fig. 19.2. The global and local encoding kernels for the two-dimensional linear network code in Example 19.8.

where $a, b, c, \cdots, l$, the entries of the matrices, are indeterminates in the base field F. Starting with

$$\mathbf{f}_{(o,s)} = \begin{bmatrix} 1 \\ 0 \end{bmatrix} \text{ and } \mathbf{f}_{(o,s)'} = \begin{bmatrix} 0 \\ 1 \end{bmatrix}, \tag{19.24}$$

we can obtain all the global encoding kernels below by applying (19.17) recursively:

$$\mathbf{f}_{(s,t)} = \begin{bmatrix} a \\ b \end{bmatrix}, \ \mathbf{f}_{(s,u)} = \begin{bmatrix} c \\ d \end{bmatrix}, \ \mathbf{f}_{(t,w)} = \begin{bmatrix} ae \\ be \end{bmatrix}, \ \mathbf{f}_{(t,y)} = \begin{bmatrix} af \\ bf \end{bmatrix}, \tag{19.25}$$

$$\mathbf{f}_{(u,w)} = \begin{bmatrix} cg \\ dg \end{bmatrix}, \ \mathbf{f}_{(u,z)} = \begin{bmatrix} ch \\ dh \end{bmatrix}, \ \mathbf{f}_{(w,x)} = \begin{bmatrix} aei + cgj \\ bei + dgj \end{bmatrix}, \tag{19.26}$$

$$\mathbf{f}_{(x,y)} = \begin{bmatrix} aeik + cgjk \\ beik + dgjk \end{bmatrix}, \ \mathbf{f}_{(x,z)} = \begin{bmatrix} aeil + cgjl \\ beil + dgjl \end{bmatrix}. \tag{19.27}$$

For example, $\mathbf{f}_{(w,x)}$ is obtained from $\mathbf{f}_{(t,w)}$ and $\mathbf{f}_{(u,w)}$ by

$$\mathbf{f}_{(w,x)} = k_{(t,w),(w,x)}\mathbf{f}_{(t,w)} + k_{(u,w),(w,x)}\mathbf{f}_{(u,w)} \tag{19.28}$$

$$= i \begin{bmatrix} ae \\ be \end{bmatrix} + j \begin{bmatrix} cg \\ dg \end{bmatrix} \tag{19.29}$$

$$= \begin{bmatrix} aei + cgj \\ bei + dgj \end{bmatrix}. \tag{19.30}$$

The local/global encoding kernels of the general linear network code are summarized in Figure 19.3.

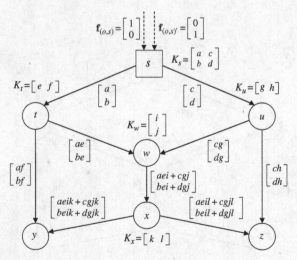

Fig. 19.3. Local/global encoding kernels of a general two-dimensional linear network code.

19.3 Desirable Properties of a Linear Network Code

We have proved in Section 18.4 that in a communication network represented by a graph G, the rate at which information is transmitted from source node s to any node t cannot exceed maxflow(t), the value of a max-flow from node s to node t. For a collection of non-source nodes T, denote by maxflow(T) the value of a max-flow from node s to T. Then it is readily seen that the rate at which information is transmitted from source node s to the collection of nodes T cannot exceed maxflow(T).

In the sequel, we adopt the conventional notation $\langle \cdot \rangle$ for the linear span of a set of vectors. For a node t, let

$$V_t = \langle \{ \mathbf{f}_e : e \in \text{In}(t) \} \rangle \tag{19.31}$$

and for a collection T of nodes, let

$$V_T = \langle \cup_{t \in T} V_t \rangle. \tag{19.32}$$

For a collection ξ of channels, let

$$V_\xi = \langle \{ \mathbf{f}_e : e \in \xi \} \rangle, \tag{19.33}$$

with the convention $V_\emptyset = \{0\}$, where 0 denotes the zero column ω-vector.

In the next theorem, we first establish a specific form of the max-flow bound which applies to linear network coding.

Theorem 19.10 (Max-Flow Bound for Linear Network Coding). *For an ω-dimensional linear network code on an acyclic network, for any collection T of non-source nodes,*

$$\dim(V_T) \leq \min\{\omega, \text{maxflow}(T)\}. \tag{19.34}$$

Proof. Let the acyclic network be $G = (V, E)$. Consider U be a cut between source node s and a collection T of non-source nodes, and let E_U be the set of edges across the cut U as in (18.7). Then V_T is a linear transformation of V_{E_U}, where

$$\dim(V_T) \leq \dim(V_{E_U}) \leq |E_U|. \tag{19.35}$$

Minimizing over all the cuts between s and T and invoking the max-flow min-cut theorem, we have

$$\dim(V_T) \leq \text{maxflow}(T). \tag{19.36}$$

On the other hand, V_T is a linear transformation of $V_s = F^\omega$. Therefore,

$$\dim(V_T) \leq \dim(V_s) = \omega. \tag{19.37}$$

Then the proof is completed by combining (19.36) and (19.37). □

For a collection of channels $\xi \subset E$ (i.e., not including the imaginary channels), we denote by $\mathrm{maxflow}(\xi)$ the value of a max-flow from source node s to ξ. Theorem 19.10 has the following straightforward corollary.

Corollary 19.11. *For an ω-dimensional linear network code on an acyclic network, for any collection of channels $\xi \subset E$,*

$$\dim(V_\xi) \leq \min\{\omega, \mathrm{maxflow}(\xi)\}. \tag{19.38}$$

Whether the max-flow bound in Theorem 19.10 or Corollary 19.11 is achievable depends on the network topology, the dimension ω, and the coding scheme. Three special classes of linear network codes are defined below by the achievement of this bound to three different extents.

Definition 19.12. *An ω-dimensional linear network code on an acyclic network qualifies as a linear multicast, a linear broadcast, or a linear dispersion, respectively, if the following hold:*

(19.39) $\dim(V_t) = \omega$ *for every non-source node t with* $\mathrm{maxflow}(t) \geq \omega$.
(19.40) $\dim(V_t) = \min\{\omega, \mathrm{maxflow}(t)\}$ *for every non-source node t.*
(19.41) $\dim(V_T) = \min\{\omega, \mathrm{maxflow}(T)\}$ *for every collection T of non-source nodes.*

For a set ξ of channels, including possibly the imaginary channels, let

$$F_\xi = \big[\, \mathbf{f}_e \,\big]_{e \in \xi} \tag{19.42}$$

be the $\omega \times |\xi|$ matrix obtained by putting $\mathbf{f}_e, e \in \xi$ in juxtaposition. For a node t, the symbols $\mathbf{x} \cdot \mathbf{f}_e$, $e \in \mathrm{In}(t)$ are received on the input channels. Equivalently, the row $|\mathrm{In}(t)|$-vector

$$\mathbf{x} \cdot F_{\mathrm{In}(t)} \tag{19.43}$$

is received. Obviously, the message $\mathbf{x}$, consisting of ω information units, can be uniquely determined at the node if and only if the rank of $F_{\mathrm{In}(t)}$ is equal to ω, i.e.,

$$\dim(V_t) = \omega. \tag{19.44}$$

The same applies to a collection T of non-source nodes.

For a linear multicast, a node t can decode the message $\mathbf{x}$ if and only if $\mathrm{maxflow}(t) \geq \omega$. For a node t with $\mathrm{maxflow}(t) < \omega$, nothing is guaranteed. An application of an ω-dimensional linear multicast is for multicasting information at rate ω to all (or some of) those non-source nodes with max-flow at least equal to ω.

For a linear broadcast, like a linear multicast, a node t can decode the message $\mathbf{x}$ if and only if $\mathrm{maxflow}(t) \geq \omega$. For a node t with $\mathrm{maxflow}(t) < \omega$, the set of all received vectors, namely

$$\{\mathbf{x} \cdot F_{\text{In}(t)} : \mathbf{x} \in F^\omega\}, \tag{19.45}$$

forms a vector subspace of F^ω with dimension equal to maxflow(t), but there is no guarantee on which such subspace is actually received.[2] An application of linear broadcast is for multicasting information on a network at a variable rate (see Problem 14). A random version of linear broadcast (to be discussed in Section 19.4) is also useful for identifying the max-flow of a non-source in an unknown network topology [352].

For a linear dispersion, a collection T of non-source nodes can decode the message $\mathbf{x}$ if and only if maxflow(T) $\geq \omega$. If maxflow(T) $< \omega$, the collection T receives a vector subspace with dimension equal to maxflow(T). Again, there is no guarantee on which such subspace is actually received. An application of linear dispersion is in a two-tier network system consisting of the backbone network and a number of local area networks (LANs), where each LAN is connected to one or more nodes on the backbone network. An information source with rate ω, generated at a node s in the backbone network, is to be transmitted to every user on the LANs. With a linear dispersion on the backbone network, every user on a LAN can receive the information source as long as the LAN acquires through the backbone network an aggregated max-flow from node s at least equal to ω. Moreover, new LANs can be established under the same criterion without modifying the linear dispersion on the backbone network.

Note that for all the three classes of linear network codes in Definition 19.12, a sink node is not explicitly identified. Also, it is immediate from the definition that every linear dispersion is a linear broadcast, and every linear broadcast is a linear multicast. The example below shows that a linear broadcast is not necessarily a linear dispersion, a linear multicast is not necessarily a linear broadcast, and a linear network code is not necessarily a linear multicast.

Example 19.13. Figure 19.4(a) shows a two-dimensional linear dispersion on an acyclic network with the global encoding kernels as prescribed. Figure 19.4(b) shows a two-dimensional linear broadcast on the same network that is not a linear dispersion because

$$\text{maxflow}(\{t, u\}) = 2 = \omega, \tag{19.46}$$

while the global encoding kernels of the channels in $\text{In}(t) \cup \text{In}(u)$ span only a one-dimensional subspace. Figure 19.4(c) shows a two-dimensional linear multicast that is not a linear broadcast since node u receives no information at all. Finally, the two-dimensional linear network code in Figure 19.4(d) is not a linear multicast.

Example 19.14. The linear network code in Example 19.8 meets all the criteria (19.39) through (19.41) in Definition 19.12. Thus it is a two-dimensional linear

---

[2] Here F^ω refers to the row vector space.

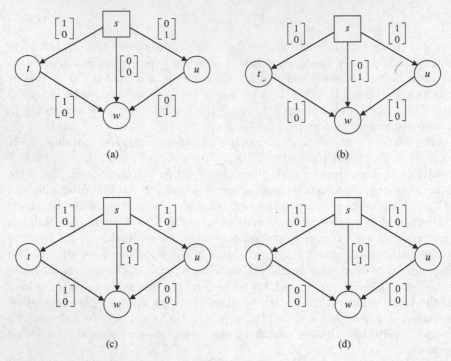

Fig. 19.4. (a) A two-dimensional linear dispersion over an acyclic network. (b) A two-dimensional linear broadcast that is not a linear dispersion. (c) A two-dimensional linear multicast that is not a linear broadcast. (d) A two-dimensional linear network code that is not a linear multicast.

dispersion, and hence also a linear broadcast and linear multicast, regardless of the choice of the base field. The same applies to the linear network code in Figure 19.4(a).

Example 19.15. The general linear network code in Example 19.9 meets the criterion (19.39) for a linear multicast when

- $\mathbf{f}_{(t,w)}$ and $\mathbf{f}_{(u,w)}$ are linearly independent;
- $\mathbf{f}_{(t,y)}$ and $\mathbf{f}_{(x,y)}$ are linearly independent;
- $\mathbf{f}_{(u,z)}$ and $\mathbf{f}_{(x,z)}$ are linearly independent.

Equivalently, the criterion says that $e, f, g, h, k, l, ad - bc, abei + adgj - baei - bcgj$, and $daei + dcgj - cbei - cdgj$ are all nonzero. Example 19.8 has been the special case with

$$a = d = e = f = g = h = i = j = k = l = 1 \qquad (19.47)$$

and

$$b = c = 0. \qquad (19.48)$$

19.3.1 Transformation of a Linear Network Code

Consider an ω-dimensional linear network code $\mathcal{C}$ on an acyclic network. Suppose source node s, instead of encoding the message $\mathbf{x}$, encodes

$$\mathbf{x}' = \mathbf{x}A, \tag{19.49}$$

where A is an *invertible* $\omega \times \omega$ matrix. Then the symbol sent on a channel $e \in E$ is given by

$$\mathbf{x}' \cdot \mathbf{f}_e = (\mathbf{x}A) \cdot \mathbf{f}_e = \mathbf{x} \cdot (A\mathbf{f}_e). \tag{19.50}$$

This gives a new linear network code $\mathcal{C}'$ with respect to the message $\mathbf{x}$ with global encoding kernels

$$\mathbf{f}'_e = \begin{cases} A\,\mathbf{f}_e & \text{if } e \in E \\ \mathbf{f}_e & \text{if } e \in \text{In}(s) \end{cases}. \tag{19.51}$$

Recall the definition of the matrix F_ξ in (19.42) for a set of channels ξ. Then (19.17) can be written in matrix form as

$$F_{\text{Out}(t)} = F_{\text{In}(t)}K_t \tag{19.52}$$

for all nodes t, where K_t is the local encoding kernel at node t. Similarly, letting

$$F'_\xi = \left[\mathbf{f}'_e\right]_{e \in \xi}, \tag{19.53}$$

we obtain from (19.51) that

$$F'_{\text{Out}(t)} = AF_{\text{Out}(t)} \tag{19.54}$$

for all nodes t,

$$F'_{\text{In}(t)} = AF_{\text{In}(t)} \tag{19.55}$$

for all nodes $t \neq s$, and

$$F'_{\text{In}(s)} = F_{\text{In}(s)}. \tag{19.56}$$

For a node $t \neq s$, from (19.54), (19.52), and (19.55),

$$F'_{\text{Out}(t)} = AF_{\text{Out}(t)} \tag{19.57}$$
$$= A(F_{\text{In}(t)}K_t) \tag{19.58}$$
$$= (AF_{\text{In}(t)})K_t \tag{19.59}$$
$$= F'_{\text{In}(t)}K_t. \tag{19.60}$$

Since $\mathbf{f}_e, e \in \text{In}(s)$ form the standard basis of F^ω,

$$F_{\text{In}(s)} = F'_{\text{In}(s)} = I, \tag{19.61}$$

the $\omega \times \omega$ identity matrix. It then follows from (19.54), (19.52), and (19.61) that

$$F'_{\text{Out}(s)} = AF_{\text{Out}(s)} \tag{19.62}$$
$$= A(F_{\text{In}(s)}K_s) \tag{19.63}$$
$$= AK_s \tag{19.64}$$
$$= F'_{\text{In}(s)}(AK_s). \tag{19.65}$$

Comparing (19.60) and (19.65) with (19.52), we see that the local encoding kernels of C' are given by

$$K'_t = \begin{cases} K_t & \text{if } t \neq s \\ AK_s & \text{if } t = s \end{cases}. \tag{19.66}$$

The network code C' is called the *transformation* of the network code C by the (invertible) matrix A. In view of Definition 19.12, the requirements of a linear multicast, a linear broadcast, and a linear dispersion are all in terms of the linear independence among the global encoding kernels. We leave it as an exercise for the reader to show that if a network code is a linear multicast, broadcast, or dispersion, then any transformation of it is also a linear multicast, broadcast, or dispersion, respectively.

Suppose C is an ω-dimensional linear multicast and let C' be a transformation of C. When the network code C' is employed, the message $\mathbf{x}$ can be decoded by any node t with maxflow$(t) \geq \omega$, because from the foregoing C' is also a linear multicast. For the purpose of multicasting, there is no difference between C and C', and they can be regarded as equivalent.

If C is an ω-dimensional linear broadcast and C' is a transformation of C, then C' is also an ω-dimensional linear broadcast. However, C as a linear broadcast may deliver to a particular node t with maxflow$(t) < \omega$ a certain subset of symbols in the message $\mathbf{x}$, while C' may not be able to achieve the same. Then whether C and C' can be regarded as equivalent depends on the specific requirements of the application. As an example, the linear network code in Figure 19.1(b) delivers b_1 to node t. However, taking a transformation of the network code with matrix

$$A = \begin{bmatrix} 1 & 0 \\ 1 & 1 \end{bmatrix}, \tag{19.67}$$

the resulting network code can no longer deliver b_1 to node t, although nodes w, v, and z can continue to decode both b_1 and b_2.

19.3.2 Implementation of a Linear Network Code

In implementation of a linear network code, be it a linear multicast, a linear broadcast, a linear dispersion, or any linear network code, in order that the code can be used as intended, the global encoding kernels $\mathbf{f}_e, e \in \text{In}(t)$ must be known by each node t if node t is to recover any useful information from the symbols received on the input channels. These global encoding kernels can be made available ahead of time if the code is already decided. Alternatively, they can be delivered through the input channels if multiple usage of the network is allowed.

One possible way to deliver the global encoding kernels to node t in a coding session of length n, where $n > \omega$, is as follows. At time $k = 1, 2, \cdots, \omega$, the source node transmits the dummy message $\mathbf{m}_k$, a row ω-vector with all the components equal to 0 except that the kth component is equal to 1. Note that

$$\begin{bmatrix} \mathbf{m}_1 \\ \mathbf{m}_2 \\ \vdots \\ \mathbf{m}_\omega \end{bmatrix} = I_\omega, \tag{19.68}$$

the $\omega \times \omega$ identity matrix. At time $k = \omega + i$, where $i = 1, 2, \cdots, n - \omega$, the source node transmits the message $\mathbf{x}_i$. Then throughout the coding session, node t receives

$$\begin{bmatrix} \mathbf{m}_1 \\ \mathbf{m}_2 \\ \vdots \\ \mathbf{m}_\omega \\ \mathbf{x}_1 \\ \mathbf{x}_2 \\ \vdots \\ \mathbf{x}_{n-\omega} \end{bmatrix} F_{\mathrm{In}(t)} = \begin{bmatrix} I_\omega \\ \mathbf{x}_1 \\ \mathbf{x}_2 \\ \vdots \\ \mathbf{x}_{n-\omega} \end{bmatrix} F_{\mathrm{In}(t)} = \begin{bmatrix} F_{\mathrm{In}(t)} \\ \mathbf{x}_1 \cdot F_{\mathrm{In}(t)} \\ \mathbf{x}_2 \cdot F_{\mathrm{In}(t)} \\ \vdots \\ \mathbf{x}_{n-\omega} \cdot F_{\mathrm{In}(t)} \end{bmatrix} \tag{19.69}$$

on the input channels. In other words, the global encoding kernels of the input channels at node t are received at the beginning of the coding session. This applies to all the sink nodes in the network simultaneously because the ω dummy messages do not depend on the particular node t. If $F_{\mathrm{In}(t)}$ has full rank, then node t can start to decode $\mathbf{x}_1$ upon receiving $\mathbf{x}_1 \cdot F_{\mathrm{In}(t)}$.

Since $n - \omega$ messages are transmitted in a coding session of length n, the utilization of the network is equal to $(n - \omega)/n$, which tends to 1 as $n \to \infty$. That is, the overhead for delivering the global encoding kernels through the network is asymptotically negligible.

19.4 Existence and Construction

For a given acyclic network, the following three factors dictate the existence of an ω-dimensional linear network code with a prescribed set of desirable properties:

- the value of ω,
- the network topology,
- the choice of the base field F.

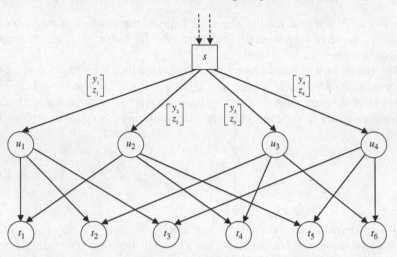

Fig. 19.5. A network with a two-dimensional ternary linear multicast but without a two-dimensional binary linear multicast.

We begin with an example illustrating the third factor.

Example 19.16. On the network in Figure 19.5, a two-dimensional ternary linear multicast can be constructed by the following local encoding kernels at the nodes:

$$K_s = \begin{bmatrix} 0 & 1 & 1 & 1 \\ 1 & 0 & 1 & 2 \end{bmatrix} \quad \text{and} \quad K_{u_i} = \begin{bmatrix} 1 & 1 & 1 \end{bmatrix} \tag{19.70}$$

for $1 \leq i \leq 4$. On the other hand, we can prove the nonexistence of a two-dimensional binary linear multicast on this network as follows. Assuming the contrary that a two-dimensional binary linear multicast exists, we will derive a contradiction. Let the global encoding kernel $\mathbf{f}_{(s,u_i)} = [\, y_i \ z_i \,]^\top$ for $1 \leq i \leq 4$. Since maxflow$(t_k) = 2$ for all $1 \leq k \leq 6$, the global encoding kernels for the two input channels to each node t_k must be linearly independent. Thus, if node t_k is at the downstream of both nodes u_i and u_j, then the two vectors $[\, y_i \ z_i \,]^\top$ and $[\, y_i \ z_i \,]^\top$ must be linearly independent. As each node t_k is at the downstream of a different pair of nodes among u_1, u_2, u_3, and u_4, the four vectors $[\, y_i \ z_i \,]^\top$, $1 \leq i \leq 4$, are pairwise linearly independent, and consequently, must be four distinct vectors in $GF(2)^2$. Then one of them must be $[\, 0 \ 0 \,]^\top$ since there are only four vectors in $GF(2)^2$. This contradicts the pairwise linear independence among the four vectors.

In order for the linear network code to qualify as a linear multicast, a linear broadcast, or a linear dispersion, it is required that certain collections of global encoding kernels span the maximum possible dimensions. This is equivalent to certain polynomial functions taking nonzero values, where the indeterminates of these polynomials are the local encoding kernels. To fix ideas, take $\omega = 3$, consider a node t with two input channels, and put the global encoding kernels of these two channels in juxtaposition to form a 3×2 matrix. Then, this matrix attains the maximum possible rank of 2 if and only if there exists a 2×2 submatrix with nonzero determinant.

According to the local description, a linear network code is specified by the local encoding kernels, and the global encoding kernels can be derived recursively in an upstream-to-downstream order. From Example 19.8, it is not hard to see that every component in a global encoding kernel is a polynomial function whose indeterminates are the local encoding kernels.

When a nonzero value of a polynomial function is required, it does not merely mean that at least one coefficient in the polynomial is nonzero. Rather, it means a way to choose scalar values for the indeterminates so that the polynomial function is evaluated to a nonzero scalar value.

When the base field is small, certain polynomial equations may be unavoidable. For instance, for any prime number p, the polynomial equation $z^p - z = 0$ is satisfied for any $z \in GF(p)$. The nonexistence of a binary linear multicast in Example 19.16 can also trace its root to a set of polynomial equations that cannot be avoided simultaneously over $GF(2)$.

However, when the base field is sufficiently large, every nonzero polynomial function can indeed be evaluated to a nonzero value with a proper choice of the values taken by the set of indeterminates involved. This is formally stated in the following elementary lemma, which will be instrumental in the proof of Theorem 19.20 asserting the existence of a linear multicast on an acyclic network when the base field is sufficiently large.

Lemma 19.17. *Let $g(z_1, z_2, \cdots, z_n)$ be a nonzero polynomial with coefficients in a field F. If $|F|$ is greater than the degree of g in every z_j for $1 \leq j \leq n$, then there exist $a_1, a_2, \cdots, a_n \in F$ such that*

$$g(a_1, a_2, \cdots, a_n) \neq 0. \tag{19.71}$$

Proof. The lemma is proved by induction on n. For $n = 0$, g is a nonzero constant in F, and the lemma is obviously true. Assume that the lemma is true for $n - 1$ for some $n \geq 1$. Express $g(z_1, z_2, \cdots, z_n)$ as a polynomial in z_n with coefficients in the polynomial ring $F[z_1, z_2, \cdots, z_{n-1}]$, i.e.,

$$g(z_1, z_2, \cdots, z_n) = h(z_1, z_2, \cdots, z_{n-1}) z_n{}^k + \cdots, \tag{19.72}$$

where k is the degree of g in z_n and the leading coefficient $h(z_1, z_2, \cdots, z_{n-1})$ is a nonzero polynomial in $F[z_1, z_2, \cdots, z_{n-1}]$. By the induction hypothesis, there exist $a_1, a_2, \cdots, a_{n-1} \in F$ such that $h(a_1, a_2, \cdots, a_{n-1}) \neq 0$. Thus $g(a_1, a_2, \cdots, a_{n-1}, z)$ is a nonzero polynomial in z with degree $k < |F|$. Since this polynomial cannot have more than k roots in F and $|F| > k$, there exists $a_n \in F$ such that

$$g(a_1, a_2, \cdots, a_{n-1}, a_n) \neq 0. \tag{19.73}$$

Corollary 19.18. *Let $g(z_1, z_2, \cdots, z_n)$ be a nonzero polynomial with coefficients in a field F with $|F| > m$, where m is the highest degree of g in z_j for*

$1 \leq j \leq n$. Let $a_1, a_2, \cdots, a_n$ be chosen independently according to the uniform distribution on F. Then

$$\Pr\{g(a_1, a_2, \cdots, a_n) \neq 0\} \geq \left(1 - \frac{m}{|F|}\right)^n. \tag{19.74}$$

In particular,

$$\Pr\{g(a_1, a_2, \cdots, a_n) \neq 0\} \to 1 \tag{19.75}$$

as $|F| \to \infty$.

Proof. The first part of the corollary is proved by induction on n. For $n = 0$, g is a nonzero constant in F, and the proposition is obviously true. Assume that the proposition is true for $n - 1$ for some $n \geq 1$. From (19.72) and the induction hypothesis, we see that

$$\Pr\{g(z_1, z_2, \cdots, z_n) \neq 0\} \tag{19.76}$$
$$= \Pr\{h(z_1, z_2, \cdots, z_{n-1}) \neq 0\} \Pr\{g(z_1, z_2, \cdots, z_n) \neq 0| \tag{19.77}$$
$$h(z_1, z_2, \cdots, z_{n-1}) \neq 0\} \tag{19.78}$$
$$\geq \left(1 - \frac{m}{|F|}\right)^{n-1} \Pr\{g(z_1, z_2, \cdots, z_n) \neq 0| \tag{19.79}$$
$$h(z_1, z_2, \cdots, z_{n-1}) \neq 0\} \tag{19.80}$$
$$\geq \left(1 - \frac{m}{|F|}\right)^{n-1} \left(1 - \frac{m}{|F|}\right) \tag{19.81}$$
$$= \left(1 - \frac{m}{|F|}\right)^n. \tag{19.82}$$

This proves the first part of the corollary. As n is fixed, the lower bound above tends to 1 as $|F| \to \infty$. This completes the proof. $\square$

Example 19.19. Recall the two-dimensional linear network code in Example 19.9 that is expressed in the 12 indeterminates $a, b, c, \cdots, l$. Place the vectors $\mathbf{f}_{(t,w)}$ and $\mathbf{f}_{(u,w)}$ in juxtaposition into the 2×2 matrix

$$L_w = \begin{bmatrix} ae & cg \\ be & dg \end{bmatrix}, \tag{19.83}$$

the vectors $\mathbf{f}_{(t,y)}$ and $\mathbf{f}_{(x,y)}$ into the 2×2 matrix

$$L_y = \begin{bmatrix} af & aeik + cgjk \\ bf & beik + dgjk \end{bmatrix}, \tag{19.84}$$

and the vectors $\mathbf{f}_{(u,z)}$ and $\mathbf{f}_{(x,z)}$ into the 2×2 matrix

$$L_z = \begin{bmatrix} aeil + cgjl & ch \\ beil + dgjl & dh \end{bmatrix}. \tag{19.85}$$

Clearly,

$$\det(L_w) \cdot \det(L_y) \cdot \det(L_z) \neq 0 \in F[a, b, c, \cdots, l]. \tag{19.86}$$

Applying Lemma 19.17 to the polynomial on the left-hand side above, we can set scalar values for the 12 indeterminates so that it is evaluated to a nonzero value in F when F is sufficiently large. This implies that the determinants on the left-hand side of (19.86) are evaluated to nonzero values in F simultaneously. Thus these scalar values yield a two-dimensional linear multicast. In fact,

$$\det(L_w) \cdot \det(L_y) \cdot \det(L_z) = 1 \tag{19.87}$$

when

$$b = c = 0 \tag{19.88}$$

and

$$a = d = e = f = \cdots = l = 1. \tag{19.89}$$

Therefore, the two-dimensional linear network code depicted in Figure 19.2 is a linear multicast, and this fact is regardless of the choice of the base field F.

Theorem 19.20. *There exists an ω-dimensional linear multicast on an acyclic network for sufficiently large base field F.*

Proof. For a directed path $P = e_1, e_2, \cdots, e_m$, define

$$K_P = \prod_{1 \leq j < m} k_{e_j, e_{j+1}}. \tag{19.90}$$

Calculating by (19.17) recursively from the upstream channels to the downstream channels, it is not hard to find that

$$\mathbf{f}_e = \sum_{d \in \text{In}(s)} \left(\Sigma_{P: \text{ a path from } d \text{ to } e} K_P \right) \mathbf{f}_d \tag{19.91}$$

for every channel e (see Example 19.23 below). Denote by $F[*]$ the polynomial ring over the field F with all the $k_{d,e}$ as indeterminates, where the total number of such indeterminates is equal to $\sum_t |\text{In}(t)| \cdot |\text{Out}(t)|$. Thus, every component of every global encoding kernel belongs to $F[*]$. The subsequent arguments in this proof actually depend on this fact alone and not on the exact form of (19.91).

Let t be a non-source node with $\text{maxflow}(t) \geq \omega$. Then there exist ω edge-disjoint paths from the ω imaginary channels to ω distinct channels in $\text{In}(t)$. Put the global encoding kernels of these ω channels in juxtaposition to form an $\omega \times \omega$ matrix L_t. We will prove that

$$\det(L_t) = 1 \tag{19.92}$$

for properly set scalar values of the indeterminates.

Toward proving this claim, we set

$$k_{d,e} = 1 \tag{19.93}$$

for all adjacent pairs of channels (d, e) along any one of the ω edge-disjoint paths, and set

$$k_{d,e} = 0 \tag{19.94}$$

otherwise. With such local encoding kernels, the symbols sent on the ω imaginary channels at source node s are routed to node t via the ω edge-disjoint paths. Thus the columns in L_t are simply the global encoding kernels of the imaginary channels, which form the standard basis of the space F^ω. Then (19.92) follows, and the claim is proved.

Consequently,

$$\det(L_t) \neq 0 \in F[*], \tag{19.95}$$

i.e., $\det(L_t)$ is a nonzero polynomial in the indeterminates $k_{d,e}$. Since this conclusion applies to every non-source node t with maxflow$(t) \geq \omega$,

$$\prod_{t:\text{maxflow}(t) \geq \omega} \det(L_t) \neq 0 \in F[*]. \tag{19.96}$$

Applying Lemma 19.17 to the above polynomial when the field F is sufficiently large, we can set scalar values in F for the indeterminates so that

$$\prod_{t:\text{maxflow}(t) \geq \omega} \det(L_t) \neq 0 \in F, \tag{19.97}$$

which in turn implies that

$$\det(L_t) \neq 0 \in F \tag{19.98}$$

for all t such that maxflow$(t) \geq \omega$. These scalar values then yield a linear network code that meets the requirement (19.39) for a linear multicast.   □

Corollary 19.21. *There exists an ω-dimensional linear broadcast on an acyclic network for sufficiently large base field F.*

Proof. For every non-source node t in the given acyclic network, install a new node t' and ω input channels to this new node, with $\min\{\omega, \text{maxflow}(t)\}$ of them from node t and the remaining $\omega - \min\{\omega, \text{maxflow}(t)\}$ from source node s. This constructs a new acyclic network. Now consider an ω-dimensional linear multicast on the new network whose existence follows from Theorem 19.20. For every node t' as described above, $\dim(V_{t'}) = \omega$ because maxflow$(t') = \omega$. Moreover, since $|\text{In}(t')| = \omega$, the global encoding kernels $\mathbf{f}_e, e \in \text{In}(t')$ are linearly independent. Therefore,

$$\dim(\langle\{\mathbf{f}_e : e \in \text{In}(t') \cap \text{Out}(t)\}\rangle) = |\text{In}(t') \cap \text{Out}(t)| \tag{19.99}$$

$$= \min\{\omega, \text{maxflow}(t)\}. \tag{19.100}$$

By (19.17),

$$\langle\{\mathbf{f}_e : e \in \operatorname{In}(t') \cap \operatorname{Out}(t)\}\rangle \subset V_t. \tag{19.101}$$

Therefore,

$$\dim(V_t) \geq \dim(\langle\{\mathbf{f}_e : e \in \operatorname{In}(t') \cap \operatorname{Out}(t)\}\rangle) \tag{19.102}$$
$$= \min\{\omega, \operatorname{maxflow}(t)\}. \tag{19.103}$$

Then by invoking Theorem 19.10, we conclude that

$$\dim(V_t) = \min\{\omega, \operatorname{maxflow}(t)\}. \tag{19.104}$$

In other words, an ω-dimensional linear multicast on the new network incorporates an ω-dimensional linear broadcast on the original network. □

Corollary 19.22. *There exists an ω-dimensional linear dispersion on an acyclic network for sufficiently large base field F.*

Proof. For every nonempty collection T of non-source nodes in the given acyclic network, install a new node u_T and $\operatorname{maxflow}(t)$ channels from every node $t \in T$ to this new node. This constructs a new acyclic network with

$$\operatorname{maxflow}(u_T) = \operatorname{maxflow}(T) \tag{19.105}$$

for every T. Now consider an ω-dimensional linear broadcast on the new network whose existence follows from Corollary 19.21. By (19.17),

$$V_{u_T} \subset V_T. \tag{19.106}$$

Then

$$\dim(V_T) \geq \dim(V_{u_T}) \tag{19.107}$$
$$= \min\{\omega, \operatorname{maxflow}(u_T)\} \tag{19.108}$$
$$= \min\{\omega, \operatorname{maxflow}(T)\}. \tag{19.109}$$

By invoking Theorem 19.10, we conclude that

$$\dim(V_T) = \min\{\omega, \operatorname{maxflow}(T)\}. \tag{19.110}$$

In other words, an ω-dimensional linear broadcast on the new network incorporates an ω-dimensional linear dispersion on the original network. □

Example 19.23. We now illustrate the formula (19.91) in the proof of Theorem 19.20 with the two-dimensional linear network code in Example 19.9 which is expressed in the 12 indeterminates $a, b, c, \cdots, l$. The local encoding kernels at the nodes are

$$K_s = \begin{bmatrix} a & c \\ b & d \end{bmatrix}, \ K_t = \begin{bmatrix} e & f \end{bmatrix}, \ K_u = \begin{bmatrix} g & h \end{bmatrix}, \tag{19.111}$$

$$K_w = \begin{bmatrix} i \\ j \end{bmatrix}, \ K_x = \begin{bmatrix} k & l \end{bmatrix}. \tag{19.112}$$

Starting with $\mathbf{f}_{(o,s)} = [\,1 \ 0\,]^\top$ and $\mathbf{f}_{(o,s)'} = [\,0 \ 1\,]^\top$, we can calculate the global encoding kernels by the formula (19.91). Take $\mathbf{f}_{(x,y)}$ as the example. There are two paths from (o,s) to (x,y) and two from $(o,s)'$ to (x,y). For these paths,

$$K_P = \begin{cases} aeik & \text{for } P = (o,s),(s,t),(t,w),(w,x),(x,y) \\ beik & \text{for } P = (o,s)',(s,t),(t,w),(w,x),(x,y) \\ cgjk & \text{for } P = (o,s),(s,u),(u,w),(w,x),(x,y) \\ dgjk & \text{for } P = (o,s)',(s,u),(u,w),(w,x),(x,y) \end{cases}. \tag{19.113}$$

Thus

$$\mathbf{f}_{(x,y)} = (aeik)\mathbf{f}_{(o,s)} + (beik)\mathbf{f}_{(o,s)'} + (cgjk)\mathbf{f}_{(o,s)} + (dgjk)\mathbf{f}_{(o,s)'} \tag{19.114}$$

$$= \begin{bmatrix} aeik + cgjk \\ beik + dgjk \end{bmatrix}, \tag{19.115}$$

which is consistent with Example 19.9.

The proof of Theorem 19.20 provides an algorithm for constructing a linear multicast that uses Lemma 19.17 as a subroutine to search for scalars $a_1, a_2, \cdots, a_n \in F$ such that $g(a_1, a_2, \cdots, a_n) \neq 0$ whenever $g(z_1, z_2, \cdots, z_n)$ is a nonzero polynomial over a sufficiently large field F. The straightforward implementation of this subroutine is exhaustive search, which is generally computationally inefficient. Nevertheless, the proof of Theorem 19.20 renders a simple method to construct a linear multicast randomly.

Corollary 19.24. *Consider an ω-dimensional linear network code on an acyclic network. By choosing the local encoding kernels $k_{d,e}$ for all adjacent pairs of channels (d,e) independently according to the uniform distribution on the base field F, a linear multicast can be constructed with probability tends to 1 as $|F| \to \infty$.*

Proof. This follows directly from Corollary 19.18 and the proof of Theorem 19.20. □

The technique described in the above theorem for constructing a linear network code is called *random network coding*. Random network coding has the advantage that the code construction can be done independent of the network topology, making it very useful when the network topology is unknown. A case study for an application of random network coding will be presented in Section 19.7.

While random network coding offers a simple construction and more flexibility, a much larger base field is usually required. In some applications, it is necessary to verify that the code randomly constructed indeed possesses the desired properties. Such a task can be computationally nontrivial.

The next algorithm constructs a linear multicast deterministically in polynomial time. Unlike the algorithm given in the proof of Theorem 19.20 that assigns values to the local encoding kernels, this algorithm assigns values to the global encoding kernels.

Algorithm 19.25 (Jaggi–Sanders Algorithm). *This algorithm constructs an ω-dimensional linear multicast over a finite field F on an acyclic network when $|F| > \eta$, the number of non-source nodes t in the network with $\mathrm{maxflow}(t) \geq \omega$. Denote these η non-source nodes by $t_1, t_2, \cdots, t_\eta$.*

A sequence of channels $e_1, e_2, \cdots, e_l$ is called a path leading to a node t_q if $e_1 \in \mathrm{In}(s)$, $e_l \in \mathrm{In}(t_q)$, and (e_j, e_{j+1}) is an adjacent pair for all $1 \leq j \leq l-1$. For each q, $1 \leq q \leq \eta$, there exist ω edge-disjoint paths $P_{q,1}, P_{q,2}, \cdots, P_{q,\omega}$ leading to t_q. All together there are $\eta\omega$ such paths. The following procedure assigns a global encoding kernel $\mathbf{f}_e$ for every channel e in the network in an upstream-to-downstream order such that $\dim(V_{t_q}) = \omega$ for $1 \leq q \leq \eta$.

```
{
                    // By definition, the global encoding kernels of the ω
                    // imaginary channels form the standard basis of Fω.
    for (q = 1; q ≤ η; q + +)
        for (i = 1; i ≤ ω; i + +)
        eq,i = the imaginary channel initiating path Pq,i;
                    // This initializes eq,i. Subsequently, eq,i will be
                    // dynamically updated by moving down path Pq,i
                    // until it finally becomes a channel in In(tq).
    for (every node t, in any upstream-to-downstream order)
    {
        for (every channel e ∈ Out(t))
        {
                    // With respect to this channel e, define a "pair" as a
                    // pair (q, i) of indices such that channel e is on the
                    // path Pq,i. Note that for each q, there exists at most
                    // one pair (q, i). Thus the number of pairs is at least 0
                    // and at most η. Since the nodes t are chosen in
                    // an upstream-to-downstream order, if (q, i) is a pair,
                    // then eq,i ∈ In(t) by induction, so that feq,i ∈ Vt. For
                    // reasons to be explained in the algorithm verification
                    // below, feq,i ∉ ⟨{feq,j : j ≠ i}⟩, and therefore
                    // feq,i ∈ Vt \ ⟨{feq,j : j ≠ i}⟩.
        Choose a vector w in Vt such that w ∉ ⟨{feq,j : j ≠ i}⟩ for
        every pair (q, i);
                    // To see the existence of such a vector w, let
                    // dim(Vt) = ν. Then, dim(Vt ∩ ⟨{feq,j : j ≠ i}⟩) ≤
                    // ν − 1 for every pair (q, i) since
                    // feq,i ∈ Vt \ ⟨{feq,j : j ≠ i}⟩. Thus
```

$\quad$ // $|V_t \cap (\cup_{(q,i):\ a\ pair}\langle\{f_{e_{q,j}} : j \neq i\}\rangle)|$
$\quad$ // $\leq \eta|F|^{\nu-1} < |F|^{\nu} = |V_t|$.
$\mathbf{f}_e = w;$
$\quad$ // This is equivalent to choosing scalar values for local
$\quad$ // encoding kernels $k_{d,e}$ for all $d \in \text{In}(t)$ such that
$\quad$ // $\sum_{d \in \text{In}(t)} k_{d,e}\mathbf{f}_d \notin \langle\{f_{e_{q,j}} : j \neq i\}\rangle$ for every pair (q,i).
for ($every\ pair\ (q,i)$)
$\qquad$ $e_{q,i} = e;$
$\quad$ }
}
}

Algorithm Verification. For $1 \leq q \leq \eta$ and $1 \leq i \leq \omega$, the channel $e_{q,i}$ is on the path $P_{q,i}$. Initially $e_{q,i}$ is an imaginary channel at source node s. Through dynamic updating, it moves downstream along the path until finally reaching a channel in $\text{In}(t_q)$.

Fix an index q, where $1 \leq q \leq \eta$. Initially, the vectors $\mathbf{f}_{e_{q,1}}, \mathbf{f}_{e_{q,2}}, \cdots, \mathbf{f}_{e_{q,\omega}}$ are linearly independent because they form the standard basis of F^ω. At the end, they need to span the vector space F^ω. Therefore, in order for the eventually constructed linear network code to qualify as a linear multicast, it suffices to show the preservation of the linear independence among $\mathbf{f}_{e_{q,1}}, \mathbf{f}_{e_{q,2}}, \cdots, \mathbf{f}_{e_{q,\omega}}$ throughout the algorithm.

We need to show the preservation in the generic step inside the "for loop" for each channel e in the algorithm. The algorithm defines a "pair" as a pair (q,i) of indices such that channel e is on path $P_{q,i}$. When no (q,i) is a pair for $1 \leq i \leq \omega$, the channels $e_{q,1}, e_{q,2}, \cdots, e_{q,\omega}$ are not changed in the generic step; neither are the vectors $\mathbf{f}_{e_{q,1}}, \mathbf{f}_{e_{q,2}}, \cdots, \mathbf{f}_{e_{q,\omega}}$. So we only need to consider the scenario that a pair (q,i) exists for some i. The only change among the channels $e_{q,1}, e_{q,2}, \cdots, e_{q,\omega}$ is that $e_{q,i}$ becomes e. Meanwhile, the only change among the vectors $\mathbf{f}_{e_{q,1}}, \mathbf{f}_{e_{q,2}}, \cdots, \mathbf{f}_{e_{q,\omega}}$ is that $\mathbf{f}_{e_{q,i}}$ becomes a vector

$$w \notin \langle\{\mathbf{f}_{e_{q,j}} : j \neq i\}\rangle. \tag{19.116}$$

This preserves the linear independence among $\mathbf{f}_{e_{q,1}}, \mathbf{f}_{e_{q,2}}, \cdots, \mathbf{f}_{e_{q,\omega}}$ as desired.

Complexity Analysis. There are a total of $|E|$ channels in the network. In the algorithm, the generic step in the "for loop" for each channel e processes at most η pairs. Throughout the algorithm, at most $|E|\eta$ such collections of channels are processed. From this, it is not hard to implement the algorithm within a polynomial time in $|E|$ for a fixed ω. The computational details can be found in [184].

Remark 1 In the Jaggi–Sanders algorithm, all nodes t in the network with $\text{maxflow}(t) \geq \omega$ serve as a sink node that receives the message $\mathbf{x}$. The algorithm can easily be modified accordingly if only a subset of such nodes need to serve as a sink node. In that case, the field size requirement is $|F| > \eta'$, where η' is the total number of sink nodes.

Remark 2 It is not difficult to see from the lower bound on the required field size in the Jaggi–Sanders algorithm that if a field much larger than sufficient is used, then a linear multicast can be constructed with high probability by randomly choosing the global encoding kernels.

Example 19.26 (Multi-Source Multicast). Consider a network coding problem on an acyclic network G with a set S of source nodes. At node $s \in S$, a message $\mathbf{x}_s$ in the form of a row vector in F^{ω_s} is generated. Let

$$\omega = \sum_{s \in S} \omega_s \tag{19.117}$$

be the total dimension of all the messages, and let

$$\mathbf{x} = (\mathbf{x}_s : s \in S) \tag{19.118}$$

be referred to as *the message*. Here, we do not impose the constraint that a node $s \in S$ has no input channels.

Expand the network G into a network G' by installing a new node 0, and ω_s channels from node 0 to node s for each $s \in S$. Denote the value of a max-flow from node 0 to node t in G' by $\text{maxflow}_{G'}(t)$.

Suppose there exists a coding scheme on G such that a node t can decode the message $\mathbf{x}$. Such a coding scheme induces a coding scheme on G' for which

1. the message $\mathbf{x}$ is generated at node 0;
2. for all $s \in S$, the message $\mathbf{x}_s$ is sent uncoded from node 0 to node s through the ω_s channels from node 0 to node s.

Applying the max-flow bound to node t with respect to this coding scheme on G', we obtain

$$\text{maxflow}_{G'}(t) \geq \omega. \tag{19.119}$$

Thus we have shown that if a node t in G can decode the message $\mathbf{x}$, then (19.119) has to be satisfied.

We now show that for a sufficiently large base field F, there exists a coding scheme on G such that a node t satisfying (19.119) can decode the message $\mathbf{x}$. Let η be the number of nodes in G that satisfies (19.119). To avoid triviality, assume $\eta \geq 1$. By Theorem 19.20, there exists an ω-dimensional linear multicast $\mathcal{C}$ on G' when the base field is sufficiently large. From the proof of Theorem 19.10, we see that for this linear multicast, the $\omega \times \omega$ matrix $F_{\text{Out}(0)}$ must be invertible, otherwise a node t satisfying (19.119) cannot possibly decode the message $\mathbf{x}$. Transforming $\mathcal{C}$ by the matrix $[F_{\text{Out}(0)}]^{-1}$, we obtain from (19.54) a linear multicast $\mathcal{C}'$ with

$$F'_{\text{Out}(0)} = \left[F_{\text{Out}(0)}\right]^{-1} F_{\text{Out}(0)} = I_\omega. \tag{19.120}$$

Accordingly, for this linear multicast, the message $\mathbf{x}_s$ is sent uncoded from node 0 to node s for all $s \in S$. Thus a coding scheme on G with the message $\mathbf{x}_s$ being generated at node s for all $s \in S$ instead of being received from node 0 is naturally induced, and this coding scheme inherits from the linear multicast $\mathcal{C}'$ that a node t satisfying (19.119) can decode the message $\mathbf{x}$.

Therefore, instead of tackling the multi-source multicast problem on G, we can tackle the single-source multicast problem on G'. This has already seen illustrated in Examples 17.1 and 17.2 for the butterfly network.

19.5 Generic Network Codes

In the last section, we have seen how to construct a linear multicast by the Jaggi–Sanders algorithm. In light of Corollaries 19.21 and 19.22, the same algorithm can be used for constructing a linear broadcast or a linear dispersion.

It is not difficult to see that if the Jaggi–Sanders algorithm is used for constructing a linear broadcast, then the computational complexity of the algorithm remains polynomial in $|E|$, the total number of channels in the network. However, if the algorithm is used for constructing a linear dispersion, the computational complexity becomes exponential because the number of channels that need to be installed in constructing the new network in Corollary 19.22 grows exponentially with the number of channels in the original network.

In this section, we introduce a class of linear network codes called *generic network codes*. As we will see, if a linear network code is generic, then it is a linear dispersion, and hence also a linear broadcast and a linear multicast. Toward the end of the section, we will present a polynomial-time algorithm that constructs a generic network code.

Imagine that in an ω-dimensional linear network code, the base field F is replaced by the real field $\Re$. Then arbitrary infinitesimal perturbation of the local encoding kernels would place the global encoding kernels at *general positions* with respect to one another in the space $\Re^\omega$. General positions of the global encoding kernels maximize the dimensions of various linear spans by avoiding linear dependence in every conceivable way. The concepts of general positions and infinitesimal perturbation do not apply to the vector space F^ω when F is a finite field. However, they can be emulated when F is sufficiently large with the effect of avoiding unnecessary linear dependence.

The following definitions of a *generic network code* captures the notion of placing the global encoding kernels in general positions. In the sequel, for a channel $e_j \in E$, let $e_j \in \text{Out}(t_j)$, and for a collection of channels $\xi = \{e_1, e_2, \cdots, e_{|\xi|}\} \subset E$, let $\xi_{\bar{k}} = \xi \backslash \{e_k\}$.

Definition 19.27 (Generic Network Code I). *An ω-dimensional linear network code on an acyclic network is generic if for any nonempty collection of channels $\xi = \{e_1, e_2, \cdots, e_m\} \subset E$ and any $1 \leq k \leq m$, if*

a) there is no directed path from t_k to t_j for $j \neq k$,
b) $V_{t_k} \not\subset V_{\xi_{\bar{k}}}$,

then $\mathbf{f}_{e_k} \notin V_{\xi_{\bar{k}}}$.

Definition 19.28 (Generic Network Code II). *An ω-dimensional linear network code on an acyclic network is generic if for any nonempty collection of channels $\xi = \{e_1, e_2, \cdots, e_m\} \subset E$ and any $1 \leq k \leq m$, if*

a) there is no directed path from t_k to t_j for $j \neq k$,
b) $V_{t_k} \not\subset V_{\xi_{\bar{k}}}$,
c) $\mathbf{f}_e, e \in \xi_{\bar{k}}$ are linearly independent,[3]

then $\mathbf{f}_{e_k} \notin V_{\xi_{\bar{k}}}$.

In Definitions 19.27 and 19.28, if (a) does not hold, then $\mathbf{f}_{e_k} \notin V_{\xi_{\bar{k}}}$ may not be possible at all as we now explain. Let $\xi = \{e_1, e_2\}$ and $k = 1$. Suppose $\text{In}(t_2) = \{e_1\}$ so that (a) is violated. Since node t_2 has only e_1 as the input channel, $\mathbf{f}_{e_1}$ cannot possibly be linear independent of $\mathbf{f}_{e_2}$.

The only difference between Definitions 19.27 and 19.28 is the additional requirement (c) in the latter. The equivalence between these two definitions of a generic network code can be seen as follows. It is obvious that if a linear network code satisfies Definition 19.27, then it also satisfies Definition 19.28. To prove the converse, suppose a linear network code satisfies Definition 19.28. Consider any collection of channels $\xi = \{e_1, e_2, \cdots, e_m\} \subset E$ such that there exists $1 \leq k \leq m$ satisfying (a) and (b) in Definition 19.27 but not necessarily (c) in Definition 19.28. Then we can always find a subset $\xi'_{\bar{k}}$ of $\xi_{\bar{k}}$ such that $\mathbf{f}_e, e \in \xi'_{\bar{k}}$ are linearly independent and $V_{\xi'_{\bar{k}}} = V_{\xi_{\bar{k}}}$. Upon letting $\xi' = \{e_k\} \cup \xi'_k$ and applying Definition 19.28 with ξ' in place of ξ, we have

$$\mathbf{f}_{e_k} \notin V_{\xi'_{\bar{k}}} = V_{\xi_{\bar{k}}}, \tag{19.121}$$

so the network code also satisfies Definition 19.27. This shows that the two definitions of a generic network code are equivalent. Note that in Definition 19.28, if ξ satisfies all the prescribed conditions, then $m \leq \omega$ because (c) and $\mathbf{f}_{e_k} \notin V_{\xi_{\bar{k}}}$ together imply that $\mathbf{f}_e, e \in \xi$ are linearly independent. Definition 19.28, which has a slightly more complicated form compared with Definition 19.27, will be instrumental in the proof of Theorem 19.32 that establishes various characterizations of a generic network code.

Proposition 19.29. *For a generic network code, for any collection of m output channels at a node t, where $1 \leq m \leq \dim(V_t)$, the global encoding kernels are linearly independent.*

---

[3] By convention, an empty collection of vectors is linearly independent.

Proof. Since the proposition becomes degenerate if $\dim(V_t) = 0$, we assume $\dim(V_t) > 0$. In Definition 19.27, let all the nodes t_j be node t and let $1 \leq m \leq \dim(V_t)$. First note that there is no directed path from node t to itself because the network is acyclic. For $m = 1$, $\xi_{\bar{1}} = \emptyset$ and $V_{\xi_{\bar{1}}} = V_\emptyset = \{0\}$. Since $\dim(V_t) > 0$, we have $V_t \not\subset V_{\xi_{\bar{1}}}$. Then $\mathbf{f}_{e_1} \notin V_{\xi_{\bar{1}}}$, which implies $\mathbf{f}_{e_1} \neq 0$. This proves the proposition for $m = 1$.

Assume that the proposition is true for $m - 1$ some $2 \leq m \leq \dim(V_t)$. We now prove that the proposition is true for m. By the induction hypothesis, $\mathbf{f}_{e_1}, \mathbf{f}_{e_2}, \cdots, \mathbf{f}_{e_{m-1}}$ are linearly independent. Since

$$\dim(\langle\{\mathbf{f}_{e_1}, \mathbf{f}_{e_2}, \cdots, \mathbf{f}_{e_{m-1}}\}\rangle) = m - 1 < \dim(V_t), \tag{19.122}$$

we have

$$V_t \not\subset \langle\{\mathbf{f}_{e_1}, \mathbf{f}_{e_2}, \cdots, \mathbf{f}_{e_{m-1}}\}\rangle. \tag{19.123}$$

Then by Definition 19.27,

$$\mathbf{f}_{e_m} \notin \langle\{\mathbf{f}_{e_1}, \mathbf{f}_{e_2}, \cdots, \mathbf{f}_{e_{m-1}}\}\rangle. \tag{19.124}$$

Hence, $\mathbf{f}_{e_1}, \mathbf{f}_{e_2}, \cdots, \mathbf{f}_{e_m}$ are linearly independent. The proposition is proved. $\square$

Corollary 19.30. *For a generic network code, if $|\text{Out}(t)| \leq \dim(V_t)$ for a node t, then the global encoding kernels of all the output channels of t are linearly independent.*

A linear dispersion on an acyclic network is not necessarily a generic network code. The following is a counterexample.

Example 19.31. The two-dimensional linear dispersion on the network in Figure 19.6 is not generic because the global encoding kernels of two of the output channels from source node s are equal to $[1\ 1]^\top$, a contradiction to Proposition 19.29. It can be shown, however, that a generic network code on an acyclic network G can be constructed through a linear dispersion on an expanded network G'. See Problem 13 for details.

Together with Example 19.13, the example above shows that the four classes of linear network codes we have discussed, namely linear multicast, linear broadcast, linear dispersion, and generic network code, achieve the max-flow bound to strictly increasing extents.

In the following theorem, we prove two characterizations of a generic network code, each can be regarded as an alternative definition of a generic network code. The reader should understand this theorem before proceeding further but may skip the proof at the first reading.

Theorem 19.32. *For an ω-dimensional linear network code on an acyclic network, the following conditions are equivalent:*

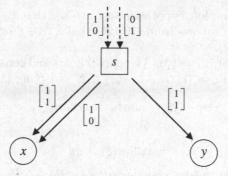

Fig. 19.6. A two-dimensional linear dispersion that is not a generic network code.

1) *The network code is generic.*
2) *For any nonempty collection of channels $\xi = \{e_1, e_2, \cdots, e_m\} \subset E$, if $V_{t_j} \not\subset V_{\xi_{\bar{j}}}$ for all $1 \leq j \leq m$, then $\mathbf{f}_e, e \in \xi$ are linearly independent.*
3) *For any nonempty collection of channels $\xi \subset E$, if*

$$|\xi| = \min\{\omega, \text{maxflow}(\xi)\}, \qquad (19.125)$$

then $\mathbf{f}_e, e \in \xi$ are linearly independent.

Proof. We will prove the theorem by showing that $(1) \Rightarrow (2) \Rightarrow (3) \Rightarrow (1)$.

We first show that $(1) \Rightarrow (2)$ by using Definition 19.27 as the definition of a generic network code. Assume (1) holds. Consider any $m \geq 1$ and any collection of channels $\xi = \{e_1, e_2, \cdots, e_m\} \subset E$, and assume $V_{t_j} \not\subset V_{\xi_{\bar{j}}}$ for all $1 \leq j \leq m$. We will show by induction on m that $\mathbf{f}_e, e \in \xi$ are linearly independent. The claim is trivially true for $m = 1$. Assume the claim is true for $m - 1$ for some $2 \leq m \leq \omega$, and we will show that it is true for m.

Consider $\xi = \{e_1, e_2, \cdots, e_m\}$ and assume $V_{t_j} \not\subset V_{\xi_{\bar{j}}}$ for all $1 \leq j \leq m$. We first prove by contradiction that there exists at least one k such that there is no directed path from t_k to t_j for all $j \neq k$, where $1 \leq j, k \leq m$. Assume that for all k, there is at least one directed path from node t_k to node t_j for some $j \neq k$. Starting at any node t_k, by traversing such directed paths, we see that there exists a directed cycle in the network because the set $\{t_k : 1 \leq k \leq m\}$ is finite. This leads to a contradiction because the network is acyclic, proving the existence of k as prescribed. Then apply Definition 19.27 to this k to see that

$$\mathbf{f}_{e_k} \notin V_{\xi_{\bar{k}}} = \langle \{\mathbf{f}_e : e \in \xi_{\bar{k}}\} \rangle. \qquad (19.126)$$

Now for any $j \neq k$, since $V_{t_j} \not\subset V_{\xi_{\bar{j}}}$ and $V_{\xi_{\bar{k}} \setminus \{e_j\}} = V_{\xi_{\bar{j}} \setminus \{e_k\}} \subset V_{\xi_{\bar{j}}}$, we have

$$V_{t_j} \not\subset V_{\xi_{\bar{k}} \setminus \{e_j\}}. \qquad (19.127)$$

Then apply the induction hypothesis to $\xi_{\bar{k}}$ to see that $\mathbf{f}_e, e \in \xi_{\bar{k}}$ are linearly independent. It then follows from (19.126) that $\mathbf{f}_e, e \in \xi$ are linearly independent. Thus (1) $\Rightarrow$ (2).

We now show that (2) $\Rightarrow$ (3). Assume (2) holds and consider any nonempty collection of channel $\xi = \{e_1, e_2, \cdots, e_m\} \subset E$ satisfying (19.125). Then

$$m = |\xi| = \min\{\omega, \text{maxflow}(\xi)\}, \tag{19.128}$$

which implies

$$\text{maxflow}(\xi) \geq m. \tag{19.129}$$

Therefore, there exist m edge-disjoint paths $P_1, P_2, \cdots, P_m$ from source node s to the channels in ξ, where the last channel on path P_j is e_j.

Denote the length of P_j by l_j and let

$$L = \sum_{j=1}^{m} l_j \tag{19.130}$$

be the total length of all the paths. We will prove the claim that $\mathbf{f}_{e_1}, \mathbf{f}_{e_2}, \cdots, \mathbf{f}_{e_m}$ are linearly independent by induction on L. For the base case $L = m$, since $m \leq \omega$ by (19.128), the claim is true by Proposition 19.29 with $t = s$. Assume that the claim is true for $L - 1$ for some $L \geq m + 1$, and we will prove that it is true for L. Let $A = \{j : l_j > 1\}$ and for $j \in A$, let $\xi'_j = \{e_1, e_2, \cdots, e_{j-1}, e'_j, e_{j+1}, \cdots, e_m\}$, where e'_j is the channel preceding e_j on P_j. Then by the induction hypothesis, $\mathbf{f}_e, e \in \xi'_j$ are linearly independent, which implies that

$$V_{t_j} \not\subset V_{\xi_{\bar{j}}}. \tag{19.131}$$

For $j \notin A$, $l_j = 1$, i.e., $t_j = s$. It follows from (19.128) that $m \leq \omega$. Then

$$V_{t_j} = V_s \not\subset V_{\xi_{\bar{j}}} \tag{19.132}$$

because $\dim(V_{\xi_{\bar{j}}}) \leq |\xi_{\bar{j}}| = m - 1 < m \leq \omega$. Therefore, (19.131) holds for all j, and hence by (2), $\mathbf{f}_e, e \in \xi$ are linearly independent. Thus (2) $\Rightarrow$ (3).

Finally, we show that (3) $\Rightarrow$ (1) by using Definition 19.28 as the definition of a generic network code. Assume (3) holds and consider any collection of channels $\xi = \{e_1, e_2, \cdots, e_m\} \subset E$, where $1 \leq m \leq \omega$, such that (a) to (c) in Definition 19.28 hold for some $1 \leq k \leq m$. Then either $t_j = s$ for all $1 \leq j \leq m$ or $t_k \neq s$, because otherwise (a) in Definition 19.28 is violated.

If $t_j = s$ for all $1 \leq j \leq m$, then

$$m = |\xi| = \text{maxflow}(\xi). \tag{19.133}$$

Since $m \leq \omega$, we have

$$|\xi| = \min\{\omega, \text{maxflow}(\xi)\}. \tag{19.134}$$

Then $\mathbf{f}_e, e \in \xi$ are linearly independent by (3), proving that $\mathbf{f}_{e_k} \notin V_{\xi_{\bar{k}}}$.

Otherwise, $t_k \neq s$. Following (b) in Definition 19.28, there exists $e'_k \in$ In$(t_k) \subset E$ such that $\mathbf{f}_{e'_k}$ and $\mathbf{f}_e, e \in \xi_{\bar{k}}$ are linearly independent. Let $\xi'_k = \{e_1, e_2, \cdots, e_{k-1}, e'_k, e_{k+1}, \cdots, e_m\}$. By Corollary 19.11,

$$\text{maxflow}(\xi'_k) \geq \dim(V_{\xi'_k}) = m, \tag{19.135}$$

so $e_1, e_2, \cdots, e_{k-1}, e'_k, e_{k+1}, e_m$ can be traced back to source node s via some edge-disjoint paths $P_1, P_2, \cdots, P_{k-1}, P'_k, P_{k+1}, \cdots, P_m$, respectively. Let P_k be obtained by appending e_k to P'_k. Since there is no directed path from t_k to t_j and $e_k \neq e_j$ for all $j \neq k$, $P_1, P_2, \cdots, P_{k-1}, P_k, P_{k+1}, \cdots, P_m$ are edge-disjoint. Therefore,

$$\text{maxflow}(\xi) \geq m. \tag{19.136}$$

On the other hand,

$$\text{maxflow}(\xi) \leq |\xi| = m. \tag{19.137}$$

Therefore,

$$m = |\xi| = \text{maxflow}(\xi), \tag{19.138}$$

i.e., (19.133). As before, we can further obtain (19.134). Then by (3), $\mathbf{f}_e, e \in \xi$, are linearly independent, and therefore $\mathbf{f}_{e_k} \notin V_{\xi_{\bar{k}}}$. Thus (3) $\Rightarrow$ (1).

Hence, the theorem is proved. □

Corollary 19.33. *An ω-dimensional generic network code on an acyclic network is an ω-dimensional linear dispersion on the same network.*

Proof. Consider an ω-dimensional generic network code on an acyclic network and let T be any collection of non-source nodes. Let

$$m = \min\{\omega, \text{maxflow}(T)\}. \tag{19.139}$$

Since maxflow$(T) \geq m$, there exists m edge-disjoint paths $P_1, P_2, \cdots, P_m$ from source node s to T. Let e_i be the last channel on path P_i, and let

$$\xi = \{e_1, e_2, \cdots, e_m\}. \tag{19.140}$$

Evidently,

$$\text{maxflow}(\xi) = m. \tag{19.141}$$

It follows from (19.139) that $m \leq \omega$. Therefore,

$$|\xi| = m = \text{maxflow}(\xi) = \min\{\omega, \text{maxflow}(\xi)\}. \tag{19.142}$$

By Theorem 19.32, $\mathbf{f}_e, e \in \xi$ are linearly independent. Then

$$\dim(V_T) \geq \dim(V_\xi) = m = \min\{\omega, \text{maxflow}(T)\}. \tag{19.143}$$

By Theorem 19.10, we conclude that

$$\dim(V_T) = \min\{\omega, \text{maxflow}(T)\}. \tag{19.144}$$

Hence, we have shown that a generic network code is a linear dispersion.   □

Theorem 19.32 renders the following important interpretation of a generic network code. Consider any linear network code and any collection of channels $\xi \subset E$. If $\mathbf{f}_e, e \in \xi$ are linearly independent, then

$$|\xi| = \dim(V_\xi). \tag{19.145}$$

By Corollary 19.11,

$$\dim(V_\xi) \leq \min\{\omega, \text{maxflow}(\xi)\}. \tag{19.146}$$

Therefore,

$$|\xi| \leq \min\{\omega, \text{maxflow}(\xi)\}. \tag{19.147}$$

On the other hand,

$$\text{maxflow}(\xi) \leq |\xi|, \tag{19.148}$$

which implies

$$\min\{\omega, \text{maxflow}(\xi)\} \leq |\xi|. \tag{19.149}$$

Combining (19.147) and (19.149), we see that

$$|\xi| = \min\{\omega, \text{maxflow}(\xi)\} \tag{19.150}$$

is a necessary condition for $\mathbf{f}_e, e \in \xi$ to be linearly independent. For a generic network code, this is also a sufficient condition for $\mathbf{f}_e, e \in \xi$ to be linearly independent. Thus for a generic network code, if a set of global encoding kernels can possibly be linearly independent, then it is linear independent. In this sense, a generic network code captures the notion of placing the global encoding kernels in general positions.

The condition (2) in Theorem 19.32 is the original definition of a generic network code given in [230]. Unlike (1) and (3), this condition is purely algebraic and does not depend upon the network topology. However, it does not suggest an algorithm for constructing such a code.

Motivated by Definition 19.28, we now present an algorithm for constructing a generic network code. The computational complexity of this algorithm is polynomial in $|E|$, the total number of channels in the network.

Algorithm 19.34 (Construction of a Generic Network Code). *This algorithm constructs an ω-dimensional generic network code over a finite field F with $|F| > \sum_{m=1}^{\omega} \binom{|E|-1}{m-1}$ by prescribing global encoding kernels that constitute a generic network code.*

{
    for (every node t, following an upstream-to-downstream order)
    {
        for (every channel e ∈ Out(t))

{

  *Choose a vector w in V_t such that $w \notin V_\zeta$, where ζ is any
  collection of $m - 1$ already processed channels, where $1 \le m \le \omega$,
  such that $\mathbf{f}_e, e \in \zeta$ are linearly independent and $V_t \not\subset V_\zeta$;*
  // To see the existence of such a vector w, denote $\dim(V_t)$
  // by ν. If ζ is any collection of $m - 1$ channels with $V_t \not\subset V_\zeta$,
  // then $\dim(V_t \cap V_\zeta) \le \nu - 1$. There are at most $\sum_{m=1}^{\omega} \binom{|E|-1}{m-1}$
  // such collections ζ. Thus
  // $|V_t \cap (\cup_\zeta V_\zeta)| \le \sum_{m=1}^{\omega} \binom{|E|-1}{m-1} |F|^{\nu-1} < |F|^\nu = |V_t|$.
  $\mathbf{f}_e = w;$
  // This is equivalent to choosing scalar values for the local
  // encoding kernels $k_{d,e}$ for all d such that $\sum_{d \in \mathrm{In}(t)} k_{d,e} \mathbf{f}_d$
  // $\notin V_\zeta$ for every collection ζ of channels as prescribed.

  }

}

}

Algorithm Verification. We will verify that the code constructed is indeed
generic by way of Condition (3) in Theorem 19.32. Consider any nonempty
collection of channels $\xi = \{e_1, e_2, \cdots, e_m\} \subset E$ satisfying (19.125). Then there
exist m edge-disjoint paths $P_1, P_2, \cdots, P_m$ from source node s to the channels
in ξ, where the last channel on path P_j is e_j. Denote the length of P_j by l_j
and let

$$L = \sum_{j=1}^{m} l_j \qquad (19.151)$$

be the total length of all the paths. We will prove the claim that $\mathbf{f}_e, e \in \xi$ are
linearly independent by induction on L.

It is easy to verify that for any set of m channels in $\mathrm{Out}(s)$, the global
encoding kernels assigned are linearly independent, so the base case $L = m$ is
verified. Assume the claim is true for $L - 1$ for some $L \ge m + 1$, and we will
prove that it is true for L. Let e_k be the channel whose global encoding kernel
is last assigned among all the channels in ξ. Note that $P_k \ge 2$ since $L \ge m+1$
and the global encoding kernels are assigned by the algorithm in an upstream-
to-downstream order. Then let e'_k be the channel preceding e_k on P_k, and let

$$\xi' = \{e_1, e_2, \cdots, e_{k-1}, e'_k, e_{k+1}, \cdots, e_m\}. \qquad (19.152)$$

By the induction hypothesis, $\mathbf{f}_e, e \in \xi'$ are linearly independent. Since $\mathbf{f}_{e'_k}$ is
linearly independent of $\mathbf{f}_e$ for $e \in \xi' \setminus \{e'_k\} = \xi_{\bar{k}}$, $V_{t_k} \not\subset V_{\xi_{\bar{k}}}$. It then follows
from the construction that $\mathbf{f}_{e_k} \notin V_{\xi_{\bar{k}}}$ because $\xi_{\bar{k}}$ is one of the collections ζ
considered when $\mathbf{f}_{e_k}$ is assigned. Hence, $\mathbf{f}_e, e \in \xi$ are linearly independent,
verifying that the network code constructed is generic.

Complexity Analysis. In the algorithm, the "for loop" for each channel e pro-
cesses at most $\sum_{m=1}^{\omega} \binom{|E|-1}{m-1}$ collections of $m - 1$ channels. The processing

includes the detection of those collections ζ as well as the computation of the set $V_t \backslash (\cup_\zeta V_\zeta)$. This can be done, for instance, by Gauss elimination. Throughout the algorithm, the total number of collections of channels processed is at most $|E| \sum_{m=1}^{\omega} \binom{|E|-1}{m-1}$, a polynomial in $|E|$ of degree ω. Thus for a fixed ω, it is not hard to implement the algorithm within a polynomial time in $|E|$.

Algorithm 19.34 constitutes a constructive proof for the next theorem.

Theorem 19.35. *There exists an ω-dimensional generic network code on an acyclic network for sufficiently large base field F.*

By noting the lower bound on the required field size in Algorithm 19.34, a generic network code can be constructed with high probability by randomly choosing the global encoding kernels provided that the base field is much larger than sufficient.

19.6 Static Network Codes

In our discussion so far, a linear network code has been defined on a network with a fixed topology, where all the channels are assumed to be available at all times. In a real network, however, a channel may fail due to various reasons, for example, hardware failure, cable cut, or natural disasters. With the failure of some subset of channels, the communication capacity of the resulting network is generally reduced.

Consider the use of, for instance, an ω-dimensional multicast on an acyclic network for multicasting a sequence of messages generated at the source node. When no channel failure occurs, a non-source node with the value of a max-flow at least equal to ω would be able to receive the sequence of messages. In case of channel failures, if the value of a max-flow of that node in the resulting network is at least ω, the sequence of messages in principle can still be received at that node. However, this would involve the deployment of a network code for the new network topology, which not only is cumbersome but may also cause a significant loss of data during the switchover.

In this section, we discuss a class of linear network codes called *static network codes* that can provide the network with maximum robustness in case of channel failures. To fix ideas, we first introduce some terminology. The status of the network is specified by a mapping $\lambda : E \to \{0,1\}$ called a *configuration*. A channel being in the set

$$\lambda^{-1}(0) = \{e \in E : \lambda(e) = 0\} \tag{19.153}$$

indicates the failure of that channel, and the subnetwork resulting from the deletion of all the channels in $\lambda^{-1}(0)$ is called the λ-subnetwork. For the λ-subnetwork, the value of a max-flow from source node s to a non-source node t is denoted by $\mathrm{maxflow}_\lambda(t)$. Likewise, the value of a max-flow from source node s to a collection T of non-source nodes is denoted by $\mathrm{maxflow}_\lambda(T)$. It is easy to see that the total number of configurations is equal to $2^{|E|}$.

Definition 19.36. *Let λ be a configuration of the network. For an ω-dimensional linear network code on the network, the λ-global encoding kernel of channel e, denoted by $\mathbf{f}_{e,\lambda}$, is the column ω-vector calculated recursively in an upstream-to-downstream order by*

(19.154) $\mathbf{f}_{e,\lambda} = \lambda(e) \sum_{d \in \mathrm{In}(t)} k_{d,e}\, \mathbf{f}_{d,\lambda}$ *for $e \in \mathrm{Out}(t)$.*

(19.155) *The λ-global encoding kernels of the ω imaginary channels are independent of λ and form the standard basis of the space F^ω.*

Note that in the above definition, the local encoding kernels $k_{d,e}$ are not changed with the configuration λ. Given the local encoding kernels, the λ-global encoding kernels can be calculated recursively by (19.154), while (19.155) serves as the boundary conditions. For a channel $e \in \mathrm{Out}(t)$ with $\lambda(e) = 0$, we see from (19.154) that

$$\mathbf{f}_{e,\lambda} = 0. \tag{19.156}$$

As before, the message generated at source node s is denoted by a row ω-vector $\mathbf{x}$. When the prevailing configuration is λ, a node t receives the symbols $\mathbf{x} \cdot \mathbf{f}_{d,\lambda}$, $d \in \mathrm{In}(t)$, from which it calculates the symbol $\mathbf{x} \cdot \mathbf{f}_{e,\lambda}$ to be sent on each channel $e \in \mathrm{Out}(t)$ via

$$\mathbf{x} \cdot \mathbf{f}_{e,\lambda} = \mathbf{x} \left[\lambda(e) \sum_{d \in \mathrm{In}(t)} k_{d,e}\, \mathbf{f}_{d,\lambda} \right] \tag{19.157}$$

$$= \lambda(e) \sum_{d \in \mathrm{In}(t)} k_{d,e}(\mathbf{x} \cdot \mathbf{f}_{d,\lambda}). \tag{19.158}$$

In particular, if $\lambda(e) = 0$, the zero symbol is sent on channel e regardless of the symbols received at node t.

In a real network, the zero symbol is not sent on a failed channel. Rather, whenever a symbol is not received on an input channel, the symbol is regarded by the receiving node as being the zero symbol.

For a configuration λ of the network, we let

$$V_{t,\lambda} = \langle \{ \mathbf{f}_{e,\lambda} : e \in \mathrm{In}(t) \} \rangle \tag{19.159}$$

for a node t,

$$V_{T,\lambda} = \langle \cup_{t \in T} V_{t,\lambda} \rangle \tag{19.160}$$

for a collection T of nodes, and

$$V_{\xi,\lambda} = \langle \{ \mathbf{f}_{e,\lambda} : e \in \xi \} \rangle, \tag{19.161}$$

for a collection ξ of channels.

Definition 19.37. *An ω-dimensional linear network code on an acyclic network qualifies as a static linear multicast, a static linear broadcast, a static linear dispersion, or a static generic network code, respectively, if the following hold:*

(19.162) $\dim(V_{t,\lambda}) = \omega$ *for every configuration λ and every non-source node t with* $\mathrm{maxflow}_\lambda(t) \geq \omega$.

(19.163) $\dim(V_{t,\lambda}) = \min\{\omega, \mathrm{maxflow}_\lambda(t)\}$ *for every configuration λ and every non-source node t.*

(19.164) $\dim(V_{T,\lambda}) = \min\{\omega, \mathrm{maxflow}_\lambda(T)\}$ *for every configuration λ and every collection T of non-source nodes.*

(19.165) *For any configuration λ and any nonempty collection of channels $\xi \subset E$, if $|\xi| = \min\{\omega, \mathrm{maxflow}_\lambda(\xi)\}$, then $\mathbf{f}_{e,\lambda}, e \in \xi$ are linearly independent.*

Here we have adopted Condition (3) in Theorem 19.32 for the purpose of defining a static generic network code. The qualifier "static" in the terms above stresses the fact that, while the configuration λ varies, the local encoding kernels remain unchanged. The advantage of using a static linear multicast, broadcast, or dispersion is that in case of channel failures, the local operation at every node in the network is affected only at the minimum level. Each receiving node in the network, however, needs to know the configuration λ before decoding can be done correctly. In implementation, this information can be provided by a separate signaling network.

For each class of static network codes in Definition 19.37, the requirement for its non-static version is applied to the λ-subnetwork for every configuration λ. Accordingly, a static linear multicast, a static linear broadcast, a static linear dispersion, and a static generic network code are increasingly stronger linear network codes as for the non-static versions.

Example 19.38. A two-dimensional linear network code over $GF(5)$ on the network in Figure 19.7 is prescribed by the local encoding kernels

$$K_s = \begin{bmatrix} 1 & 0 & 1 \\ 0 & 1 & 1 \end{bmatrix} \tag{19.166}$$

and

$$K_x = \begin{bmatrix} 1 & 3 \\ 3 & 2 \\ 1 & 1 \end{bmatrix} . \tag{19.167}$$

We claim that this is a static generic network code. Denote the three channels in $\mathrm{In}(x)$ by $c, d,$ and e and the two channels in $\mathrm{Out}(x)$ by g and h. The vectors $\mathbf{f}_{g,\lambda}$ and $\mathbf{f}_{h,\lambda}$ for all possible configurations λ are tabulated in Table 19.1, from which it is straightforward to verify the condition (19.165).

The following is an example of a generic network code that does not qualify even as a static linear multicast.

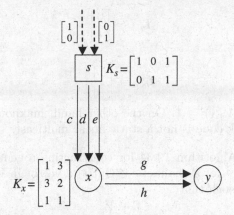

Fig. 19.7. A two-dimensional $GF(5)$-valued static generic network code.

Example 19.39. On the network in Figure 19.7, a two-dimensional generic network code over $GF(5)$ is prescribed by the local encoding kernels

$$K_s = \begin{bmatrix} 1 & 0 & 1 \\ 0 & 1 & 1 \end{bmatrix} \tag{19.168}$$

and

$$K_x = \begin{bmatrix} 2 & 1 \\ 1 & 2 \\ 0 & 0 \end{bmatrix}. \tag{19.169}$$

For a configuration λ such that

$$\lambda(c) = 0 \tag{19.170}$$

and

$$\lambda(d) = \lambda(e) = 1, \tag{19.171}$$

we have the λ-global encoding kernels

Table 19.1. The vectors $\mathbf{f}_{g,\lambda}$ and $\mathbf{f}_{h,\lambda}$ for all possible configurations λ in Example 19.38.

$\lambda(c)$	0	0	0	1	1	1	1
$\lambda(d)$	0	1	1	0	0	1	1
$\lambda(e)$	1	0	1	0	1	0	1
$\mathbf{f}_{g,\lambda}$	$\lambda(g)\begin{bmatrix}1\\1\end{bmatrix}$	$\lambda(g)\begin{bmatrix}0\\3\end{bmatrix}$	$\lambda(g)\begin{bmatrix}1\\4\end{bmatrix}$	$\lambda(g)\begin{bmatrix}1\\0\end{bmatrix}$	$\lambda(g)\begin{bmatrix}2\\1\end{bmatrix}$	$\lambda(g)\begin{bmatrix}1\\3\end{bmatrix}$	$\lambda(g)\begin{bmatrix}2\\4\end{bmatrix}$
$\mathbf{f}_{h,\lambda}$	$\lambda(h)\begin{bmatrix}1\\1\end{bmatrix}$	$\lambda(h)\begin{bmatrix}0\\2\end{bmatrix}$	$\lambda(h)\begin{bmatrix}1\\3\end{bmatrix}$	$\lambda(h)\begin{bmatrix}3\\0\end{bmatrix}$	$\lambda(h)\begin{bmatrix}4\\1\end{bmatrix}$	$\lambda(h)\begin{bmatrix}3\\2\end{bmatrix}$	$\lambda(h)\begin{bmatrix}4\\3\end{bmatrix}$

$$\mathbf{f}_{g,\lambda} = \begin{bmatrix} 0 \\ 1 \end{bmatrix} \tag{19.172}$$

and

$$\mathbf{f}_{h,\lambda} = \begin{bmatrix} 0 \\ 2 \end{bmatrix}, \tag{19.173}$$

and therefore $\dim(V_{y,\lambda}) = 1$. On the other hand, $\mathrm{maxflow}_\lambda(y) = 2$. Hence, this generic network code is not a static linear multicast.

Recall that in Algorithm 19.34 for constructing a generic network code, the key step chooses for a channel $e \in \mathrm{Out}(t)$ a vector in V_t to be the global encoding kernel $\mathbf{f}_e$ such that

$$\mathbf{f}_e \notin V_\zeta, \tag{19.174}$$

where ζ is any collection of $m - 1$ channels as prescribed with $1 \le m \le \omega$. This is equivalent to choosing scalar values for the local encoding kernels $k_{d,e}$ for all $d \in \mathrm{In}(t)$ such that

$$\sum_{d \in \mathrm{In}(t)} k_{d,e}\,\mathbf{f}_d \notin V_\zeta. \tag{19.175}$$

Algorithm 19.34 is adapted below for the construction of a static generic network code.

Algorithm 19.40 (Construction of a Static Generic Network Code).
This algorithm constructs an ω-dimensional static generic network code over a finite field F on an acyclic network with $|F| > 2^{|E|} \sum_{m=1}^{\omega} \binom{|E|-1}{m-1}$.

{
    for (every node t, following an upstream-to-downstream order)
    {
        for (every channel $e \in \mathrm{Out}(t)$)
        {
            Choose scalar values for $k_{d,e}$ for all $d \in \mathrm{In}(t)$ such that for
            any configuration λ, $\sum_{d \in \mathrm{In}(t)} k_{d,e}\mathbf{f}_d \notin V_{\zeta,\lambda}$, where ζ is any
            collection of $m - 1$ already processed channels such that
            $\mathbf{f}_{e,\lambda}, e \in \zeta$ are linearly independent and $V_{t,\lambda} \not\subset V_{\zeta,\lambda}$;
            // To see the existence of such values $k_{d,e}$, denote
            // $\dim(V_{t,\lambda})$ by ν. For any collection ζ of channels
            // with $V_{t,\lambda} \not\subset V_{\zeta,\lambda}$, $\dim(V_{t,\lambda} \cap V_{\zeta,\lambda}) < \nu$. Consider
            // the linear mapping $[k_{d,e}]_{d \in \mathrm{In}(t)} \mapsto \sum_{d \in \mathrm{In}(t)} k_{d,e}\,\mathbf{f}_{d,\lambda}$
            // from $F^{|\mathrm{In}(t)|}$ to F^ω. The nullity of this linear
            // mapping is $|\mathrm{In}(t)| - \nu$, so the pre-image of
            // the space $(V_{t,\lambda} \cap V_{\zeta,\lambda})$ has dimension less than
            // $|\mathrm{In}(t)|$. Thus the pre-image of $\cup_{\lambda,\zeta}(V_{t,\lambda} \cap V_{\zeta,\lambda})$
            // contains at most $2^{|E|} \sum_{m=1}^{\omega} \binom{|E|-1}{m-1}|F|^{|\mathrm{In}(t)|-1}$

```
    // elements, which are fewer than |F|^{|In(t)|} if
    // |F| > 2^{|E|} \sum_{m=1}^{\omega} \binom{|E|-1}{m-1}.
    for (every configuration λ)
        f_{e,λ} = λ(e) \sum_{d∈In(t)} k_{d,e} f_{d,λ};
    }
  }
}
```

Algorithm Verification. The explanation for the code constructed by Algorithm 19.40 being a static generic network code is exactly the same as that given for Algorithm 19.34. The details are omitted.

Algorithm 19.40 constitutes a constructive proof for the next theorem. By noting the lower bound on the required field size in the algorithm, we see that a generic network code can be constructed with high probability by randomly choosing the local encoding kernels provided that the base field is much larger than sufficient.

Theorem 19.41. *There exist an ω-dimensional static linear multicast, a static linear broadcast, a static linear dispersion, and a static generic network code on an acyclic network for sufficiently large base field F.*

The requirements (19.162) through (19.165) in Definition 19.37 refer to all the $2^{|E|}$ possible configurations. Conceivably, a practical application may only need to deal with a certain collection $\{\lambda_1, \lambda_2, \cdots, \lambda_\kappa\}$ of configurations, where $\kappa \ll 2^{|E|}$. Thus we may define, for instance, an $\{\lambda_1, \lambda_2, \cdots, \lambda_\kappa\}$-*static linear multicast* and an $\{\lambda_1, \lambda_2, \cdots, \lambda_\kappa\}$-*static linear broadcast* by replacing the conditions (19.162) and (19.163), respectively, by

(19.176) $\dim(V_{t,\lambda}) = \omega$ for every configuration $\lambda \in \{\lambda_1, \lambda_2, \cdots, \lambda_\kappa\}$ and every non-source node t with $\text{maxflow}_\lambda(t) \geq \omega$.

(19.177) $\dim(V_{t,\lambda}) = \min\{\omega, \text{maxflow}_\lambda(t)\}$ for every configuration $\lambda \in \{\lambda_1, \lambda_2, \cdots, \lambda_\kappa\}$ and every non-source node t.

Algorithm 19.34 has been converted into Algorithm 19.40 by modifying the key step in the former. In a similar fashion, Algorithm 19.25 can be adapted for the construction of an $\{\lambda_1, \lambda_2, \cdots, \lambda_\kappa\}$-static linear multicast or broadcast. This will lower the threshold for the sufficient size of the base field as well as the computational complexity. The details are left as an exercise.

19.7 Random Network Coding: A Case Study

We have seen in Corollary 19.24 that if the local encoding kernels of a linear network code are randomly chosen, a linear multicast can be obtained with high probability provided that the base field is sufficiently large. Since the code construction is independent of the network topology, the network code so constructed can be used when the network topology is unknown. In this section, we study an application of random network coding in *peer-to-peer* (P2P) networks. The system we will analyze is based on a prototype for large-scale content distribution on such networks proposed in [133].

19.7.1 How the System Works

A file originally residing on a single server is to be distributed to a large number of users through a network. The server divides the file into k data blocks, $B_1, B_2, \cdots, B_k$, and uploads coded versions of these blocks to different users according to some protocol. These users again help distributing the file by uploading blocks to other users in the network. By means of such repeated operations, a logical network called an *overlay network* is formed by the users as the process evolves. On this logical network, henceforth referred to as the network, information can be dispersed very rapidly, and the file is eventually delivered to every user in the network. Note that the topology of the network is not known ahead of time.

In the system, new users can join the network as a node at any time as long as the distribution process is active. Upon arrival, a new user will contact a designated node called the *tracker* that provides a subset of the other users already in the system forming the set of neighboring nodes of the new user. Subsequent information flow in the network is possible only between neighboring nodes.

For the purpose of coding, the data blocks $B_1, B_2, \cdots, B_k$ are represented as symbols in a large finite field F referred to as the base field.[4] At the beginning of the distribution process, a Client A contacts the server and receives a number of *encoded blocks*. For example, the server uploads two encoded blocks E_1 and E_2 to Client A, where for $i = 1, 2$,

$$E_i = c_1^i B_1 + c_2^i B_2 + \cdots + c_k^i B_k, \qquad (19.178)$$

with c_j^i, $1 \leq j \leq k$ being chosen randomly from the base field F. Note that each E_1 and E_2 is some random linear combination of $B_1, B_2, \cdots, B_k$.

In general, whenever a node needs to upload an encoded block to a neighboring node, the block is formed by taking a random linear combination of all the blocks possessed by that node. Continuing with the above example, when Client A needs to upload an encoded block E_3 to a neighboring Client B, we have

$$E_3 = c_1^3 E_1 + c_2^3 E_2, \qquad (19.179)$$

where c_1^3 and c_2^3 are randomly chosen from F. Substituting (19.178) into (19.179), we obtain

---

[4] In the system proposed in [133], the size of the base field is of the order 2^{16}.

$$E_3 = \sum_{j=1}^{k} (c_1^3 c_j^1 + c_2^3 c_j^2) B_j. \tag{19.180}$$

Thus E_3 and in general every encoded block subsequently uploaded by a node in the network is some random linear combination of the data blocks $B_1, B_2, \cdots, B_k$.

The exact strategy for downloading encoded blocks from the neighboring nodes so as to avoid receiving redundant information depends on the implementation. The main idea is that downloading from a neighboring node is necessary only if the neighboring node has at least one block not in the linear span of all the blocks possessed by that particular node. Upon receiving enough linearly independent encoded blocks, a node is able to decode the whole file.

Compared with store-and-forward, the application of network coding as described in the above system can reduce the file download time because an encoded block uploaded by a node contains information about every block possessed by that node. Moreover, in case some nodes leave the system before the end of the distribution process, it is more likely that the remaining nodes have the necessary information to recover the whole file if network coding is used. In the following, we will give a quantitative analysis to substantiate these claimed advantages of network coding.

19.7.2 Model and Analysis

Let V be the set of all the nodes in the system. In implementation, blocks of data are transmitted between neighboring nodes in an asynchronous manner, and possibly at different speeds. To simplify the analysis, we assume that every transmission from one node to a neighboring node is completed in an integral number of time units. Then we can unfold the network of nodes in discrete time into a graph $G^* = (V^*, E^*)$ with the node set

$$V^* = \{i_t : i \in V \text{ and } t \geq 0\}, \tag{19.181}$$

where node $i_t \in V^*$ corresponds to node $i \in V$ at time t. The edge set E^* specified below is determined by the strategy adopted for the server as well as for all the other nodes in V to request uploading of data blocks from the neighboring nodes. Specifically, there are two types of edges in E^*:

1. There is an edge with capacity m from node i_t to node $j_{t'}$, where $t < t'$, if m blocks are transmitted from node i to node j, starting at time t and ending at time t'.
2. For each $i \in V$ and $t \geq 0$, there is an edge with infinite capacity from node i_t to node i_{t+1}.

An edge of the second type models the assumption that the blocks, once possessed by a node, are retained in that node indefinitely over time. Without loss of generality, we may assume that all the blocks possessed by nodes $i_l, l \leq t$ are transmitted uncoded on the edge from node i_t to node i_{t+1}.

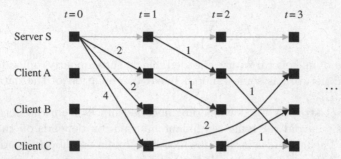

Fig. 19.8. A illustration of the graph G^*.

An illustration of the graph G^* up to $t = 3$ with V consisting of the server S and three clients A, B, and C is given in Figure 19.8, where the edges with infinite capacities are lightened for clarity. Note that the graph G^* is acyclic because each edge is pointed in the positive time direction and hence a cycle cannot be formed.

Denote the server S by node $s \in V$ and regard node s_0 in G^* as the source node generating the whole file consisting of k data blocks and multicasting it to all the other nodes in G^* via random linear network coding, with the coefficients in the random linear combinations forming the encoded blocks being the local encoding kernels of the network code. Note that random network coding is applied on G^*, not the logical network formed by the user nodes.

Also note that in order to simplify our description of the system, we have omitted the necessity of delivering the global encoding kernels to the nodes for the purpose of decoding. We refer the reader to the discussion toward the end of Section 19.3 for this implementation detail.

We are now ready to determine the time it takes for a particular node $i \in V$ to receive the whole file. Denote the value of a max-flow from node s_0 to a node $v \in G^*$ other than s_0 by maxflow(v). When the base field is sufficiently large, by Corollary 19.24, with probability close to 1, the network code generated randomly during the process is a linear multicast, so that those nodes i_t with

$$\text{maxflow}(i_t) \geq k \tag{19.182}$$

can receive the whole file. In other words, with high probability, the time it takes a node $i \in V$ to receive the whole file is equal to t^*, the minimum t that satisfies (19.182). Obviously, this is a lower bound on the time it takes a node $i \in V$ to receive the whole file, and it is achievable with high probability by the system under investigation. In the rare event that node i cannot decode at time t^*, it can eventually decode upon downloading some additional encoded blocks from the neighboring nodes.

When some nodes leave the system before the end of the distribution process, an important question is whether the remaining nodes have the necessary information to recover the whole file. To be specific, assume that a subset of users $U^c \subset V$ leave the system after time t, and we want to know whether the users in $U = V \setminus U^c$ have sufficient information to recover the whole file. If they do, by further exchanging information among themselves, every user in U can eventually receive the whole file (provided that no more nodes leave the system). Toward this end, again consider the graph G^*. Let

$$U_t = \{u_t : u \in U\} \tag{19.183}$$

and denote the value of a max-flow from node s_0 to the set of nodes U_t by maxflow(U_t). If

$$\text{maxflow}(U_t) \geq k, \tag{19.184}$$

then the users in U with high probability would have the necessary information to recover the whole file. This is almost the best possible performance one can expect from such a system, because if

$$\text{maxflow}(U_t) < k, \tag{19.185}$$

it is simply impossible for the users in U to recover the whole file even if they are allowed to exchange information among themselves.

Thus we see that random network coding provides the system with both maximum bandwidth efficiency and maximum robustness. However, additional computational resource is required compared with store-and-forward. These are engineering tradeoffs in the design of such systems.

We conclude this section by an example further demonstrating the advantage of random network coding when it is applied to packet networks with packet loss.

Example 19.42. The random network coding scheme discussed in this section can be applied to packet networks with packet loss. Consider the network depicted in Figure 19.9 consisting of three nodes, s, t, and u. Data packets are sent from node s to node u via node t. Let the packet loss rates of channels (s,t) and (t,u) be γ, i.e., a fraction γ of packets are lost during their transmission through the channel. Then the fraction of packets sent by node s that are eventually received at node u is $(1-\gamma)^2$.

To fix idea, assume the packet size is sufficiently large and one packet is sent on each channel per unit time. To remedy the problem of packet loss, a fountain code [50] can be employed at node s. This would allow data packets to be sent from node s to node u reliably at an effective rate equal to $(1-\gamma)^2$. In such a scheme, node t simply forwards to node u the packets it receives

Fig. 19.9. A simple packet network.

from node s. On the other hand, by using the random network coding scheme we have discussed, data packets can be sent from node s to node u reliably at an effective rate equal to $1-\gamma$, which is strictly higher than $(1-\gamma)^2$ whenever $\gamma > 0$. This can be proved by means of the analysis presented in this section. The details are left as an exercise.

While a fountain code can remedy the problem of packet loss between the source node and the sink node, it cannot prevent the packet loss rate from accumulating when packets are routed through the network. On the other hand, the use of random network coding allows information to be transmitted from the source node to the sink node at the maximum possible rate, namely the min-cut between the source node and the sink node after the packet loss in the channels has been taken into account.

Chapter Summary

Linear Network Code:

- $k_{d,e}$ is the local encoding kernel of the adjacent pair of channels (d, e).
- $\mathbf{f}_e$ is the global encoding kernel of channel e.
- $\mathbf{f}_e = \sum_{d \in \text{In}(t)} k_{d,e} \mathbf{f}_d$ for $e \in \text{Out}(t)$.
- $\mathbf{f}_e, e \in \text{In}(s)$ form the standard basis of F^ω.
- Channel e transmits the symbol $\mathbf{x} \cdot \mathbf{f}_e$, where $\mathbf{x} \in F^\omega$ is the message generated at source node s.

Linear Multicast, Broadcast, and Dispersion: An ω-dimensional linear network code is a

linear multicast if $\dim(V_t) = \omega$ for every node $t \neq s$ with $\text{maxflow}(t) \geq \omega$.
linear broadcast if $\dim(V_t) = \min\{\omega, \text{maxflow}(t)\}$ for every node $t \neq s$.
linear dispersion if $\dim(V_T) = \min\{\omega, \text{maxflow}(T)\}$ for every collection T of non-source nodes.

Generic Network Code: An ω-dimensional linear network code is generic if for any nonempty collection of channels $\xi = \{e_1, e_2, \cdots, e_m\} \subset E$, where $e_j \in \text{Out}(t_j)$, and any $1 \leq k \leq m$, if

a) there is no directed path from t_k to t_j for $j \neq k$,
b) $V_{t_k} \not\subset V_{\xi_{\bar{k}}}$, where $\xi_{\bar{k}} = \xi \backslash \{e_k\}$,

then $\mathbf{f}_{e_k} \notin V_{\xi_{\bar{k}}}$. A generic network code is a linear dispersion and hence a linear broadcast and a linear multicast.

Characterizations of Generic Network Code: Each of the following is a necessary and sufficient condition for an ω-dimensional linear network code to be generic:

1) For any nonempty collection of channels $\xi = \{e_1, e_2, \cdots, e_m\} \subset E$, where $e_j \in \text{Out}(t_j)$, if $V_{t_j} \not\subset V_{\xi_j}$ for all $1 \leq j \leq m$, then $\mathbf{f}_e, e \in \xi$ are linearly independent.

2) For any nonempty collection of channels $\xi \subset E$, if $|\xi| = \min\{\omega, \text{maxflow}(\xi)\}$, then $\mathbf{f}_e, e \in \xi$ are linearly independent.

Static Network Code: For a given linear network code and a configuration λ of the network, the λ-global encoding kernel $\mathbf{f}_{e,\lambda}$ of channel e is calculated recursively by

- $\mathbf{f}_{e,\lambda} = \lambda(e) \sum_{d \in \text{In}(t)} k_{d,e}\, \mathbf{f}_{d,\lambda}$ for $e \in \text{Out}(t)$.
- $\mathbf{f}_{e,\lambda}, e \in \text{In}(s)$ are independent of λ and form the standard basis of F^ω.

An ω-dimensional linear network code is a static

linear multicast if $\dim(V_{t,\lambda}) = \omega$ for every λ and every node $t \neq s$ with $\text{maxflow}_\lambda(t) \geq \omega$.

linear broadcast if $\dim(V_{t,\lambda}) = \min\{\omega, \text{maxflow}_\lambda(t)\}$ for every λ and every node $t \neq s$.

linear dispersion if $\dim(V_{T,\lambda}) = \min\{\omega, \text{maxflow}_\lambda(T)\}$ for every λ and every collection T of non-source nodes.

generic network code if for every λ and any nonempty collection of channels $\xi \subset E$, if $|\xi| = \min\{\omega, \text{maxflow}_\lambda(\xi)\}$, then $\mathbf{f}_{e,\lambda}, e \in \xi$ are linearly independent.

Lemma: Let $g(z_1, z_2, \cdots, z_n)$ be a nonzero polynomial with coefficients in a field F. If $|F|$ is greater than the degree of g in every z_j for $1 \leq j \leq n$, then there exist $a_1, a_2, \cdots, a_n \in F$ such that $g(a_1, a_2, \cdots, a_n) \neq 0$.

Existence and Construction: All the linear network codes defined in this chapter exist and can be constructed either deterministically or randomly (with high probability) when the base field is sufficiently large.

Problems

In the following, let $G = (V, E)$ be the underlying directed acyclic network on which the linear network code is defined, and let s be the unique source node in the network.

1. Show that in a network with the capacities of all the edges equal to 1, the number of edge-disjoint paths from source node s to a non-source node t is equal to $\text{maxflow}(t)$.

2. For the network code in Definitions 19.4 and 19.6, show that if the global encoding mappings are linear, then so are the local encoding mappings. (Yeung *et al.* [408].)

3. *Network transfer matrix.* Consider an ω-dimensional linear network code.

a) Prove (19.91).

b) Fix an upstream-to-downstream order for the channels in the network and let K be the $|E| \times |E|$ matrix with the (d, e)th element equal to $k_{d,e}$ if (d, e) is an adjacent pair of channels and equal to 0 otherwise. Let A be the $\omega \times |E|$ matrix obtaining by appending $|E| - |\mathrm{Out}(s)|$ columns of zeroes to K_s, and B_e be the $|E|$-column vector with all the components equal to 0 except that the eth component is equal to 1. Show that

$$\mathbf{f}_e = A(I - K)^{-1} B_e$$

for all $e \in E$. The matrix $M = (I - K)^{-1}$ is called the *network transfer matrix*.

(Koetter and Médard [202].)

4. Apply Lemma 19.17 to obtain a lower bound on the field size for the existence of a two-dimensional linear multicast on the butterfly network.

5. Show that $|E| \sum_{m=1}^{\omega} \binom{|E|-1}{m-1}$ is a polynomial in $|E|$ of degree ω. This is the lower bound on the required field size in Algorithm 19.34.

6. Verify that the network code in Example 19.38 is a generic network code.

7. *Simplified characterization of a generic network code.* Consider an ω-dimensional generic network code on a network for which $|\mathrm{Out}(s)| \geq \omega$.

   a) Show that Condition (3) in Theorem 19.32 can be modified to restricting the cardinality of ξ to ω. Hint: If $|\xi| < \omega$, expand ξ by including a certain subset of the channels in $\mathrm{Out}(s)$.

   b) Simplify Algorithm 19.34 and tighten the lower bound on the required field size accordingly.

   (Tan et al. [348].)

8. For the network below, prove the nonexistence of a two-dimensional binary generic network code.

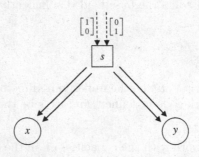

9. Show that for $\eta \geq 2$, a linear multicast can be constructed by the Jaggi–Sanders algorithm provided that $|F| \geq \eta$. Hint: Two vector subspaces intersect at the origin.

10. Modify the Jaggi–Sanders algorithm for the construction of a static linear multicast.

11. Obtain a lower bound on the required field size and determine the computational complexity when Algorithm 19.40 is adapted for the construction of an $\{\lambda_1, \lambda_2, \cdots, \lambda_\kappa\}$-static generic network code.

12. Show that a transformation of a static generic network code is also a static generic network code.

13. *A generic network code as a linear dispersion.* Expand the network G into a network $G' = (V', E')$ as follows. For an edge $e \in E$, let the edge be from node v_e to node w_e. Install a new node t_e and replace edge e by two new edges e' and e'', where e' is from node v_e to node t_e and e'' is from node t_e to node w_e. Show that a linear dispersion on G' is equivalent to a generic network code on G. Hint: Use Theorem 19.32. (Kwok and Yeung [217], Tan et al. [348].)

14. *Multi-rate linear broadcast.* Consider a network on which an ω-dimensional linear network code over a base field F is defined. For all $e \in E$, let

$$\mathbf{f}'_e = [\; I \;\; \mathbf{b}\;] \, \mathbf{f}_e,$$

where I is the $(\omega - 1) \times (\omega - 1)$ identity matrix and $\mathbf{b}$ is an $(\omega - 1)$-column vector.

   a) Show that $\mathbf{f}'_e, e \in E$ constitute the global encoding kernels of an $(\omega - 1)$-dimensional linear network code on the same network.

   b) Show that the $(\omega - 1)$-dimensional linear network code in (a) and the original ω-dimensional linear network code have the same local encoding kernels for all the non-source nodes.

   It was shown in Fong and Yeung [115] that an $(\omega - 1)$-dimensional linear broadcast can be constructed from any ω-dimensional linear broadcast by choosing a suitable vector $\mathbf{b}$, provided $|F| \geq |V|$. This implies that multi-rate linear multicast/broadcast can be supported on a network without changing the local encoding kernels of the non-source nodes.

15. Let a message $\mathbf{x} \in F^\omega$ be generated at source node s in a network for which $\mathrm{maxflow}(t) \geq \omega$ for all non-source nodes t. Show that $\mathbf{x}$ can be multicast to all the non-source nodes by store-and-forward. In other words, for this special case, network coding has no advantage over store-and-forward if complete information on the network topology is known ahead of time. This result is implied by a theorem on *directed spanning tree packing* by Edmonds [100] (see also Wu et al. [386]).

16. Let L be the length of the message $\mathbf{x}$ generated at source node s, where L is divisible by $\mathrm{maxflow}(t)$ for all non-source nodes t. Allowing multiple usage of the network, devise a linear network coding scheme such that each non-source node t can receive $\mathbf{x}$ in $L/\mathrm{maxflow}(t)$ units of time. Such a scheme enables each non-source node in the network to receive the message within the shortest possible time.

17. Consider distributing a message of 5 data blocks on a P2P network with 4 nodes, Server S and Clients A, B, and C, by the system discussed in Section 19.7. Assume each data block is sufficiently large. The following transmissions take place during the process.

From	To	Start Time	End Time	# Blocks
S	A	0	1	2
S	B	0	1	3
S	C	0	1	2
B	A	1	2	1
C	B	1	3	2
S	B	2	3	1
B	C	2	3	2

a) Which client is the first to receive the whole message?
b) If Client B leaves the system after $t = 3$, do Clients A and C have sufficient information to reconstruct the whole message?
c) Suppose the hard disk of Client B crashes at $t = 1.5$ and loses 2 blocks of data. Repeat (b) by making the assumption that the transmissions by Client B starting at $t \leq 1$ are not affected by the disk failure.

18. Prove the claim in Example 19.42 that by using random network coding, data packets can be sent from node s to node u at an effective rate equal to $1 - \gamma$.

Historical Notes

The achievability of the max-flow bound by linear network codes was proved by Li et al. [230] using a vector space approach and then by Koetter and Médard [202] using a matrix approach. These two approaches correspond, respectively, to the notions of global encoding kernel and local encoding kernel discussed here. Neither the construction in [230] for a generic network code nor the construction in [202] for a linear multicast is a polynomial-time algorithm. Jaggi and Sanders et al. [184] obtained a polynomial-time algorithm for constructing a linear multicast by modifying the construction of a generic network code in [230]. A polynomial-time algorithm for constructing a generic network code was subsequently obtained in Yeung et al. [408].

In [202], static network code was introduced and its existence was proved. An explicit construction of such codes was given in [408].

The optimality of random network coding was proved in Ahlswede et al. [6]. Ho et al. [162] proved the optimality of random linear network coding and proposed the use of such codes on networks with unknown topology. A tight upper bound on the probability of decoding error for random linear network coding has recently been obtained by Balli et al. [23].

Implementation issues of network coding were discussed in Chou et al. [69]. The application of random network coding in peer-to-peer networks discussed in Section 19.7 is due to Gkantsidis and Rodriguez [133].

Cai and Yeung have generalized the theory of single-source network coding on acyclic networks to network error correction [52][405][53] and secure network coding [51][406]. Network error correction subsumes classical algebraic coding, while secure network coding subsumes secret sharing in cryptography.

The presentation in this chapter is largely based on the tutorial paper by Yeung et al. [408]. The various characterizations of a generic network code are due to Tan et al. [348]. The analysis of a large-scale content distribution system with network coding is due to Yeung [403].

Single-Source Linear Network Coding: Cyclic Networks

A directed network is *cyclic* if it contains at least one directed cycle. In Chapter 19, we have discussed network coding over an acyclic network, for which there exists an upstream-to-downstream order on the nodes. Following such an order, whenever a node encodes, all the information needed would have already been received on the input channels of that node. For a cyclic network, such an order of the nodes does not exist. This makes network coding over a cyclic network substantially different from network coding over an acyclic network.

20.1 Delay-Free Cyclic Networks

When we discussed network coding over an acyclic network in Chapter 19, we assumed that there is no propagation delay in the network. Based on this assumption, a linear network code can be specified by either the local description in Definition 19.6 or the global description in Definition 19.7. The local and global descriptions of a linear network code are equivalent over an acyclic network because given the local encoding kernels, the global encoding kernels can be calculated recursively in any upstream-to-downstream order. In other words, equation (19.17) has a unique solution for the global encoding kernels in terms of the local encoding kernels, while (19.18) serves as the boundary conditions.

If these descriptions are applied to a cyclic network, it is not clear whether for any given set of local encoding kernels, there exists a unique solution for the global encoding kernels. In the following, we give one example with a unique solution, one with no solution, and one with multiple solutions.

Example 20.1. Consider the cyclic network in Figure 20.1. Let (s, t) precede (v, t) in the ordering among the channels. Similarly, let (s, t') precede (v, t'). Given the local encoding kernels

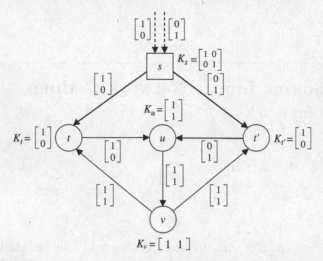

Fig. 20.1. A two-dimensional linear broadcast on a cyclic network.

$$K_s = \begin{bmatrix} 1 & 0 \\ 0 & 1 \end{bmatrix}, \; K_t = K_{t'} = \begin{bmatrix} 1 \\ 0 \end{bmatrix}, \; K_u = \begin{bmatrix} 1 \\ 1 \end{bmatrix}, \; K_v = \begin{bmatrix} 1 & 1 \end{bmatrix}, \qquad (20.1)$$

equation (19.17) yields the following unique solution for the global encoding kernels:

$$\mathbf{f}_{(s,t)} = \mathbf{f}_{(t,u)} = \begin{bmatrix} 1 \\ 0 \end{bmatrix}, \; \mathbf{f}_{(s,t')} = \mathbf{f}_{(t',u)} = \begin{bmatrix} 0 \\ 1 \end{bmatrix} \qquad (20.2)$$

$$\mathbf{f}_{(u,v)} = \mathbf{f}_{(v,t)} = \mathbf{f}_{(v,t')} = \begin{bmatrix} 1 \\ 1 \end{bmatrix}. \qquad (20.3)$$

These global encoding kernels are shown in Figure 20.1, and they in fact define a two-dimensional linear broadcast regardless of the choice of the base field. Since

$$k_{(v,t),(t,u)} = 0 \qquad (20.4)$$

and

$$k_{(v,t'),(t',u)} = 0, \qquad (20.5)$$

information looping in the directed cycles (t, u), (u, v), (v, t) and (t', u), (u, v), (v, t') is prevented.

Example 20.2. An arbitrarily prescribed set of local encoding kernels on a cyclic network is unlikely to be compatible with any global encoding kernels. In Figure 20.2(a), a local encoding kernel is prescribed at each node in a cyclic network. Had a global encoding kernel $\mathbf{f}_e$ existed for each channel e, the requirement (19.17) would imply the equations

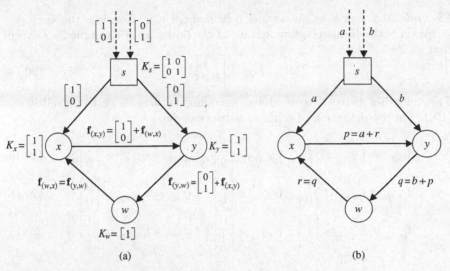

(a)   (b)

Fig. 20.2. An example of a cyclic network and local encoding kernels that do not render a solution for the global encoding kernels.

$$\mathbf{f}_{(x,y)} = \begin{bmatrix} 1 \\ 0 \end{bmatrix} + \mathbf{f}_{(w,x)} \tag{20.6}$$

$$\mathbf{f}_{(y,w)} = \begin{bmatrix} 0 \\ 1 \end{bmatrix} + \mathbf{f}_{(x,y)} \tag{20.7}$$

$$\mathbf{f}_{(w,x)} = \mathbf{f}_{(y,w)}, \tag{20.8}$$

which sum up to

$$\begin{bmatrix} 1 \\ 0 \end{bmatrix} = \begin{bmatrix} 0 \\ 1 \end{bmatrix}, \tag{20.9}$$

a contradiction.

The nonexistence of compatible global encoding kernels can also be interpreted in terms of the message transmission. Let the message $\mathbf{x} = [\, a\, b\,]$ be a generic vector in F^2, where F denotes the base field. The symbol transmitted on channel e, given by $\mathbf{x} \cdot \mathbf{f}_e$, are shown in Figure 20.2(b). In particular, the symbols transmitted on channels (x,y), (y,w), and (w,x), namely p, q, and r, are related through

$$p = a + r \tag{20.10}$$
$$q = b + p \tag{20.11}$$
$$r = q. \tag{20.12}$$

These equalities imply that

$$a + b = 0, \tag{20.13}$$

a contradiction to the independence between the two components a and b of a generic message.

Example 20.3. Let F be an extension field of $GF(2)$.[1] Consider the same prescription of the local encoding kernels at the nodes as in Example 20.2 except that

$$K_S = \begin{bmatrix} 1 & 1 \\ 0 & 0 \end{bmatrix}. \tag{20.14}$$

The following three sets of global encoding kernels meet the requirement (19.17) in the definition of a linear network code:

$$\mathbf{f}_{(s,x)} = \mathbf{f}_{(s,y)} = \begin{bmatrix} 1 \\ 0 \end{bmatrix}, \ \mathbf{f}_{(x,y)} = \begin{bmatrix} 0 \\ 0 \end{bmatrix}, \ \mathbf{f}_{(y,w)} = \mathbf{f}_{(w,x)} = \begin{bmatrix} 1 \\ 0 \end{bmatrix}; \tag{20.15}$$

$$\mathbf{f}_{(s,x)} = \mathbf{f}_{(s,y)} = \begin{bmatrix} 1 \\ 0 \end{bmatrix}, \ \mathbf{f}_{(x,y)} = \begin{bmatrix} 1 \\ 0 \end{bmatrix}, \ \mathbf{f}_{(y,w)} = \mathbf{f}_{(w,x)} = \begin{bmatrix} 0 \\ 0 \end{bmatrix}; \tag{20.16}$$

$$\mathbf{f}_{(s,x)} = \mathbf{f}_{(s,y)} = \begin{bmatrix} 1 \\ 0 \end{bmatrix}, \ \mathbf{f}_{(x,y)} = \begin{bmatrix} 0 \\ 1 \end{bmatrix}, \ \mathbf{f}_{(y,w)} = \mathbf{f}_{(w,x)} = \begin{bmatrix} 1 \\ 1 \end{bmatrix}. \tag{20.17}$$

20.2 Convolutional Network Codes

In a real network, the propagation delay, which includes the processing delay at the nodes and the transmission delay over the channels, cannot be zero. For a cyclic network, this renders the implementation non-physical because the transmission on an output channel of a node can only depend on the information received on the input channels of that node. Besides, technical difficulties as described in the last section arise even with the ideal assumption that there is no propagation delay.

In this section, we introduce the *unit-delay network* as a model for network coding on a cyclic network $G = (V, E)$, where V and E are the sets of nodes and channels of the network, respectively. In this model, a symbol is transmitted on every channel in the network at every discrete-time index, with the transmission delay equal to exactly one time unit. Intuitively, this assumption on the transmission delay over a channel ensures no information looping in the network even in the presence of a directed cycle. The results to be developed in this chapter, although discussed in the context of cyclic networks, apply equally well to acyclic networks.

As a time-multiplexed network in the combined space–time domain, a unit-delay network can be unfolded with respect to the time dimension into an indefinitely long network called a *trellis network*. Corresponding to a physical node t is a sequence of nodes $t_0, t_1, t_2, \cdots$ in the trellis network, with the subscripts being the time indices. A channel e_j in the trellis network represents the transmission on the physical channel e between times j and $j+1$. When the physical channel e is from node t to node u, the channel e_j in the trellis network

---

[1] In an extension field of $GF(2)$, the arithmetic on the symbols 0 and 1 are modulo 2 arithmetic.

is from node t_j to node u_{j+1}. Note that the trellis network is acyclic regardless of the topology of the physical network, because all the channels are pointing in the forward time direction so that a directed cycle cannot be formed.

Example 20.4. Regard the network in Figure 20.2 as a unit-delay network. For each channel e in the network, the scalar values in the base field F transmitted on the channels e_j, $j \geq 0$ in the corresponding trellis network are determined by the local encoding kernels. This is illustrated in Figure 20.3. For instance, the channels $(x, y)_j$, $j \geq 0$ carry the scalar values

$$0, 0, a_0, a_1, a_2 + b_0, a_0 + a_3 + b_1, a_1 + a_4 + b_2, \cdots, \qquad (20.18)$$

respectively. This constitutes an example of a convolutional network code to be formally defined in Definition 20.6.

Let c_j be the scalar value in F transmitted on a particular channel in the network at time j. A succinct mathematical expression for the sequence of scalars $c_0, c_1, c_2, \cdots$ is the z-transform

$$\sum_{j=0}^{\infty} c_j z^j = c_0 + c_1 z + c_2 z^2 + \cdots, \qquad (20.19)$$

where the power j of the *dummy variable* z represents discrete time. The pipelining of scalars transmitted over a time-multiplexed channel can thus be regarded as the transmission of a power series over the channel. For example,

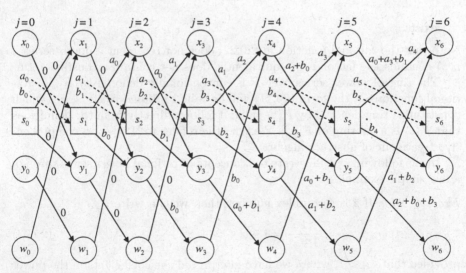

Fig. 20.3. The trellis network depicting a convolutional network code defined on the physical network in Figure 20.2.

the transmission of a scalar value on the channel $(x, y)_j$ for each $j \geq 0$ in the trellis network in Figure 20.3 translates into the transmission of the power series

$$a_0 z^2 + a_1 z^3 + (a_2 + b_0) z^4 + (a_0 + a_3 + b_1) z^5 + (a_1 + a_4 + b_2) z^6 + \cdots$$
(20.20)

over the channel (x, y) in the network in Figure 20.2.

The z-transform in (20.19) is a power series in the dummy variable z, which would be regarded as either a real number or a complex number in the context of signal analysis. However, in the context of convolutional coding, such a power series should not be regarded as anything more than a representation of the sequence of scalars $c_0, c_1, c_2, \cdots$. Specifically, the dummy variable z is not associated with any value, and there is no notion of convergence. Such power series are called *formal power series*.

Given a field F, consider rational functions of a dummy variable z of the form

$$\frac{p(z)}{1 + zq(z)},$$
(20.21)

where $p(z)$ and $q(z)$ are polynomials. The following properties of such a function are relevant to our subsequent discussion:

1. The denominator has a constant term, so the function can be expanded into a power series by long division (see Example 20.5).
2. If $p(z)$ is not the zero polynomial, the inverse function, namely

$$\frac{1 + zq(z)}{p(z)},$$
(20.22)

exists.

Note that the rational function in (20.22) does not represent a power series if $p(z)$ contains the factor z, or equivalently, does not contain a constant term.

The ring of power series over F is conventionally denoted by $F[[z]]$. Rational functions of the form (20.21) will be called *rational power series* which constitute a ring denoted by $F\langle z \rangle$ [408]. It follows directly from the definitions that $F\langle z \rangle$ is a subring of $F[[z]]$. We refer the reader to [124] for a comprehensive treatment of abstract algebra.

In the following, we illustrate the concepts of rational power series through a few simple examples.

Example 20.5. If z is a complex number, then we can write

$$\frac{1}{1 - z} = 1 + z + z^2 + z^3 + \cdots$$
(20.23)

provided that $|z| < 1$, where we have interpreted the coefficients in the power series on the right-hand side as real (or complex) numbers. If $|z| > 1$, the above expression is not meaningful because the power series diverges.

However, if we do not associate z with a value but regard the coefficients in the power series as elements in a commutative ring, we can always write

$$(1 - z)(1 + z + z^2 + z^3 + \cdots)$$
$$= (1 + z + z^2 + z^3 + \cdots) - (z + z^2 + z^3 + \cdots) \qquad (20.24)$$
$$= 1. \qquad (20.25)$$

In this sense, we say that $1 - z$ is the reciprocal of the power series $1 + z + z^2 + z^3 + \cdots$ and write

$$\frac{1}{1 - z} = 1 + z + z^2 + z^3 + \cdots. \qquad (20.26)$$

We also say that $1 + z + z^2 + z^3 + \cdots$ is the power series expansion of $1/1 - z$. In fact, the power series on the right-hand side can be readily obtained by dividing 1 by $1 - z$ using long division.

Alternatively, we can seek the inverse of $1 - z$ by considering the identity

$$(1 - z)(a_0 + a_1 z + a_2 z^2 + \cdots) = 1. \qquad (20.27)$$

By equating the powers of z on both sides, we have

$$a_0 = 1 \qquad (20.28)$$
$$-a_0 + a_1 = 0 \qquad (20.29)$$
$$-a_1 + a_2 = 0 \qquad (20.30)$$
$$\vdots \qquad (20.31)$$

Then by forward substitution, we immediately obtain

$$1 = a_0 = a_1 = a_2 = \cdots, \qquad (20.32)$$

which gives exactly the power series obtained by long division. The reader can easily verify that long division indeed mimics the process of forward substitution.

For polynomials $p(z)$ and $q(z)$ where $q(z)$ is not the zero polynomial, we can always expand the rational function $p(z)/q(z)$ into a series. However, such a series is not always a power series. For example,

$$\frac{1}{z - z^2} = \frac{1}{z}\left(\frac{1}{1 - z}\right) \qquad (20.33)$$

$$= \frac{1}{z}(1 + z + z^2 + \cdots) \qquad (20.34)$$

$$= z^{-1} + 1 + z + z^2 + \cdots. \qquad (20.35)$$

The above is not a power series because of the term involving a negative power of z. In fact, the identity

$$(z - z^2)(a_0 + a_1 z + a_2 z^2 + \cdots) = 1 \qquad (20.36)$$

has no solution for $a_0, a_1, a_2, \cdots$ since there is no constant term on the left-hand side. Therefore, $1/(z - z^2)$ indeed does not have a power series expansion.

From the above example, we see that $p(z)/q(z)$ represents a rational power series if and only if $q(z)$ has a nonzero constant term, or equivalently, does not contain the factor z.

Definition 20.6 (Convolutional Network Code). *An ω-dimensional convolutional network code on a unit-delay network over a base field F consists of an element $k_{d,e}(z) \in F\langle z \rangle$ for every adjacent pair of channels (d, e) in the network as well as a column ω-vector $\mathbf{f}_e(z)$ over $F\langle z \rangle$ for every channel e such that*

(20.37) $\mathbf{f}_e(z) = z \sum_{d \in In(t)} k_{d,e}(z) \mathbf{f}_d(z)$ *for $e \in \mathrm{Out}(t)$.*

(20.38) *The vectors $\mathbf{f}_e(z)$ for the imaginary channels $e \in \mathrm{In}(s)$ consist of scalar components that form the standard basis of the vector space F^ω.*

The vector $\mathbf{f}_e(z)$ is called the global encoding kernel for channel e, and $k_{d,e}(z)$ is called the local encoding kernel for the adjacent pair of channels (d, e). The $|\mathrm{In}(t)| \times |\mathrm{Out}(t)|$ matrix

$$K_t(z) = [k_{d,e}(z)]_{d \in In(t), e \in Out(t)} \qquad (20.39)$$

is called the local encoding kernel at node t.

The constraint (20.37) is the time-multiplexed version of (19.17), with the factor z in the equation indicating a unit-time delay that represents the transmission delay over a channel. In the language of electronic circuit theory, for an adjacent pair of channels (d, e), the "gain" from channel d to channel e is given by $zk_{d,e}(z)$.

A convolutional network code over a unit-delay network can be viewed as a discrete-time *linear time-invariant* (LTI) system defined by the local encoding kernels, where the local encoding kernel $k_{d,e}(z)$ specifies the impulse response of an LTI filter from channel d to channel e. The requirement that $k_{d,e}(z)$ is a power series corresponds to the causality of the filter. The additional requirement that $k_{d,e}(z)$ is rational ensures that the filter is implementable by a finite circuitry of shift registers. Intuitively, once the local encoding kernels are given, the global encoding kernels are uniquely determined. This is explained as follows. Write

$$\mathbf{f}_e(z) = \sum_{j=0}^{\infty} \mathbf{f}_{e,j} z^j = \mathbf{f}_{e,0} + \mathbf{f}_{e,1} z + \mathbf{f}_{e,2} z^2 + \cdots \qquad (20.40)$$

and

$$k_{d,e}(z) = \sum_{j=0}^{\infty} k_{d,e,j} z^j = k_{d,e,0} + k_{d,e,1} z + k_{d,e,2} z^2 + \cdots, \tag{20.41}$$

where $\mathbf{f}_{e,j}$ is a column ω-vector in F^ω and $k_{d,e,j}$ is a scalar in F. Then (20.37) can be written in time domain as the convolutional equation

$$\mathbf{f}_{e,j} = \sum_{d \in \mathrm{In}(t)} \left(\sum_{u=0}^{j-1} k_{d,e,u} \, \mathbf{f}_{d,j-1-u} \right) \tag{20.42}$$

for $j \geq 0$, with the boundary conditions provided by (20.38):

- The vectors $\mathbf{f}_{e,0}, e \in \mathrm{In}(t)$ form the standard basis of the vector space F^ω.
- The vectors $\mathbf{f}_{e,j}, e \in \mathrm{In}(t)$ are the zero vector for all $j \geq 1$.

For $j = 0$, the summation in (20.42) is empty, so that $\mathbf{f}_{e,0}$ vanishes. For $j \geq 0$, the right-hand side of (20.42) involves the vectors $\mathbf{f}_{d,i}$ only for $0 \leq i \leq j-1$. Thus the vectors $\mathbf{f}_{e,j}, j \geq 1$ can be calculated recursively via (20.42) with the boundary condition

$$\mathbf{f}_{d,0} = 0 \quad \text{for all } d \in E. \tag{20.43}$$

Together with $\mathbf{f}_{e,0} = 0$, the global encoding kernel $\mathbf{f}_e(z)$ is determined (cf. (20.40)). In other words, in a convolutional network code over a unit-delay network, the global encoding kernels are determined once the local encoding kernels are given. From (20.40), we see that the components of $\mathbf{f}_e(z)$ are power series in z, so $\mathbf{f}_e(z)$ is a column ω-vector over $F[[z]]$. In Theorem 20.9, we will further establish that the components of the global encoding kernels are in fact rational functions in z, proving that $\mathbf{f}_e(z)$ is indeed a column ω-vector over $f\langle z \rangle$ as required in Definition 20.6 for a convolutional network code.

Example 20.7. In Figure 20.2, denote the two imaginary channels by (o, s) and $(o, s)'$. A convolutional network code is specified by the prescription of a local encoding kernel at every node as shown in the figure:

$$K_s(z) = \begin{bmatrix} 1 & 0 \\ 0 & 1 \end{bmatrix}, \ K_x(z) = K_y(z) = \begin{bmatrix} 1 \\ 1 \end{bmatrix}, \ K_w(z) = \begin{bmatrix} 1 \end{bmatrix}, \tag{20.44}$$

and a global encoding kernel for every channel:

$$\mathbf{f}_{(o,s)}(z) = \begin{bmatrix} 1 \\ 0 \end{bmatrix}, \ \mathbf{f}_{(o,s)'}(z) = \begin{bmatrix} 0 \\ 1 \end{bmatrix} \tag{20.45}$$

$$\mathbf{f}_{(s,x)}(z) = z \begin{bmatrix} 1 & 0 \\ 0 & 1 \end{bmatrix} \begin{bmatrix} 1 \\ 0 \end{bmatrix} = \begin{bmatrix} z \\ 0 \end{bmatrix} \tag{20.46}$$

$$\mathbf{f}_{(s,y)}(z) = z \begin{bmatrix} 1 & 0 \\ 0 & 1 \end{bmatrix} \begin{bmatrix} 0 \\ 1 \end{bmatrix} = \begin{bmatrix} 0 \\ z \end{bmatrix} \tag{20.47}$$

$$\mathbf{f}_{(x,y)}(z) = \begin{bmatrix} z^2/(1-z^3) \\ z^4/(1-z^3) \end{bmatrix}, \tag{20.48}$$

$$\mathbf{f}_{(y,w)}(z) = \begin{bmatrix} z^3/(1-z^3) \\ z^2/(1-z^3) \end{bmatrix}, \tag{20.49}$$

$$\mathbf{f}_{(w,x)}(z) = \begin{bmatrix} z^4/(1-z^3) \\ z^3/(1-z^3) \end{bmatrix}, \tag{20.50}$$

where the last three global encoding kernels have been solved from the following equations:

$$\mathbf{f}_{(x,y)}(z) = z \left[\mathbf{f}_{(s,x)}(z) \ \mathbf{f}_{(w,x)}(z) \right] \begin{bmatrix} 1 \\ 1 \end{bmatrix} = z^2 \begin{bmatrix} 1 \\ 0 \end{bmatrix} + z \, \mathbf{f}_{(w,x)}(z), \tag{20.51}$$

$$\mathbf{f}_{(y,w)}(z) = z \left[\mathbf{f}_{(s,y)}(z) \ \mathbf{f}_{(x,y)}(z) \right] \begin{bmatrix} 1 \\ 1 \end{bmatrix} = z^2 \begin{bmatrix} 0 \\ 1 \end{bmatrix} + z \, \mathbf{f}_{(x,y)}(z), \tag{20.52}$$

$$\mathbf{f}_{(w,x)}(z) = z(\mathbf{f}_{(y,w)}(z)) \left[1 \right] = z \, \mathbf{f}_{(y,w)}(z). \tag{20.53}$$

These local and global encoding kernels of a two-dimensional convolutional network code are summarized in Figure 20.4.

Represent the message generated at source node s at time j, where $j \geq 0$, by a row ω-vector $\mathbf{x}_j \in F^\omega$. Equivalently, source node s generates the message pipeline represented by the z-transform

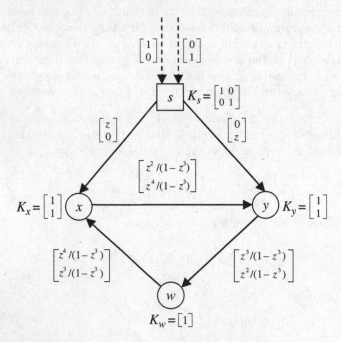

Fig. 20.4. The local and global encoding kernels of the convolutional network code in Example 20.7.

$$\mathbf{x}(z) = \sum_{j=0}^{\infty} \mathbf{x}_j z^j, \tag{20.54}$$

which is a row ω-vector over $F[[z]]$, the ring of power series over F. Here, $\mathbf{x}(z)$ is not necessarily rational.

Through a convolutional network code, each channel e carries the power series $\mathbf{x}(z)\mathbf{f}_e(z)$. Write

$$\mathbf{x}(z)\mathbf{f}_e(z) = \sum_{j=0}^{\infty} m_{e,j} z^j, \tag{20.55}$$

where

$$m_{e,j} = \sum_{u=0}^{j} \mathbf{x}_u \mathbf{f}_{e,j-u}. \tag{20.56}$$

For $e \in \text{Out}(t)$, from (20.37), we obtain

$$\mathbf{x}(z)\mathbf{f}_e(z) = \mathbf{x}(z)\left[z \sum_{d \in \text{In}(t)} k_{d,e}(z)\mathbf{f}_d(z) \right] \tag{20.57}$$

$$= z \sum_{d \in \text{In}(t)} k_{d,e}(z)\left[\mathbf{x}(z)\mathbf{f}_d(z) \right], \tag{20.58}$$

or equivalently in time domain,

$$m_{e,j} = \sum_{d \in \text{In}(t)} \left(\sum_{u=0}^{j-1} k_{d,e,u}\, m_{d,j-1-u} \right). \tag{20.59}$$

The reader should compare (20.59) with (20.42). Note that the scalar values $m_{e,j}$, $j \geq 1$ can be calculated recursively via (20.59) with the boundary condition

$$m_{d,0} = 0 \quad \text{for all } d \in E. \tag{20.60}$$

Thus a node t calculates the scalar value $m_{e,j}$ for transmitting on each output channel e at time j from the cumulative information it has received on all the input channels up to time $j-1$. The convolutional equation (20.59) can be implemented by a finite circuit of shift-registers in a causal manner because the local encoding kernels belong to $F\langle z\rangle$, the ring of rational power series over F (cf. Definition 20.6).

Example 20.8. Consider the convolutional network code in Example 20.7. Let source node s pipelines the message

$$\mathbf{x}(z) = \left[\sum_{j=0}^{\infty} a_j z^j \quad \sum_{j=0}^{\infty} b_j z^j \right]. \tag{20.61}$$

Then the five channels $(s, x), (s, y), (x, y), (y, w),$ and (w, x) carry the following power series, respectively:

$$\mathbf{x}(z)\, \mathbf{f}_{(s,x)}(z) = \sum_{j=0}^{\infty} a_j z^{j+1} \tag{20.62}$$

$$\mathbf{x}(z)\, \mathbf{f}_{(s,y)}(z) = \sum_{j=0}^{\infty} b_j z^{j+1} \tag{20.63}$$

$$\mathbf{x}(z)\, \mathbf{f}_{(x,y)}(z) = \left(\sum_{j=0}^{\infty} a_j z^{j+2} + \sum_{j=0}^{\infty} b_j z^{j+4} \right) / (1 - z^3) \tag{20.64}$$

$$= \left(\sum_{j=0}^{\infty} a_j z^{j+2} + \sum_{j=0}^{\infty} b_j z^{j+4} \right) \sum_{j=0}^{\infty} z^{3j} \tag{20.65}$$

$$= a_0 z^2 + a_1 z^3 + (a_2 + b_0) z^4$$
$$+ (a_0 + a_3 + b_1) z^5 + \cdots \tag{20.66}$$

$$\mathbf{x}(z)\, \mathbf{f}_{(y,w)}(z) = \left(\sum_{j=0}^{\infty} a_j z^{j+3} + \sum_{j=0}^{\infty} b_j z^{j+2} \right) / (1 - z^3) \tag{20.67}$$

$$\mathbf{x}(z)\, \mathbf{f}_{(w,x)}(z) = \left(\sum_{j=0}^{\infty} a_j z^{j+4} + \sum_{j=0}^{\infty} b_j z^{j+3} \right) / (1 - z^3). \tag{20.68}$$

At each time $j \geq 0$, the source generates a message $\mathbf{x}_j = [\, a_j\, b_j\,]$. Thus channel (s, x) carries the scalar 0 at time 0 and the scalar a_{j-1} at time $j \geq 1$. Similarly, channel (s, y) carries the scalar 0 at time 0 and the scalar b_{j-1} at time $j \geq 1$. For every channel e, write

$$\mathbf{x}(z)\, \mathbf{f}_e(z) = \sum_{j=0}^{\infty} m_{e,j} z^j \tag{20.69}$$

as in (20.55). The actual encoding process at node x is as follows. At time j, node x has received the sequence $m_{d,0}, m_{d,1}, \cdots, m_{d,j-1}$ for $d = (s, x)$ and (w, x). Accordingly, at time $j \geq 1$, channel (x, y) transmits the scalar value

$$m_{(x,y),j} = \sum_{u=0}^{j-1} k_{(s,x),(x,y),u}\, m_{(s,x),j-1-u}$$

$$+ \sum_{u=0}^{j-1} k_{(w,x),(x,y),u}\, m_{(w,x),j-1-u} \tag{20.70}$$

$$= m_{(s,x),j-1} + m_{(w,x),j-1}. \tag{20.71}$$

Similarly, channels (y, w) and (w, x) transmit the scalar values

$$m_{(y,w),j} = m_{(s,y),j-1} + m_{(x,y),j-1} \tag{20.72}$$

and

$$m_{(w,x),j} = m_{(y,w),j-1}, \tag{20.73}$$

respectively. The values $m_{(x,y),j}$, $m_{(y,w),j}$, and $m_{(w,x),j}$ for $j \geq 1$ can be calculated recursively by the above formulas with the boundary condition

$$m_{e,0} = 0 \quad \text{for all } e \in E, \tag{20.74}$$

and they are shown in the trellis network in Figure 20.3 for small values of j. For instance, the channel (x, y) carries the scalar values

$$m_{(x,y),0} = 0, \ m_{(x,y),1} = 0, \ m_{(x,y),2} = a_0, \ m_{(x,y),3} = a_1,$$
$$m_{(x,y),4} = a_2 + b_0, \ m_{(x,y),5} = a_0 + a_3 + b_1, \ \cdots. \tag{20.75}$$

The z-transform of this sequence is

$$x(z) \, \mathbf{f}_{(x,y)}(z) = \left(\sum_{j=0}^{\infty} a_j z^{j+2} + \sum_{j=0}^{\infty} b_j z^{j+4} \right) / (1 - z^3), \tag{20.76}$$

as calculated in (20.66).

In the discussion following Definition 20.6, we have shown that once the local encoding kernels of a convolutional network code over a unit-delay network are given, the global encoding kernels are determined. The proof of the next theorem further provides a simple closed-form expression for the global encoding kernels $\mathbf{f}_e(z)$, from which it follows that the entries in $\mathbf{f}_e(z)$ indeed belong to $F\langle z \rangle$ as required in Definition 20.6.

Theorem 20.9. *Let F be the base field and $k_{d,e}(z) \in F\langle z \rangle$ be given for every adjacent pair of channels (d, e) on a unit-delay network. Then there exists a unique ω-dimensional convolutional network code over F with $k_{d,e}(z)$ as the local encoding kernel for every (d, e).*

Proof. Let the unit-delay network be represented by a directed graph $G = (V, E)$. Let $[k_{d,e}(z)]$ be the $|E| \times |E|$ matrix in which both the rows and columns are indexed by E, with the (d, e)th entry equal to the given $k_{d,e}(z)$ if (d, e) is an adjacent pair of channels, and equal to zero otherwise. Denote the global encoding kernel of channel e by $\mathbf{f}_e(z)$ if exists. Let $[\mathbf{f}_e(z)]$ be the $\omega \times |E|$ matrix obtained by putting the global encoding kernels $\mathbf{f}_e(z), e \in E$ in juxtaposition. Let $H_s(z)$ be the $\omega \times |E|$ matrix obtained by appending $|E| - |\text{Out}(s)|$ columns of zeroes to the local encoding kernel $K_s(z)$. The requirements (20.37) and (20.38) in Definition 20.6 can be written as

$$[\mathbf{f}_e(z)] = z[\mathbf{f}_e(z)] \, [k_{d,e}(z)] + zI H_s(z), \tag{20.77}$$

where I in the above equation denotes the $\omega \times \omega$ identity matrix representing the global encoding kernels $\mathbf{f}_e(z)$, $e \in \text{In}(s)$ in juxtaposition. Rearranging the terms in (20.77), we obtain

$$[\mathbf{f}_e(z)](I - z[k_{d,e}(z)]) = zH_s(z). \tag{20.78}$$

In the matrix $z[k_{d,e}(z)]$, the diagonal elements are equal to zero because (e, e) does not form an adjacent pair of channels for all $e \in E$, while the nonzero off-diagonal elements all contain the factor z. Therefore, $\det(I - z[k_{d,e}(z)])$ has the form

$$1 + zq(z), \tag{20.79}$$

where $q(z) \in F\langle z \rangle$, so that it is invertible inside $F\langle z \rangle$ because

$$[\det(I - z[k_{d,e}(z)])]^{-1} = \frac{1}{1 + zq(z)} \tag{20.80}$$

is a rational power series. It follows that

$$(I - z[k_{d,e}(z)])^{-1} \tag{20.81}$$

exists and is a matrix over $F\langle z \rangle$. Then the unique solution for $[\mathbf{f}_e(z)]$ in (20.78) is given by

$$[\mathbf{f}_e(z)] = zH_s(z)(I - z[k_{d,e}(z)])^{-1}. \tag{20.82}$$

With the two matrices $[k_{d,e}(z)]$ and $H_s(z)$ representing the given local encoding kernels and the matrix $[\mathbf{f}_e(z)]$ representing the global encoding kernels, (20.82) is a closed-form expression for the global encoding kernels in terms of the local encoding kernels. In particular, $[\mathbf{f}_e(z)]$ is a matrix over $F\langle z \rangle$ because all the matrices on the right-hand side of (20.82) are over $F\langle z \rangle$. Thus we conclude that all the components of the global encoding kernels are in $F\langle z \rangle$. Hence, the given local encoding kernels $k_{d,e}(z)$ for all adjacent pairs (d, e) together with the associated global encoding kernels $\mathbf{f}_e(z)$, $e \in \text{In}(s) \cup E$ constitute a unique convolutional network code over the unit-delay network G. $\square$

In view of Definition 19.7 for the global description of a linear network code over an acyclic network, Definition 20.6 can be regarded as the global description of a convolutional network code over a unit-delay network, while Theorem 20.9 renders a local description by specifying the local encoding kernels only.

20.3 Decoding of Convolutional Network Codes

For a node t, let

$$F_t(z) = [\mathbf{f}_e(z)]_{e \in \text{In}(t)} \tag{20.83}$$

be the $\omega \times |\text{In}(t)|$ matrix obtained by putting the global encoding kernels $\mathbf{f}_e(z)$, $e \in \text{In}(t)$ in juxtaposition. In the following, we define a *convolutional multicast*, the counterpart of a linear multicast defined in Chapter 19, for a unit-delay cyclic network. The existence of a convolutional multicast will also be established.

Definition 20.10 (Convolutional Multicast). *An ω-dimensional convolutional network code on a unit-delay network qualifies as an ω-dimensional convolutional multicast if for every non-source node t with* $\mathrm{maxflow}(t) \geq \omega$*, there exists an* $|\mathrm{In}(t)| \times \omega$ *matrix $D_t(z)$ over $F\langle z \rangle$ and a positive integer τ such that*

$$F_t(z)\, D_t(z) = z^\tau I, \tag{20.84}$$

where $\tau > 0$ depends on node t and I is the $\omega \times \omega$ identity matrix. The matrix $D_t(z)$ and the integer τ are called the decoding kernel and the decoding delay at node t, respectively.

Source node s generates the message pipeline

$$\mathbf{x}(z) = \sum_{j=0}^{\infty} \mathbf{x}_j z^j, \tag{20.85}$$

where $\mathbf{x}_j$ is a row ω-vector in F^ω and $\mathbf{x}(z)$ is a row ω-vector over $F[[z]]$. Through the convolutional network code, a channel e carries the power series $\mathbf{x}(z)\, \mathbf{f}_e(z)$. The power series $\mathbf{x}(z)\, \mathbf{f}_e(z)$ received by a node t from the input channels $e \in \mathrm{In}(t)$ forms the row $|\mathrm{In}(t)|$-vector $\mathbf{x}(z)\, F_t(z)$ over $F[[z]]$. If the convolutional network code is a convolutional multicast, node t can use the decoding kernel $D_t(z)$ to calculate

$$(\mathbf{x}(z)\, F_t(z))D_t(z) = \mathbf{x}(z)(F_t(z)D_t(z)) \tag{20.86}$$
$$= \mathbf{x}(z)(z^\tau I) \tag{20.87}$$
$$= z^\tau \mathbf{x}(z). \tag{20.88}$$

The row ω-vector $z^\tau \mathbf{x}(z)$ of power series represents the message pipeline generated by source node s delayed by τ time units. Note that $\tau > 0$ because the message pipeline $\mathbf{x}(z)$ is delayed by one time unit at node s.

Example 20.11. Consider the network in Figure 20.4. Again let source node s pipeline the message

$$\mathbf{x}(z) = \left[\sum_{j=0}^{\infty} a_j z^j \quad \sum_{j=0}^{\infty} b_j z^j \right]. \tag{20.89}$$

For node x, we have

$$F_x(z) = \begin{bmatrix} z & z^4/(1-z^3) \\ 0 & z^3/(1-z^3) \end{bmatrix}. \tag{20.90}$$

Let

$$D_x(z) = \begin{bmatrix} z^2 & -z^3 \\ 0 & 1 - z^3 \end{bmatrix}. \tag{20.91}$$

Then

$$F_x(z)D_t(z) = z^3 I_2 \tag{20.92}$$

(I_2 is the 2×2 identity matrix). From channels (s, x) and (w, x), node x receives the row vector

$$\mathbf{x}(z)F_x(z) = \begin{bmatrix} \sum_{j=0}^{\infty} a_j z^{j+1} & \sum_{j=0}^{\infty} \dfrac{a_j z^{j+4} + b_j z^{j+3}}{1 - z^3} \end{bmatrix} \tag{20.93}$$

and decodes the message pipeline as

$$z^3 \mathbf{x}(z) = \begin{bmatrix} \sum_{j=0}^{\infty} a_j z^{j+1} & \sum_{j=0}^{\infty} \dfrac{a_j z^{j+4} + b_j z^{j+3}}{1 - z^3} \end{bmatrix} \begin{bmatrix} z^2 & -z^3 \\ 0 & 1 - z^3 \end{bmatrix}. \tag{20.94}$$

Decoding at node y is similar. Thus the two-dimensional convolutional network code is a convolutional multicast.

Toward proving the existence of a convolutional multicast, we first observe that Lemma 19.17 can be strengthened as follows with essentially no change in the proof.

Lemma 20.12. *Let $g(y_1, y_2, \cdots, y_m)$ be a nonzero polynomial with coefficients in a field $\tilde{F}$. For any subset $\tilde{E}$ of $\tilde{F}$, if $|\tilde{E}|$ is greater than the degree of g in every y_j, then there exist $a_1, a_2, \cdots, a_m \in \tilde{E}$ such that*

$$g(a_1, a_2, \cdots, a_m) \neq 0. \tag{20.95}$$

In the above lemma, the values $a_1, a_2, \cdots, a_m$ can be found by exhaustive search in $\tilde{E}$ provided that $\tilde{E}$ is finite. If $\tilde{E}$ is infinite, simply replace $\tilde{E}$ by a sufficiently large finite subset of $\tilde{E}$.

Theorem 20.13. *There exists an ω-dimensional convolutional multicast over any base field F. Furthermore, the local encoding kernels of the convolutional multicast can be chosen in any sufficiently large subset Φ of $F\langle z \rangle$.*

Proof. Recall equation (20.82) in the proof of Theorem 20.9:

$$[\mathbf{f}_e(z)] = zH_s(z)(I - z[k_{d,e}(z)])^{-1}. \tag{20.96}$$

In this equation, the $\omega \times |E|$ matrix $[\mathbf{f}_e(z)]$ on the left-hand side represents the global encoding kernels, while the $\omega \times |E|$ matrix $H_s(z)$ and the $|E| \times |E|$ matrix $[k_{d,e}(z)]$ on the right-hand side represent the local encoding kernels. Analogous to the proof of Theorem 19.20, denote by $(F\langle z \rangle)[*]$ the polynomial ring over $F\langle z \rangle$ with all $k_{d,e}(z)$ as indeterminates.

Let t be a non-source node with maxflow$(t) \geq \omega$. Then there exist ω edge-disjoint paths from the ω imaginary channels to ω distinct channels in In(t). Put the global encoding kernels of these ω channels in juxtaposition to form the $\omega \times \omega$ matrix $L_t(z)$ over $(F\langle z \rangle)[*]$. We will show that

$$\det(L_t(z)) \neq 0 \in (F\langle z \rangle)[*]. \tag{20.97}$$

Toward proving (20.97), it suffices to show that

$$\det(L_t(z)) \neq 0 \in F\langle z \rangle \tag{20.98}$$

when the determinant is evaluated at some particular values for the indeterminates $k_{d,e}(z)$. Analogous to the proof of Theorem 19.20, we set

$$k_{d,e}(z) = 1 \tag{20.99}$$

for all adjacent pairs of channels (d, e) along any one of the ω edge-disjoint paths, and set

$$k_{d,e}(z) = 0 \tag{20.100}$$

otherwise. Then with a suitable indexing of the columns, the matrix $L_t(z)$ becomes diagonal with all the diagonal entries being powers of z. Hence, $\det(L_t(z))$ is equal to some positive power of z, proving (20.98) for this particular choice of the indeterminates $k_{d,e}(x)$ and hence proving (20.97). As the conclusion (20.97) applies to every non-source node t with maxflow$(t) \geq \omega$, it follows that

$$\prod_{t:\text{maxflow}(t)\geq\omega} \det(L_t(z)) \neq 0 \in (F\langle z \rangle)[*]. \tag{20.101}$$

Let $F(z)$ be the conventional notation for the field of rational functions in z over the given base field F. The ring $F\langle z \rangle$ of rational power series is a subset of $F(z)$. Then any subset Φ of $F\langle z \rangle$ is also a subset of $F(z)$. Note that the ring $F\langle z \rangle$ is infinite. Then for any sufficiently large subset Φ of $F\langle z \rangle$, we can apply Lemma 20.12 to the polynomial in (20.101) with $\tilde{F} = F(z)$ and $\tilde{E} = \Phi$ to see that we can choose a value $a_{d,e}(z) \in F\langle z \rangle$ for each of the indeterminates $k_{d,e}(z)$ so that

$$\prod_{t:\text{maxflow}(t)\geq\omega} \det(L_t(z)) \neq 0 \in F\langle z \rangle \tag{20.102}$$

when evaluated at $k_{d,e}(z) = a_{d,e}(z)$ for all (d, e), which in turn implies that

$$\det(L_t(z)) \neq 0 \in F\langle z \rangle \tag{20.103}$$

for all nodes t such that maxflow$(t) \geq \omega$.

Henceforth, the local encoding kernel $k_{d,e}(z)$ will be fixed at the appropriately chosen value $a_{d,e}(z)$ for all (d, e) as prescribed above. Without loss of generality, we assume that $L_t(z)$ consists of the first ω columns of $F_t(z)$. From (20.103), we can write

$$\det(L_t(z)) = z^\tau \left[\frac{1 + zq(z)}{p(z)} \right], \tag{20.104}$$

where $p(z)$ and $q(z)$ are polynomials over F, and $p(z)$ is not the zero polynomial. Note that the right-hand side of (20.104) is the general form for a nonzero rational function in z. In this particular context, since the columns of $L_t(z)$ are global encoding kernels as prescribed by (20.82), each containing the factor z in the numerator, we see that $\tau > 0$.

Denote by $J_t(z)$ the adjoint matrix[2] of $L_t(z)$. Take the $\omega \times \omega$ matrix

$$\left[\frac{p(z)}{1 + zq(z)} \right] J_t(z) \tag{20.105}$$

and append to it $|\mathrm{In}(t)| - \omega$ rows of zeroes to form an $|\mathrm{In}(t)| \times \omega$ matrix $D_t(z)$. Then

$$F_t(z)D_t(z) = \begin{bmatrix} L_t(z)\, 0 \end{bmatrix} \begin{bmatrix} \left[\frac{p(z)}{1+zq(z)} \right] J_t(z) \\ 0 \end{bmatrix} \tag{20.106}$$

$$= \left[\frac{p(z)}{1 + zq(z)} \right] L_t(z) J_t(z) \tag{20.107}$$

$$= \left[\frac{p(z)}{1 + zq(z)} \right] \det(L_t(z)) I \tag{20.108}$$

$$= z^\tau I, \tag{20.109}$$

where the last equality follows from (20.104). Hence, the matrix $D_t(z)$ qualifies as a decoding kernel at node t in Definition 20.10. This proves the existence of the convolutional multicast as required. □

The proof of Theorem 20.13 constitutes an algorithm for constructing a convolutional multicast. By noting the lower bound on the size of $\tilde{E}$ in Lemma 20.12, a convolutional multicast can be constructed with high probability by randomly choosing the local encoding kernels in the subset Φ of $F\langle z \rangle$ provided that Φ is much larger than sufficient.

Example 20.14. When the base field F is sufficiently large, Theorem 20.13 can be applied with $\Phi = F$ so that the local encoding kernels of the convolutional multicast can be chosen to be scalars. This special case is the convolutional counterpart of Theorem 19.20 for the existence of a linear multicast over an acyclic network. In this case, the local encoding kernels can be found by exhaustive search over F.

More generally, by virtue of Lemma 20.12, the same exhaustive search applies to any large enough subset Φ of $F\langle z \rangle$. For example, F can be $GF(2)$ and Φ can be the set of all binary polynomials up to a sufficiently large degree.

---

[2] For a matrix B whose entries are elements in a ring, denote by $\mathrm{Adj}(B)$ the adjoint matrix of B. Then $\mathrm{Adj}(B)B = B\mathrm{Adj}(B) = \det(A)I$.

Chapter Summary

Algebraic Structures:

- $F[[z]]$ denotes the ring of power series in z over F.
- $F\langle z \rangle$ denotes the ring of rational power series in z over F.
- $F(z)$ denotes the field of rational functions in z over F.

Convolutional Network Code:

- $k_{d,e}(z) \in F\langle z \rangle$ is the local encoding kernel of the adjacent pair of channels (d, e).
- $\mathbf{f}_e(z) \in (F\langle z \rangle)^\omega$ is the global encoding kernel of channel e.
- $\mathbf{f}_e(z) = z \sum_{d \in In(t)} k_{d,e}(z) \mathbf{f}_d(z)$ for $e \in \text{Out}(t)$.
- $\mathbf{f}_e(z), e \in \text{In}(s)$ forms the standard basis of F^ω.
- Channel e carries the power series $\mathbf{x}(z)\, \mathbf{f}_e(z)$, where $\mathbf{x}(z) \in (F[[z]])^\omega$ is the message pipeline generated at source node s.

Uniqueness of Convolutional Network Code: For given $k_{d,e}(z)$ for every adjacent pair of channels (d, e) on a unit-delay network, there exists a unique convolutional network code with $k_{d,e}(z)$ as the local encoding kernel for every (d, e). The global encoding kernels $\mathbf{f}_e(z)$ can be expressed in terms of the local encoding kernels $k_{d,e}(z)$ as

$$[\mathbf{f}_e(z)] = z H_s(z)(I - z[k_{d,e}(z)])^{-1},$$

where $H_s(z)$ is determined by the local encoding kernel at source node s.

Convolutional Multicast: A convolutional network code is a convolutional multicast if for every node $t \neq s$ with $\text{maxflow}(t) \geq \omega$, there exists an $|In(t)| \times \omega$ matrix $D_t(z)$ over $F\langle z \rangle$ and a positive integer τ such that

$$F_t(z)\, D_t(z) = z^\tau I,$$

where $F_t(z) = [\mathbf{f}_e(z)]_{e \in \text{In}(t)}$ and $\tau > 0$ depends on node t. The matrix $D_t(z)$ and the integer τ are called the decoding kernel and the decoding delay at node t, respectively.

Existence and Construction: A convolutional multicast on a unit-delay network with the local encoding kernels chosen in a subset Φ of $F\langle z \rangle$ exists and can be constructed randomly (with high probability) when Φ is sufficiently large. For example, Φ can be the base field F provided that F is sufficiently large or Φ can be the set of all binary polynomials up to a sufficiently large degree.

Problems

1. Show that the right-hand side of (20.104) is the general form for a nonzero rational function in z.

2. A formal Laurent series over a field F has the form

$$a_{-m}z^{-m} + a_{-(m-1)}z^{-(m-1)} + \cdots + a_{-1}z^{-1} + a_0 + a_1 z + a_2 z^2 + \cdots,$$

where m is a nonnegative integer. Show that for any formal Laurent series $f(z)$ over F, there exists a unique formal Laurent series $g(z)$ over F such that $f(z)g(z) = 1$.

3. Verify the following series expansion:

$$\frac{1}{1-z} = -z^{-1} - z^{-2} - z^{-3} - \cdots.$$

Can you obtain this series by long division?

4. Construct a finite circuit of shift-registers that implements a discrete-time LTI system with transfer function

$$\frac{a_0 + a_1 z + \cdots + a_n z^n}{b_0 + b_1 z + \cdots + b_n z^n},$$

where a_i and b_i are elements in a finite field and $b_0 \neq 0$.

5. Consider the convolutional network code in Figure 20.4.
   a) Is it a convolutional multicast?
   b) If your answer in (a) is positive, give the decoding kernel at node y with minimum decoding delay.
   c) Change K_s to $\begin{bmatrix} 1 & 1 \\ 0 & 1 \end{bmatrix}$ and determine the corresponding global encoding kernels.
   d) Instead of a convolutional multicast, can you construct a linear multicast on the network?

Historical Notes

The asymptotic achievability of the max-flow bound for unit-delay cyclic networks was proved by Ahlswede et al. [6], where an example of a convolutional network code achieving this bound was given. Li et al. [230] conjectured the existence of convolutional multicasts on such networks. This conjecture was subsequently proved by Koetter and Médard [202]. Construction and decoding of convolutional multicast have been studied by Erez and Feder [105, 106], Fragouli and Soljanin [122], and Barbero and Ytrehus [24]. The unifying treatment of convolutional codes here is based on Li and Yeung [229] (see also Yeung et al. [408]). Li and Ho [228] recently obtained a general abstract formulation of convolutional network codes based on ring theory.

Multi-source Network Coding

In Chapters 19 and 20, we have discussed single-source network coding in which an information source is multicast in a point-to-point communication network. The maximum rate at which information can be multicast has a simple characterization in terms of the maximum flows in the graph representing the network. In this chapter, we consider the more general multi-source network coding problem in which more than one *mutually independent* information sources are generated at possibly different nodes, and each of the information sources is multicast to a specific set of nodes.

The *achievable information rate region* of a multi-source network coding problem, which will be formally defined in Section 21.4, refers to the set of all possible rates at which multiple information sources can be multicast simultaneously on a network. In a single-source network coding problem, we are interested in characterizing the maximum rate at which information can be multicast from the source node to all the sink nodes. In a multi-source network coding problem, we are interested in characterizing the achievable information rate region.

As discussed in Section 17.3, source separation is not necessarily optimal for multi-source network coding. It is therefore *not* a simple extension of single-source network coding. Unlike the single-source network coding problem which has an explicit solution, the multi-source network coding problem has not been completely solved. In this chapter, by making use of the tools we have developed for information inequalities in Chapters 13–15, we will develop an implicit characterization of the achievable information rate region for multi-source network coding on acyclic networks.

21.1 The Max-Flow Bounds

The max-flow bound, which fully characterizes the maximum rate of an information source that can be multicast in a network, plays a central role in single-source network coding. We now revisit this bound in the context of

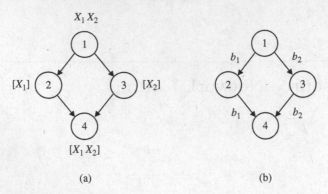

Fig. 21.1. A network which achieves the max-flow bound.

multi-source network coding. In the following discussion, the unit of information is the bit.

Consider the graph in Figure 21.1(a). The capacity of each edge is equal to 1. Two independent information sources X_1 and X_2 with rates ω_1 and ω_2, respectively, are generated at node 1. Suppose we want to multicast X_1 to nodes 2 and 4 and multicast X_2 to nodes 3 and 4. In the figure, an information source in square brackets is one which is to be received at that node.

It is easy to see that the values of a max-flow from node 1 to node 2, from node 1 to node 3, and from node 1 to node 4 are, respectively, 1, 1, and 2. At node 2 and node 3, information is received at rates ω_1 and ω_2, respectively. At node 4, information is received at rate $\omega_1 + \omega_2$ because X_1 and X_2 are independent. Applying the max-flow bound at nodes 2, 3, and 4, we have

$$\omega_1 \leq 1, \tag{21.1}$$
$$\omega_2 \leq 1, \tag{21.2}$$

and

$$\omega_1 + \omega_2 \leq 2, \tag{21.3}$$

respectively. We refer to (21.1) to (21.3) as the max-flow bounds. Figure 21.2 is an illustration of all (ω_1, ω_2) which satisfy these bounds, where ω_1 and ω_2 are obviously nonnegative.

We now show that the rate pair $(1, 1)$ is achievable. Let b_1 be a bit generated by X_1 and b_2 be a bit generated by X_2. In the scheme in Figure 21.1(b), b_1 is received at node 2, b_2 is received at node 3, and both b_1 and b_2 are received at node 4. Thus the multicast requirements are satisfied, and the information rate pair $(1, 1)$ is achievable. This implies that all (ω_1, ω_2) which satisfy the max-flow bounds are achievable because they are all inferior to $(1, 1)$ (see Figure 21.2). In this sense, we say that the max-flow bounds are achievable.

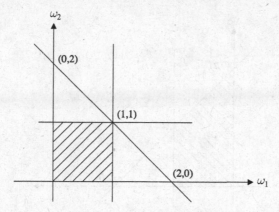

Fig. 21.2. The max-flow bounds for the network in Figure 21.1.

Suppose we now want to multicast X_1 to nodes 2, 3, and 4 and multicast X_2 to node 4 as illustrated in Figure 21.3. Applying the max-flow bound at either node 2 or node 3 gives

$$\omega_1 \leq 1, \tag{21.4}$$

and applying the max-flow bound at node 4 gives

$$\omega_1 + \omega_2 \leq 2. \tag{21.5}$$

Figure 21.4 is an illustration of all (ω_1, ω_2) which satisfy these bounds.

We now show that the information rate pair $(1, 1)$ is not achievable. Suppose we need to send a bit b_1 generated by X_1 to nodes 2, 3, and 4 and send a bit b_2 generated by X_2 to node 4. Since b_1 has to be recovered at node 2, the bit sent to node 2 must be an invertible transformation of b_1. This implies that the bit sent to node 2 cannot depend on b_2. Similarly, the bit sent to node 3 also cannot depend on b_2. Therefore, it is impossible for node 4 to recover b_2 because both the bits received at nodes 2 and 3 do not depend on

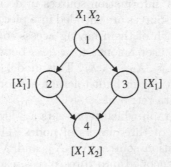

Fig. 21.3. A network which does not achieve the max-flow bounds.

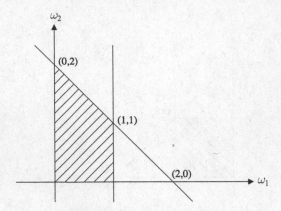

Fig. 21.4. The max-flow bounds for the network in Figure 21.3.

b_2. Thus the information rate pair $(1, 1)$ is not achievable, which implies that the max-flow bounds (21.4) and (21.5) are not achievable.

From this example, we see that the max-flow bounds do not always fully characterize the achievable information rate region. We leave it as an exercise for the reader to show that for this example, source separation is in fact optimal.

21.2 Examples of Application

Multi-source network coding is a very rich model which encompasses many communication situations arising from fault-tolerant network communication, disk array, satellite communication, etc. In this section, we discuss some applications of the model.

21.2.1 Multilevel Diversity Coding

Let $X_1, X_2, \cdots, X_K$ be K information sources in decreasing order of importance. These information sources are encoded into pieces of information. There are a number of users, each of them having access to a certain subset of the information pieces. Each user belongs to a *level* between 1 and K, where a Level k user can decode $X_1, X_2, \cdots, X_k$. This model, called *multilevel diversity coding*, finds applications in fault-tolerant network communication, disk array, and distributed data retrieval.

Figure 21.5 shows a graph which represents a 3-level diversity coding system. The graph consists of three layers of nodes. The top layer consists of a node at which information sources X_1, X_2, and X_3 are generated. These information sources are encoded into three pieces, each of which is stored in a distinct node in the middle layer. A dummy node is associated with such a

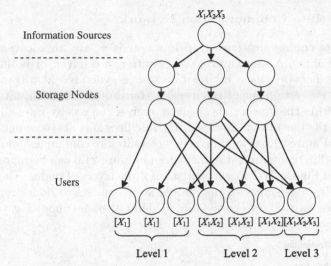

Fig. 21.5. A 3-level diversity coding system.

node to model the effect that the same information is retrieved every time the node is accessed (see the discussion in Section 17.2). The nodes in the bottom layer represent the users, each of them belonging to one of the three levels. Each of the three Level 1 users has access to a distinct node in the second layer (through the associated dummy node) and decodes X_1. Similarly, each of the three Level 2 users has access to a distinct set of two nodes in the second layer and decodes X_1 and X_2. There is only one Level 3 user, who has access to all the three nodes in the second layer and decodes X_1, X_2, and X_3.

The model represented by the graph in Figure 21.5 is called *symmetrical* 3-level diversity coding because the model is unchanged by permuting the nodes in the middle layer. By degenerating information sources X_1 and X_3, the model is reduced to the diversity coding model discussed in Section 18.2.

In the following, we describe two applications of symmetrical multilevel diversity coding:

Fault-Tolerant Network Communication In a computer network, a data packet can be lost due to buffer overflow, false routing, breakdown of communication links, etc. Suppose the packet carries K messages, $X_1, X_2, \cdots, X_K$, in decreasing order of importance. For improved reliability, the packet is encoded into K sub-packets, each of which is sent over a different channel. If any k sub-packets are received, then the messages $X_1, X_2, \cdots, X_k$ can be recovered.

Disk Array Consider a disk array which consists of K disks. The data to be stored in the disk array are segmented into K pieces, $X_1, X_2, \cdots, X_K$, in decreasing order of importance. Then $X_1, X_2, \cdots, X_K$ are encoded into K pieces, each of which is stored on a separate disk. When any k out of the K disks are functioning, the data $X_1, X_2, \cdots, X_k$ can be recovered.

21.2.2 Satellite Communication Network

In a satellite communication network, a user is at any time covered by one
or more satellites. A user can be a transmitter, a receiver, or both. Through
the satellite network, each information source generated at a transmitter is
multicast to a certain set of receivers. A transmitter can transmit to all the
satellites within the line of sight, while a receiver can receive from all the satellites within the line of sight. Neighboring satellites may also communicate with
each other. Figure 21.6 is an illustration of a satellite communication network.

The satellite communication network in Figure 21.6 can be represented by
the graph in Figure 21.7 which consists of three layers of nodes. The top layer
represents the transmitters, the middle layer consists of nodes representing
the satellites as well as the associated dummy nodes modeling the broad-

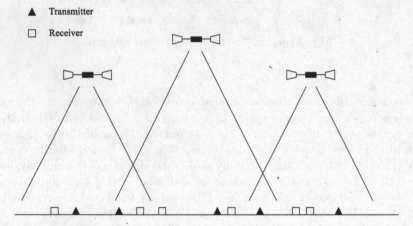

Fig. 21.6. A satellite communication network.

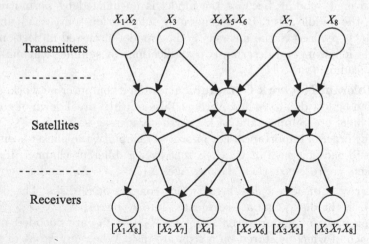

Fig. 21.7. A graph representing a satellite communication network.

cast nature of the satellites, and the bottom layer represents the receivers. If a satellite is within the line of sight of a transmitter, then the corresponding pair of nodes are connected by a directed edge. Likewise, if a receiver is within the line of sight of a satellite, then the corresponding pair nodes are connected by a directed edge. An edge between two nodes in the middle layer represents the communication links between two neighboring satellites. Each information source is multicast to a specified set of receiving nodes as shown.

21.3 A Network Code for Acyclic Networks

Let $G = (V, E)$ denote an acyclic point-to-point communication network, where V and E are the set of nodes and the set of channels, respectively. We assume that each channel $e \in E$ is error-free with rate constraint R_e. As in our previous discussions, we let $\text{In}(t)$ and $\text{Out}(t)$ be the set of input channels and the set of output channels of node t, respectively.

Let $S \subset V$ be the set of source nodes and $T \subset V$ be the set of sink nodes. Without loss of generality, we assume G has the structure that a source node has no input channel and a sink node has no output channel. Accordingly, S and T are disjoint subsets of V.

An information source represented by a random variable X_s is generated at a source node $s \in S$, where X_s takes values in

$$\mathcal{X}_s = \{1, 2, \cdots, \lceil 2^{n\tau_s} \rceil\} \tag{21.6}$$

according to the uniform distribution, where τ_s is the rate of the information source. The information sources $X_s, s \in S$ are assumed to be mutually independent.

To simplify the notation, we will denote $(X_s : s \in \mathcal{A})$ by $X_{\mathcal{A}}$, $\prod_{s \in \mathcal{A}} \mathcal{X}_s$ by $\mathcal{X}_{\mathcal{A}}$, etc. At a sink node $t \in T$, the set of information sources $X_{\beta(t)}$, where $\beta(t) \subset S$, is received. We assume that each information source is received at at least one sink node, i.e., for every $s \in S$, $s \in \beta(t)$ for some $t \in T$. In the case when $\beta(t) = S$ for all $t \in T$, the problem is reduced to the single-source network coding problem.

Definition 21.1. *An $(n, (\eta_e : e \in E), (\tau_s : s \in S))$ block code of length n on a given communication network is defined as follows:*

1) *for all source node $s \in S$ and all channels $e \in \text{Out}(s)$, a local encoding function*

$$k_e : \mathcal{X}_s \to \{0, 1, \cdots, \eta_e\}; \tag{21.7}$$

2) *for all node $i \in V \setminus (S \cup T)$ and all channels $e \in \text{Out}(i)$, a local encoding function*

$$k_e : \prod_{d \in \text{In}(i)} \{0, 1, \cdots, \eta_d\} \to \{0, 1, \cdots, \eta_e\}; \tag{21.8}$$

3) *for all sink node* $t \in T$, *a decoding function*

$$g_t : \prod_{d \in \text{In}(t)} \{0, 1, \cdots, \eta_d\} \to \mathcal{X}_{\beta(t)}. \qquad (21.9)$$

The nodes in V are assumed to be ordered in an upstream-to-downstream manner as prescribed in Proposition 19.1. This defines a coding order among the nodes such that whenever a node encodes, all the information needed would have already been received on the input channels of that node.

For all sink node $t \in T$, define

$$\Delta_t = \Pr \left\{ \tilde{g}_t(X_S) \neq X_{\beta(t)} \right\}, \qquad (21.10)$$

where $\tilde{g}_t(X_S)$ denotes the value of g_t as a function of X_S. Δ_t is the probability that the set of information sources $X_{\beta(t)}$ is decoded incorrectly at sink node t.

Throughout this chapter, all the logarithms are in the base 2.

Definition 21.2. *An information rate tuple* $\boldsymbol{\omega} = (\omega_s : s \in S)$, *where* $\boldsymbol{\omega} \geq 0$ *(componentwise) is asymptotically achievable if for any* $\epsilon > 0$, *there exists for sufficient large* n *an* $(n, (\eta_e : e \in E), (\tau_s : s \in S))$ *code such that*

$$n^{-1} \log \eta_e \leq R_e + \epsilon, \ e \in E, \qquad (21.11)$$

$$\tau_s \geq \omega_s - \epsilon, \ s \in S, \qquad (21.12)$$

$$\Delta_t \leq \epsilon, \ t \in T. \qquad (21.13)$$

For brevity, an asymptotically achievable information rate tuple will be referred to as an achievable information rate tuple.

21.4 The Achievable Information Rate Region

In this section, we define the achievable information rate region and give a characterization of this region.

Definition 21.3. *The achievable information rate region, denoted by* $\mathcal{R}$, *is the set of all achievable information rate tuples* $\boldsymbol{\omega}$.

Remark It follows from the definition of the achievability of an information rate vector that if $\boldsymbol{\omega}$ is achievable, then $\boldsymbol{\omega}'$ is achievable for all $0 \leq \boldsymbol{\omega}' \leq \boldsymbol{\omega}$. Also, if $\boldsymbol{\omega}^{(k)}$, $k \geq 1$ are achievable, then it can be proved by techniques similar to those in the proof of Theorem 8.12 that

$$\boldsymbol{\omega} = \lim_{k \to \infty} \boldsymbol{\omega}^{(k)} \qquad (21.14)$$

is also achievable, i.e., $\mathcal{R}$ is closed. The details are omitted here.

Consider the set of all information rate tuples $\boldsymbol{\omega}$ such that there exist auxiliary random variables $\{Y_s, s \in S\}$ and $\{U_e, e \in E\}$ which satisfy the following conditions:

$$H(Y_s) \geq \omega_s, \ s \in S, \tag{21.15}$$

$$H(Y_S) = \sum_{s \in S} H(Y_s), \tag{21.16}$$

$$H(U_{\text{Out}(s)}|Y_s) = 0, \ s \in S, \tag{21.17}$$

$$H(U_{\text{Out}(i)}|U_{\text{In}(i)}) = 0, \ i \in V \setminus (S \cup T), \tag{21.18}$$

$$H(U_e) \leq R_e, \ e \in E, \tag{21.19}$$

$$H(Y_{\beta(t)}|U_{\text{In}(t)}) = 0, \ t \in T, \tag{21.20}$$

where Y_S denotes $(Y_s : s \in S)$, $U_{\text{Out}(s)}$ denotes $(U_e : e \in \text{Out}(s))$, etc. Here, Y_s is an auxiliary random variable associated with the information source X_s, and U_e is an auxiliary random variable associated with the codeword sent on channel e. The interpretations of (21.15)–(21.20) are as follows. The inequality in (21.15) says that the entropy of Y_s is greater than or equal to ω_s, the rate of the information source X_s. The equality in (21.16) says that $Y_s, s \in S$ are mutually independent, which corresponds to the assumption that the information sources $X_s, s \in S$ are mutually independent. The equality in (21.17) says that $U_{\text{Out}(s)}$ is a function of Y_s for $s \in S$, and the equality in (21.18) says that $U_{\text{Out}(i)}$ is a function of $U_{\text{In}(i)}$ for $i \in V \setminus (S \cup T)$. These correspond to the requirement that the codewords sent out by a source node s are functions of the information source X_s, and that the codewords sent out by a non-source node i are functions of the codewords received by node i. The inequality in (21.19) says that the entropy of U_e less than or equal to R_e, the rate constraint for channel e. The equality in (21.20) says that $Y_{\beta(t)}$ is a function of $U_{\text{In}(t)}$ for $t \in T$, which corresponds to the requirement that the information sources to be received at a sink node t can be decoded from the codewords received at node t.

For a given multi-source network coding problem, let

$$\mathcal{N} = \{Y_s : s \in S; U_e : e \in E\} \tag{21.21}$$

be a collection of discrete random variables whose joint distribution is unspecified, and let

$$\mathcal{Q}_{\mathcal{N}} = 2^{\mathcal{N}} \setminus \{\phi\} \tag{21.22}$$

with cardinality $2^{|\mathcal{N}|} - 1$. Let $\mathcal{H}_{\mathcal{N}}$ be the $|\mathcal{Q}_{\mathcal{N}}|$-dimensional Euclidean space with the coordinates labeled by $h_A, A \in \mathcal{Q}_{\mathcal{N}}$. A vector

$$.\mathbf{h} = (h_A : A \in \mathcal{Q}_{\mathcal{N}}) \tag{21.23}$$

in $\mathcal{H}_{\mathcal{N}}$ is said to be *finitely entropic* if there exists a joint distribution for all $X \in \mathcal{N}$, where $|\mathcal{X}| < \infty$ for all $X \in \mathcal{N}$ and

$$h_A = H(X : X \in A) \tag{21.24}$$

for all $A \in \mathcal{Q}_{\mathcal{N}}$. Note that $\mathbf{h} \in \mathcal{H}_{\mathcal{N}}$ is entropic if it is finitely entropic, but not vice versa. We then define the region

$$\Gamma_{\mathcal{N}}^{**} = \{\mathbf{h} \in \mathcal{H}_{\mathcal{N}} : \mathbf{h} \text{ is finitely entropic}\}. \tag{21.25}$$

To simplify notation, for any nonempty $A, A' \in \mathcal{Q}_{\mathcal{N}}$, define

$$h_{A|A'} = h_{AA'} - h_{A'}, \tag{21.26}$$

where we have used juxtaposition to denote the union of two sets. In using the above notation, we do not distinguish elements and singletons of $\mathcal{N}$, i.e., for a random variable $Z \in \mathcal{N}$, h_Z is the same as $h_{\{Z\}}$. We now define the following regions in $\mathcal{H}_{\mathcal{N}}$:

$$\mathcal{L}_1 = \left\{ \mathbf{h} \in \mathcal{H}_{\mathcal{N}} : h_{Y_S} = \sum_{s \in S} h_{Y_s} \right\}, \tag{21.27}$$

$$\mathcal{L}_2 = \left\{ \mathbf{h} \in \mathcal{H}_{\mathcal{N}} : h_{U_{\text{Out}(s)}|Y_s} = 0, s \in S \right\}, \tag{21.28}$$

$$\mathcal{L}_3 = \left\{ \mathbf{h} \in \mathcal{H}_{\mathcal{N}} : h_{U_{\text{Out}(i)}|U_{\text{In}(i)}} = 0, i \in V \setminus (S \cup T) \right\}, \tag{21.29}$$

$$\mathcal{L}_4 = \left\{ \mathbf{h} \in \mathcal{H}_{\mathcal{N}} : h_{U_e} \leq R_e, e \in E \right\}, \tag{21.30}$$

$$\mathcal{L}_5 = \left\{ \mathbf{h} \in \mathcal{H}_{\mathcal{N}} : h_{Y_{\beta(t)}|U_{\text{In}(t)}} = 0, t \in T \right\}. \tag{21.31}$$

Evidently, (21.27)–(21.31) are the regions in $\mathcal{H}_{\mathcal{N}}$ corresponding to (21.16)–(21.20), respectively. We further denote $\bigcap_{i \in \alpha} \mathcal{L}_i$ by $\mathcal{L}_\alpha$ for $\alpha \subset \{1, 2, 3, 4, 5\}$.

We now introduce a few notations. For a vector $\mathbf{h} \in \mathcal{H}_{\mathcal{N}}$, let $\mathbf{h}_{Y_S} = (h_{Y_s} : s \in S)$. For a subset $\mathcal{B}$ of $\mathcal{H}_{\mathcal{N}}$, let

1. $\text{proj}_{Y_S}(\mathcal{B}) = \{\mathbf{h}_{Y_S} : \mathbf{h} \in \mathcal{B}\}$ be the projection of the set $\mathcal{B}$ on the coordinates $h_{Y_s}, s \in S$;
2. $\Lambda(\mathcal{B}) = \{\mathbf{h} \in \mathcal{H}_{\mathcal{N}} : 0 \leq \mathbf{h} \leq \mathbf{h}' \text{ for some } \mathbf{h}' \in \mathcal{B}\}$;
3. $\text{con}(\mathcal{B})$ be the convex hull of $\mathcal{B}$;
4. $\overline{\mathcal{B}}$ be the closure of $\mathcal{B}$.

Note that a vector $\mathbf{h} \geq 0$ is in $\Lambda(\mathcal{B})$ if and only if it is inferior to some vector $\mathbf{h}'$ in $\mathcal{B}$. The following theorem gives a characterization of the achievable information rate region $\mathcal{R}$ in terms of the region $\Gamma_{\mathcal{N}}^{**}$.

Definition 21.4. *Define the region*

$$\mathcal{R}' = \Lambda \left(\text{proj}_{Y_S} \left(\overline{\text{con}(\Gamma_{\mathcal{N}}^{**} \cap \mathcal{L}_{123})} \cap \mathcal{L}_4 \cap \mathcal{L}_5 \right) \right). \tag{21.32}$$

Theorem 21.5. $\mathcal{R} = \mathcal{R}'$.

This theorem, which characterizes the achievable information rate region $\mathcal{R}$, will be proved in Sections 21.6 and 21.7. In the next section, we first discuss how more explicit inner and outer bounds on $\mathcal{R}$ can be obtained.

21.5 Explicit Inner and Outer Bounds

Theorem 21.5 gives a characterization of the achievable information rate region $\mathcal{R}$ in terms of the region $\overline{\Gamma}_{\mathcal{N}}^{**}$. However, so far there exists no complete characterization of $\overline{\Gamma}_{\mathcal{N}}^{**}$. Therefore, the region $\mathcal{R}$ cannot be evaluated explicitly.

In the definition of $\mathcal{R}'$ in (21.32), if $\overline{\Gamma}_{\mathcal{N}}^{**}$ is replaced by an inner bound (outer bound) on $\overline{\Gamma}_{\mathcal{N}}^{**}$, then an inner bound (outer bound) on $\mathcal{R}$ is obtained. The results in [261] and [259], which are beyond the scope of our discussion here, provide explicit constructions of inner bounds on $\overline{\Gamma}_{\mathcal{N}}^{**}$.

We now discuss how an explicit outer bound on $\mathcal{R}$ can be obtained. To facilitate our discussion, we further define

$$i_{A;A'} = h_A - h_{A|A'} \tag{21.33}$$

and

$$i_{A;A'|A''} = h_{A|A''} - h_{A|A'A''} \tag{21.34}$$

for $A, A', A'' \in \mathcal{Q}_{\mathcal{N}}$. Let $\Gamma_{\mathcal{N}}$ be the set of $\mathbf{h} \in \mathcal{H}_{\mathcal{N}}$ such that $\mathbf{h}$ satisfies all the basic inequalities involving some or all of the random variables in $\mathcal{N}$, i.e., for all $A, A', A'' \in \mathcal{Q}_{\mathcal{N}}$,

$$h_A \geq 0, \tag{21.35}$$

$$h_{A|A'} \geq 0, \tag{21.36}$$

$$i_{A;A'} \geq 0, \tag{21.37}$$

$$i_{A;A'|A''} \geq 0. \tag{21.38}$$

We know from Section 14.2 that $\overline{\Gamma}_{\mathcal{N}}^{**} \subset \Gamma_{\mathcal{N}}$. Then upon replacing $\overline{\Gamma}_{\mathcal{N}}^{**}$ by $\Gamma_{\mathcal{N}}$ in (21.32), we obtain an outer bound on $\mathcal{R}_{out}$. This outer bound, called the *LP bound* (LP for linear programming), is given by

$$\mathcal{R}_{LP} = \Lambda\left(\text{proj}_{Y_S}\left(\overline{\text{con}(\Gamma_{\mathcal{N}} \cap \mathcal{L}_{123})} \cap \mathcal{L}_4 \cap \mathcal{L}_5\right)\right) \tag{21.39}$$

$$= \Lambda\left(\text{proj}_{Y_S}(\Gamma_{\mathcal{N}} \cap \mathcal{L}_{12345})\right), \tag{21.40}$$

where the last equality follows because both $\Gamma_{\mathcal{N}}$ and $\mathcal{L}_{123}$ are closed and convex. Since $\mathcal{R}_{LP}$ involves only a finite number of linear constraints, $\mathcal{R}_{LP}$ can be evaluated explicitly.

Using the technique in [411], it can be proved that $\mathcal{R}_{LP}$ is tight for most special cases of multi-source network coding on an acyclic network for which the achievable information region is known. In addition to single-source network coding, these include the models described in [160] [400] [308] [410] [411] [341]. Since $\mathcal{R}_{LP}$ encompasses all Shannon-type information inequalities and the converse proofs of the achievable information rate region for all these special cases do not involve non-Shannon-type inequalities, the tightness of $\mathcal{R}_{LP}$ for all these cases is expected.

However, there exist multi-source network coding problems that require non-Shannon-type inequalities for the characterization of the achievable information rate region [98] [62]. As new non-Shannon-type inequalities are discovered from time to time, improved outer bounds on $\mathcal{R}$ can be obtained by incorporating these inequalities.

21.6 The Converse

In this section, we establish the converse part of Theorem 21.5, namely

$$\mathcal{R} \subset \Lambda\left(\text{proj}_{Y_S}\left(\overline{\text{con}(\Gamma_N^{**} \cap \mathcal{L}_{123})} \cap \mathcal{L}_4 \cap \mathcal{L}_5\right)\right) = \mathcal{R}'. \tag{21.41}$$

Let ϵ_k be a sequence such that $0 < \epsilon_k < 1$ for all k, and ϵ_k monotonically decreases to 0 as $k \to \infty$. Consider an achievable information rate tuple $\boldsymbol{\omega} \in \mathcal{R}$. Then for all k, for all sufficiently large n, there exists an

$$\left(n, (\eta_e^{(k)} : e \in E), (\tau_s^{(k)} : s \in S)\right) \tag{21.42}$$

code satisfying

$$n^{-1}\log \eta_e^{(k)} \le R_e + \epsilon_k, \ e \in E \tag{21.43}$$

$$\tau_s^{(k)} \ge \omega_s - \epsilon_k, \ s \in S \tag{21.44}$$

$$\Delta_t^{(k)} \le \epsilon_k, \ t \in T, \tag{21.45}$$

where $\Delta_t^{(k)}$ denotes the decoding error probability at sink node t (cf. (21.10)).

We now fix k to be any positive integer and temporarily suppress all the superscripts involving k. For all $e \in E$, let U_e be the codeword sent on channel e and denote the alphabet of U_e by $\mathcal{U}_e$. The following lemma, whose proof will be deferred to the end of the section, is a consequence of Fano's inequality.

Lemma 21.6. *For all n and k, for all $t \in T$,*

$$H(X_{\beta(t)}|U_{\text{In}(t)}) \le n\phi_t(n, \epsilon_k), \tag{21.46}$$

where

1. *$\phi_t(n, \epsilon_k)$ is bounded;*
2. *$\phi_t(n, \epsilon_k) \to 0$ as $n, k \to \infty$;*
3. *$\phi_t(n, \epsilon_k)$ is monotonically decreasing in both n and k.*

Since the information source $X_s, s \in S$ are mutually independent,

$$H(X_S) = \sum_{s \in S} H(X_s). \tag{21.47}$$

For any $s \in S$, since $U_{\text{Out}(s)}$ is a function of X_s,

$$H(U_{\text{Out}(s)}|X_s) = 0. \tag{21.48}$$

Similarly, for all $i \in V \setminus (S \cup T)$,

$$H(U_{\text{Out}(i)}|U_{\text{In}(i)}) = 0. \tag{21.49}$$

For all $e \in E$,

$$H(U_e) \leq \log |\mathcal{U}_e| \tag{21.50}$$
$$= \log(\eta_e + 1) \tag{21.51}$$
$$\leq n(R_e + 2\epsilon_k) \tag{21.52}$$

where (21.52) follows from (21.43) assuming that n is sufficiently large. For all $t \in T$, from Lemma 21.6, we have

$$H(X_{\beta(t)}|U_{\text{In}(t)}) \leq n\phi_t(n, \epsilon_k). \tag{21.53}$$

For all $s \in S$, from (21.44),

$$H(X_s) = \log |\mathcal{X}_s| = \log \lceil 2^{n\tau_s} \rceil \geq n\tau_s \geq n(\omega_s - \epsilon_k). \tag{21.54}$$

By letting $Y_s = X_s$ for all $s \in S$, we then obtain from (21.47) to (21.49) and (21.52) to (21.54) that

$$H(Y_S) = \sum_{s \in S} H(Y_s) \tag{21.55}$$
$$H(U_{\text{Out}(s)}|Y_s) = 0, \ s \in S \tag{21.56}$$
$$H(U_{\text{Out}(i)}|U_{\text{In}(i)}) = 0, \ i \in V \setminus (S \cup T) \tag{21.57}$$
$$H(U_e) \leq n(R_e + 2\epsilon_k), \ e \in E \tag{21.58}$$
$$H(Y_{\beta(t)}|U_{\text{In}(t)}) \leq n\phi_t(n, \epsilon_k), \ t \in T \tag{21.59}$$
$$H(Y_s) \geq n(\omega_s - \epsilon_k), \ s \in S. \tag{21.60}$$

Now define the following two regions in $\mathcal{H}_\mathcal{N}$:

$$\mathcal{L}_{4,\epsilon_k}^n = \{\mathbf{h} \in \mathcal{H}_\mathcal{N} : h_{U_e} \leq n(R_e + 2\epsilon_k), e \in E\}, \tag{21.61}$$
$$\mathcal{L}_{5,\epsilon_k}^n = \{\mathbf{h} \in \mathcal{H}_\mathcal{N} : h_{Y_{\beta(t)}|U_{\text{In}(t)}} \leq n\phi_t(n, \epsilon_k), t \in T\}. \tag{21.62}$$

Note that all the auxiliary random variables $Y_s, s \in S$ and $U_e, e \in E$ have finite alphabets, because

$$|\mathcal{Y}_s| = |\mathcal{X}_s| = \lceil 2^{n\tau_s} \rceil < \infty \tag{21.63}$$

and

$$\log |\mathcal{U}_e| \leq n(R_e + 2\epsilon_k) < \infty \tag{21.64}$$

(cf. (21.50) through (21.52)). Then we see from (21.55) to (21.60) that there exists

$$\mathbf{h}^{(k)} \in \Gamma_{\mathcal{N}}^{**} \tag{21.65}$$

such that

$$\mathbf{h}^{(k)} \in \mathcal{L}_{123} \cap \mathcal{L}_{4,\epsilon_k}^n \cap \mathcal{L}_{5,\epsilon_k}^n \tag{21.66}$$

and

$$h_{Y_s}^{(k)} \geq n(\omega_s - \epsilon_k) \tag{21.67}$$

for all $s \in S$. From (21.65) and (21.66), we obtain

$$\mathbf{h}^{(k)} \in \Gamma_{\mathcal{N}}^{**} \cap \mathcal{L}_{123} \cap \mathcal{L}_{4,\epsilon_k}^n \cap \mathcal{L}_{5,\epsilon_k}^n. \tag{21.68}$$

Upon dividing by n, (21.67) becomes

$$n^{-1} h_{Y_s}^{(k)} \geq \omega_s - \epsilon_k. \tag{21.69}$$

Since $\Gamma_{\mathcal{N}}^{**} \cap \mathcal{L}_{123}$ contains the origin in $\mathcal{H}_{\mathcal{N}}$, we see that

$$n^{-1} \mathbf{h}^{(k)} \in \overline{\mathrm{con}(\Gamma_{\mathcal{N}}^{**} \cap \mathcal{L}_{123})} \cap \mathcal{L}_{4,\epsilon_k} \cap \mathcal{L}_{5,\epsilon_k}, \tag{21.70}$$

where

$$\mathcal{L}_{4,\epsilon_k} = \{\mathbf{h} \in \mathcal{H}_{\mathcal{N}} : h_{U_e} \leq R_e + 2\epsilon_k, e \in E\}, \tag{21.71}$$
$$\mathcal{L}_{5,\epsilon_k} = \{\mathbf{h} \in \mathcal{H}_{\mathcal{N}} : h_{Y_{\beta(t)}|U_{\mathrm{In}(t)}} \leq \phi_t(n, \epsilon_k), t \in T\}. \tag{21.72}$$

Note that the region $\mathcal{L}_{5,\epsilon_k}$ depends on n though it is not indicated explicitly. For all n and k, define the set

$$\mathcal{B}^{(n,k)} = \{\mathbf{h} \in \overline{\mathrm{con}(\Gamma_{\mathcal{N}}^{**} \cap \mathcal{L}_{123})} \cap \mathcal{L}_{4,\epsilon_k} \cap \mathcal{L}_{5,\epsilon_k} :$$
$$h_{Y_s} \geq \omega_s - \epsilon_k \text{ for all } s \in S\}. \tag{21.73}$$

Lemma 21.7. *For all n and k, the set $\mathcal{B}^{(n,k)}$ is compact.*[1]

Again, the proof of this lemma is deferred to the end of the section. Now from Lemma 21.6, $\phi_t(n, \epsilon_k)$ is monotonically decreasing in both n and k, so for all n and k,

$$\mathcal{B}^{(n+1,k)} \subset \mathcal{B}^{(n,k)} \tag{21.74}$$

and

$$\mathcal{B}^{(n,k+1)} \subset \mathcal{B}^{(n,k)}. \tag{21.75}$$

For any fixed k and all sufficiently large n, we see from (21.69) and (21.70) that $\mathcal{B}^{(n,k)}$ is nonempty. Since $\mathcal{B}^{(n,k)}$ is compact by Lemma 21.7,

$$\lim_{n \to \infty} \mathcal{B}^{(n,k)} = \bigcap_{n=1}^{\infty} \mathcal{B}^{(n,k)} \tag{21.76}$$

---

[1] A subset of the Euclidean space is compact if and only if it is closed and bounded.

is both compact and nonempty. By the same argument, we conclude that

$$\lim_{k\to\infty} \lim_{n\to\infty} \mathcal{B}^{(n,k)} = \bigcap_{k=1}^{\infty} \bigcap_{n=1}^{\infty} \mathcal{B}^{(n,k)} \qquad (21.77)$$

is also nonempty. Now the set

$$\lim_{k\to\infty} \lim_{n\to\infty} \mathcal{B}^{(n,k)} \qquad (21.78)$$

is equal to

$$\left\{ \mathbf{h} \in \overline{\mathrm{con}(\Gamma_{\mathcal{N}}^{**} \cap \mathcal{L}_{123})} \cap \mathcal{L}_4 \cap \mathcal{L}_5 : h_{Y_s} \geq \omega_s \text{ for all } s \in S \right\}. \qquad (21.79)$$

Hence, there exists $\mathbf{h}'$ satisfying

$$\mathbf{h}' \in \overline{\mathrm{con}(\Gamma_{\mathcal{N}}^{**} \cap \mathcal{L}_{123})} \cap \mathcal{L}_4 \cap \mathcal{L}_5 \qquad (21.80)$$

and

$$h'_{Y_s} \geq \omega_s, \quad s \in S. \qquad (21.81)$$

Let $\mathbf{r} = \mathrm{proj}_{Y_S}(\mathbf{h}')$. Then we have

$$\mathbf{r} \in \mathrm{proj}_{Y_S}\left(\overline{\mathrm{con}(\Gamma_{\mathcal{N}}^{**} \cap \mathcal{L}_{123})} \cap \mathcal{L}_4 \cap \mathcal{L}_5 \right) \qquad (21.82)$$

and

$$\mathbf{r} \geq \omega \qquad (21.83)$$

componentwise. By (21.82) and (21.83), we finally conclude that

$$\omega \in \Lambda\left(\mathrm{proj}_{Y_S}\left(\overline{\mathrm{con}(\Gamma_{\mathcal{N}}^{**} \cap \mathcal{L}_{123})} \cap \mathcal{L}_4 \cap \mathcal{L}_5 \right) \right). \qquad (21.84)$$

This completes the proof of the converse part of Theorem 21.5.

Proof of Lemma 21.6. For any $t \in T$, by Fano's inequality, we have

$$H(X_{\beta(t)}|U_{\mathrm{In}(t)}) \leq 1 + \Delta_t \log |\mathcal{X}_{\beta(t)}| \qquad (21.85)$$

$$= 1 + \Delta_t H(X_{\beta(t)}) \qquad (21.86)$$

$$\leq 1 + \epsilon_k H(X_{\beta(t)}), \qquad (21.87)$$

where (21.86) follows because X_s is distributed uniformly on $\mathcal{X}_s$ and X_s, $s \in S$ are mutually independent, and (21.87) follows from (21.45). Then

$$H(X_{\beta(t)}) = I(X_{\beta(t)}; U_{\mathrm{In}(t)}) + H(X_{\beta(t)}|U_{\mathrm{In}(t)}) \qquad (21.88)$$

$$\overset{a)}{\leq} I(X_{\beta(t)}; U_{\mathrm{In}(t)}) + 1 + \epsilon_k H(X_{\beta(t)}) \qquad (21.89)$$

$$\leq H(U_{\mathrm{In}(t)}) + 1 + \epsilon_k H(X_{\beta(t)}) \qquad (21.90)$$

$$\overset{b)}{\leq} \left(\sum_{e \in \mathrm{In}(t)} \log \eta_e \right) + 1 + \epsilon_k H(X_{\beta(t)}) \qquad (21.91)$$

$$\overset{c)}{\leq} \left(\sum_{e \in \mathrm{In}(t)} n(R_e + \epsilon_k) \right) + 1 + \epsilon_k H(X_{\beta(t)}), \qquad (21.92)$$

where

(a) follows from (21.87);
(b) follows from Theorem 2.43;
(c) follows from (21.43).

Rearranging the terms in (21.92), we obtain

$$H(X_{\beta(t)}) \leq \frac{n}{1 - \epsilon_k} \left(\sum_{e \in \text{In}(t)} (R_e + \epsilon_k) + \frac{1}{n} \right). \tag{21.93}$$

Substituting (21.93) into (21.87), we have

$$H(X_{\beta(t)} | U_{\text{In}(t)}) < n \left[\frac{1}{n} + \frac{\epsilon_k}{1 - \epsilon_k} \left(\sum_{e \in \text{In}(t)} (R_e + \epsilon_k) + \frac{1}{n} \right) \right] \tag{21.94}$$

$$= n\phi_t(n, \epsilon_k), \tag{21.95}$$

where

$$\phi_t(n, \epsilon_k) = \frac{1}{n} + \frac{\epsilon_k}{1 - \epsilon_k} \left(\sum_{e \in \text{In}(t)} (R_e + \epsilon_k) + \frac{1}{n} \right). \tag{21.96}$$

Invoking the assumption that $0 < \epsilon_k < 1$ for all k and ϵ_k monotonically decreases to 0 as $k \to \infty$, it is evident that

1. $\phi_t(n, \epsilon_k)$ is bounded for all n and k;
2. $\phi_t(n, \epsilon_k) \to 0$ as $n, k \to \infty$;
3. $\phi_t(n, \epsilon_k)$ is monotonically non-increasing in both n and k.

The lemma is proved. $\square$

Proof of Lemma 21.7. We need to show that the set $\mathcal{B}^{(n,k)}$ is both closed and bounded. The closedness of $\mathcal{B}^{(n,k)}$ is immediate from its definition. To establish the boundedness of $\mathcal{B}^{(n,k)}$, we need to show that for any $\mathbf{h} \in \mathcal{B}^{(n,k)}$, all the components of $\mathbf{h}$ are bounded. Consider any $\mathbf{h} \in \mathcal{B}^{(n,k)}$. Since

$$\mathcal{B}^{(n,k)} \subset \mathcal{L}_{4,\epsilon_k}, \tag{21.97}$$

we see from (21.71) that h_{U_e} are bounded for all $e \in E$. Since

$$\mathcal{B}^{(n,k)} \subset \mathcal{L}_{5,\epsilon_k}, \tag{21.98}$$

we see from (21.72) that for every $t \in T$,

$$h_{Y_{\beta(t)}} \leq h_{Y_{\beta(t)} U_{\text{In}(t)}} \tag{21.99}$$

$$= h_{Y_{\beta(t)} | U_{\text{In}(t)}} + h_{U_{\text{In}(t)}} \tag{21.100}$$

$$\leq \phi_t(n, \epsilon_k) + h_{U_{\text{In}(t)}} \tag{21.101}$$

$$\leq \phi_t(n, \epsilon_k) + \sum_{e \in \text{In}(t)} h_{U_e}, \tag{21.102}$$

where (21.100) and the boundedness of $\phi_t(n, \epsilon_k)$ follow from Lemma 21.6. This shows that $h_{Y_{\beta(t)}}$ is bounded for all $t \in T$.

In our model, for every $s \in S$, there exists at least one $t \in T$ such that $s \in \beta(t)$. Then the boundedness of $h_{Y_{\beta(t)}}$ for all $t \in T$ implies the boundedness of h_{Y_s} for all $s \in S$. Finally, the boundedness of all the other components of $\mathbf{h}$ is established by invoking the independence bound for entropy. The lemma is proved. $\square$

21.7 Achievability

In this section, we establish the direct part of Theorem 21.5, namely

$$\mathcal{R}' = \Lambda \left(\text{proj}_{Y_S} \left(\overline{\text{con}(\Gamma_{\mathcal{N}}^{**} \cap \mathcal{L}_{123})} \cap \mathcal{L}_4 \cap \mathcal{L}_5 \right) \right) \subset \mathcal{R}. \tag{21.103}$$

Before we proceed, we first prove an alternative form of $\mathcal{R}'$ that will be used in constructing the random code. For a subset $\mathcal{B}$ of $\mathcal{H}_{\mathcal{N}}$, let

$$D(\mathcal{B}) = \{\alpha \mathbf{h} : \mathbf{h} \in \mathcal{B} \text{ and } 0 \leq \alpha \leq 1\}. \tag{21.104}$$

Define the two subsets

$$\mathcal{A}_1 = \overline{\text{con}(\Gamma_{\mathcal{N}}^{**} \cap \mathcal{L}_{123})} \tag{21.105}$$

and

$$\mathcal{A}_2 = \overline{D(\Gamma_{\mathcal{N}}^{**} \cap \mathcal{L}_{123})} \tag{21.106}$$

of $\mathcal{H}_{\mathcal{N}}$.

Lemma 21.8. $\mathcal{A}_1 = \mathcal{A}_2$.

Proof. Since the origin of $\mathcal{H}_{\mathcal{N}}$ is in $\Gamma_{\mathcal{N}}^{**}$, it is also in $\Gamma_{\mathcal{N}}^{**} \cap \mathcal{L}_{123}$ because $\mathcal{L}_{123}$ is a linear subspace of $\mathcal{H}_{\mathcal{N}}$. Upon observing that for $0 \leq \alpha \leq 1$,

$$\alpha \mathbf{h} = (1 - \alpha)\mathbf{0} + \alpha \mathbf{h} \tag{21.107}$$

is a convex combination of $\mathbf{0}$ and $\mathbf{h}$, we obtain

$$D(\Gamma_{\mathcal{N}}^{**} \cap \mathcal{L}_{123}) \subset \text{con}(\Gamma_{\mathcal{N}}^{**} \cap \mathcal{L}_{123}). \tag{21.108}$$

It follows that $\mathcal{A}_2 \subset \mathcal{A}_1$.

To prove that $\mathcal{A}_1 \subset \mathcal{A}_2$, it suffices to show that $\mathcal{A}_2$ is convex because

1. $(\Gamma_{\mathcal{N}}^{**} \cap \mathcal{L}_{123}) \subset \mathcal{A}_2$, where $\mathcal{A}_2$ is closed;
2. $\mathcal{A}_1$ is the smallest closed convex set containing $\Gamma_{\mathcal{N}}^{**} \cap \mathcal{L}_{123}$.

Toward this end, consider any $\mathbf{h}_1, \mathbf{h}_2 \in \mathcal{A}_2$ and any $0 \leq \lambda \leq 1$. We will show that

$$\mathbf{h} = \lambda \mathbf{h}_1 + (1 - \lambda)\mathbf{h}_2 \in \mathcal{A}_2. \tag{21.109}$$

Here, we can assume without loss of generality that $\mathbf{h}_1, \mathbf{h}_2 \neq 0$, because otherwise (21.109) holds by the definition of $\mathcal{A}_2$. Since $\mathbf{h}_1, \mathbf{h}_2 \in \mathcal{A}_2$, there exist $\mathbf{h}_1^k, \mathbf{h}_2^k \in D(\Gamma_{\mathcal{N}}^{**} \cap \mathcal{L}_{123})$ such that $\mathbf{h}_1^k \to \mathbf{h}_1$ and $\mathbf{h}_2^k \to \mathbf{h}_2$. Again, we can assume without loss of generality that $\mathbf{h}_1^k, \mathbf{h}_2^k \neq 0$ for all k because $\mathbf{h}_1, \mathbf{h}_2 \neq 0$. Since $\mathbf{h}_1^k, \mathbf{h}_2^k \in D(\Gamma_{\mathcal{N}}^{**} \cap \mathcal{L}_{123})$, we can write

$$\mathbf{h}_1^k = \alpha_1^k \hat{\mathbf{h}}_1^k \tag{21.110}$$

and

$$\mathbf{h}_2^k = \alpha_2^k \hat{\mathbf{h}}_2^k, \tag{21.111}$$

where $\hat{\mathbf{h}}_1^k, \hat{\mathbf{h}}_2^k \in \Gamma_{\mathcal{N}}^{**} \cap \mathcal{L}_{123}$ and $0 < \alpha_1^k, \alpha_2^k \leq 1$. Note that α_1^k and α_2^k are strictly positive because $\mathbf{h}_1^k, \mathbf{h}_2^k \neq 0$. Now let n_1^k and n_2^k be integer sequences such that $n_1^k, n_2^k \to \infty$ and

$$\frac{n_1^k \alpha_2^k}{n_2^k \alpha_1^k} \to \frac{\lambda}{1 - \lambda}, \tag{21.112}$$

and let

$$\hat{\mathbf{h}}^k = n_1^k \hat{\mathbf{h}}_1^k + n_2^k \hat{\mathbf{h}}_2^k. \tag{21.113}$$

It can be seen from Lemma 15.3 and Corollary 15.4 that

$$\hat{\mathbf{h}}^k \in \Gamma_{\mathcal{N}}^{**}. \tag{21.114}$$

Furthermore, since $\hat{\mathbf{h}}_1^k, \hat{\mathbf{h}}_2^k \in \mathcal{L}_{123}$ and $\mathcal{L}_{123}$ is a linear subspace, $\hat{\mathbf{h}}^k \in \mathcal{L}_{123}$. Therefore,

$$\hat{\mathbf{h}}^k \in \Gamma_{\mathcal{N}}^{**} \cap \mathcal{L}_{123}. \tag{21.115}$$

Let

$$\mathbf{h}^k = \frac{\alpha_1^k \alpha_2^k}{n_1^k \alpha_2^k + n_2^k \alpha_1^k} \hat{\mathbf{h}}^k. \tag{21.116}$$

Since $\alpha_1^k, \alpha_2^k \leq 1$ and $n_1^k, n_2^k \to \infty$, for sufficiently large k,

$$\frac{\alpha_1^k \alpha_2^k}{n_1^k \alpha_2^k + n_2^k \alpha_1^k} \leq 1, \tag{21.117}$$

and therefore

$$\mathbf{h}^k \in D(\Gamma_{\mathcal{N}}^{**} \cap \mathcal{L}_{123}) \subset \overline{D(\Gamma_{\mathcal{N}}^{**} \cap \mathcal{L}_{123})} = \mathcal{A}_2. \tag{21.118}$$

Substituting (21.113), (21.110), and (21.111) into (21.116), we obtain

$$\mathbf{h}^k = \frac{n_1^k \alpha_2^k}{n_1^k \alpha_2^k + n_2^k \alpha_1^k} \mathbf{h}_1^k + \frac{n_2^k \alpha_1^k}{n_1^k \alpha_2^k + n_2^k \alpha_1^k} \mathbf{h}_2^k. \tag{21.119}$$

It can readily be seen from (21.112) that

$$\frac{n_1^k \alpha_2^k}{n_1^k \alpha_2^k + n_2^k \alpha_1^k} \to \lambda \tag{21.120}$$

and

$$\frac{n_2^k \alpha_1^k}{n_1^k \alpha_2^k + n_2^k \alpha_1^k} \to 1 - \lambda. \tag{21.121}$$

Since $\mathbf{h}_1^k \to \mathbf{h}_1$ and $\mathbf{h}_2^k \to \mathbf{h}_2$, we see from (21.119) and (21.109) that $\mathbf{h}^k \to \mathbf{h}$. Finally, since $\mathbf{h}^k \in \mathcal{A}_2$ and $\mathcal{A}_2$ is closed, we conclude that $\mathbf{h} \in \mathcal{A}_2$. Therefore, $\mathcal{A}_2$ is convex, and hence $\mathcal{A}_1 \subset \mathcal{A}_2$. The lemma is proved. $\square$

By virtue of this lemma, we can write

$$\mathcal{R}' = \Lambda \left(\text{proj}_{Y_S} \left(\overline{D(\Gamma_{\mathcal{N}}^{**} \cap \mathcal{L}_{123})} \cap \mathcal{L}_4 \cap \mathcal{L}_5 \right) \right), \tag{21.122}$$

and we will establish $\mathcal{R}' \subset \mathcal{R}$ by proving that

$$\Lambda \left(\text{proj}_{Y_S} \left(\overline{D(\Gamma_{\mathcal{N}}^{**} \cap \mathcal{L}_{123})} \cap \mathcal{L}_4 \cap \mathcal{L}_5 \right) \right) \subset \mathcal{R}. \tag{21.123}$$

By the remark following Definition 21.3, we only need to show the achievability of the region

$$\text{proj}_{Y_S} \left(\overline{D(\Gamma_{\mathcal{N}}^{**} \cap \mathcal{L}_{123})} \cap \mathcal{L}_4 \cap \mathcal{L}_5 \right). \tag{21.124}$$

Consider any ω in this region. Then there exists

$$\mathbf{h} \in \overline{D(\Gamma_{\mathcal{N}}^{**} \cap \mathcal{L}_{123})} \cap \mathcal{L}_4 \cap \mathcal{L}_5 \tag{21.125}$$

such that

$$\omega = \text{proj}_{Y_S}(\mathbf{h}). \tag{21.126}$$

Since

$$\mathbf{h} \in \overline{D(\Gamma_{\mathcal{N}}^{**} \cap \mathcal{L}_{123})}, \tag{21.127}$$

there exists a sequence

$$\mathbf{h}^{(k)} \in D(\Gamma_{\mathcal{N}}^{**} \cap \mathcal{L}_{123}) \tag{21.128}$$

such that

$$\mathbf{h} = \lim_{k \to \infty} \mathbf{h}^{(k)}. \tag{21.129}$$

Let

$$\omega^{(k)} = \text{proj}_{Y_S}(\mathbf{h}^{(k)}). \tag{21.130}$$

It then follows from (21.129) that

$$\lim_{k \to \infty} \omega^{(k)} = \omega. \tag{21.131}$$

By (21.128),

$$\mathbf{h}^{(k)} = \alpha^{(k)} \hat{\mathbf{h}}^{(k)} \tag{21.132}$$

where $\hat{\mathbf{h}}^{(k)} \in \Gamma_\mathcal{N}^{**} \cap \mathcal{L}_{123}$ and

$$0 \le \alpha^{(k)} \le 1. \tag{21.133}$$

Note that $\hat{\mathbf{h}}^{(k)}$ is an entropy function because it is in $\Gamma_\mathcal{N}^{**}$, but $\mathbf{h}^{(k)}$ and $\mathbf{h}$ are not necessarily entropy functions. Since $\hat{\mathbf{h}}^{(k)} \in \Gamma_\mathcal{N}^{**} \cap \mathcal{L}_{123}$, there exists a collection of random variables with finite alphabets

$$\mathcal{N}^{(k)} = \left\{ (Y_s^{(k)} : s \in S), (U_e^{(k)} : e \in E) \right\} \tag{21.134}$$

such that

$$\alpha^{(k)} H\left(Y_s^{(k)}\right) = \omega_s^{(k)}, \; s \in S, \tag{21.135}$$

$$H\left(Y_S^{(k)}\right) = \sum_{s \in S} H\left(Y_s^{(k)}\right), \tag{21.136}$$

$$H\left(U_{\text{Out}(s)}^{(k)} \middle| Y_s^{(k)}\right) = 0, \; s \in S, \tag{21.137}$$

$$H\left(U_{\text{Out}(i)}^{(k)} \middle| U_{\text{In}(i)}^{(k)}\right) = 0, \; i \in V \setminus (S \cup T), \tag{21.138}$$

where (21.135) is implied by (21.130). Furthermore, since $\mathbf{h} \in \mathcal{L}_4 \cap \mathcal{L}_5$, it follows from (21.129) and (21.132) that

$$\alpha^{(k)} H\left(U_e^{(k)}\right) \le R_e + \mu^{(k)}, \; e \in E, \tag{21.139}$$

$$\alpha^{(k)} H\left(Y_{\beta(t)}^{(k)} \middle| U_{\text{In}(t)}^{(k)}\right) \le \gamma^{(k)}, \; t \in T, \tag{21.140}$$

where $\mu^{(k)}, \gamma^{(k)} \to 0$ as $k \to \infty$. In the rest of the section, we will prove the achievability of $\boldsymbol{\omega}^{(k)}$ for all sufficiently large k. Then the closedness of $\mathcal{R}$ implies the achievability of $\boldsymbol{\omega}$ by the remark following Definition 21.3.

21.7.1 Random Code Construction

Fix k and $\epsilon > 0$, and let δ be a small positive quantity to be specified later. We first construct a random

$$(n, (\eta_e^{(k)} : e \in E), (\tau_s^{(k)} : s \in S)) \tag{21.141}$$

code with

$$\eta_e^{(k)} \le 2^{n(\alpha^{(k)} H(U_e^{(k)}) + \psi_e^{(k)})} \tag{21.142}$$

for all $e \in E$ and

$$\omega_s^{(k)} - \frac{\epsilon}{2} \le \tau_s^{(k)} \le \omega_s^{(k)} - \frac{\epsilon}{3}, \tag{21.143}$$

where $\psi_e^{(k)} > 0$ and $\psi_e^{(k)} \to 0$ as $\delta \to 0$, by the steps below. For the sake of simplicity, we temporarily suppress all the superscripts involving k. In light of the heavy notation, we list in Table 21.1 the symbols involved in the description of the random code construction.

Table 21.1. The list of symbols involved in the random code construction.

n	The block length of the random code
$\hat{n}$	The length of a sequence in the typical sets used in the code construction
X_s	The information source generated at source node s
$\mathcal{X}_s$	The alphabet of information source X_s, equal to $\{1, 2, \cdots, \theta_s\}$
Y_s	The auxiliary random variable associated with X_s
$\mathcal{C}_s$	The codebook for information source X_s consisting of the codewords $\mathbf{Y}_s(1), \mathbf{Y}_s(2), \cdots, \mathbf{Y}_s(\theta_s) \in \mathcal{Y}_s^{\hat{n}}$
$\mathbf{Y}_s(0)$	An arbitrary constant sequence in $\mathcal{Y}_s^{\hat{n}}$
θ_s	The common size of the alphabet $\mathcal{X}_s$ and the codebook $\mathcal{C}_s$
$T_{[U_e]\delta}^{\hat{n}}$	Equal to $\{\mathbf{U}_e(1), \mathbf{U}_e(2), \cdots, \mathbf{U}_e(\zeta_e)\}$
$\mathbf{U}_e(0)$	An arbitrary constant sequence in $\mathcal{U}_e^{\hat{n}}$
k_e	The local encoding function for channel e
$\hat{u}_e$	The function defined by $U_e = \hat{u}_e(Y_s)$ for $e \in \text{Out}(s)$ and $U_e = \hat{u}_e(U_{\text{In}(i)})$ for $e \in \text{Out}(i)$
C_e	The index sent on channel e, i.e., the value taken by the local encoding function k_e
g_t	The decoding function at sink node t

1. Let

$$\hat{n} = \lceil n\alpha \rceil. \tag{21.144}$$

Here n is the block length of the random code we will construct, while $\hat{n}$ is the length of a sequence of the typical sets that we will use for constructing the random code. For each source $s \in S$, let

$$\theta_s = \lceil 2^{n\tau_s} \rceil \tag{21.145}$$

and construct a codebook $\mathcal{C}_s$ by generating θ_s codewords in $\mathcal{Y}_s^{\hat{n}}$ randomly and independently according to $p^{\hat{n}}(y_s)$. Denote these sequences by $\mathbf{Y}_s(1), \mathbf{Y}_s(2), \cdots, \mathbf{Y}_s(\theta_s)$, and let $\mathbf{Y}_s(0)$ be an arbitrary constant sequence in $\mathcal{Y}_s^{\hat{n}}$.
2. Reveal the codebook $\mathcal{C}_s, s \in S$ to all the nodes in the network.
3. At a source node $s \in S$, the information source X_s is generated according to the uniform distribution on

$$\mathcal{X}_s = \{1, 2, \cdots, \theta_s\}. \tag{21.146}$$

4. Let $T_{[U_e]\delta}^{\hat{n}}$ denote the set of strongly typical sequences[2] with respect to the distribution $p(u_e)$. Let

$$\zeta_e = |T_{[U_e]\delta}^{\hat{n}}|. \tag{21.147}$$

---

[2] Strong typicality applies because all the random variables in $\mathcal{N}^{(k)}$ have finite alphabets.

By the strong AEP and (21.144),

$$\zeta_e \leq 2^{\hat{n}(H(U_e)+\psi_e/(2\alpha))} \leq 2^{n(\alpha H(U_e)+\psi_e/2)}, \tag{21.148}$$

where $\psi_e \to 0$ as $\delta \to 0$. For all channels $e \in E$, choose an η_e satisfying

$$2^{n(\alpha H(U_e)+\psi_e/2)} \leq \eta_e \leq 2^{n(\alpha H(U_e)+\psi_e)}. \tag{21.149}$$

Denote the sequences in $T^{\hat{n}}_{[U_e]\delta}$ by $\mathbf{U}_e(1), \mathbf{U}_e(2), \cdots, \mathbf{U}_e(\zeta_e)$, and let $\mathbf{U}_e(0)$ be an arbitrary constant sequence in $\mathcal{U}_e^{\hat{n}}$.

a) Let the outcome of X_s be x_s for a source node s. For a channel $e \in \text{Out}(s)$, define the local encoding function

$$k_e : \mathcal{X}_s \to \{0, 1, \cdots, \eta_e\} \tag{21.150}$$

as follows. By (21.137), for each channel $e \in \text{Out}(s)$, there exists a function $\hat{u}_e$ such that

$$U_e = \hat{u}_e(Y_s), \tag{21.151}$$

i.e.,

$$\Pr\{U_e = \hat{u}_e(y) | Y_s = y\} = 1 \tag{21.152}$$

for all $y \in \mathcal{Y}_s$. By the preservation property of strong typicality (Theorem 6.8), if

$$\mathbf{Y}_s(x_s) \in T^{\hat{n}}_{[Y_s]\delta}, \tag{21.153}$$

then

$$\hat{u}_e(\mathbf{Y}_s(x_s)) \in T^{\hat{n}}_{[U_e]\delta}, \tag{21.154}$$

where in $\hat{u}_e(\mathbf{Y}_s(x_s))$, the function $\hat{u}_e$ is applied to $\mathbf{Y}_s(x_s)$ componentwise. If so, let $k_e(x_s)$ be the index of $\hat{u}_e(\mathbf{Y}_s(x_s))$ in $T^{\hat{n}}_{[U_e]\delta}$, i.e.,

$$\mathbf{U}_e(k_e(x_s)) = \hat{u}_e(\mathbf{Y}_s(x_s)). \tag{21.155}$$

Otherwise, let $k_e(x_s)$ be 0. Note that k_e is well defined because

$$\zeta_e \leq \eta_e \tag{21.156}$$

by (21.148) and (21.149).
b) For a channel $e \in \text{Out}(i)$, where $i \in V \backslash (S \cup T)$, define the local encoding function

$$k_e : \prod_{d \in \text{In}(i)} \{0, 1, \cdots, \eta_d\} \to \{0, 1, \cdots, \eta_e\} \tag{21.157}$$

as follows. By (21.138), there exists a function $\hat{u}_e$ such that

$$U_e = \hat{u}_e(U_{\text{In}(i)}). \tag{21.158}$$

Let C_e be the index sent on channel e, i.e., the value taken by the local encoding function k_e. With a slight abuse of notation, we write

$$\mathbf{U}_{E'}(C_{E'}) = (\mathbf{U}_d(C_d) : d \in E') \tag{21.159}$$

for $E' \subset E$, and

$$\mathbf{Y}_{S'}(x_{S'}) = (\mathbf{Y}_s(x_s) : s \in S') \tag{21.160}$$

for $S' \subset S$. By the preservation property of strong typicality, if

$$\mathbf{U}_{\text{In}(i)}(C_{\text{In}(i)}) \in T_{[U_{\text{In}(i)}]\delta}^{\hat{n}}, \tag{21.161}$$

then

$$\hat{u}_e(\mathbf{U}_{\text{In}(i)}(C_{\text{In}(i)})) \in T_{[U_e]\delta}^{\hat{n}}. \tag{21.162}$$

If so, let $k_e(C_{\text{In}(i)})$ be the index of $\hat{u}_e(\mathbf{U}_{\text{In}(i)}(C_{\text{In}(i)}))$ in $T_{[U_e]\delta}^{\hat{n}}$, i.e.,

$$\mathbf{U}_e(k_e(C_{\text{In}(i)})) = \hat{u}_e(\mathbf{U}_{\text{In}(i)}(C_{\text{In}(i)})). \tag{21.163}$$

Otherwise, let $k_e(C_{\text{In}(i)})$ be 0. Again, k_e is well defined because (21.156) holds.

5. For a sink node $t \in T$, define the decoding function

$$g_t : \prod_{d \in \text{In}(t)} \{0, 1, \cdots, \eta_d\} \to \mathcal{X}_{\beta(t)} \tag{21.164}$$

as follows. If the received index C_d on channel d is nonzero for all $d \in \text{In}(t)$ and there exists a unique tuple

$$x_{\beta(t)} \in \mathcal{X}_{\beta(t)} \tag{21.165}$$

such that

$$(\mathbf{Y}_{\beta(t)}(x_{\beta(t)}), \mathbf{U}_{\text{In}(i)}(C_{\text{In}(t)})) \in T_{[U_{\text{In}(t)}Y_{\beta(t)}]\delta}^{\hat{n}}, \tag{21.166}$$

then let $g_t(C_{\text{In}(t)})$ be $x_{\beta(t)}$. Otherwise, declare a decoding error.

21.7.2 Performance Analysis

Let us reinstate all the superscripts involving k that were suppressed when we described the construction of the random code. Our task is to show that for any sufficiently large k and any $\epsilon > 0$, the random code we have constructed satisfies the following when n is sufficiently large:

$$n^{-1} \log \eta_e^{(k)} \leq R_e + \epsilon, \ e \in E, \tag{21.167}$$

$$\tau_s^{(k)} \geq \omega_s^{(k)} - \epsilon, \ s \in S, \tag{21.168}$$

$$\Delta_t^{(k)} \leq \epsilon, \ t \in T. \tag{21.169}$$

For $e \in E$, consider

$$n^{-1} \log \eta_e^{(k)} \leq \alpha^{(k)} H(U_e^{(k)}) + \psi_e^{(k)} \tag{21.170}$$

$$\leq R_e + \mu^{(k)} + \psi_e^{(k)}, \tag{21.171}$$

where the first inequality follows from the upper bound in (21.149) and the second inequality follows from (21.139). Since $\mu^{(k)} \to 0$ as $k \to \infty$, we can let k be sufficiently large so that

$$\mu^{(k)} < \epsilon. \tag{21.172}$$

With k fixed, since $\psi_e^{(k)} \to 0$ as $\delta \to 0$, by letting δ be sufficiently small, we have

$$\mu^{(k)} + \psi_e^{(k)} \leq \epsilon, \tag{21.173}$$

and (21.167) follows from (21.171). For $s \in S$, from the lower bound in (21.143), we have

$$\tau_s^{(k)} \geq \omega_s^{(k)} - \epsilon, \tag{21.174}$$

proving (21.168).

The proof of (21.169), which is considerably more involved, will be organized into a few lemmas. For the sake of presentation, the proofs of these lemmas will be deferred to the end of the section.

For $i \in S$ and $i \in V \backslash (S \cup T)$, the function $\hat{u}_e$, where $e \in \text{Out}(i)$, has been defined in (21.151) and (21.158), respectively. Since the network is acyclic, we see by induction that all the auxiliary random variables $U_e, e \in E$ are functions of the auxiliary random variables Y_S. Thus there exists a function $\tilde{u}_e$ such that

$$U_e = \tilde{u}_e(Y_S). \tag{21.175}$$

Equating the above with (21.151) and (21.158), we obtain

$$\hat{u}_e(Y_s) = \tilde{u}_e(Y_S) \tag{21.176}$$

and

$$\hat{u}_e(U_{\text{In}(i)}) = \tilde{u}_e(Y_S), \tag{21.177}$$

respectively. These relations will be useful subsequently.

In the rest of the section, we will analyze the probabilities of decoding error for the random code we have constructed for a fixed k, namely $\Delta_t^{(k)}$, for $t \in T$. With a slight abuse of notation, we write

$$\tilde{u}_{E'}(\cdot) = (\tilde{u}_d(\cdot) : d \in E') \tag{21.178}$$

for $E' \subset E$. Again, we temporarily suppress all the superscripts invoking k.

Lemma 21.9. *Let*

$$X_S = x_S, \tag{21.179}$$

$$\mathbf{Y}_S(x_S) = \mathbf{y}_S \in T_{[Y_S]\delta}^{\hat{n}}, \tag{21.180}$$

and for $e \in E$, let C_e take the value c_e, which by the code construction is a function of x_S and $\mathbf{y}_S$. Then

$$\mathbf{U}_{\text{In}(t)}(c_{\text{In}(t)}) = \tilde{u}_{\text{In}(t)}(\mathbf{y}_S) \tag{21.181}$$

and

$$(\mathbf{y}_S, \mathbf{U}_{\text{In}(t)}(c_{\text{In}(t)})) \in T_{[Y_S U_{\text{In}(t)}]\delta}^{\hat{n}} \tag{21.182}$$

for all $t \in T$.

Let

$$Err_t = \{g_t(C_{\text{In}(t)}) \neq X_{\beta(t)}\} = \{\tilde{g}_t(X_S) \neq X_{\beta(t)}\} \tag{21.183}$$

be the event of a decoding error at sink node t, i.e.,

$$\Pr\{Err_t\} = \Delta_t \tag{21.184}$$

(cf. (21.10)). In the following, we will obtain an upper bound on $\Pr\{Err_t\}$. Consider

$$\Pr\{Err_t\} = \sum_{x_{\beta(t)} \in \mathcal{X}_{\beta(t)}} \Pr\{Err_t | X_{\beta(t)} = x_{\beta(t)}\} \Pr\{X_{\beta(t)} = x_{\beta(t)}\}, \tag{21.185}$$

and for $S' \subset S$, let

$$1_{S'} = \underbrace{(1, 1, \cdots, 1)}_{|S'|}. \tag{21.186}$$

Since $\Pr\{Err_t | X_{\beta(t)} = x_{\beta(t)}\}$ are identical for all $x_{\beta(t)}$ by symmetry in the code construction, from (21.185), we have

$$\Pr\{Err_t\}$$
$$= \Pr\{Err_t | X_{\beta(t)} = 1_{\beta(t)}\} \sum_{x_{\beta(t)} \in \mathcal{X}_{\beta(t)}} \Pr\{X_{\beta(t)} = x_{\beta(t)}\} \tag{21.187}$$
$$= \Pr\{Err_t | X_{\beta(t)} = 1_{\beta(t)}\}. \tag{21.188}$$

In other words, we can assume without loss of generality that $X_{\beta(t)} = 1_{\beta(t)}$. To facilitate our discussion, define the event

$$E_S = \{\mathbf{Y}_S(1_S) \in T_{[Y_S]\delta}^{\hat{n}}\}. \tag{21.189}$$

Following (21.188), we have

$$\Pr\{Err_t\}$$
$$= \Pr\{Err_t | X_{\beta(t)} = 1_{\beta(t)}, E_S\} \Pr\{E_S | X_{\beta(t)} = 1_{\beta(t)}\}$$
$$\quad + \Pr\{Err_t | X_{\beta(t)} = 1_{\beta(t)}, E_S^c\} \Pr\{E_S^c | X_{\beta(t)} = 1_{\beta(t)}\} \tag{21.190}$$
$$= \Pr\{Err_t | X_{\beta(t)} = 1_{\beta(t)}, E_S\} \Pr\{E_S\}$$
$$\quad + \Pr\{Err_t | X_{\beta(t)} = 1_{\beta(t)}, E_S^c\} \Pr\{E_S^c\} \tag{21.191}$$
$$\leq \Pr\{Err_t | X_{\beta(t)} = 1_{\beta(t)}, E_S\} \cdot 1 + 1 \cdot \Pr\{E_S^c\} \tag{21.192}$$
$$\leq \Pr\{Err_t | X_{\beta(t)} = 1_{\beta(t)}, E_S\} + \lambda, \tag{21.193}$$

where the last inequality follows from the strong AEP and $\lambda \to 0$ as $\delta \to 0$. Upon defining the event

$$E'_S = \{X_{\beta(t)} = 1_{\beta(t)}\} \cap E_S, \tag{21.194}$$

we have

$$\Pr\{Err_t\} \le Pr\{Err_t|E'_S\} + \lambda. \tag{21.195}$$

We now further analyze the conditional probability in (21.195). For $x_{\beta(t)} \in \mathcal{X}_{\beta(t)}$, define the event

$$E_t(x_{\beta(t)}) = \left\{ (\mathbf{Y}_{\beta(t)}(x_{\beta(t)}), \mathbf{U}_{\mathrm{In}(t)}(C_{\mathrm{In}(t)})) \in T^{\hat{n}}_{[Y_{\beta(t)}U_{\mathrm{In}(t)}]\delta} \right\}. \tag{21.196}$$

Since $X_{\beta(t)} = 1_{\beta(t)}$, decoding at sink node t is correct if the received indices $C_{\mathrm{In}(t)}$ are decoded to $1_{\beta(t)}$. This is the case if and only if $E_t(1_{\beta(t)})$ occurs but $E_t(x_{\beta(t)})$ does not occur for all $x_{\beta(t)} \ne 1_{\beta(t)}$. It follows that

$$Err_t^c = E_t(1_{\beta(t)}) \cap \left(\cap_{x_{\beta(t)} \ne 1_{\beta(t)}} E_t(x_{\beta(t)})^c \right) \tag{21.197}$$

or

$$Err_t = E_t(1_{\beta(t)})^c \cup \left(\cup_{x_{\beta(t)} \ne 1_{\beta(t)}} E_t(x_{\beta(t)}) \right), \tag{21.198}$$

which implies

$$\Pr\{Err_t|E'_S\} = \Pr\left\{ E_t(1_{\beta(t)})^c \cup \left(\cup_{x_{\beta(t)} \ne 1_{\beta(t)}} E_t(x_{\beta(t)}) \right) \middle| E'_S \right\}. \tag{21.199}$$

By the union bound, we have

$$\Pr\{Err_t|E'_S\}$$
$$\le \Pr\{E_t(1_{\beta(t)})^c|E'_S\} + \sum_{x_{\beta(t)} \ne 1_{\beta(t)}} \Pr\{E_t(x_{\beta(t)})|E'_S\} \tag{21.200}$$

$$= \sum_{x_{\beta(t)} \ne 1_{\beta(t)}} \Pr\{E_t(x_{\beta(t)})|E'_S\}, \tag{21.201}$$

where the last step follows because $\Pr\{E_t(1_{\beta(t)})^c|E'_S\}$ in (21.200) vanishes by Lemma 21.9. The next two lemmas will be instrumental in obtaining an upper bound on $\Pr\{E_t(x_{\beta(t)})|E'_S\}$ in (21.201).

For any proper subset Ψ of $\beta(t)$, let

$$\Lambda_\Psi = \{x_{\beta(t)} \ne 1_{\beta(t)} : x_s = 1 \text{ if and only if } s \in \Psi\}. \tag{21.202}$$

Note that $\{\Lambda_\Psi\}$ is a partition of the set $\mathcal{X}_{\beta(t)} \backslash \{1_{\beta(t)}\}$. For $x_{\beta(t)} \in \Lambda_\Psi$, $x_{\beta(t)}$ and $1_{\beta(t)}$ are identical for exactly the components indexed by Ψ.

Lemma 21.10. *For $x_{\beta(t)} \in \Lambda_\Psi$, where Ψ is a proper subset of $\beta(t)$,*

$$\Pr\{E_t(x_{\beta(t)})|E'_S\} \le 2^{-n\alpha(H(Y_{\beta(t) \backslash \Psi}) - H(Y_{\beta(t)}|U_{\mathrm{In}(t)}) - \varphi_t)}, \tag{21.203}$$

where $\varphi_t \to 0$ as $n \to \infty$ and $\delta \to 0$.

Lemma 21.11. *For all sufficiently large* n,

$$|\Lambda_\Psi| \leq 2^{n(\alpha H(Y_{\beta(t)\setminus\Psi}) - \epsilon/4)}. \tag{21.204}$$

We now reinstate all the superscripts involving k that have been suppressed. By Lemma 21.10, Lemma 21.11, and (21.201),

$$\Pr\{Err_t | E'_S\}$$

$$\leq \sum_{x_{\beta(t)} \neq 1_{\beta(t)}} \Pr\{E_t(x_{\beta(t)}) | E'_S\} \tag{21.205}$$

$$\leq \sum_\Psi \sum_{x_{\beta(t)} \in \Psi} \Pr\{E_t(x_{\beta(t)}) | E'_S\} \tag{21.206}$$

$$\leq 2^{|E|} 2^{-n\alpha^{(k)} \left[\left(H\left(Y^{(k)}_{\beta(t)\setminus\Psi}\right) - H\left(Y^{(k)}_{\beta(t)} | U^{(k)}_{In(t)}\right)\right) - \varphi_t\right]}$$

$$\cdot 2^{n\left[\alpha^{(k)} H\left(Y^{(k)}_{\beta(t)\setminus\Psi}\right) - \epsilon/4\right]} \tag{21.207}$$

$$= 2^{|E|} 2^{-n\left[\epsilon/4 - \alpha^{(k)} H\left(Y^{(k)}_{\beta(t)} | U^{(k)}_{In(t)}\right) - \alpha^{(k)}\varphi_t\right]} \tag{21.208}$$

$$\leq 2^{|E|} 2^{-n\left(\epsilon/4 - \gamma^{(k)} - \alpha^{(k)}\varphi_t\right)} \tag{21.209}$$

$$\leq 2^{|E|} 2^{-n\left(\epsilon/4 - \gamma^{(k)} - \varphi_t\right)}, \tag{21.210}$$

where (21.209) follows from (21.140) and (21.210) follows from (21.133). Then from (21.184), (21.195), and (21.210), we have

$$\Delta_t^{(k)} \leq 2^{|E|} 2^{-n(\epsilon/4 - \gamma^{(k)} - \varphi_t)} + \lambda. \tag{21.211}$$

We now choose k, n, and δ to make the upper bound above smaller than any prescribed $\epsilon > 0$. Since $\gamma^{(k)} \to 0$ as $k \to \infty$, we can let k to be sufficiently large so that

$$\gamma^{(k)} < \epsilon/4. \tag{21.212}$$

Then with k fixed, since $\varphi_t \to 0$ as $n \to \infty$ and $\delta \to 0$, and $\lambda \to 0$ as $\delta \to 0$, by letting n be sufficiently large and δ be sufficiently small, we have

1. $\gamma^{(k)} + \varphi_t < \epsilon/4$, so that $2^{|E|} 2^{-n(\epsilon/4 - \gamma^{(k)} - \varphi_t)} \to 0$ as $n \to \infty$;
2. $\Delta_t^{(k)} \leq \epsilon$.

This completes the proof of (21.169).

Hence, we have proved the achievability of $\omega^{(k)}$ for all sufficiently large k. Then the closedness of $\mathcal{R}$ implies the achievability of $\omega = \lim_{k\to\infty} \omega^{(k)}$, where $\omega \in \mathcal{R}'$. The achievability of $\mathcal{R}'$ is established.

Proof of Lemma 21.9. We first prove that given

$$X_S = x_S \tag{21.213}$$

and

$$\mathbf{Y}_S(x_S) = \mathbf{y}_S \in T^{\hat{n}}_{[Y_S]\delta}, \tag{21.214}$$

the following hold for all non-source nodes i (i.e., $i \in V \backslash S$):

i) $\mathbf{U}_{\text{In}(i)}(c_{\text{In}(i)}) \in T^{\hat{n}}_{[U_{\text{In}(i)}]\delta}$;
ii) $k_e(c_{\text{In}(i)}) \neq 0$, $e \in \text{Out}(i)$;
iii) $\mathbf{U}_e(k_e(c_{\text{In}(i)})) = \tilde{u}_e(\mathbf{y}_S)$, $e \in \text{Out}(i)$.

Note that for $i \in T$, $\text{Out}(i) = \emptyset$ in (ii) and (iii). By the consistency of strong typicality (Theorem 6.7),

$$\mathbf{y}_s \in T^{\hat{n}}_{[Y_s]\delta} \tag{21.215}$$

for all $s \in S$. Then according to the construction of the code, for all $e \in \text{Out}(s)$,

$$k_e(x_s) \neq 0 \tag{21.216}$$

and

$$\mathbf{U}_e(k_e(x_s)) = \hat{u}_e(\mathbf{y}_s). \tag{21.217}$$

We now prove (i)–(iii) by induction on the non-source nodes according to any given coding order. Let i_1 be the first non-source node to encode. Since

$$\text{In}(i_1) \subset S, \tag{21.218}$$

for all $d \in \text{In}(i_1)$, $d \in \text{Out}(s)$ for some $s \in S$. Then

$$\mathbf{U}_d(c_d) = \mathbf{U}_d(k_d(x_s)) \tag{21.219}$$
$$= \hat{u}_d(\mathbf{y}_s) \tag{21.220}$$
$$= \tilde{u}_d(\mathbf{y}_S), \tag{21.221}$$

where (21.220) and (21.221) follow from (21.217) and (21.176), respectively. Thus

$$\mathbf{U}_{\text{In}(i_1)}(c_{\text{In}(i_1)}) = \tilde{u}_{\text{In}(i_1)}(\mathbf{y}_S). \tag{21.222}$$

Since $U_{\text{In}(i_1)}$ is a function of Y_S, in light of (21.180),

$$\mathbf{U}_{\text{In}(i)}(c_{\text{In}(i)}) \in T^{\hat{n}}_{[U_{\text{In}(i)}]\delta} \tag{21.223}$$

by the preservation property of strong typicality, proving (i). According to the code construction, this also implies (ii). Moreover,

$$\mathbf{U}_e(k_e(c_{\text{In}(i_1)})) = \hat{u}_e(\mathbf{U}_{\text{In}(i_1)}(c_{\text{In}(i_1)})) \tag{21.224}$$
$$= \tilde{u}_e(\mathbf{y}_S), \tag{21.225}$$

where the last equality is obtained by replacing in (21.177) the random variable U_d by the sequence $\mathbf{U}_d(c_d)$ and the random variable Y_s by the sequence $\mathbf{Y}_S(x_S) = \mathbf{y}_S$, proving (iii).

We now consider any non-source node i in the network. Assume that (i)–(iii) are true for all the nodes upstream to node i. For $d \in \text{In}(i)$, if $d \in \text{Out}(s)$ for some $s \in S$, we have already proved in (21.221) that

$$\mathbf{U}_d(c_d) = \tilde{u}_d(\mathbf{y}_S). \tag{21.226}$$

Otherwise, $d \in \text{Out}(i')$, where node i' is upstream to node i. Then

$$\mathbf{U}_d(c_d) = \mathbf{U}_d(k_d(c_{\text{In}(i')})) \tag{21.227}$$
$$= \hat{u}_d(\mathbf{U}_{\text{In}(i')}(c_{\text{In}(i')})) \tag{21.228}$$
$$= \tilde{u}_d(\mathbf{y}_S). \tag{21.229}$$

In the above, (21.228) follows from (ii) for node i' by the induction hypothesis and the code construction, and (21.229) follows from (21.177). Therefore, (21.226) is valid for all $d \in \text{In}(i)$. Hence,

$$\mathbf{U}_{\text{In}(i)}(c_{\text{In}(i)}) = \tilde{u}_{\text{In}(i)}(\mathbf{y}_S), \tag{21.230}$$

which is exactly the same as (21.222) except that i_1 is replaced by i. Then by means of the same argument, we conclude that (i), (ii), and (iii) hold for node i.

As (21.230) holds for any non-source node i, it holds for any sink node t. This proves (21.181) for all $t \in T$. Furthermore, since $U_{\text{In}(t)}$ is a function of Y_S, $(Y_S, U_{\text{In}(t)})$ is also a function of Y_S. Then in view of (21.181), (21.182) follows from the preservation property of strong typicality. This completes the proof of the lemma. $\square$

Proof of Lemma 21.10. Consider

$$\Pr\{E_t(x_{\beta(t)})|E_S'\}$$
$$= \sum_{\mathbf{y}_S \in T_{[Y_S]\delta}^{\hat{n}}} \Pr\{E_t(x_{\beta(t)})|\mathbf{Y}_S(1_S) = \mathbf{y}_S, E_S'\} \Pr\{\mathbf{Y}_S(1_S) = \mathbf{y}_S|E_S'\}.$$
$$\tag{21.231}$$

To analyze $\Pr\{E_t(x_{\beta(t)})|\mathbf{Y}_S(1_S) = \mathbf{y}_S, E_S'\}$ in the above summation, let us condition on the event $\{\mathbf{Y}_S(1_S) = \mathbf{y}_S, E_S'\}$, where $\mathbf{y}_S \in T_{[Y_S]\delta}^{\hat{n}}$. It then follows from (21.181) in Lemma 21.9 that

$$\mathbf{U}_{\text{In}(t)}(c_{\text{In}(t)}) = \tilde{u}_{\text{In}(t)}(\mathbf{y}_S). \tag{21.232}$$

Therefore, the event $E_t(x_{\beta(t)})$ is equivalent to

$$(\mathbf{y}_\Psi, \mathbf{Y}_{\beta(t)\setminus\Psi}(x_{\beta(t)\setminus\Psi}), \tilde{u}_{\text{In}(t)}(\mathbf{y}_S)) \in T_{[Y_\Psi Y_{\beta(t)\setminus\Psi} U_{\text{In}(t)}]\delta}^{\hat{n}} \tag{21.233}$$

(cf. (21.196)) or

$$\mathbf{Y}_{\beta(t)\setminus\Psi}(x_{\beta(t)\setminus\Psi}) \in T_{[Y_{\beta(t)\setminus\Psi}|Y_\Psi U_{\text{In}(t)}]\delta}^{\hat{n}}(\mathbf{y}_\Psi, \tilde{u}_{\text{In}(t)}(\mathbf{y}_S)). \tag{21.234}$$

Thus

$$\Pr\{E_t(x_{\beta(t)})|\mathbf{Y}_S(1_S) = \mathbf{y}_S, E'_S\}$$
$$= \sum_{\mathbf{y}_{\beta(t)\backslash\Psi} \in T^{\hat{n}}_{[Y_{\beta(t)\backslash\Psi}|Y_\Psi U_{\mathrm{In}(t)}]\delta}(\mathbf{y}_\Psi, \tilde{u}_{\mathrm{In}(t)}(\mathbf{y}_S))}$$
$$\Pr\{\mathbf{Y}_{\beta(t)\backslash\Psi}(x_{\beta(t)\backslash\Psi}) = \mathbf{y}_{\beta(t)\backslash\Psi}|\mathbf{Y}_S(1_S) = \mathbf{y}_S, E'_S\}. \qquad (21.235)$$

Since $x_s \neq 1$ for $s \in \beta(t)\backslash\Psi$, $\mathbf{Y}_{\beta(t)\backslash\Psi}(x_{\beta(t)\backslash\Psi})$ is independent of the random sequences $\mathbf{Y}_S(1_S)$ and the event E'_S by construction. Therefore,

$$\Pr\{\mathbf{Y}_{\beta(t)\backslash\Psi}(x_{\beta(t)\backslash\Psi}) = \mathbf{y}_{\beta(t)\backslash\Psi}|\mathbf{Y}_S(1_S) = \mathbf{y}_S, E'_S\}$$
$$= \Pr\{\mathbf{Y}_{\beta(t)\backslash\Psi}(x_{\beta(t)\backslash\Psi}) = \mathbf{y}_{\beta(t)\backslash\Psi}\}. \qquad (21.236)$$

By the consistency of strong typicality, if

$$\mathbf{y}_{\beta(t)\backslash\Psi} \in T^{\hat{n}}_{[Y_{\beta(t)\backslash\Psi}|Y_\Psi U_{\mathrm{In}(t)}]\delta}(\mathbf{y}_\Psi, \tilde{u}_{\mathrm{In}(t)}(\mathbf{y}_S)), \qquad (21.237)$$

then

$$\mathbf{y}_{\beta(t)\backslash\Psi} \in T^{\hat{n}}_{[Y_{\beta(t)\backslash\Psi}]\delta}. \qquad (21.238)$$

Since $\mathbf{Y}_{\beta(t)\backslash\Psi}(x_{\beta(t)\backslash\Psi})$ are generated i.i.d. according to the distribution of $Y_{\beta(t)\backslash\Psi}$, by the strong AEP,

$$\Pr\{\mathbf{Y}_{\beta(t)\backslash\Psi}(x_{\beta(t)\backslash\Psi}) = \mathbf{y}_{\beta(t)\backslash\Psi}\} \leq 2^{-\hat{n}(H(Y_{\beta(t)\backslash\Psi})-\rho)}, \qquad (21.239)$$

where $\rho \to 0$ as $\delta \to 0$. Combining (21.236) and (21.239), we have

$$\Pr\{\mathbf{Y}_{\beta(t)\backslash\Psi}(x_{\beta(t)\backslash\Psi}) = \mathbf{y}_{\beta(t)\backslash\Psi}|\mathbf{Y}_S(1_S) = \mathbf{y}_S, E'_S\} \leq 2^{-\hat{n}(H(Y_{\beta(t)\backslash\Psi})-\rho)}.$$
$$(21.240)$$

By the strong conditional AEP,

$$|T^{\hat{n}}_{[Y_{\beta(t)\backslash\Psi}|Y_\Psi U_{\mathrm{In}(t)}]\delta}(\mathbf{y}_\Psi, \tilde{u}_{\mathrm{In}(t)}(\mathbf{y}_S))| \leq 2^{\hat{n}(H(Y_{\beta(t)\backslash\Psi}|Y_\Psi U_{\mathrm{In}(t)})+\sigma)}, \qquad (21.241)$$

where $\sigma \to 0$ as $\hat{n} \to \infty$ and $\delta \to 0$. It then follows from (21.235), (21.240), and (21.241) that

$$\Pr\{E_t(x_{\beta(t)})|\mathbf{Y}_S(1_S) = \mathbf{y}_S, E'_S\}$$
$$\leq 2^{\hat{n}(H(Y_{\beta(t)\backslash\Psi}|Y_\Psi U_{\mathrm{In}(t)})+\sigma)} 2^{-\hat{n}(H(Y_{\beta(t)\backslash\Psi})-\rho)} \qquad (21.242)$$
$$= 2^{-\hat{n}(H(Y_{\beta(t)\backslash\Psi})-H(Y_{\beta(t)\backslash\Psi}|Y_\Psi U_{\mathrm{In}(t)})-\sigma-\rho)} \qquad (21.243)$$
$$\leq 2^{-\hat{n}(H(Y_{\beta(t)\backslash\Psi})-H(Y_{\beta(t)}|U_{\mathrm{In}(t)})-\sigma-\rho)} \qquad (21.244)$$
$$\leq 2^{-n\alpha(H(Y_{\beta(t)\backslash\Psi})-H(Y_{\beta(t)}|U_{\mathrm{In}(t)})-\varphi_t)}, \qquad (21.245)$$

where (21.244) is justified by

$$H(Y_{\beta(t)\backslash\Psi}|Y_\Psi U_{\text{In}(t)}) \leq H(Y_{\beta(t)\backslash\Psi}|Y_\Psi U_{\text{In}(t)}) + H(Y_\Psi|U_{\text{In}(t)}) \tag{21.246}$$
$$= H(Y_{\beta(t),\backslash\Psi}, Y_\Psi|U_{\text{In}(t)}) \tag{21.247}$$
$$= H(Y_{\beta(t)}|U_{\text{In}(t)}), \tag{21.248}$$

(21.245) follows from (21.144), and $\varphi_t \to 0$ as $n \to \infty$ and $\delta \to 0$.

In (21.231),

$$\Pr\{\mathbf{Y}_S(1_S) = \mathbf{y}_S|E'_S\}$$
$$= \Pr\{\mathbf{Y}_S(1_S) = \mathbf{y}_S|X_{\beta(t)} = 1_{\beta(t)}, E_S\} \tag{21.249}$$
$$= \Pr\{\mathbf{Y}_S(1_S) = \mathbf{y}_S|E_S\} \tag{21.250}$$
$$= \Pr\{\mathbf{Y}_S(1_S) = \mathbf{y}_S|\mathbf{Y}_S(1_S) \in T^{\hat{n}}_{[Y_S]\delta}\}. \tag{21.251}$$

Hence, it follows from (21.231) and (21.245) that

$$\Pr\{E_t(x_{\beta(t)})|E'_S\}$$
$$\leq 2^{-n\alpha(H(Y_{\beta(t)\backslash\Psi})-H(Y_{\beta(t)}|U_{\text{In}(t)})-\varphi_t)} \cdot$$
$$\sum_{\mathbf{y}_S \in T^{\hat{n}}_{[Y_S]\delta}} \Pr\{\mathbf{Y}_S(1_S) = \mathbf{y}_S|\mathbf{Y}_S(1_S) \in T^{\hat{n}}_{[Y_S]\delta}\} \tag{21.252}$$
$$= 2^{-n\alpha(H(Y_{\beta(t)\backslash\Psi})-H(Y_{\beta(t)}|U_{\text{In}(t)})-\varphi_t)} \cdot 1 \tag{21.253}$$
$$= 2^{-n\alpha(H(Y_{\beta(t)\backslash\Psi})-H(Y_{\beta(t)}|U_{\text{In}(t)})-\varphi_t)}. \tag{21.254}$$

The lemma is proved. $\square$

Proof of Lemma 21.11. Let n be sufficiently large. Consider

$$|\Lambda_\Psi| = \prod_{s\in\beta(t)\backslash\Psi} |\mathcal{X}_s| \tag{21.255}$$
$$\stackrel{a)}{=} \prod_{s\in\beta(t)\backslash\Psi} \theta_s \tag{21.256}$$
$$\stackrel{b)}{=} \prod_{s\in\beta(t)\backslash\Psi} \lceil 2^{n\tau_s} \rceil \tag{21.257}$$
$$\stackrel{c)}{\leq} \prod_{s\in\beta(t)\backslash\Psi} \lceil 2^{n(\omega_s-\epsilon/3)} \rceil \tag{21.258}$$
$$\leq \prod_{s\in\beta(t)\backslash\Psi} 2^{n(\omega_s-\epsilon/4)} \tag{21.259}$$
$$\stackrel{d)}{=} \prod_{s\in\beta(t)\backslash\Psi} 2^{n(\alpha H(Y_s)-\epsilon/4)} \tag{21.260}$$
$$= 2^{n\sum_{s\in\beta(t)\backslash\Psi}(\alpha H(Y_s)-\epsilon/4)} \tag{21.261}$$
$$= 2^{n\left[\alpha\sum_{s\in\beta(t)\backslash\Psi} H(Y_s)-(|\beta(t)|-|\Psi|)\epsilon/4\right]} \tag{21.262}$$

$$= 2^{n\left[\alpha H(Y_{\beta(t)\setminus\Psi})-(|\beta(t)|-|\Psi|)\epsilon/4\right]} \tag{21.263}$$

$$\overset{e)}{\leq} 2^{n(\alpha H(Y_{\beta(t)\setminus\Psi})-\epsilon/4)}, \tag{21.264}$$

where

(a) follows from (21.146);
(b) follows from (21.145);
(c) follows from (21.143);
(d) follows from (21.135);
(e) follows because Ψ is a proper subset of $\beta(t)$.

The lemma is proved.   □

Chapter Summary

A Multi-source Network Coding Problem: A point-to-point communication network is represented by an acyclic graph consisting of a set of nodes V and a set of channels E. The set of source nodes is denoted by S. At a source node s, an information source X_s is generated. The rate constraint on a channel e is R_e. The set of sink nodes is denoted by T. At a sink node t, the information sources $X_s, s \in \beta(t)$ are received, where $\beta(t) \subset S$ depends on t.

Achievable Information Rate Region $\mathcal{R}$: An information rate tuple $\omega = (\omega_s : s \in S)$ is achievable if for a sufficiently large block length n, there exists a network code such that

1. at a source node s, the rate of the information source X_s is at least $\omega_s - \epsilon$;
2. the rate of the network code on a channel e is at most $R_e + \epsilon$;
3. at a sink node t, the information sources $X_s, s \in \beta(t)$ can be decoded with negligible probability of error.

The achievable information rate region $\mathcal{R}$ is the set of all achievable information rate tuples ω.

Characterization of $\mathcal{R}$: Let $\mathcal{N} = \{Y_s : s \in S; U_e : e \in E\}$ and $\mathcal{H}_{\mathcal{N}}$ be the entropy space for the random variables in $\mathcal{N}$. Then

$$\mathcal{R} = \Lambda\left(\text{proj}_{Y_S}\left(\overline{\text{con}(\Gamma_{\mathcal{N}}^{**} \cap \mathcal{L}_{123})} \cap \mathcal{L}_4 \cap \mathcal{L}_5\right)\right),$$

where

$$\Gamma_{\mathcal{N}}^{**} = \{\mathbf{h} \in \mathcal{H}_{\mathcal{N}} : \mathbf{h} \text{ is finitely entropic}\}$$

and

$$\mathcal{L}_1 = \left\{ \mathbf{h} \in \mathcal{H}_\mathcal{N} : h_{Y_S} = \sum_{s \in S} h_{Y_s} \right\},$$

$$\mathcal{L}_2 = \left\{ \mathbf{h} \in \mathcal{H}_\mathcal{N} : h_{U_{\mathrm{Out}(s)}|Y_s} = 0, s \in S \right\},$$

$$\mathcal{L}_3 = \left\{ \mathbf{h} \in \mathcal{H}_\mathcal{N} : h_{U_{\mathrm{Out}(i)}|U_{\mathrm{In}(i)}} = 0, i \in V \setminus (S \cup T) \right\},$$

$$\mathcal{L}_4 = \left\{ \mathbf{h} \in \mathcal{H}_\mathcal{N} : h_{U_e} \leq R_e, e \in E \right\},$$

$$\mathcal{L}_5 = \left\{ \mathbf{h} \in \mathcal{H}_\mathcal{N} : h_{Y_{\beta(t)}|\dot{U}_{\mathrm{In}(t)}} = 0, t \in T \right\}.$$

An Explicit Outer Bound on $\mathcal{R}$ (LP Bound):

$$\mathcal{R}_{LP} = \Lambda \left(\mathrm{proj}_{Y_S} \left(\overline{\mathrm{con}(\Gamma_\mathcal{N} \cap \mathcal{L}_{123})} \cap \mathcal{L}_4 \cap \mathcal{L}_5 \right) \right).$$

Problems

1. Show that source separation is optimal for the networking problem depicted in Figure 21.3.
2. By letting $S = \{s\}$ and $\beta(t) = \{s\}$ for all $t \in T$, the multi-source network coding problem described in Section 21.3 becomes a single-source network coding problem. Write $\omega = \omega_s$.
   a) Write out the achievable information rate region $\mathcal{R}$.
   b) Show that if $\omega_s \in \mathcal{R}$, then $\omega_s \leq \mathrm{maxflow}(t)$ for all $t \in T$.
3. Consider the following network.

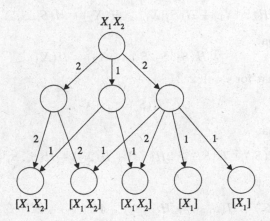

a) Let ω_i be the rate of information source X_i. Determine and illustrate the max-flow bounds.
b) Are the max-flow bounds achievable?
c) Is source separation always optimal?

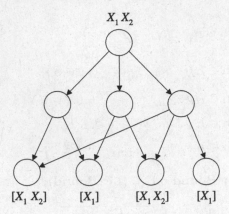

$$X_1 X_2$$

$$[X_1 X_2] \qquad [X_1] \qquad [X_1 X_2] \qquad [X_1]$$

4. Repeat Problem 3 for the above network in which the capacities of all the edges are equal to 1.

5. Consider a disk array with three disks. Let X_1, X_2, and X_3 be three mutually independent pieces of information to be retrieved from the disk array, and let S_1, S_2, and S_3 be the data to be stored separately in the three disks. It is required that X_1 can be retrieved from S_i, $i = 1, 2, 3$, X_2 can be retrieved from (S_i, S_j), $1 \le i < j \le 3$, and X_3 can be retrieved from (S_1, S_2, S_3).

a) Prove that for $i = 1, 2, 3$,

$$H(S_i) = H(X_1) + H(S_i|X_1).$$

b) Prove that for $1 \le i < j \le 3$,

$$H(S_i|X_1) + H(S_j|X_1) \ge H(X_2) + H(S_i, S_j|X_1, X_2).$$

c) Prove that

$$H(S_1, S_2, S_3|X_1, X_2) = H(X_3).$$

d) Prove that for $i = 1, 2, 3$,

$$H(S_i) \ge H(X_1).$$

e) Prove that

$$H(S_i) + H(S_j) \ge 2H(X_1) + H(X_2) + H(S_i, S_j|X_1, X_2).$$

f) Prove that

$$H(2S_i) + H(S_{i\oplus 1}) + H(S_{i\oplus 2}) \ge 4H(X_1) + 2H(X_2) + H(X_3),$$

where $i = 1, 2, 3$ and

$$i \oplus j = \begin{cases} i + j & \text{if } i + j \le 3 \\ i + j - 3 & \text{if } i + j > 3 \end{cases}$$

for $1 \le i, j \le 3$.

g) Prove that

$$H(S_1) + H(S_2) + H(S_3) \geq 3H(X_1) + \frac{3}{2}H(X_2) + H(X_3).$$

Parts (d)–(g) give constraints on $H(S_1)$, $H(S_2)$, and $H(S_3)$ in terms of $H(X_1)$, $H(X_2)$, and $H(X_3)$. It was shown in Roche et al. [308] that these constraints are the tightest possible.

6. Generalize the setup in Problem 5 to K disks and show that

$$\sum_{i=1}^{K} H(S_i) \geq K \sum_{\alpha=1}^{K} \frac{H(X_\alpha)}{\alpha}.$$

Hint: Use the inequalities in Problem 18 in Chapter 2 to prove that for $s = 0, 1, \cdots, K - 1$,

$$\sum_{i=1}^{K} H(S_i) \geq nK \sum_{\alpha=1}^{s} \frac{H(X_\alpha)}{\alpha} + \frac{K}{\binom{K}{s+1}}$$
$$\times \sum_{T:|T|=s+1} \frac{H(S_T|X_1, X_2, \cdots, X_s)}{s+1}$$

by induction on s, where T is a subset of $\{1, 2, \cdots, K\}$.

7. Write out the achievable information rate region $\mathcal{R}$ for the network in Problem 3.

8. Show that if there exists an

$$(n, (\eta_{ij} : (i, j) \in E), (\tau_s : s \in S))$$

code which satisfies (21.11) and (21.13), then there always exists an

$$(n, (\eta_{ij} : (i, j) \in E), (\tau_s' : s \in S))$$

code which satisfies (21.11) and (21.13), where $\tau_s' \leq \tau_s$ for all $s \in S$. Hint: use a random coding argument.

Historical Notes

Multilevel diversity coding was studied by Yeung [400], where it was shown that source separation is not always optimal. Roche et al. [308] showed that source separation is optimal for symmetrical 3 level diversity coding. This result was extended to any level by Yeung and Zhang [410] with a painstaking proof. Hau [160] studied all the 100 configurations of a three-encoder diversity coding systems and found that source separation is optimal for 86 configurations.

Yeung and Zhang [411] introduced the distributed source coding model discussed in Section 21.2.2 which subsumes multilevel diversity coding. The region of all entropy functions previously introduced by Yeung [401] for studying information inequalities enabled them to obtain inner and outer bounds on the achievable information rate region for a variety of networks.

Distributed source coding is equivalent to multi-source network coding on a special class of acyclic networks. The inner and outer bounds on the achievable information rate region in [411] were generalized to arbitrary acyclic networks by Song et al. [341]. The gap between these bounds was finally closed by Yan et al. [391].

The insufficiency of specific forms of linear coding for multi-source network coding were demonstrated and discussed by Riis [303], Rasala Lehman and Lehman [299], and Medard et al. [268]. The insufficiency of very general forms of linear coding has been proved by Dougherty et al. [96]. Chan [61] proved the sufficiency of a class of network codes constructed by groups when all the information sources are generated at the same node in the network.

Even though the achievable information rate region for multi-source network coding is characterized by all information inequalities (Shannon-type and non-Shannon-type), it is not clear whether there exists a multi-source network coding problem for which the characterization of the achievable information rate region necessarily involves non-Shannon-type inequalities. This important question was resolved by Dougherty et al. [98]. In this work, they constructed a multi-source network coding problem from matroids and demonstrated that a tighter outer bound on the achievable information rate region can be obtained by invoking the unconstrained non-Shannon-type inequality discovered by Zhang and Yeung [416]. Chan and Grant [62] recently proved that for every non-Shannon-type inequality that exists, there is a multi-source network coding problem for which the characterization of the achievable information rate region necessarily involves that particular inequality.

Bibliography

1. J. Abrahams, "Code and parse trees for lossless source encoding," *Comm. Info. and Syst.*, 1: 113–146, 2001 (http://www.ims.cuhk.edu.hk/~cis/).
2. N. Abramson, *Information Theory and Coding,* McGraw-Hill, New York, 1963.
3. Y. S. Abu-Mostafa, Ed., *Complexity in Information Theory*, Springer-Verlag, New York, 1988.
4. J. Aczél and Z. Daróczy, *On Measures of Information and Their Characterizations*, Academic Press, New York, 1975.
5. R. Ahlswede, B. Balkenhol, and L. Khachatrian, "Some properties of fix-free codes," preprint 97–039, Sonderforschungsbereich 343, Universität Bielefeld, Bielefeld, Germany 1997.
6. R. Ahlswede, N. Cai, S.-Y. R. Li, and R. W. Yeung, "Network information flow," *IEEE Trans. Info. Theory*, IT-46: 1204–1216, 2000.
7. R. Ahlswede and I. Csiszár, "Common randomness in information theory and cryptography – Part I: Secret sharing," *IEEE Trans. Info. Theory*, IT-39: 1121–1132, 1993.
8. R. Ahlswede and I. Csiszár, "Common randomness in information theory and cryptography – Part II: CR capacity," *IEEE Trans. Info. Theory*, IT-44: 225–240, 1998.
9. R. Ahlswede, P. Gács, and J. Körner, "Bounds of conditional probabilities with applications in multi-user communication," *Z. Wahrscheinlichketisheorie u. verw. Geb.*, 34: 157–177, 1976.
10. R. Ahlswede and J. Körner, "Source coding with side information and a converse for degraded broadcast channels," *IEEE Trans. Info. Theory*, IT-21: 629–637, 1975.
11. R. Ahlswede and I. Wegener, *Suchprobleme*, Teubner Studienbcher. B. G. Teubner, Stuttgart, 1979 (in German). English translation: *Search Problems*, Wiley, New York, 1987.
12. R. Ahlswede and J. Wolfowitz, "The capacity of a channel with arbitrarily varying cpf's and binary output alphabet," *Zeitschrift für Wahrscheinlichkeitstheorie und verwandte Gebiete*, 15: 186–194, 1970.
13. P. Algoet and T. M. Cover, "A sandwich proof of the Shannon–McMillan–Breiman theorem," *Ann. Prob.*, 16: 899–909, 1988.
14. S. Amari, *Differential-Geometrical Methods in Statistics*, Springer-Verlag, New York, 1985.

15. V. Anantharam and S. Verdú, "Bits through queues," *IEEE Trans. Info. Theory*, IT-42: 4–18, 1996.

16. J. B. Anderson and S. Mohan, *Source and Channel Coding: An Algorithmic Approach*, Kluwer Academic Publishers, Boston, 1991.

17. A. Argawal and M. Charikar, "On the advantage of network coding for improving network throughput," 2004 IEEE Information Theory Workshop, San Antonio, TX, Oct. 25–29, 2004.

18. S. Arimoto, "Encoding and decoding of p-ary group codes and the correction system," *Inf. Process. Japan*, 2: 321–325, 1961 (in Japanese).

19. S. Arimoto, "An algorithm for calculating the capacity of arbitrary discrete memoryless channels," *IEEE Trans. Info. Theory*, IT-18: 14–20, 1972.

20. S. Arimoto, "On the converse to the coding theorem for discrete memoryless channels," *IEEE Trans. Info. Theory*, IT-19: 357–359, 1973.

21. R. B. Ash, *Information Theory*, Interscience, New York, 1965.

22. E. Ayanoglu, R. D. Gitlin, C.-L. I, and J. Mazo, "Diversity coding for transparent self-healing and fault-tolerant communication networks," 1990 IEEE International Symposium on Information Theory, San Diego, CA, Jan. 1990.

23. H. Balli, X. Yan, and Z. Zhang, "Error correction capability of random network error correction codes," submitted to *IEEE Trans. Info. Theory*.

24. Á. I. Barbero and Ø. Ytrehus, "Cycle-logical treatment for "cyclopathic" networks," joint special issue of *IEEE Trans. Info. Theory* and *IEEE/ACM Trans. Networking* on Networking and Information Theory, IT-52: 2795–2804, 2006.

25. A. R. Barron, "The strong ergodic theorem for densities: Generalized Shannon–McMillan–Breiman theorem," *Ann. Prob.*, 13: 1292–1303, 1985.

26. L. A. Bassalygo, R. L. Dobrushin, and M. S. Pinsker, "Kolmogorov remembered," *IEEE Trans. Info. Theory*, IT-34: 174–175, 1988.

27. T. Berger, *Rate Distortion Theory: A Mathematical Basis for Data Compression*, Prentice-Hall, Englewood Cliffs, New Jersey, 1971.

28. T. Berger, "Multiterminal source coding," in *The Information Theory Approach to Communications*, G. Longo, Ed., CISM Courses and Lectures #229, Springer-Verlag, New York, 1978.

29. T. Berger and R. W. Yeung, "Multiterminal source coding with encoder breakdown," *IEEE Trans. Info. Theory*, IT-35: 237–244, 1989.

30. T. Berger, Z. Zhang, and H. Viswanathan, "The CEO problem," *IEEE Trans. Info. Theory*, IT-42, 887–902, May 1996.

31. E. R. Berlekamp, "Block coding for the binary symmetric channel with noiseless, delayless feedback," in *Error Correcting Codes*, H. B. Mann, Ed., Wiley, New York, 1968.

32. E. R. Berlekamp, Ed., *Key Papers in the Development of Coding Theory*, IEEE Press, New York, 1974.

33. C. Berrou, A. Glavieux, and P. Thitimajshima, "Near Shannon limit error-correcting coding and decoding: Turbo codes," Proceedings of the 1993 International Conferences on Communications, 1064–1070, 1993.

34. J. Berstel and D. Perrin, *Theory of Codes*, Academic Press, Orlando, 1985.

35. D. Bertsekas and R. Gallager, *Data Networks*, 2nd ed., Prentice-Hall, Englewood Cliffs, New Jersey, 1992.

36. D. Blackwell, L. Breiman, and A. J. Thomasian, "The capacities of certain channel classes under random coding," *Ann. Math. Stat.*, 31: 558–567, 1960.

37. R. E. Blahut, "Computation of channel capacity and rate distortion functions," *IEEE Trans. Info. Theory*, IT-18: 460–473, 1972.

38. R. E. Blahut, "Information bounds of the Fano-Kullback type," *IEEE Trans. Info. Theory*, IT-22: 410–421, 1976.

39. R. E. Blahut, *Theory and Practice of Error Control Codes*, Addison-Wesley, Reading, Massachusetts, 1983.

40. R. E. Blahut, *Principles and Practice of Information Theory*, Addison-Wesley, Reading, Massachusetts, 1987.

41. R. E. Blahut, D. J. Costello, Jr., U. Maurer, and T. Mittelholzer, Ed., *Communications and Cryptography: Two Sides of One Tapestry*, Kluwer Academic Publishers, Boston, 1994.

42. G. R. Blakley, "Safeguarding cryptographic keys," in *Proceedings of the National Computer Conference*, 48: 313–317, 1979.

43. C. Blundo, A. De Santis, R. De Simone, and U. Vaccaro, "Tight bounds on the information rate of secret sharing schemes," *Designs, Codes and Cryptography*, 11: 107–110, 1997.

44. B. Bollobás, *Graph Theory: An Introductory Course*, Springer-Verlag, New York, 1979.

45. J. A. Bondy and U. S. R. Murty, *Graph Theory with Applications*, North Holland, New York, 1976.

46. S. Borade, "Network information flow: Limits and achievability," 2002 IEEE International Symposium on Information Theory, Lausanne, Switzerland, Jun. 30–Jul. 5, 2002.

47. R. C. Bose and D. K. Ray-Chaudhuri, "On a class of error correcting binary group codes," *Info. Contr.*, 3: 68–79, Mar. 1960.

48. L. Breiman, "The individual ergodic theorems of information theory," *Ann. Math. Stat.*, 28: 809–811, 1957.

49. M. Burrows and D. J. Wheeler, "A block-sorting lossless data compression algorithm," Technical Report 124, Digital Equipment Corporation, 1994.

50. J. Byers, M. Luby, M. Mitzenmacher, "A digital fountain approach to asynchronous reliable multicast," *IEEE J. Selected Areas Comm.*, 20: 1528–1540, 2002.

51. N. Cai and R. W. Yeung, "Secure network coding," 2002 IEEE International Symposium on Information Theory, Lausanne, Switzerland, Jun. 30–Jul. 5, 2002.

52. N. Cai and R. W. Yeung, "Network coding and error correction," 2002 IEEE Information Theory Workshop, Bangalore, India, Oct. 20–25, 2002.

53. N. Cai and R. W. Yeung, "Network error correction, Part II: Lower bounds," *Comm. Info. and Syst.*, 6: 37–54, 2006 (http://www.ims.cuhk.edu.hk/~cis/).

54. G. Caire and S. Shamai, "On the achievable throughput of a multiantenna Gaussian broadcast channel," *IEEE Trans. Info. Theory*, IT-49: 1691–1706, 2003.

55. R. Calderbank and N. J. A. Sloane, "Obituary: Claude Shannon (1916–2001)," *Nature*, 410: 768, April 12, 2001.

56. R. M. Capocelli, A. De Santis, L. Gargano, and U. Vaccaro, "On the size of shares for secret sharing schemes," *J. Cryptology*, 6: 157–168, 1993.

57. H. L. Chan (T. H. Chan), "Aspects of information inequalities and its applications," M.Phil. thesis, The Chinese University of Hong Kong, 1998.

58. T. H. Chan, "A combinatorial approach to information inequalities," *Comm. Info. and Syst.*, 1: 241–253, 2001 (http://www.ims.cuhk.edu.hk/~cis/).

59. H. L. Chan (T. H. Chan), "New results in probabilistic modeling," Ph.D. thesis, The Chinese University of Hong Kong, 2001.

60. T. H. Chan, "Balanced information inequalities," *IEEE Trans. Info. Theory*, IT-49: 3261–3267, 2003.

61. T. H. Chan, "On the optimality of group network codes," 2005 IEEE International Symposium on Information Theory, Adelaide, South Australia, Australia, Sept. 4–9, 2005.

62. T. Chan and A. Grant, "Entropy vectors and network codes," 2007 IEEE International Symposium on Information Theory, Nice, France, Jun. 24–29, 2007.

63. T. H. Chan and R. W. Yeung, "On a relation between information inequalities and group theory," *IEEE Trans. Info. Theory*. IT-48:1992–1995, 2002.

64. G. J. Chatin, *Algorithmic Information Theory*, Cambridge University Press, Cambridge, 1987.

65. C. Chekuri, C. Fragouli, and E. Soljanin, "On average throughput benefits and alphabet size for network coding," joint special issue of *IEEE Trans. Info. Theory* and *IEEE/ACM Trans. Networking* on Networking and Information Theory, IT-52: 2410–2424, 2006.

66. H. Chernoff, "A measure of the asymptotic efficiency of test of a hypothesis based on a sum of observations," *Ann. Math. Stat.*, 23: 493–507, 1952.

67. D. M. Chiu, R. W. Yeung, J. Huang, and B. Fan, "Can network coding help in P2P networks?" NetCod 2006, Boston, MA, Apr. 3–7, 2006.

68. P. A. Chou and Y. Wu, "Network coding for the Internet and wireless networks," *IEEE Signal Processing Magazine*, 77–85, Sept. 2007.

69. P. A. Chou, Y. Wu, and K. Jain, "Practical network coding," 41st Annual Allerton Conference on Communication, Control, and Computing, Monticello, IL, Oct. 2003.

70. K. L. Chung, "A note on the ergodic theorem of information theory," *Ann. Math. Stat.*, 32: 612–614, 1961.

71. A. S. Cohen and A. Lapidoth, "The Gaussian watermarking game," *IEEE Trans. Info. Theory*, IT-48: 1639–1667, 2002.

72. M. H. M. Costa, "Writing on dirty paper," *IEEE Trans. Info. Theory*, IT-29: 439–441, 1983.

73. T. M. Cover, "A proof of the data compression theorem of Slepian and Wolf for ergodic sources," *IEEE Trans. Info. Theory*, IT-21: 226–228, 1975.

74. T. M. Cover, "An algorithm for maximizing expected log investment return," *IEEE Trans. Info. Theory*, IT-30: 369–373, 1984.

75. T. M. Cover and M. Chiang, "Duality between channel capacity and rate distortion with two-sided state information," *IEEE Trans. Info. Theory*, IT-48: 1629–1638, 2002.

76. T. M. Cover, P. Gács, and R. M. Gray, "Kolmogorov's contribution to information theory and algorithmic complexity," *Ann. Prob.*, 17: 840–865, 1989.

77. T. M. Cover and R. King, "A convergent gambling estimate of the entropy of English," *IEEE Trans. Info. Theory*, IT-24: 413–421, 1978.

78. T. M. Cover and S. K. Leung, "Some equivalences between Shannon entropy and Kolmogorov complexity," *IEEE Trans. Info. Theory*, IT-24: 331–338, 1978.

79. T. M. Cover and J. A. Thomas, *Elements of Information Theory*, Wiley, 1991, 2nd ed., Wiley-Interscience, 2006.

80. I. Csiszár, "Information type measures of difference of probability distributions and indirect observations," *Studia Sci. Math. Hungar.*, 2: 229–318, 1967.

81. I. Csiszár, "On the computation of rate-distortion functions," *IEEE Trans. Info. Theory*, IT-20: 122–124, 1974.

82. I. Csiszár, "The method of types," *IEEE Trans. Info. Theory*, IT-44: 2505–2523, 1998.

83. I. Csiszár and J. Körner, *Information Theory: Coding Theorems for Discrete Memoryless Systems*, Academic Press, New York, 1981.

84. I. Csiszár and P. Narayan, "Arbitrarily varying channels with constrained inputs and states," *IEEE Trans. Info. Theory*, IT-34: 27–34, 1988.

85. I. Csiszár and P. Narayan, "The capacity of the arbitrarily varying channel revisited: Positivity, constraints," *IEEE Trans. Info. Theory*, IT-34: 181–193, 1988.

86. I. Csiszár and P. Narayan, "Secrecy capacities for multiple terminals," *IEEE Trans. Info. Theory*, IT-50: 3047–3061, 2004.

87. I. Csiszár and P. Narayan, "Secrecy capacities for multiterminal channel models," special issue of *IEEE Trans. Info. Theory* on Information Theoretic Security, to appear, Jun. 2008.

88. I. Csiszár and G. Tusnády, "Information geometry and alternating minimization procedures," *Statistics and Decisions*, Supplement Issue 1: 205–237, 1984.

89. A. F. Dana, R. Gowaikar, R. Palanki, B. Hassibi, and M. Effros, "Capacity of wireless erasure networks," joint special issue of *IEEE Trans. Info. Theory* and *IEEE/ACM Trans. Networking* on Networking and Information Theory, IT-52: 789–804, 2006.

90. G. B. Dantzig, *Linear Programming and Extensions*, Princeton University Press, Princeton, New Jersey, 1962.

91. L. D. Davisson, "Universal noiseless coding," *IEEE Trans. Info. Theory*, IT-19: 783–795, 1973.

92. A. P. Dawid, "Conditional independence in statistical theory (with discussion)," *J. Roy. Statist. Soc., Series B*, 41: 1–31, 1979.

93. A. P. Dempster, N. M. Laird, and D. B. Rubin, "Maximum likelihood form incomplete data via the EM algorithm," *Journal Royal Stat. Soc., Series B*, 39: 1–38, 1977.

94. S. N. Diggavi and T. M. Cover, "The worst additive noise under a covariance constraint," *IEEE Trans. Info. Theory*, IT-47: 3072–3081, 2001.

95. R. L. Dobrushin, "General formulation of Shannon's main theorem in information theory," *Uspekhi Mat. Nauk*, 14: 3–104; Translated in AMS Transl. Ser. 2, 33: 323–438, 1963.

96. R. Dougherty, C. Freiling, and K. Zeger, "Insufficiency of linear coding in network information flow," *IEEE Trans. Info. Theory*, IT-51: 2745–2759, 2005.

97. R. Dougherty, C. Freiling, and K. Zeger, "Six new non-Shannon information inequalities," 2006 IEEE International Symposium on Information Theory, Seattle, WA, Jul. 9–14, 2006.

98. R. Dougherty, C. Freiling, and K. Zeger, "Networks, matriods, and non-Shannon information inequalities," *IEEE Trans. Info. Theory*, IT-53: 1949–1969, 2007.

99. G. Dueck and J. Körner, "Reliability function of a discrete memoryless channel at rates above capacity," *IEEE Trans. Info. Theory*, IT-25: 82–85, 1979.

100. J. Edmonds, "Edge-disjoint branchings," in *Combinatorial Algorithms*, R. Rustin, Ed., 91–96, Algorithmics Press, New York, 1973.

101. P. Elias, "Universal codeword sets and representations of the integers," *IEEE Trans. Info. Theory*, IT-21: 194–203, 1975.

102. P. Elias, A. Feinstein, and C. E. Shannon, "A note on maximum flow through a network," *IRE Trans. Info. Theory*, IT-2: 117–119, 1956.

103. *Encyclopedia Britannica*, http://www.britannica.com/

104. E. Erez and M. Feder, "Capacity region and network codes for two receivers multicast with private and common data," Workshop on Coding, Cryptography and Combinatorics, Huangshen City, China, 2003.

105. E. Erez and M. Feder, "Convolutional network codes," 2004 IEEE International Symposium on Information Theory, Chicago, IL, Jun. 27–Jul. 2, 2004.

106. E. Erez and M. Feder, "Convolutional network codes for cyclic networks," NetCod 2005, Riva del Garda, Italy, Apr. 7, 2005.

107. R. M. Fano, Class notes for Transmission of Information, Course 6.574, MIT, Cambridge, Massachusetts, 1952.

108. R. M. Fano, *Transmission of Information: A Statistical Theory of Communication*, Wiley, New York, 1961.

109. M. Feder, N. Merhav, and M. Gutman, "Universal prediction of individual sequences," *IEEE Trans. Info. Theory*, IT-38: 1258–1270, 1992.

110. A. Feinstein, "A new basic theorem of information theory," *IRE Trans. Info. Theory*, IT-4: 2–22, 1954.

111. A. Feinstein, *Foundations of Information Theory*, McGraw-Hill, New York, 1958.

112. J. Feldman, T. Malkin, C. Stein, and R. A. Servedio, "On the capacity of secure network coding", 42nd Annual Allerton Conference on Communication, Control, and Computing, Monticello, IL, Sept. 29–Oct. 1, 2004.

113. W. Feller, *An Introduction to Probability Theory and Its Applications*, Vol. 1, Wiley, New York, 1950.

114. B. M. Fitingof, "Coding in the case of unknown and changing message statistics," *PPI* 2: 3–11, 1966 (in Russian).

115. S. L. Fong and R. W. Yeung, "Variable-rate linear network coding," 2006 IEEE Information Theory Workshop, Chengdu, China, Oct. 22–26, 2006.

116. L. K. Ford, Jr. and D. K. Fulkerson, *Flows in Networks*, Princeton University Press, Princeton, New Jersey, 1962.

117. G. D. Forney, Jr., "Convolutional codes I: Algebraic structure," *IEEE Trans. Info. Theory*, IT-16: 720–738, 1970.

118. G. D. Forney, Jr., Information Theory, unpublished course notes, Stanford University, 1972.

119. G. D. Forney, Jr., "The Viterbi algorithm," *Proc. IEEE*, 61: 268–278, 1973.

120. C. Fragouli, J.-Y. Le Boudec, and J. Widmer, "Network coding: An instant primer," *ACM SIGCOMM Comp. Comm. Review*, 36: 63–68, 2006.

121. C. Fragouli and E. Soljanin, "A connection between network coding and convolutional codes," IEEE International Conference on Communications, Paris, France, Jun. 20–24, 2004.

122. C. Fragouli and E. Soljanin, "Information flow decomposition for network coding," *IEEE Trans. Info. Theory*, IT-52: 829–848, 2006.

123. C. Fragouli and E. Soljanin, "Network coding fundamentals," *Foundations and Trends in Networking*, 2(1): 1–133, 2007.

124. J. B. Fraleigh, *A First Course in Abstract Algebra*, 7th ed., Addison Wesley, 2002.

125. F. Fu and R. W. Yeung, "On the rate-distortion region for multiple descriptions," *IEEE Trans. Info. Theory*, IT-48: 2012–2021, 2002.

126. S. Fujishige, "Polymatroidal dependence structure of a set of random variables," *Info. Contr.*, 39: 55–72, 1978.

127. R. G. Gallager, "Low-density parity-check codes," *IEEE Trans. Info. Theory*, IT-8: 21–28, Jan. 1962.

128. R. G. Gallager, "A simple derivation of the coding theorem and some applications," *IEEE Trans. Info. Theory*, IT-11: 3–18, 1965.

129. R. G. Gallager, *Information Theory and Reliable Communication*, Wiley, New York, 1968.

130. R. G. Gallager, "Variations on a theme by Huffman," *IEEE Trans. Info. Theory*, IT-24: 668–674, 1978.

131. Y. Ge and Z. Ye, "Information-theoretic characterizations of lattice conditional independence models," unpublished.

132. A. Gersho and R. M. Gray, *Vector Quantization and Signal Compression*, Kluwer Academic Publishers, Boston, 1992.

133. C. Gkantsidis and P. R. Rodriguez, "Network coding for large scale content distribution," IEEE INFOCOM 2005, Miami, FL, Mar. 13–17, 2005.

134. S. Goldman, *Information Theory*, Prentice-Hall, Englewood Cliffs, New Jersey, 1953.

135. A. Goldsmith and P. P. Varaiya, "Capacity of fading channels with channel side information," *IEEE Trans. Info. Theory*, IT-43: 1986–1992, 1997.

136. A. Goldsmith, *Wireless Communications*, Cambridge University Press, 2006.

137. J. Dj. Golić, "Noiseless coding for multiple channels," 1994 International Symposium on Information Theory and Its Applications, Sydney, Australia, 1994.

138. S. W. Golomb, R. E. Peile, and R. A. Scholtz, *Basic Concepts in Information Theory and Coding : The Adventures of Secret Agent 00111*, Plenum Press, New York, 1994.

139. R. M. Gray, "On the asymptotic eigenvalue distribution of Toeplitz matrices," *IEEE Trans. Info. Theory*, IT-18: 725–730, 1972.

140. R. M. Gray, *Entropy and Information Theory*, Springer-Verlag, New York, 1990.

141. S. Guiasu, *Information Theory with Applications*, McGraw-Hill, New York, 1976.

142. J. Hadamard, "Résolution d'une question relative aux déterminans," *Bull. Sci. Math. Sér. 2*, 17: 240–246, 1893.

143. B. Hajek and T. Berger, "A decomposition theorem for binary Markov random fields," *Ann. Prob.*, 15: 1112–1125, 1987.

144. D. Hammer, A. Romashchenko, A. Shen, and N. Vereshchagin, "Inequalities for Shannon Entropy and Kolmogorov Complexity," *J. Comp. Syst. Sci.*, 60: 442–464, 2000.

145. R. V. Hamming, "Error detecting and error correcting codes," *Bell Sys. Tech. J.*, 29: 147–160, 1950.

146. T. S. Han, "Linear dependence structure of the entropy space," *Info. Contr.*, 29: 337–368, 1975.

147. T. S. Han, "Nonnegative entropy measures of multivariate symmetric correlations," *Info. Contr.*, 36: 133–156, 1978.

148. T. S. Han, "A uniqueness of Shannon's information distance and related non-negativity problems," *J. Comb., Info., Syst. Sci.*, 6: 320–321, 1981.

149. T. S. Han, "An information-spectrum approach to source coding theorems with a fidelity criterion," *IEEE Trans. Info. Theory*, IT-43: 1145–1164, 1997.

150. T. S. Han and K. Kobayashi, "A unified achievable rate region for a general class of multiterminal source coding systems," *IEEE Trans. Info. Theory*, IT-26: 277–288, 1980.

151. T. S. Han and K. Kobayashi, *Mathematics of Information and Coding*, American Mathematical Society, 2003.

152. T. S. Han and H. Koga, *Information-Spectrum Methods in Information Theory*, Springer, Berlin, 2003.

153. T. S. Han and S. Verdú, "Generalizing the Fano inequality," *IEEE Trans. Info. Theory*, IT-40: 1247–1251, 1994.

154. G. H. Hardy, J. E. Littlewood, and G. Polya, *Inequalities*, 2nd ed., Cambridge University Press, London, 1952.

155. P. Harremoës, "Information topologies with applications," in *Entropy, Search, Complexity (Bolyai Society Mathematical Studies)*, I. Csiszár, G. O. H. Katona, and G. Tardos, Ed., Springer, Berlin, 2007.

156. P. Harremoës and F. Topsøe, "Inequalities between entropy and index of coincidence derived from information diagrams," *IEEE Trans. Info. Theory*, IT-47: 2944–2960, 2001.

157. N. Harvey, R. Kleinberg, R. Nair, and Y. Wu, "A "chicken and egg" network coding problem," 2007 IEEE International Symposium on Information Theory, Nice, France, Jun. 24–29, 2007.

158. B. Hassibi, "Normalized entropy vectors, network information theory and convex optimization," IEEE Information Theory Workshop on Information Theory for Wireless Networks, Solstrand, Norway, Jul. 1–6, 2007.

159. B. Hassibi and S. Shadbakht, "On a construction of entropic vectors using lattice-generated distributions," 2007 IEEE International Symposium on Information Theory, Nice, France, Jun. 24–29, 2007.

160. K. P. Hau, "Multilevel diversity coding with independent data streams," M.Phil. thesis, The Chinese University of Hong Kong, Jun. 1995.

161. C. Heegard and S. B. Wicker, *Turbo Coding*, Kluwer Academic Publishers, Boston, 1999.

162. T. Ho, R. Koetter, M. Médard, D. R. Karger, and M. Effros, "The benefits of coding over routing in a randomized setting," 2003 IEEE International Symposium on Information Theory, Yokohama, Japan, Jun. 29–Jul. 4, 2003.

163. T. Ho, B. Leong, R. Koetter, M. Médard, M. Effros, and D. R. Karger, "Byzantine modification detection in multicast networks using randomized network coding", 2004 IEEE International Symposium on Information Theory, Chicago, IL, Jun. 27–Jul. 2, 2007.

164. T. Ho and D. S. Lun, *Network Coding: An Introduction*, Cambridge University Press, 2008.

165. S.-W. Ho, "The interplay between entropy and variational distance, Part II: Applications," submitted to *IEEE Trans. Info. Theory*.

166. S.-W. Ho and R. W. Yeung, "On the discontinuity of the Shannon information measures," 2005 IEEE International Symposium on Information Theory, Adelaide, South Australia, Australia, Sept. 4–9, 2005.

167. S.-W. Ho and R. W. Yeung, "On information divergence measures and a unified typicality," 2006 IEEE International Symposium on Information Theory, Seattle, WA, Jul. 9–14, 2006.

168. S.-W. Ho and R. W. Yeung, "The interplay between entropy and variational distance," 2007 IEEE International Symposium on Information Theory, Nice, France, Jun. 24–29, 2007.

169. S.-W. Ho and R. W. Yeung, "The interplay between entropy and variational distance, Part I: Basic concepts and bounds," submitted to *IEEE Trans. Info. Theory*.

170. A. Hocquenghem, "Codes correcteurs d'erreurs," *Chiffres*, 2: 147–156, 1959.

171. Y. Horibe, "An improved bound for weight-balanced tree," *Info. Contr.*, 34: 148–151, 1977.

172. Hu Guo-Ding, "Three converse theorems for the Shannon theorem in information theory," *Acta Mathematica Sinica (Chinese Edition)*, 11: 260–293, 1961.

173. Hu Guo-Ding, "On the amount of Information," *Teor. Veroyatnost. i Primenen.*, 4: 447–455, 1962 (in Russian).

174. Hu Guo-Ding, "On Shannon theorem and its converse for sequences of communication schemes in the case of abstract random variables," *Trans. Third Prague Conference on Information Theory etc.*, 285–332, 1964.

175. D. A. Huffman, "A method for the construction of minimum redundancy codes," *Proc. IRE*, 40: 1098–1101, 1952.

176. J. Y. Hui, *Switching and Traffic Theory for Integrated Broadband Networks*, Springer, Boston, 1990.

177. L. P. Hyvarinen, *Information Theory for Systems Engineers*, Springer-Verlag, Berlin, 1968.

178. B. Ibinson, N. Linden, and A. Winter, "All inequalities for the relative entropy," *Comm. Math. Phys.*, 269: 223–238, 2006.

179. S. Ihara, "On the capacity of channels with additive non-Gaussian noise," *Info. Contr.*, 37: 34–39, 1978.

180. S. Ihara, *Information Theory for Continuous Systems*, World Scientific, Singapore, 1993.

181. A. W. Ingleton, "Representation of matroids," in *Combinatorial Mathematics and Its Applications*, D. J. A. Welsh, Ed., 149–167, Academic Press, London, 1971.

182. C. Intanagonwiwat, R. Govindan, and D. Estrin, "Directed diffusion: A scalable and robust communication paradigm for sensor networks," 6th Annual International Conference on Mobile Computing and Networking (Mobicom 2000), Boston, MA, Aug. 6–11, 2000.

183. P. Jacquet and W. Szpankowski, "Entropy computations via analytic depoissonization," *IEEE Trans. Info. Theory*, IT-45: 1072–1081, 1999.

184. S. Jaggi, P. Sanders, P. A. Chou, M. Effros, S. Egner, K. Jain, and L. Tolhuizen, "Polynomial time algorithms for multicast network code construction," *IEEE Trans. Info. Theory*, IT-51: 1973–1982, 2005.

185. S. Jaggi, M. Langberg, S. Katti, D. Katabi, M. Médard, and M. Effros, "Resilient network coding in the presence of Byzantine adversaries," IEEE INFOCOM 2007, Anchorage, AK, May 6–12, 2007.

186. E. T. Jaynes, "On the rationale of maximum entropy methods," *Proc. IEEE*, 70: 939–052, 1982.

187. E. T. Jaynes, *Probability Theory: The Logic of Science*, Cambridge University Press, Cambridge, 2003.

188. F. Jelinek, *Probabilistic Information Theory*, McGraw-Hill, New York, 1968.

189. J. L. W. V. Jensen, "Sur les fonctions convexes et les inégalités entre les valeurs moyennes," *Acta Mathematica*, 30: 175–193, 1906.

190. V. D. Jerohin, "ε-entropy of discrete random objects," *Teor. Veroyatnost. i Primenen*, 3: 103–107, 1958.

191. N. Jindal, S. Viswanath, and A. Goldsmith, "On the duality of Gaussian multiple-access and broadcast channels," *IEEE Trans. Info. Theory*, IT-50: 768–783, 2004.

192. O. Johnsen, "On the redundancy of binary Huffman codes," *IEEE Trans. Info. Theory*, IT-26: 220–222, 1980.

193. G. A. Jones and J. M. Jones, *Information and Coding Theory*, Springer, London, 2000.

194. Y. Kakihara, *Abstract Methods in Information Theory*, World-Scientific, Singapore, 1999.

195. J. Karush, "A simple proof of an inequality of McMillan," *IRE Trans. Info. Theory*, 7: 118, 1961.

196. T. Kawabata, "Gaussian multiterminal source coding," Master thesis, Math. Eng., Univ. of Tokyo, Japan, Feb. 1980.

197. T. Kawabata and R. W. Yeung, "The structure of the I-Measure of a Markov chain," *IEEE Trans. Info. Theory*, IT-38: 1146–1149, 1992.

198. A. I. Khinchin, *Mathematical Foundations of Information Theory*, Dover, New York, 1957.

199. J. C. Kieffer, "A survey of the theory of source coding with a fidelity criterion," *IEEE Trans. Info. Theory*, IT-39: 1473–1490, 1993.

200. J. C. Kieffer and E.-h. Yang, "Grammar-based codes: A new class of universal lossless source codes," *IEEE Trans. Info. Theory*, IT-46: 737–754, 2000.

201. R. Kindermann and J. Snell, *Markov Random Fields and Their Applications*, American Mathematical Society, Providence, Rhode Island, 1980.

202. R. Koetter and M. Médard, "An algebraic approach to network coding," *IEEE/ACM Trans. Networking*, 11: 782–795, 2003.

203. R. Koetter and F. Kschischang, "Coding for errors and erasures in random network coding," 2007 IEEE International Symposium on Information Theory, Nice, France, Jun. 24–29, 2007.

204. A. N. Kolmogorov, "On the Shannon theory of information transmission in the case of continuous signals," *IEEE Trans. Info. Theory*, IT-2: 102–108, 1956.

205. A. N. Kolmogorov, "Three approaches to the quantitative definition of information," *Prob. Info. Trans.*, 1: 4–7, 1965.

206. A. N. Kolmogorov, "Logical basis for information theory and probability theory," *IEEE Trans. Info. Theory*, IT-14: 662–664, 1968.

207. L. G. Kraft, "A device for quantizing, grouping and coding amplitude modulated pulses," M.S. thesis, Dept. of Elec. Engr., MIT, 1949.

208. G. Kramer, "Directed information for channels with feedback," Ph.D. thesis, Swiss Federal Institute of Technology, Zurich, 1998.

209. G. Kramer and S. A. Savari, "Cut sets and information flow in networks of two-way channels," 2004 IEEE International Symposium on Information Theory, Chicago, IL, Jun. 27–Jul. 2, 2004.

210. F. R. Kschischang, B. J. Frey, and H.-A. Loeliger, "Factor graphs and the sum-product algorithm," *IEEE Trans. Info. Theory*, IT-47: 498–519, 2001.

211. H. W. Kuhn and A. W. Tucker, "Nonlinear programming," Proceedings of 2nd Berkeley Symposium: 481–492, University of California Press, 1951.

212. S. Kullback, *Information Theory and Statistics*, Wiley, New York, 1959.

213. S. Kullback, *Topics in Statistical Information Theory*, Springer-Verlag, Berlin, 1987.

214. S. Kullback and R. A. Leibler, "On information and sufficiency," *Ann. Math. Stat.*, 22: 79–86, 1951.

215. J. F. Kurose and K. W. Ross, *Computer Networking: A Top-Down Approach Featuring the Internet*, 2007, 4th ed., Addison Wesley, Boston, 2004.

216. E. Kushilevitz and N. Nisan, *Communication Complexity*, Cambridge University Press, Cambridge, 2006.

217. P.-W. Kwok and R. W. Yeung, "On the relation between linear dispersion and generic network code," 2006 IEEE Information Theory Workshop, Chengdu, China, Oct. 22–26, 2006.

218. H. J. Landau and H. O. Pollak, "Prolate spheroidal wave functions, Fourier analysis, and uncertainty-II," *Bell Sys. Tech. J.*, 40: 65–84, 1961.

219. H. J. Landau and H. O. Pollak, "Prolate spheroidal wave functions, Fourier analysis, and uncertainty-III," *Bell Sys. Tech. J.*, 41: 1295–1336, 1962.

220. M. Langberg, A. Sprintson, and J. Bruck, "The encoding complexity of network coding," joint special issue of *IEEE Trans. Info. Theory* and *IEEE/ACM Trans. Networking* on Networking and Information Theory, IT-52: 2386–2397, 2006.

221. G. G. Langdon, "An introduction to arithmetic coding," *IBM J. Res. Devel.*, 28: 135–149, 1984.

222. A. Lapidoth, "Mismatched decoding and the multiple access channels," *IEEE Trans. Info. Theory*, IT-42: 1520–1529, 1996.

223. A. Lapidoth and P. Narayan, "Reliable communication under channel uncertainty," *IEEE Trans. Info. Theory, Commemorative Issue: 1948–1998*, IT-44: 2148–2177, 1998.

224. S. L. Lauritzen, *Graphical Models*, Oxford Science Publications, Oxford, 1996.

225. J. Li, P. A. Chou, and C. Zhang, "Mutualcast: An efficient mechanism for one-to many content distribution," ACM SIGCOMM Asia Workshop, Beijing, China, Apr. 11–13, 2005.

226. M. Li and P. Vitányi, *An Introduction to Kolmogorov Complexity and Its Applications*, 2nd ed., Springer, New York, 1997.

227. S.-Y. R. Li, *Algebraic Switching Theory and Broadband Applications*, Academic Press, San Diego, California, 2000.

228. S.-Y. R. Li and S. T. Ho, "Ring-theoretic foundation of convolutional network coding," NetCod 2008, Hong Kong, Jan. 3–4, 2008.

229. S.-Y. R. Li and R. W. Yeung, "On convolutional network coding," 2006 IEEE International Symposium on Information Theory, Seattle, WA, Jul. 9–14, 2006.

230. S.-Y. R. Li, R. W. Yeung and N. Cai, "Linear network coding," *IEEE Trans. Info. Theory*, IT-49: 371–381, 2003.

231. Z. Li, B. Li, and L. C. Lau, "On achieving optimal multicast throughput in undirected networks," joint special issue of *IEEE Trans. Info. Theory* and *IEEE/ACM Trans. Networking* on Networking and Information Theory, IT-52: 2410–2424, 2006.

232. X.-B. Liang, "Matrix games in the multicast networks: maximum information flows with network switching," joint special issue of *IEEE Trans. Info. Theory* and *IEEE/ACM Trans. Networking* on Networking and Information Theory, IT-52: 2433–2466, 2006.

233. E. H. Lieb and M. B. Ruskai, "Proof of the strong subadditivity of quantum-mechanical entropy," *J. Math. Phys.*, 14: 1938–1941, 1973.

234. S. Lin and D. J. Costello, Jr., *Error Control Coding: Fundamentals and Applications*, Prentice-Hall, Englewood Cliffs, New Jersey, 1st ed., 1983, Upper Saddle River, New Jersey, 2nd ed., 2004.

235. N. Linden and A. Winter, "A new inequality for the von Neumann entropy," *Comm. Math. Phys.*, 259: 129–138, 2005.

236. T. Linder, V. Tarokh, and K. Zeger, "Existence of optimal codes for infinite source alphabets," *IEEE Trans. Info. Theory*, IT-43: 2026–2028, 1997.

237. R. Lněnička, "On the tightness of the Zhang-Yeung inequality for Gaussian vectors," *Comm. Info. and Syst.*, 6: 41–46, 2003 (http://www.ims.cuhk.edu.hk/∼cis/).

238. L. Lovász, "On the Shannon capacity of a graph," *IEEE Trans. Info. Theory*, IT-25: 1–7, 1979.

239. D. S. Lun, N. Ratnakar, M. Médard, R. Koetter, D. R. Karger, T. Ho, E. Ahmed, and F. Zhao, "Minimum-cost multicast over coded packet networks," joint special issue of *IEEE Trans. Info. Theory* and *IEEE/ACM Trans. Networking* on Networking and Information Theory, IT-52: 2608–2623, 2006.

240. D. J. C. MacKay, "Good error-correcting codes based on very sparse matrices," *IEEE Trans. Info. Theory*, IT-45: 399–431, Mar. 1999.

241. D. J. C. MacKay, *Information Theory, Inference, and Learning Algorithms*, Cambridge University Press, Cambridge, 2003.

242. K. Makarychev, Y. Makarychev, A. Romashchenko, and N. Vereshchagin, "A new class of non-Shannon-type inequalities for entropies," *Comm. Info. and Syst.*, 2: 147–166, 2002 (http://www.ims.cuhk.edu.hk/∼cis/).

243. F. M. Malvestuto, "A unique formal system for binary decompositions of database relations, probability distributions, and graphs," *Info. Sci.*, 59: 21–52, 1992; with Comment by F. M. Malvestuto and M. Studený, *Info. Sci.*, 63: 1–2, 1992.

244. M. Mansuripur, *Introduction to Information Theory*, Prentice-Hall, Englewood Cliffs, New Jersey, 1987.

245. H. Marko, "The bidirectional communication theory – A generalization of information theory," *IEEE Trans. Comm.*, 21: 1345–1351, 1973.

246. A. W. Marshall and I. Olkin, *Inequalities: Theory of Majorization and Its Applications*, Academic Press, New York, 1979.

247. K. Marton, "Error exponent for source coding with a fidelity criterion," *IEEE Trans. Info. Theory*, IT-20: 197–199, 1974.

248. K. Marton, "A simple proof of the blowing-up lemma," *IEEE Trans. Info. Theory*, IT-32: 445–446, 1986.

249. J. L. Massey, "Shift-register synthesis and BCH decoding," *IEEE Trans. Info. Theory*, IT-15: 122–127, 1969.

250. J. L. Massey, "Causality, feedback and directed information," in *Proc. 1990 Int. Symp. on Info. Theory and Its Applications*, 303–305, 1990.

251. J. L. Massey, "Contemporary cryptology: An introduction," in *Contemporary Cryptology: The Science of Information Integrity*, G. J. Simmons, Ed., IEEE Press, Piscataway, New Jersey, 1992.

252. J. L. Massey, "Conservation of mutual and directed information," 2005 IEEE International Symposium on Information Theory, Adelaide, South Australia, Australia, Sept. 4–9, 2005.

253. A. M. Mathai and P. N. Rathie, *Basic Concepts in Information Theory and Statistics: Axiomatic Foundations and Applications*, Wiley, New York, 1975.

254. F. Matúš, "Probabilistic conditional independence structures and matroid theory: Background," *Int. J. of General Syst.*, 22: 185–196, 1994.

255. F. Matúš, "Conditional independences among four random variables II," *Combinatorics, Probability and Computing*, 4: 407–417, 1995.

256. F. Matúš, "Conditional independences among four random variables III: Final conclusion," *Combinatorics, Probability and Computing*, 8: 269–276, 1999.

257. F. Matúš, "Inequalities for Shannon entropies and adhesivity of polyma-troids," 9th Canadian Workshop on Information Theory, McGill University, Montréal, Québec, Canada, 2005.

258. F. Matúš, "Piecewise linear conditional information inequalities," *IEEE Trans. Info. Theory*, IT-52: 236–238, 2006.

259. F. Matúš, "Two constructions on limits of entropy functions," *IEEE Trans. Info. Theory*, IT-53: 320–330, 2007.

260. F. Matúš, "Infinitely many information inequalities," 2007 IEEE International Symposium on Information Theory, Nice, France, Jun. 24–29, 2007.

261. F. Matúš and M. Studený, "Conditional independences among four random variables I," *Combinatorics, Probability and Computing*, 4: 269–278, 1995.

262. U. M. Maurer, "Secret key agreement by public discussion from common information," *IEEE Trans. Info. Theory*, IT-39: 733–742, 1993.

263. R. J. McEliece, *The Theory of Information and Coding*, Addison-Wesley, Reading, Massachusetts, 1977.

264. R. J. McEliece, *Finite Fields for Computer Scientists and Engineers*, Kluwer Academic Publishers, Boston, 1987.

265. W. J. McGill, "Multivariate information transmission," *Transactions PGIT, 1954 Symposium on Information Theory*, PGIT-4: 93–111, 1954.

266. B. McMillan, "The basic theorems of information theory," *Ann. Math. Stat.*, 24: 196–219, 1953.

267. B. McMillan, "Two inequalities implied by unique decipherability," *IRE Trans. Info. Theory*, 2: 115–116, 1956.

268. M. Médard, M. Effros, T. Ho, and D. Karger, "On coding for nonmulticast networks," 41st Annual Allerton Conference on Communication, Control, and Computing, Monticello, IL, Oct. 2003.

269. K. Menger, "Zur allgemeinen Kurventhoerie," *Fund. Math.*, 10: 96–115, 1927.

270. M. Mitzenmacher, "Digital fountains: A survey and look forward," 2004 IEEE Information Theory Workshop, San Antonio, TX, Oct. 24–29, 2004.

271. P. Moulin and J. A. O'Sullivan, " Information-theoretic analysis of information hiding," *IEEE Trans. Info. Theory*, IT-49: 563–593, 2003.

272. S. C. Moy, "Generalization of the Shannon–McMillan theorem," *Pacific J. Math.*, 11: 705–714, 1961.

273. Network Coding Homepage, http://www.networkcoding.info

274. M. A, Nielsen and I. L. Chuang, *Quantum Computation and Quantum Information*, Cambridge University Press, Cambridge, 2000.

275. H. Nyquist, "Certain factors affecting telegraph speed," *Bell Sys. Tech. J.*, 3: 324, 1924.

276. J. K. Omura, "A coding theorem for discrete-time sources," *IEEE Trans. Info. Theory*, IT-19: 490–498, 1973.

277. J. M. Ooi, *Coding for Channels with Feedback*, Kluwer Academic Publishers, Boston, 1998.

278. E. Ordentlich and M. J. Weinberger, "A distribution dependent refinement of Pinsker's inequality," *IEEE Trans. Info. Theory*, IT-51: 1836–1840, 2005.

279. A. Orlitsky, "Worst-case interactive communication I: Two messages are almost optimal," *IEEE Trans. Info. Theory*, IT-36: 1111–1126, 1990.

280. A. Orlitsky, "Worst-case interactive communication II: Two messages are not optimal," *IEEE Trans. Info. Theory*, IT-37: 995–1005, 1991.

554   Bibliography

281. A. Orlitsky, N. P. Santhanam, and J. Zhang, "Universal compression of memoryless sources over unknown alphabets," *IEEE Trans. Info. Theory*, IT-50: 1469–1481, 2004.

282. D. S. Ornstein, "Bernoulli shifts with the same entropy are isomorphic," *Adv. Math.*, 4: 337–352, 1970.

283. J. G. Oxley, *Matroid Theory*, Oxford University Press, Oxford, 1992.

284. L. H. Ozarow and A. D. Wyner, "Wire-tap channel II," *AT&T Bell Labs Tech. Journal*, 63: 2135–2157, 1984.

285. C. H. Papadimitriou and K. Steiglitz, *Combinatorial Optimization: Algorithms and Complexity*, Prentice-Hall, Englewood Cliffs, New Jersey, 1982.

286. A. Papoulis, *Probability, Random Variables and Stochastic Processes*, 2nd ed., McGraw-Hill, New York, 1984.

287. J. Pearl, *Probabilistic Reasoning in Intelligent Systems*, Morgan Kaufman, San Meteo, California, 1988.

288. A. Perez, "Extensions of Shannon–McMillan's limit theorem to more general stochastic processes," in Trans. Third Prague Conference on Information Theory, Statistical Decision Functions and Random Processes, 545–574, Prague, 1964.

289. J. R. Pierce, *An Introduction to Information Theory: Symbols, Signals and Noise*, 2nd rev. ed., Dover, New York, 1980.

290. J. T. Pinkston, "An application of rate-distortion theory to a converse to the coding theorem," *IEEE Trans. Info. Theory*, IT-15: 66–71, 1969.

291. M. S. Pinsker, "Calculation of the rate of information transmission of stationary random processes and the capacity of stationary channels," *Dokl. Akad. Nauk SSSR*, 111: 753–756 (in Russian).

292. M. S. Pinsker, *Information and Information Stability of Random Variables and Processes*, vol. 7 of the series *Problemy Peredači Informacii*, AN SSSR, Moscow, 1960 (in Russian). English translation: Holden-Day, San Francisco, 1964.

293. M. S. Pinsker, "Gaussian sources," *Prob. Info. Trans.*, 14: 59–100, 1963 (in Russian).

294. N. Pippenger, "What are the laws of information theory?" 1986 Special Problems on Communication and Computation Conference, Palo Alto, CA, Sept. 3–5, 1986.

295. N. Pippenger, "The inequalities of quantum information theory," *IEEE Trans. Info. Theory*, IT-49: 773–789, 2003.

296. C. Preston, *Random Fields*, Springer-Verlag, New York, 1974.

297. M. O. Rabin, "Efficient dispersal of information for security, load balancing, and fault-tolerance," *J. ACM*, 36: 335–348, 1989.

298. A. Rasala Lehman, "Network coding," Ph.D. thesis, MIT, Dept. of Elec. Engr. and Comp. Sci., Feb. 2005.

299. A. Rasala Lehman and E. Lehman, "Complexity classification of network information flow problems," 41st Annual Allerton Conference on Communication, Control, and Computing, Monticello, IL, Oct. 2003.

300. I. S. Reed and G. Solomon, "Polynomial codes over certain finite fields," *SIAM Journal Appl. Math.*, 8: 300–304, 1960.

301. F. M. Reza, *An Introduction to Information Theory*, McGraw-Hill, New York, 1961.

302. A. Rényi, *Foundations of Probability*, Holden-Day, San Francisco, 1970.

303. S. Riis, "Linear versus nonlinear boolean functions in network flows," 38th Annual Conference on Information Sciences and Systems (CISS), Princeton, NJ, Mar. 17–19, 2004.

304. J. Rissanen, "Generalized Kraft inequality and arithmetic coding," *IBM J. Res. Devel.*, 20: 198, 1976.

305. J. Rissanen, "Universal coding, information, prediction, and estimation," *IEEE Trans. Info. Theory*, IT-30: 629–636, 1984.

306. J. R. Roche, "Distributed information storage," Ph.D. thesis, Stanford University, Mar. 1992.

307. J. R. Roche, A. Dembo, and A. Nobel, "Distributed information storage," 1988 IEEE International Symposium on Information Theory, Kobe, Japan, Jun. 1988.

308. J. R. Roche, R. W. Yeung, and K. P. Hau, "Symmetrical multilevel diversity coding," *IEEE Trans. Info. Theory*, IT-43: 1059–1064, 1997.

309. R. T. Rockafellar, *Convex Analysis*, Princeton University Press, Princeton, New Jersey, 1970.

310. S. Roman, *Coding and Information Theory*, Springer-Verlag, New York, 1992.

311. A. Romashchenko, A. Shen, and N. Vereshchagin, "Combinatorial interpretation of Kolmogorov complexity," *Electronic Colloquium on Computational Complexity*, vol. 7, 2000.

312. K. Rose, "A mapping approach to rate-distortion computation and analysis," *IEEE Trans. Info. Theory*, IT-40: 1939–1952, 1994.

313. W. Rudin, *Principles of Mathematical Analysis*, 3rd ed., McGraw Hill, New York, 1976.

314. W. Rudin, *Real and Complex Analysis*, 3rd ed., McGraw Hill, New York, 1987.

315. F. Ruskey, "A survey of Venn diagrams," http://www.combinatorics.org/Surveys/ds5/VennEJC.html

316. S. A. Savari, "Redundancy of the Lempel-Ziv incremental parsing rule," *IEEE Trans. Info. Theory*, IT-43: 9–21, 1997.

317. S. A. Savari and R. G. Gallager, "Generalized Tunstall codes for sources with memory," *IEEE Trans. Info. Theory*, IT-43: 658–668, 1997.

318. S. Shamai and I. Sason, "Variations on the Gallager bounds, connections, and applications," *IEEE Trans. Info. Theory*, IT-48: 3029–3051, 2002.

319. S. Shamai and S. Verdú, "The empirical distribution of good codes," *IEEE Trans. Info. Theory*, IT-43: 836–846, 1997.

320. S. Shamai, S. Verdú, and R. Zamir, "Systematic lossy source/channel coding," *IEEE Trans. Info. Theory*, IT-44: 564–579, 1998.

321. A. Shamir, "How to share a secret," *Comm. ACM*, 22: 612–613, 1979.

322. C. E. Shannon, "A Mathematical Theory of Communication," *Bell Sys. Tech. J.*, 27: 379–423, 623–656, 1948.

323. C. E. Shannon, "Communication theory of secrecy systems," *Bell Sys. Tech. J.*, 28: 656–715, 1949.

324. C. E. Shannon, "Communication in the presence of noise," *Proc. IRE*, 37: 10–21, 1949.

325. C. E. Shannon, "Prediction and entropy of printed English," *Bell Sys. Tech. J.*, 30: 50–64, 1951.

326. C. E. Shannon, "The zero-error capacity of a noisy channel," *IRE Trans. Info. Theory*, IT-2: 8–19, 1956.

327. C. E. Shannon, "Coding theorems for a discrete source with a fidelity criterion," *IRE National Convention Record, Part 4*, 142–163, 1959.

328. C. E. Shannon, R. G. Gallager, and E. R. Berlekamp, "Lower bounds to error probability for coding in discrete memoryless channels," *Info. Contr.*, 10: 65–103 (Part I), 522–552 (Part II), 1967.

329. C. E. Shannon and W. W. Weaver, *The Mathematical Theory of Communication*, University of Illinois Press, Urbana, Illinois, 1949.

330. A. Shen, "Multisource information theory," Electronic Colloquium on Computational Complexity, Report No. 6, 2006.

331. Shen Shi-Yi, "The necessary and sufficient condition for the validity of information criterion in Shannon theory," *Acta Mathematica Sinica (Chinese Edition)*, 12: 389–407, 1962.

332. Shen Shi-Yi, "The basic theorem of the stationary channel," *Trans. Third Prague Conference on Information Theory etc.*, 637–649, 1964.

333. P. C. Shields, *The Ergodic Theory of Discrete Sample Paths*, American Mathematical Society, Providence, Rhode Island, 1996.

334. J. E. Shore and R. W. Johnson, "Axiomatic derivation of the principle of maximum entropy and the principle of minimum cross-entropy," *IEEE Trans. Info. Theory*, IT-26: 26–37, 1980.

335. I. Shunsuke, *Information theory for continuous systems*, World Scientific, Singapore, 1993.

336. M. Simonnard, *Linear Programming*, translated by William S. Jewell, Prentice-Hall, Englewood Cliffs, New Jersey, 1966.

337. D. Slepian, Ed., *Key Papers in the Development of Information Theory*, IEEE Press, New York, 1974.

338. D. Slepian and H. O. Pollak, "Prolate spheroidal wave functions, Fourier analysis, and uncertainty-I," *Bell Sys. Tech. J.*, 40: 43–64.

339. D. Slepian and J. K. Wolf, "Noiseless coding of correlated information sources," *IEEE Trans. Info. Theory*, IT-19: 471–480, 1973.

340. N. J. A. Sloane and A. D. Wyner, Ed., *Claude Elwood Shannon Collected Papers*, IEEE Press, New York, 1993.

341. L. Song, R. W. Yeung, and N. Cai, "Zero-error network coding for acyclic networks," *IEEE Trans. Info. Theory*, IT-49: 3129–3139, 2003.

342. L. Song, R. W. Yeung and N. Cai, "A separation theorem for single-source network coding," *IEEE Trans. Info. Theory*, IT-52: 1861–1871, 2006.

343. F. Spitzer, "Random fields and interacting particle systems," M. A. A. Summer Seminar Notes, 1971.

344. D. R. Stinson, "An explication of secret sharing schemes," *Designs, Codes and Cryptography*, 2: 357–390, 1992.

345. D. R. Stinson, "New general lower bounds on the information rate of secret sharing schemes," in *Adv. in Cryptology – CRYPTO '92, Lecture Notes in Comput. Sci.*, vol. 740, 168–182, 1993.

346. M. Studený, "Multiinformation and the problem of characterization of conditional-independence relations," *Prob. Contr. Info. Theory*, 18: 3–16, 1989.

347. W. Szpankowski, "Asymptotic average redundancy of Huffman (and other) block codes," *IEEE Trans. Info. Theory*, IT-46: 2434–2443, 2000.

348. M. Tan, R. W. Yeung, and S.-T. Ho, "A unified framework for linear network codes," NetCod 2008, Hong Kong, Jan. 3–4, 2008.

349. I. J. Taneja, *Generalized Information Measures and Their Applications*, http://www.mtm.ufsc.br/∼taneja/book/book.html

350. S. Tatikonda and S. Mitter, "Channel coding with feedback," 38th Annual Allerton Conference on Communication, Control, and Computing, Monticello, IL, Oct. 2000.

351. İ. E. Telatar, "Capacity of multi-antenna Gaussian channels," *Euro. Trans. Telecom.*, 10: 585–595, 1999.

352. A. L. Toledo and X. Wang, "Efficient multipath in sensor networks using diffusion and network coding," 40th Annual Conference on Information Sciences and Systems, Princeton University, Princeton, NJ, Mar. 22–24, 2006.

353. F. Topsøe, "Information theoretical optimization techniques," *Kyberneticka*, 15: 8–27, 1979.

354. F. Topsøe, "Some inequalities for information divergence and related measures of discrimination," *IEEE Trans. Info. Theory*, IT-46: 1602–1609, 2000.

355. F. Topsøe, "Basic concepts, identities and inequalities – the toolkit of information theory," *Entropy*, 3: 162–190, 2001.

356. F. Topsøe, "Information theory at the service of science," in *Entropy, Search, Complexity (Bolyai Society Mathematical Studies)*, I. Csiszár, G. O. H. Katona, and G. Tardos, Ed., Springer, Berlin, 2007.

357. D. N. C. Tse and S. V. Hanly, "Linear multiuser receivers: effective interference, effective bandwidth and user capacity," *IEEE Trans. Info. Theory*, IT-45: 641–657, 1999.

358. D. Tse and P. Viswanath, *Fundamentals of Wireless Communication*, Cambridge University Press, Cambridge, 2005.

359. B. P. Tunstall, "Synthesis of noiseless compression codes," Ph.D. dissertation, Georgia Institute of Technology, Atlanta, GA, 1967.

360. J. C. A. van der Lubbe, *Information Theory*, Cambridge University Press, Cambridge, 1997 (English translation).

361. E. C. van der Meulen, "A survey of multi-way channels in information theory: 1961–1976," *IEEE Trans. Info. Theory*, IT-23: 1–37, 1977.

362. E. C. van der Meulen, "Some reflections on the interference channel," in *Communications and Cryptography: Two Side of One Tapestry*, R. E. Blahut, D. J. Costello, Jr., U. Maurer, and T. Mittelholzer, Ed., Kluwer Academic Publishers, Boston, 1994.

363. M. van Dijk, "On the information rate of perfect secret sharing schemes," *Designs, Codes and Cryptography*, 6: 143–169, 1995.

364. M. van Dijk, "Secret key sharing and secret key generation," Ph.D. thesis, Eindhoven University of Technology, Dec. 1997.

365. S. Vembu, S. Verdú, and Y. Steinberg, "The source–channel separation theorem revisited," *IEEE Trans. Info. Theory*, IT-41: 44–54, 1995.

366. S. Verdú and T. S. Han, "A general formula for channel capacity," *IEEE Trans. Info. Theory*, IT-40: 1147–1157, 1994.

367. S. Verdú and T. S. Han, "The role of the asymptotic equipartition property in noiseless source coding," *IEEE Trans. Info. Theory*, IT-43: 847–857, 1997.

368. S. Verdú and S. W. McLaughlin, Ed., *Information Theory : 50 Years of Discovery*, IEEE Press, New York, 2000.

369. A. J. Viterbi, "Error bounds for convolutional codes and an asymptotically optimum decoding algorithm," *IEEE Trans. Info. Theory*, IT-13: 260–269, 1967.

370. A. J. Viterbi and J. K. Omura, *Principles of Digital Communications and Coding*, McGraw-Hill, New York, 1979.

371. J. von Neumann, *Mathematical Foundations of Quantum Mechanics*, Princeton University Press, Princeton, New Jersey, 1996 (translation from German edition, 1932).

372. A. Wald, "Sequential tests of statistical hypothesis," *Ann. Math. Stat.*, 16: 117–186, 1945.

373. V. K. Wei, "Generalized Hamming weight for linear codes," *IEEE Trans. Info. Theory*, IT-37: 1412–1418, 1991.

374. H. Weingarten, Y. Steinberg, and S. Shamai, "The capacity region of the Gaussian multiple-input multiple-output broadcast channel," *IEEE Trans. Info. Theory*, IT-52: 3936–3964, 2006.

375. T. Weissman, E.Ordentlich, G. Seroussi, S. Verdú, and M. J. Weinberger, "Universal discrete denoising: Known channel," *IEEE Trans. Info. Theory*, IT-51: 5–28, 2005.

376. T. A. Welch, "A technique for high-performance data compression," *Computer*, 17: 8–19, 1984.

377. E. T. Whittaker, "On the functions which are represented by the expansions of the interpolation theory", *Proc. Royal Soc. Edinburgh*, Sec. A, 35: 181–194, 1915.

378. S. B. Wicker, *Error Control Systems for Digital Communication and Storage*, Prentice-Hall, Englewood Cliffs, New Jersey, 1995.

379. S. B. Wicker and V. K. Bhargava, Ed., *Reed–Solomon Codes and Their Applications*, IEEE Press, Piscataway, New Jersey, 1994.

380. F. M. J. Willems, "Universal data compression and repetition times," *IEEE Trans. Info. Theory*, IT-35: 54–58, 1989.

381. F. M. J. Willems, Y. M. Shtarkov, and T. J. Tjalkens, "The context-tree weighting method: basic properties," *IEEE Trans. Info. Theory*, IT-41: 653–664, 1995.

382. F. M. J. Willems, Y. M. Shtarkov, and T. J. Tjalkens, "Context weighting for general finite-context sources," *IEEE Trans. Info. Theory*, IT-42: 1514–1520, 1996.

383. J. Wolfowitz, "The coding of messages subject to chance errors," *Illinois Journal of Mathematics*, 1: 591–606, 1957.

384. J. Wolfowitz, *Coding Theorems of Information Theory*, Springer, Berlin-Heidelberg, 2nd ed., 1964, 3rd ed., 1978.

385. P. M. Woodard, *Probability and Information Theory with Applications to Radar*, McGraw-Hill, New York, 1953.

386. Y. Wu, K. Jain, and S.-Y. Kung, "A unification of network coding and tree-packing (routing) theorems," joint special issue of *IEEE Trans. Info. Theory* and *IEEE/ACM Trans. Networking* on Networking and Information Theory, IT-52: 2398–2409, 2006.

387. A. D. Wyner, "The capacity of the band-limited Gaussian channel," *Bell Syst. Tech. J.*, 45: 359–371, 1966.

388. A. D. Wyner, "On source coding with side information at the decoder," *IEEE Trans. Info. Theory*, IT-21: 294–300, 1975.

389. A. D. Wyner, "The wiretap channel," *Bell Sys. Tech. Journal*, 54: 1355–1387, 1975.

390. A. D. Wyner and J. Ziv, "The rate-distortion function for source coding with side information at the decoder," *IEEE Trans. Info. Theory*, IT-22: 1–10, 1976.

391. X. Yan, R. W. Yeung, and Z. Zhang, "The capacity region for multi-source multi-sink network coding," 2007 IEEE International Symposium on Information Theory, Nice, France, Jun. 24–29, 2007.

392. E.-h. Yang and J. C. Kieffer, "Efficient universal lossless data compression algorithms based on a greedy sequential grammar transform – Part one: Without context models," *IEEE Trans. Info. Theory*, IT-46: 755–777, 2000.

393. S. Yang, R. W. Yeung, and Z. Zhang, "Weight properties of network codes," submitted to *Euro. Trans. Telecom.*

394. A. C.-C. Yao, "Some complexity questions related to distributive computing (Preliminary Report)," The Eleventh Annual ACM Symposium on Theory of Computing, Atlanta, GA, Apr. 30-May 02, 1979.

395. C. Ye and R. W. Yeung, "Some basic properties of fix-free codes," *IEEE Trans. Info. Theory*, IT-47: 72–87, 2001.

396. C. Ye and R. W. Yeung, "A simple upper bound on the redundancy of Huffman codes," *IEEE Trans. Info. Theory*, IT-48: 2132–2138, 2002.

397. Z. Ye and T. Berger, *Information Measures for Discrete Random Fields*, Science Press, Beijing/New York, 1998.

398. R. W. Yeung, "A new outlook on Shannon's information measures," *IEEE Trans. Info. Theory*, IT-37: 466–474, 1991.

399. R. W. Yeung, "Local redundancy and progressive bounds on the redundancy of a Huffman code," *IEEE Trans. Info. Theory*, IT-37: 687–691, 1991.

400. R. W. Yeung, "Multilevel diversity coding with distortion," *IEEE Trans. Info. Theory*, IT-41: 412–422, 1995.

401. R. W. Yeung, "A framework for linear information inequalities," *IEEE Trans. Info. Theory*, IT-43: 1924–1934, 1997.

402. R. W. Yeung, *A First Course in Information Theory*, Kluwer Academic/Plenum Publishers, New York, 2002.

403. R. W. Yeung, "Avalanche: A network coding analysis," *Comm. Info. and Syst.*, 7: 353–358, 2007 (http://www.ims.cuhk.edu.hk/~cis/).

404. R. W. Yeung and T. Berger, "Multi-way alternating minimization," 1995 IEEE International Symposium on Information Theory, Whistler, British Columbia, Canada, Sept. 17–22, 1995.

405. R. W. Yeung and N. Cai, "Network error correction, Part I: Basic concepts and upper bounds," *Comm. Info. and Syst.*, 6: 19–36, 2006 (http://www.ims.cuhk.edu.hk/~cis/).

406. R. W. Yeung and N. Cai, "On the optimality of a construction of secure network codes," 2008 International Symposium on Information Theory, Toronto, Canada, Jul. 6–11, 2008.

407. R. W. Yeung, T. T. Lee, and Z. Ye, "Information-theoretic characterization of conditional mutual independence and Markov random fields," *IEEE Trans. Info. Theory*, IT-48: 1996–2011, 2002.

408. R. W. Yeung, S.-Y. R. Li, N. Cai, and Z. Zhang, "Network coding theory," *Foundations and Trends in Comm. and Info. Theory*, 2(4–5): 241–381, 2005.

409. R. W. Yeung and Y.-O. Yan, Information-Theoretic Inequality Prover (ITIP), http://user-www.ie.cuhk.edu.hk/~ITIP/

410. R. W. Yeung and Z. Zhang, "On symmetrical multilevel diversity coding," *IEEE Trans. Info. Theory*, IT-45: 609–621, 1999.

411. R. W. Yeung and Z. Zhang, "Distributed source coding for satellite communications," *IEEE Trans. Info. Theory*, IT-45: 1111–1120, 1999.

412. R. W. Yeung and Z. Zhang, "A class of non-Shannon-type information inequalities and their applications," *Comm. Info. and Syst.*, 1: 87–100, 2001 (http://www.ims.cuhk.edu.hk/~cis/).

413. Z. Zhang, "On a new non-Shannon-type information inequality," *Comm. Info. and Syst.*, 3: 47–60, 2003 (http://www.ims.cuhk.edu.hk/~cis/).

414. Z. Zhang, "Linear network error correction codes in packet networks," *IEEE Trans. Info. Theory*, IT-54: 209–218, 2008.

415. Z. Zhang and R. W. Yeung, "A non-Shannon-type conditional inequality of information quantities," *IEEE Trans. Info. Theory*, IT-43: 1982–1986, 1997.

416. Z. Zhang and R. W. Yeung, "On characterization of entropy function via information inequalities," *IEEE Trans. Info. Theory*, IT-44: 1440–1452, 1998.

417. L. Zheng and D. N. C. Tse, "Communication on the Grassmann manifold: A geometric approach to the noncoherent multiple-antenna channel," *IEEE Trans. Info. Theory*, IT-48: 359–383, 2002.

418. S. Zimmerman, "An optimal search procedure," *Am. Math. Monthly*, 66: 8, 690–693, 1959.

419. K. Sh. Zigangirov, "Number of correctable errors for transmission over a binary symmetrical channel with feedback," *Prob. Info. Trans.*, 12: 85–97, 1976. Translated from *Problemi Peredachi Informatsii*, 12: 3–19 (in Russian).

420. J. Ziv and A. Lempel, "A universal algorithm for sequential data compression," *IEEE Trans. Info. Theory*, IT-23: 337–343, 1977.

421. J. Ziv and A. Lempel, "Compression of individual sequences via variable-rate coding," *IEEE Trans. Info. Theory*, IT-24: 530–536, 1978.

Index

Information Technology: Transmission, Processing, and Storage